POLLUTION
Engineering and Scientific Solutions

Environmental Science Research

Editorial Board

Alexander Hollaender
Department of Biomedical Sciences
University of Tennessee
Oak Ridge Graduate School
Oak Ridge, Tennessee
and
Director, Archives of Radiation and Biology
University of Tennessee
Knoxville, Tennessee

Ronald F. Probstein
Department of Mechanical Engineering
Massachusetts Institute of Technology
Cambridge, Massachusetts

David P. Rall
Director, National Institute of Environmental Health Sciences
Research Triangle Park, North Carolina

E. S. Starkman
General Motors Environmental Activities Staff
General Motors Technical Center
Warren, Michigan

Bruce L. Welch
Director, Environmental Neurobiology
Friends Medical Science Research Center, Inc.
and
Department of Psychiatry & Behavioral Sciences
The Johns Hopkins University School of Medicine
Baltimore, Maryland

Volume 1 — INDICATORS OF ENVIRONMENTAL QUALITY
Edited by William A. Thomas • 1972

Volume 2 — POLLUTION: ENGINEERING AND SCIENTIFIC SOLUTIONS
Edited by Euval S. Barrekette • 1972

POLLUTION
Engineering and Scientific Solutions

Proceedings of the First International Meeting of the
Society of Engineering Science held in Tel Aviv, Israel
June 12-17, 1972

Edited by
Euval S. Barrekette
Program Director of Advanced Engineering
International Business Machines Corporation
Armonk, New York

℘ **PLENUM PRESS • NEW YORK-LONDON • 1973**

Editorial Committee

Ervin Y. Rodin
Department of Applied Mathematics
Washington University
St. Louis, Missouri

Alberto M. Wachs
Dean, Civil Engineering
L. Sherman Center for Environmental Engineering Research
Technion-Israel Institute of Technology
Haifa, Israel

Americo R. DiPietro
Program Director Solid State Technology
International Business Machines Corp. Corporate Headquarters
Armonk, New York

Library of Congress Catalog Card Number 72-91328
ISBN 0-306-36302-X

© 1973 Plenum Press, New York
A Division of Plenum Publishing Corporation
227 West 17th Street, New York, N.Y. 10011

United Kingdom edition published by Plenum Press, London
A Division of Plenum Publishing Company, Ltd.
Davis House (4th Floor), 8 Scrubs Lane, Harlesden, London, NW10 6SE, England

All rights reserved

No part of this publication may be reproduced in any form
without written permission from the publisher

Printed in the United States of America

PREFACE

The rapid deterioration of the environment in many countries around the world, or of segments and aspects of the environment in specific locations, made it necessary that immediate - even if only short term - solutions be found to as many of these problems as possible. Nevertheless, in the long run, long range and long term solutions must be found taking into account the effects of one country or region on another as well as of the inter-action between the different types of pollution over extended periods of time.

It was the purpose of the Tel Aviv meeting on Pollution: Engineering and Scientific Solutions, to address presently known or foreseeable "environmental insults;" that is, to focus on those aspects of air, noise, land, water or any other environmental quality for which there already exist engineering, scientific, legal or other solutions. Consequently, people from all disciplines which are relevant to environmental problems and their solutions were invited to participate. Their contributions were organized into more than a score of sessions as follows:

Agricultural Pollution
Air Chemistry and Pollution Abatement
Canadian Pollution Problems and Solutions
Computers in the Environment
Effects of Pollutants on Human Health
Hazardous Materials
Interfaces of Various Pollution Phenomena
    with Water Pollution
Large-Scale Air Chemistry and Pollution
Mathematical Environmental Models
Monitoring and Regional Studies
Pollution Abatement in the Manufacture and
    Loading of Propellants and Explosives
Pollution Measurement Instrumentation
Pollution and Space
Pests and Their Control

Problems of Ecopolitics
Remote Sensing of Pollution
Scientific and Engineering Solutions
    to Noise Pollution
Solid Waste and Air Pollution
Sulfur Dioxide Removal from Waste Gases
The Built Environment
The Diverse Problems of the Environment
The Role of Engineering Schools in
    Environmental Education
The Role of Law in Air Pollution Control

The list of disciplines included in these sessions was unusually long. Nevertheless, several groups were unrepresented. This was due in part to the temporal and geographical proximity of this meeting to the U. N. Meeting on the Human Environment in Stockholm and the Sixth International Conference on Water Pollution Research in Jerusalem.

This volume contains a representative sample of the papers presented at the conference. Three classes of papers were included: Plenary Addresses, Invited Presentations and Contributed Papers. Of the first two categories, only those papers are included which the presenters wished to have published here. In the last category, only those papers are included which were selected by a postmeeting review committee. This group includes two student papers.

The following is a list of those, who in some measure, contributed to the success of the meeting:

GENERAL CHAIRMAN
    Professor E. Y. Rodin, Applied Mathematics,
        Washington University, St. Louis, Mo., U.S.A.
        and Director Society of Engineering Science

INTERNATIONAL ADVISORY COMMITTEE
    Professor A. C. Eringen, Chairman, Princeton, N. J., U.S.A.
        and President Society of Engineering Science
    Dr. E. E. Goldstein, European Coordinator, Paris, France
    Professor A. Akinsete, Lagos, Nigeria
    Professor D. D. Ang, Saigon, Vietnam
    Professor M. Bank, Mexico City, Mexico
    Professor H. S. Green, Adelaide, Australia
    Professor G. Hanisch, Stuttgart, Germany
    Dr. W. N. Hess, Boulder, Colorado, U.S.A., and Director
        of Society of Engineering Science
    Dr. J. P. Lodge, Boulder, Colorado, U.S.A.
    Professor L. G. Napolitano, Naples, Italy

Professor S. G. Nomachi, Muroran, Japan
Professor N. Ozdas, Istanbul, Turkey
Professor E. Y. Rodin, St. Louis, Mo., U.S.A.
Professor I. N. Sneddon, Glasgow, Scotland, U.K.
Professor L. Sobrero, Udine, Italy
Professor A. Wachs, Haifa, Israel

U.S. PROGRAM COMMITTEE
Professor G. L. Esterson, Chairman, Chemical Engineering
Professor F. Rosenbaum, Chairman, Contributed Papers, Electrical Engineering
Mr. J. Itzikowitz, Student Affairs Chairman
Professor D. K. Ai, Aerodynamics
Dean G. Anselivicius, Architecture
Dr. E. Barrekette, Computer Science
Professor E. Edgerley, Environmental Engineering
Professor I. N. Katz, Applied Mathematics
Professor D. R. Mandelker, Law
Professor S. P. Sutera, Chairman, Invited Papers, Mechanical Engineering

ISRAELI HOST COMMITTEE
Professor A. Wachs, Technion, Haifa, Israel, Chairman
Professor S. Frankenthal, Tel Aviv University, Israel
Professor A. Inselberg, Technion, Haifa, Israel
Dr. U. Marinov, Prime Minister's Office
Professor N. Narkis, Technion, Haifa, Israel
Mr. E. Pelles, C. E., Israel Association of Architects and Engineers
Mr. G. Rivlin, KENES, Ltd.
Professor A. Rubinstein, Tel Aviv University, Israel
Dr. Y. Shalhevet, Volcani Institute

SPONSORS OF THE MEETING
Society of Engineering Science
National Oceanic and Atmospheric Administration, U.S.A.
Environmental Protection Agency, U.S.A.
Washington University, St. Louis, Mo., U.S.A.
Princeton University, U.S.A.
The Association of Engineers and Architects in Israel
Ministry of the Interior, Government of Israel
Ministry of Housing, Government of Israel
The Technion - Israel Institute of Technology
    in association with the
Israel Institute of Chemical Engineers

PLENARY SPEAKERS
Alfred Eggers, U. S. National Science Foundation,
    Keynote Speaker, Director Society of Engineering Science
George B. Dantzig, Stanford University, California

James A. Fay, Massachusetts Institute of Technology
Buckminster Fuller, Southern Illinois University
Stanley Greenfield, Environmental Protection Agency
P. M. Higgins, Environment, Canada
D. O. Jordan, University of Adelaide, Australia

CONVENORS OF SESSIONS
A. V. Slack, Tennessee Valley Authority, U.S.A.
H. M. Lieberstein, Newcastle University, Australia
A. Wachs, Technion, Haifa, Israel
E. S. Barrekette, IBM Corporation, U.S.A.
I. Forsten, U. S. Army
R. Levy, McDonnell-Douglas Research Laboratories
H. E. von Gierke, Wright-Patterson Air Force Base, Ohio
W. Hess, National Oceanic and Atmospheric Administration
T. D. Sterling, Washington University
L. Machta, U.S. National Oceanic and Atmospheric Administration
G. L. Esterson, Washington University
I. Volgyes, University of Nebraska
Y. Shalhevet, Volcani Institute
G. Morley, Environment, Canada
G. Anselivicius, Washington University
D. R. Mandelker, Columbia University
K. Barbehenn, Washington University

The entire meeting was conceived in such a way as to make it conducive for an expert from one field to get an overview of many other fields and gain an appreciation of multi-disciplinary interaction. It is the hope of the Editor and the Editorial Board that these proceedings will also further this goal.

>   Euval S. Barrekette
>   Editor
>   IBM Corporation
>   Armonk, New York, U.S.A.
>
>   Ervin Y. Rodin
>   General Chairman
>   Washington University
>   St. Louis, Missouri, U.S.A.

CONTENTS

1. Climatic Aspects of Pollution

    1.1 Global Effects of Contaminants
        in the Upper Atmosphere   1
           Lester Machta

    1.2 Climatic Aspects of Waste Heat 10
           James T. Peterson

    1.3 Cooling Towers and Weather Modification 19
           John R. Hummel

2. Pollution and Space

    2.1 Man-Made Alterations of the Near
        Earth Space Environment 26
           Wilmot N. Hess

    2.2 Orbital Pollution Control 38
           Ervin Y. Rodin

    2.3 Global Air Pollution Monitoring from
        Satellites: Prospects and Problems 54
           George Ohring, Joseph Otterman, and
           Joachim Joseph

    2.4 Environmental Observations from Space -
        The Promise and the Challenge   70
           E. S. Schaller

3. Air Pollution

    3.1 A Model for Fluid Mechanical Studies
        of Air Pollution 77
           C. Y. Liu and K. G. Agopian

3.2 A Systematic Method for Evaluating the Potential Environmental Impact of New Industrial Chemicals 95
   E. J. Sowinski and I. H. Suffet

3.3 Effect of Additives on Boiler Cleanliness and Particulate Emissions 105
   Ira Kukin

3.4 High Temperature Dust Control - Application of a Combination Particle Collector and Heat Exchanger 122
   H. Gregor Rigo and James J. Stukel

4. Agricultural and Pest Control Aspects of Pollution

   4.1 Solutions to Problems of Soil Pollution by Agricultural Chemicals 133
      I. J. Graham-Bryce

   4.2 Control of Air Pollution Affecting or Caused by Agriculture 148
      John Middleton and Ellis F. Darley

   4.3 The Creation and Control of Pest Situations Through Engineering: Past, Present and Future 158
      Kyle R. Barbehenn

   4.4 The Neo-Technological Landscape Degradation and Its Ecological Restoration 168
      Zev Naveh

   4.5 Some Aspects of the Role of Engineering in the Competition Between Insects and Man 182
      D. Schröder

5. Sulfur Dioxide

   5.1 Sulfur Dioxide Removal from Gases, U. S.: Lime-Limestone 192
      I. A. Raben

   5.2 Special Problems of the Smelter Industry 215
      G. Bridgstock

CONTENTS

5.3 Sulfur Dioxide Removal from Waste Gases:
A Status Report from United States:
Recovery Processes 224
R. E. Harrington

5.4 Sulphur Dioxide Removal from Waste Gases:
A Status Report from Japan 235
Jumpei Ando and Shoichiro Hori

5.5 Sulfur Dioxide Removal from Waste Gases:
A Status Report — Europe 253
Werner Brocke

6. Propellants and Explosives

6.1 Pollution Abatement in the
Manufacture and Loading of Propellants
and Explosives 268
Irving Forsten

6.2 Investigations Related to Prevention
and Control of Water Pollution in the
U. S. TNT Industry 272
David H. Rosenblatt

6.3 Control of Nitrogen Oxide Emissions for
Nitric Acid Plants 278
H. G. Rigo, W. J. Mikucki, and
M. L. Davis

6.4 Abatement of Nitrobodies in Aqueous
Effluents from TNT Production and
Finishing Plants 288
Leo A. Spano, Ronald A. Chalk,
John T. Walsh, and Carmine DiPietro

6.5 Explosive Incineration 298
Irving Forsten

6.6 The U. S. Navy PEPPARD Program 304
Bernard E. Drimmer

6.7 Elimination of Styphnic Acid from
Plant Effluents 310
E. Kurz

7. Solid Waste, Land Use

　7.1　Environmental Problems and Solutions
　　　Associated with the Development of
　　　the World's Largest Lead Mining District 320
　　　　J. Charles Jennet, Bobby G. Wixson,
　　　　Ernst Bolter, and James O. Pierce

　7.2　A Study of the Effectiveness of
　　　Backfilling in Controlling Mine Drainage 331
　　　　Keith G. Kirk

　7.3　Land Disposal of Septage (Septic Tank
　　　Pumpings) 346
　　　　J. J. Kolega, A. W. Dewey,
　　　　R. L. Leonard, and B. J. Cosenza

　7.4　Resource Recovery from Municipal Water -
　　　A Review and Analysis of Existing and
　　　Emerging Technology 357
　　　　David Bendersky, William R. Park,
　　　　Larry J. Shannon, and William E. Franklin

　7.5　Removal of Trace Metals from Wastewater
　　　by Lime and Ozonation 380
　　　　A. Netzer, A. Bowers, and J. D. Norman

　7.6　A Floating Settler for Low Cost
　　　Clarification 387
　　　　S. C. Reed, T. Buzzell, and S. Buda

8. Sensing, Instrumentation, and Measurement

　8.1　Measurement and Collaborative Testing
　　　for Implementation of Air Quality 398
　　　　A. T. Altshuller

　8.2　The Use of Airborne Sensor Systems for
　　　Environmental Monitoring 415
　　　　Howard A. Friedman and Howard J. Mason, Jr.

　8.3　Radio Wave Monitoring of the Depth and
　　　Salinity of the Water Table 434
　　　　E. Bahar

　8.4　Remote Measurement of Air Pollutants
　　　Utilizing the Raman Effect 438
　　　　S. Lederman and M. H. Bloom

- 8.5 Remote Sensing of Pollutants by Means of Stereo Analysis 453
  W. Z. Sadeh and J. F. Ruff

- 8.6 Profiles of the Natural Contaminant Radon 222 as a Measure of Vertical Diffusivity 469
  Amiram Roffman

- 8.7 Spectrophotometric Determination of Indole and Skatole Using Lignine Extractions 477
  M. Davidson, J. Kendler, and A. E. Donagi

9. Models

- 9.1 The Steady-State Demand-Output-Waste Economy 487
  H. Melvin Lieberstein

- 9.2 Simulation of Populations, with Particular Reference to the Grain Beetle, Tribolium 504
  N. W. Taylor

- 9.3 Pollution by Diffusive Processes 510
  H. S. Green

- 9.4 Computer Control of Physical-Chemical Wastewater Treatment 522
  Dolluff F. Bishop, Walter W. Schuck, Robert B. Samworth, Ralph Bernstein, and Elliott D. Fein

- 9.5 Computation and Mapping of the Dispersion and Herbage Uptake of Gaseous Effluents from Industrial Plants 548
  A. J. H. Goddard, R. E. Holmes, and H. Apsimon

- 9.6 Particle Collection Efficiencies on Cylindrical Wires 564
  George A. Sehmel

10. Noise

- 10.1 Industrial Noise Pollution 572
  L. S. Goodfriend and F. M. Kessler

10.2 Motor Vehicle Noise 587
William J. Galloway

10.3 Engineering and Scientific
Implications of Noise Control
Legislation 594
Alvin F. Meyer, Jr.

10.4 Noise Pollution in Developing
Countries and in Israel in Particular 602
F. Michael Strumpf

11. National and Local Problems and Solutions

11.1 Environmental Protection in the U. S. 606
Stanley M. Greenfield

11.2 Pollution Problems in Australia — A
Large, Sparsely Populated, Rapidly
Developing Country 616
D. O. Jordan

11.3 The Canadian Pollution Problem — A
Need for Broader Perceptions 620
P. M. Higgins

11.4 A Cooperative Approach to Pollution
Problems in Canada 633
C. G. Morley

11.5 Some Pollution Problems in Nigeria 647
Alaba Akinesete

11.6 Legal Control of Industrial Air
Pollution in Israel 651
Gerald M. Adler

11.7 Air Pollution Trends in Tel Aviv, Israel 664
A. E. Donagi, J. Mamane, and T. Anavi

11.8 A Dual-Purpose Air Pollution Alert and
Implementation System for the Greater
Tel-Aviv Area 671
A. E. Donagi, M. Naveh, A. Manes,
M. Rindsberger, Y. Gat, and A. Friedland

12. Political, Social, Educational, and Industrial Aspects of Pollution

12.1 Politics and Pollution: Political Solutions to a Deteriorating Environment 680
Ivan Völgyes

12.2 Some Economic, Spatial, Social, and Political Indicators of Environmental Quality of Urban Life 689
Chester Rapkin

12.3 Teratological Hazards due to Phenoxy Herbicides and Dioxin Contaminants 708
Samuel S. Epstein

12.4 Strategy for Maintaining Environmental Quality in Developing Technological Societies 730
Anthony Peranio

12.5 Environment's Most Dangerous Pest: Man 738
Donald C. Royse

12.6 Environmental Education as a Means of Creating an Awareness of Pollution by Tomorrow's Youth 750
Phillip Bedein

12.7 Pollution and Public Information 758
Teri Aaronson and Daniel H. Kohl

12.8 Are Industry and Government Fulfilling Their Responsibilities for Pollution Control? 765
Mitchell R. Zavon

12.9 Industrial Zoning 773
A. E. Donagi and I. Nizan

SUBJECT INDEX 779

GLOBAL EFFECTS OF CONTAMINANTS IN THE UPPER ATMOSPHERE

Lester Machta

Air Resources Laboratories, NOAA Silver Springs, Maryland 20910

INTRODUCTION

Although direct injections of contaminants into the upper atmosphere may be derived from both rockets and aircraft, this paper will treat only emissions from aircraft flying in the upper troposphere and lower stratosphere. It must be noted that there are also "indirect" sources of upper atmospheric pollutants from upward mixing of ground level emissions. Further, each man-made pollutant also occurs naturally.

AIRCRAFT EMISSIONS

Table 1 lists the emissions from modern subsonic aircraft both relative to fuel consumption and as hourly emissions for a typical large 4 engine jet aircraft during cruise mode (Forney, 1969). Also shown in Table 1, for comparison purposes, are the estimated emission rates of man-made pollutants (Robinson and Moser, 1970; Petersen and Junge, 1971; Machta, 1971a). One may convert from the emission of an individual aircraft to the world jet fleet by multiplying the second column by about 1,000.

It is evident that the first two products, carbon dioxide and water vapor, greatly exceed all others. These emissions are inevitable end-products in producing the energy for flight.

All forms of combustion of fossil fuels create carbon dioxide. The relative contribution of carbon dioxide from the aviation industry, even if it grows faster than other users of fossil fuel, will be small, of the order of a few percent or less in the next

Table 1

Emission from 4 Engine Intercontinental Subsonic
Jet Aircraft in Cruise Mode

Total fuel consumed per hr per aircraft (JP-4 or JP-5 fuel) ~14,600 kg/hr

Man-made Global Pollutant Emissions

Emission Product	kg Product per kg fuel	Emission rate kg/hr	Emission rate kg/hr
Carbon Dioxide	2.60	38×10^3	14×10^8
Water	1.34	20×10^3	see text
Carbon Monoxide	0.0010	15	32×10^6
Nitrogen Oxides	0.0015	22	60×10^5
Hydrocarbons	0.0049	72	10×10^6
Sulfur Dioxide	0.0004-0.0010	6-15	17×10^6
Particles	0.0009	13	47×10^6

20 years. Further, the average lifetime of a carbon dioxide molecule in the atmosphere is at least two years (Machta, 1971b). During this long period, the carbon dioxide from ground based activity will intermingle with that from aircraft. The altitude of input will not make the aviation industry a special culprit in potential climate change due to carbon dioxide growth in the atmosphere. If society exhibits concern about the increase in atmospheric carbon dioxide, the aviation industry contributes only a small part to the total concern.

Water vapor from aviation sources in recent years adds only about 10^{-7} of the amount evaporated from ocean surfaces over the globe. Unfortunately, the location of the aircraft insertion in the upper troposphere, may give this relatively small injection undue significance. But in comparison to another man-made source, its contribution is small.

As far as can be judged, the trace gases (as gases) such as carbon monoxide, oxides of nitrogen, and sulfur dioxide from aircraft flying below about 40,000 feet present no known impact on the weather or climate. Although they possess absorption bands in the infrared part of the electromagnetic spectrum, their steady state concentrations are too small to be of concern. Further, the tropospheric content of these gases from man-made based sources will far exceed that from aviation as seen in Table 1 (with column 2 multiplied by 10^3).

The sulfur dioxide can convert to sulfuric acid droplets or particles of ammonium sulfate, which may act as condensation nuclei, and this deserves further study. Unburned gaseous hydrocarbons may also partially convert to particles in the presence of sunlight. The direct formation of carbon particles or soot, while aesthetically undesirable at lower altitudes, has no known importance in the quantities to be released by civil aviation in the foreseeable future.

CONTRAILS

It is a matter of everyday experience that persistent contrails form behind many aircraft flying in the upper troposphere. Many of us can report examples of real or apparent cloud decks created during the passage of jet aircraft. Dean C.L. Hosler of Pennsylvania State University lives beneath one of the busiest airlanes in the world, the route between New York and such major air terminals as Chicago and Cleveland. He reports that on some days, after the flights of many aircraft, especially in weather conditions ahead of an upper level trough (airflow from the southwest), the sky will become overcast due to aircraft activities (Hosler, 1971). The cirrostratus deck has become thick enough to reduce the normal incidence solar radiation to one half of its expected cloudless value. The clouds appear from 10 to 20 hours before natural cirrostratus move in. Hosler believes that ice crystals falling into lower layers which are saturated with respect to ice can also create lower clouds although the explanation for this observed phenomenon is not clear.

Both heat and moisture are emitted from an aircraft. The warming of the air tends to inhibit contrails by giving the air a greater capacity to hold moisture in an unsaturated state. Appleman (1953) in Figure 1 shows in which parts of the atmosphere contrails are admitted. The heavy dashed line labelled U.S. Standard Atmosphere shows the average, temperate latitude vertical profile of temperature; it has a variability which lies between the two thin dashed curves. It is evident that the altitude of the subsonic jet cruise, 30-40,000 feet, has the greatest likelihood of contrail formation.

Table 1 shows that a large aircraft will emit about 2×10^7 grams of moisture for each hour of flight. As seen in Figure 2,

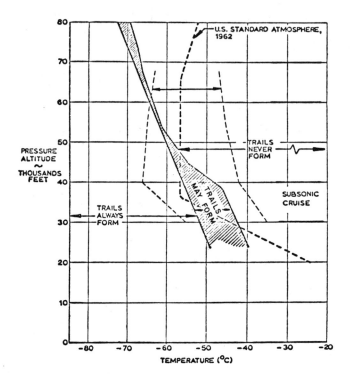

Figure 1. Contrail formation criteria for a U.S. Standard Atmosphere (temperate latitude), thin dashed lines indicate variability about the standard (Appelman, 1953).

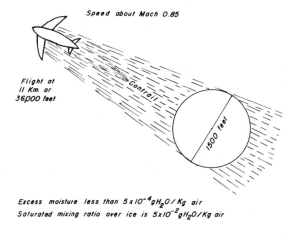

Figure 2. Schematic presentation of the dimensions of an atmospheric contrail induced by aircraft emitting 2×10^7 grams of moisture per hour.

by the time the aircraft wake and natural turbulence spread this moisture into a circular area of about 1500 feet diameter at a pressure of 225 mb at 36,000 feet the added aircraft moisture concentration will decrease to less than 5×10^{-4} g/kg of air. This is one hundredth of the saturated mixing ratio at 36,000 feet of 5×10^{-2} g/kg of air. Thus, the relative humidity would have to be in excess of 99% for the added aircraft moisture to saturate the air. Only very small volumes of the atmosphere will have air this close to saturation. Whey then do persistent contrails occur so frequently? Two possible explanations are offered.

First, the incidence of natural cirrus cloudiness is much more common than might be evident to a ground based observer. Pilots do detect high clouds when they are invisible to a ground observer. Some aircraft flying through air already saturated will thicken and make visible the clouds along the flight line. Even though natural clouds may be present, the aircraft can alter the optical and perhaps radiative properties of the upper troposphere.

Second, there is reason to believe that ice crystals will form when the air is saturated with respect to water but, once formed, will persist so long as the air is saturated with respect to ice. This means that air may have a relative humidity of between 60 and 100% with respect to water at 36,000 feet or 11 km (see Figure 3) and still be saturated with respect to ice. No

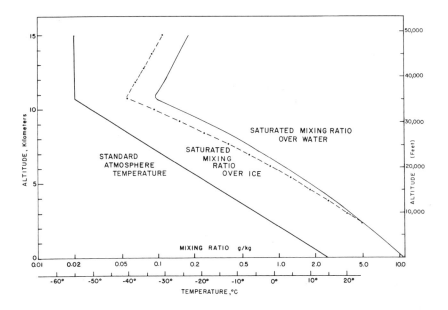

Figure 3. Saturated humidity mixing ratio over water and ice in a standard atmosphere to the lower stratosphere.

natural clouds may be present in this range (the region between the two saturation curves of Figure 3) since the air is unsaturated with respect to water. But if the air is brought to saturation with respect to water even momentarily as in the wake of the aircraft or by cooling it to dry ice temperature, then ice crystals will form and persist even when the relative humidity with respect to water falls below 100%. This now greatly enlarges the volume of air which may be affected by the small additions of aircraft moisture.

OBSERVATIONAL EVIDENCE FOR CHANGING CLOUDINESS

The most direct evidence of significant cloud modification by aircraft operations would be a parallel increase in the amount of high clouds with the increase in commercial jet aircraft operations since the late 1950's. The ordinary records of three-hourly observations of clouds at seven regular National Weather Service stations have been examined for trends in cloudiness between 1949 and 1969. It must be emphasized that observations of cloud amount and type, especially for high clouds, are very uncertain. One will expect to find considerable variability in the annual amount of high clouds due to random errors of observation as well as natural fluctuations in cloudiness.

Figure 4 shows a picture of the change in high cloudiness with time of two independent sets of data. The upper solid line applies to the average high clouds during observations with no low or middle clouds while the lower solid line is based on observations with 0.1 to 0.3 lower or middle clouds. The dashed line, read on the right hand scale, gives the time history of jet fuel consumption in the U.S. which is taken as proportional to the number of jet flight hours. The period 1965-69 has a greater cloud cover than any previous period on either curve. This recent period contains the greatest number of jet flights. The data show, then, that the last available 5 years have more high cloudiness; this is to be expected if the jet aircraft were the cause of increased clouds. So much for the argument favoring an association between cloudiness and aircraft increases. But two features in Figure 4 are difficult to explain. First, the lack of an upward trend in cloudiness between 1958 and 1964. Perhaps the smaller magnitude of the jet aircraft cloudiness was overbalanced by the noise which is evident even before the jet era. Second, the period 1951-55 has a negative anomaly in both curves which, statistically, is almost as significant as the anomaly in 1965-69. Insofar as is known, this smaller than average cloudiness is of natural origin. This negative anomaly suggests that natural causes such as year to year circulation changes rather than contrails might also be responsible for the positive anomaly in 1965-69.

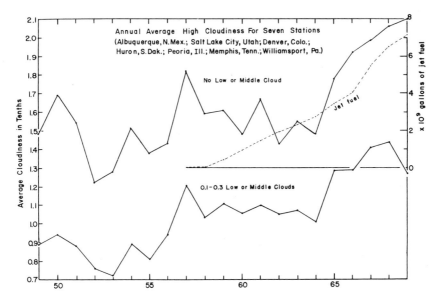

Figure 4. The history of the annual average high cloudiness for seven stations with zero low or middle clouds (upper curve) and 0.1-0.3 low or middle clouds (lower curve). The dashed line shows the growth on jet fuel consumption by U.S. domestic commercial jet aircraft.

EFFECTS OF CIRRUS CLOUDS ON THE WEATHER AND CLIMATE

There have been very few attempts to find the direct consequences of introducing artificial cirrus cloudiness. However, in a sense, once cirrus clouds form from the activity of jet aircraft, the artificial cirrus clouds are no different than naturally formed cirrus clouds. These effects generally fall into the following categories: 1) altering (probably increasing) the albedo or reflecting power of the planet Earth to solar radiation; 2) reducing the solar energy received at the earth's surface and consequently reducing the daytime temperatures; 3) reducing the outgoing long wave radiation so that at nighttime, temperatures in the lower atmosphere would be warmer than they would be with clear skies; and 4) seeding lower clouds by falling ice crystals.

The albedo change is likely to be very small since all the natural cirrus cloudiness contributes only a small fraction to the earth's albedo. However, since the climatic effects of even small changes in albedo are significant, the effect of increased cirrus cloudiness cannot be dismissed as unimportant. The reduction of the solar energy at the ground depends strongly on the cirrus character and other environmental factors. A study by Kuhn (1970) showed a 15% decrease of solar radiation due to a 500 m thick contrail sheet. Nicodemus and McQuigg (1969) have shown that the maximum temperature in the summertime over a continental area may be reduced by as much as 2°C if a cirrus deck replaces an otherwise clear sky. Since cirrus clouds are normally very cold compared to ground temperatures, the warming due to a reduction in outgoing long wave radiation will not be large. The reduction in outgoing terrestrial radiation found in Kuhn's study was about 2.4%. Finally, while seeding of lower clouds by falling ice crystals from cirrus clouds has been occasionally observed, the overall significance of this due to either natural cirrus or contrails is largely unevaluated.

SUMMARY

In summary, there is little doubt that jet aircraft have increased the amount of high clouds; all of us can attest to persistent contrails in the sky. But whether the increase is significant or negligible is not yet resolved. The weather station records consistently indicate more high clouds in the post jet period but natural variability precludes identifying the increase from 1965-69 as necessarily due to aircraft operations rather than due to natural causes. There are a variety of weather and climate effects which might follow from an increase in high clouds but further research will be required to make a case for or against a weather change.

REFERENCES

Appleman, H., The formation of exhaust condensation trails by jet aircraft, Bull. Amer. Meteor. Soc., 34, 14-20, 1953.

Forney, A.K., (FAA), Letter, "Combustion products from turbojet powered subsonic commercial aircraft," to T.C. Council (NOAA), January 17, 1969.

Hosler, C.L., Private communication, 1971.

Kuhn, P.M., Airborne observations of contrail effects on the thermal radiation budget, J. Atmos. Sci., 937-942, 1970.

Machta, L., Civil aviation and the environment, World Meteorological Organization, Commission for Aeronautical Meteorology, 5th Session, Geneva, CAeM-V/Doc. 13, ADD.1, October 1971a.

Machta, L., The role of the oceans and biosphere in the carbon dioxide cycle, Nobel Symposium, 20, "Changing chemistry of the oceans," Gothenburg, Sweden, August 1971b. (To be published.)

Nicodemus, M.L. and J.D. McQuigg, A simulation model for studying possible modification of surface temperature, J. Appl. Meteor. 8, 199-204, 1969.

Peterson, J.T. and C.E. Junge, Sources of particulate matter in the atmosphere, Man's Impact on the Climate, M.I.T. Press, Cambridge, Mass., 310-320, 1971.

Robinson, E. and C.E. Moser, Global gaseous pollutant emissions and removal mechanisms, Presented at the Second International Clean Air Congress of the International Union of Air Pollution Prevention Associations, Washington, D.C., December 6-11, 1970.

CLIMATIC ASPECTS OF WASTE HEAT

James T. Peterson[*]

Environmental Protection Agency, National Environmental Research Center, Division of Meteorology, Research Triangle Park, North Carolina

I. INTRODUCTION

This morning you have heard about two forms of man-made pollution which are capable of causing large scale climatic change, namely CO_2 and aerosols in the atmosphere. They have been receiving increasing attention in the scientific literature for about a decade. I will talk about a third pollutant, namely waste heat (or thermal pollution), and how its introduction into the environment can alter climate on large as well as small scales. Relatively little study has heretofore been devoted to this problem.

Waste heat is simply the heat or thermal energy generated essentially by all of man's varied activities. Combustion is by far the most important of these activities. Automobiles, home heating, industrial processes, and power generation all produce heat by combustion, heat that eventually is emitted into the atmosphere.

The ultimate sink of this energy, of course is outer space. Just as in the case of all solar energy absorbed by the earth, waste heat is eventually emitted by the earth and atmosphere as thermal infrared radiation to space. Since the infrared emission of energy directly depends on the fourth power of the temperature, the warmer earth-atmosphere will emit more radiation to space to achieve a long term balance between energy income and outgo.

---

[*]On assignment from the National Oceanic and Atmospheric Administration, U.S. Department of Commerce

Therefore, in sufficient quantity waste heat can significantly warm the earth and atmosphere and be climatically important. Moreover, in specific situations it can enhance convection, i.e., the vertical motion so important in precipitation processes.

Before going on to discuss specific meteorological applications, I would like to give you some background information on this general problem. There has recently been a considerable abount of publicity about the so-called energy crisis. Basically, this crisis stems from the fact that we are using energy faster than we can discover the fossil fuels used to generate it. Moreover, the demand for electricity is growing faster than the plants for generating it are being built. The energy crisis is receiving growing attention throughout government and private sectors as evidenced by many technical reports and popular articles. One small part of this total crisis is the effect of waste heat on weather and climate.

The energy crisis emphasizes the rapid rate of growth of energy use not only in the United States, but throughout the world. From 1950 to 1968 the rate of growth of energy consumption in the world exceeded 5 percent per year (see Table 1). During these years the annual amount of global energy used increased from 77 to 100×10^{15} Btu/yr (2.6 to 6.3×10^{12} watts). In the United States alone the use of electrical energy has increased by about 7 percent per year during the last five years. Darmstadter's (1972) forecast of future energy use amounts to 830×10^{15} Btu/yr by the year 2000. This is based on a 5.1 percent growth rate to 1980 and a 4.5 percent rate thereafter. This increase in global energy use by a factor of nearly 4.4 is naturally based on many assumptions. However, prediction based on simple extrapolation of current trends does not yield substantially different results by 2000. Thus, at a rate of increase of about 5 percent per year, which in itself does not seem very large, in 32 years an increase by a factor of about 4 1/2 times results!

An interesting sidelight to this problem is that nuclear generation of electrical power is often heralded as the solution or at least the partial solution, to the energy crises. From a waste heat viewpoint, however, it will only make matters worse. The reason for this is that fossil fuel generation facilities are about 35 percent efficient on the average. In other words, only 35 percent of all heat generated is converted to electrical power, the remainder goes out the stack, into cooling water, etc. Present-day nuclear plants are even less efficient. For a given amount of electricity generated, a nuclear facility has about 2/3 more heat loss than does a corresponding fossil fuel plant. However, future technology may significantly alter this comparison. Another suggestion to ease the energy crisis is the use of waste heat to increase agricultural or aquacultural production. Such schemes might simply redistribute the heat somewhat, but would not tend to reduce the total amount added to the environment.

TABLE 1. Past and predicted future total energy consumption for the world, 10^{15} Btu/yr. From Darmstadter (1972).

Year	World Energy Consumption, 10^{15} Btu/yr
1925	44
1950	77
1955	100
1960	124
1965	161
1968	190
1980	345
2000	831

Compounding the expected energy growth is the trend toward more and larger urban areas. More and more of the populace continue to move from rural to urban locations. In the United States demographers predict the existence of four major metropolitan centers by the year 2000: the North Atlantic Seaboard, the Lower Great Lakes, Florida, and California. Sixty percent of the nation's population (200 million people) will live on 8 percent of the land. Consequently, the distribution of waste heat will likely also be clustered in these areas.

Today, man-made heat emissions together with a surface changed from vegetation to building materials are altering the climate on a local scale. Urban areas are noticeably warmer than their rural surroundings. On the other extreme -- the global scale -- waste heat is insignificant compared to the natural energy balance. Our concern for the near future centers on the regional scale where population and energy use will continue to cluster. Over areas of hundreds of thousands of square kilometers emission of waste heat will be several percent of the naturally absorbed solar radiation. Such emissions may be sufficient to warm the regional climate, to initiate, intensify, or alter the tracks of storms, or to enhance convection and precipitation.

Before discussing some specific climatic problems, I will present some data on the current magnitude of waste heat emissions. Averaged over the entire globe the absorption of solar energy by the earth equals about 200 watts/meter^2 (W/m^2) while the total (solar and infrared) net (incoming minus outgoing) radiative energy balance is about 80 W/m^2. Obviously these figures vary markedly from north to south. Rates of energy use by man for several geographical areas are shown in Table 2. These vary from the intense energy use in Manhattan, New York City, to that for the world, which is insignificant in comparison to the natural energy balance. On an intermediate scale, in Los Angeles County with an area of 10,000 km^2 the consumption of energy is already several percent of that naturally received from the sun.

II. SMALL SCALE EFFECTS

Over certain rather small areas of the earth, man has concentrated his use of energy. Cooling towers, especially the large (approximately 100 meters tall) natural draft units primarily used to cool water at electrical generating plants, are prime examples. They are intense sources of heat and moisture and may liberate 10,000 W/m^2 over a small area. Another paper presented at this conference by Hummell is devoted solely to the environmental problems associated with cooling towers. Thus, they will be only briefly discussed here.

Cumulus clouds have been observed to form immediately downwind of cooling towers, and their heat and water emissions are suspected of occasionally enhancing convective precipitation by perhaps triggering a locally unstable weather situation. In addition, ground fog is slightly more prevalent in their vicinity, but this is more frequent with smaller mechanical draft towers than large natural draft untis. However, the meteorological significance of cooling towers will not be clarified until more studies are made, such as establishment of raingage networks in the vicinity of cooling towers or detailed analysis of cloud radar echos.

The most notable alteration of climate on a small scale occurs over urban areas. The intense human activity and use of energy in cities is largely responsible for urban-rural climatic differences, but the replacement of natural vegetation by building materials is also significant (Peterson, 1969). The most evident feature of city climate is its excess warmth, which is commonly referred to as the urban heat island. Occasionally urban-rural temperature differences reach 10°C, but on the average cities are 1.0 to 2.0°C warmer. This temperature effect is maximum at night, especially during weather dominated by high pressure with low winds and clear skies. The heat island can be measured in cities of all sizes and in all parts of any given city, from the center out through the outlying suburbs. Other major features of urban climates which are related to energy use include: (1) a longer frost-free growing season which, for example, in Washington, D.C., is more than a month longer than that of nearby rural areas; (2) less snowfall because it melts while falling through the warmer urban atmosphere; (3) lower relative humidity; (4) less fog because of the lower humidity, a feature which may be offset, however, by more particulate matter that serves as condensation nuclei; (5) about 5 to 10 percent more precipitation downwind of cities, a phenomenon likely at least partially to be due to increased convection (vertical motion); and (6) a slight component of the wind directed toward the city center as a result of the horizontal temperature contrast.

TABLE 2. Energy Consumption Density (ECD), W/m^2, for selected areas of the world, for the late 1960's. From Inadvertent Climate Modification (1971).

	Area (km^2)	ECD (W/m^2)
Manhattan, New York City	59	630.
Moscow	880	127.
West Berlin	234[a]	21.3
Los Angeles	3,500[a]	21.0
Northrhein-Westphalia (industrial area)	10,300	10.2
Los Angeles County	10,000	7.5
Northrhein-Westphalia	34,000	4.2
Benelux	73,000	1.66
United Kingdom	242,000	1.21
14 Eastern U.S. States	932,000	1.11
Central-Western Europe	1,665,000	0.74
World	510,000,000	0.012

[a]Building area only.

The most important non-meteorological effect is the personal discomfort for those who live in urban centers, especially during humid, summer evenings. The city simply does not cool as rapidly as the outlying areas. During prolonged hot weather, increased urban mortality frequently occurs as a result of the heat island (Clarke, 1972). Studies in such cities as New York and St. Louis have shown that the death rate noticeably increased during heat waves lasting several days.

The effect of waste heat on urban areas can be summed up by stating that there are climatic effects and there are health effects, but they do appear to be restricted to the local area. These effects are being tolerated now, and it is likely that they will continue to be tolerated. There has been much talk and discussion of urban and regional planning, but thus far the design of only a few new developments has included climatic influences.

III. GLOBAL EFFECTS

On a local scale the climatic effects of energy use are significant and well documented. On the other extreme -- the global scale -- climatic effects today and for the near future are insignificant. In 1970, world-wide energy consumption was about 7 to 8 x 10^{12} watts. Averaged over the entire globe the higher estimate equals 0.016 W/m^2. Distributed over the continental areas only, it yields an energy use density of 0.054 W/m^2. Thus, even if the global energy use should increase by tenfold, the average energy densities would be less than one percent of the natural radiation balance. Since the global energy use is likely to increase by four to five times in the next 30 years, from this viewpoint no climatic problems are likely before the year 2000.

For those people who like to play games with numbers, energy use growth rates can be extended beyond 2000. However, little reliability can be attached to such figures. For example, an annual growth rate of 5.5 percent compounded for 80 years (to about) the year 2050) results in a seventy fold increase. If such an energy use growth pattern should acutally occur, but no such prediction has been made with confidence, there would likely be a noticeable warming on a world-wide basis.

IV. INTERMEDIATE SCALE

In between the intense energy use on a local scale and the insignificant use on a global basis, is the intermediate or regional scale where, in the near future, man's generation of heat may become climatically important as urban areas expand and merge. Current forecasts of energy use, extrapolations of past trends, etc., indicate roughly a 5 percent growth rate to the year 2000. Thus, the order of a four-fold increase in global energy use can be expected by 2000. In some countries or areas this rate will be exceeded, but these figures can be used as an overall estimate. Using Table 2 as a guide to current energy statistics, it is reasonable to expect that by 2000 there will be several areas in the world of 100,000 to perhaps a million km^2 with waste heat of 5 to 10 W/m^2, i.e., several percent of the energy naturally received from the sun.

What are the likely climatic consequences of such an energy use density? Today, meteorologists don't know what all the specific consequences will be. The answer must wait until better mathematical models of the atmopshere are developed that can describe weather and climate patterns. However, some first steps have been made toward solution of this problem.

Sellers (1969) at the University of Arizona has developed a highly parameterized global climatic model in which he applied an average of about 20 W/m^2 of thermal energy over the entire globe, but weighted by latitude in accordance with the distribution of large cities. The results of this extreme example were an increase in average global temperature of about 15°C, ranging from 11°C near the the equator to 27°C at the North Pole.

Washington (1971) has applied the problem of waste heat to the more sophisticated model of the atmospheric general circulation at NCAR in Boulder. He applied 25 W/m^2 of thermal energy to all land areas of the globe which again is an extreme case. The root-mean-square temperature difference between his experimental and control results was 5°C averaged over the globe. Washington (1972) reran his model for a more realistic, but still extreme example. He applied about 40 times more thermal energy than is currently being used on a global basis, distributed in the same proportion as the present population. The results were inconclusive. The model may not have been sensitive enough to handle the experiment.

The work of Sellers and Washington suggest that extreme waste heat inputs (i.e., that are very unlikely to be reached in the foreseeable future) will likely result in extreme climatic changes. What is now needed is more refinement and further application of such models.

These few examples are the only results of quantitative work yet to be published on this problem. There are, however, some qualitative arguments that also suggest that the waste heat will have climatic importance on a regional scale. First, various climatic models (e.g., Budyko, 1969) show that when an anomaly of one percent of the absorbed solar energy occurs, significant alterations of climate result. For example, on a global scale, a one percent change in solar input or surface albedo will lead to about a 1°C change in mean surface temperature.

Second, during the 1960's the eastern half of the United States experienced below normal average temperatures of up to several degrees in some places. Namias (1969) has suggested that this temperature change was due to an anomalously warm North Pacific Ocean. In fact, throughout most of the 1960's the surface temperature of the entire Pacific Ocean north of about 20°N averaged above normal. Most of the ocean averaged only slightly above normal but about 3×10^6 km^2 averaged about 1.5°F (almost 1°C) above normal. In the not too distant future areas of one million or more km^2 (e.g., the northeast United States and West Central Europe) will have a waste heat input similar to that of some current cities, which have an average heat island of nearly 1°C. A sea surface and an urban surface are hardly similar in terms of heat and moisture exchange with the atmosphere. But, if a 1°C temperature anomaly in the Pacific can cause significant temperature changes throughout half of North America, we should determine what such a temperature anomaly over land will cause.

Third, those of us who live on the eastern coast of the United States have experienced winter storms which generally track up the eastern seaboard deriving much of their energy from the cold land -- warm sea temperature contrast. Will the climatology of such storms be affected by the expected change in thermal contrast along the northeast United States?

V. CONCLUSION

Perhaps, just perhaps, the projected 5 percent energy growth rate will not materialzie because of increased costs of energy or of some reordering of national priorities or life styles. In the United States, the birth rate is sharply declining. Statistics for the first quarter of 1972 show the lowest number of births/1000 people since tabulations began 70 years ago.

There is, however, a pressing need to develop and apply climatic models to the type of regional problems discussed above. If large scale climatic consequences are likely, many years lead time will be necessary for adjustments in regional planning. We must determine the level of waste heat input that will significantly affect regional and global climate!

REFERENCES

Budyko, M.I., 1969. The Effects of Solar Radiation Variations on the Climate of the Earth. Tellus 21 (5): 611.

Clarke, J.F., 1972. Some Effects of the Urban Structure on Heat Mortality. Envr. Res. 5(1):93-104.

Darmstadter, J., 1972. in Energy, Economic Growth and the Environment, ed. S.H. Schurr. Johns Hopkins Press.

Inadvertent Climate Modification, 1971. Report of the Study of Man's Impact on Climate. MIT Press, Cambridge, 308 p.

Namias, J., 1969. Seasonal Interactions Between the North Pacific Ocean and the Atmosphere during the 1960's. Mon. Weather Rev. 97, 173-192.

Peterson, J.T., 1969. The Climate of Cities: A Survey of Recent Literature. Nat. Air Poll. Ctr. Admin., Pub. No. AP-59. Raleigh, N.C.

Sellers, W.D., 1969. A Global Climate Model based on the Energy Balance of the Earth-Atmosphere System. J. Appl. Meteor. 8(3): 392-400.

Washington, W., 1971. On the Possible Uses of Global Atmosphere Models for the Study of Air and Thermal Pollution. in Man's Impact on the Climate, ed. W.H. Mathews et al, MIT Press, Cambridge. 594.

Washington, W., 1972. Personal communication.

COOLING TOWERS AND WEATHER MODIFICATION

John R. Hummel

Department of Engineering Science, Pennsylvania State University

Electric power production has reached astronomical proportions worldwide. Most of this power is produced by either coal-fired or nuclear steam turbines, and because of the inefficiencies involved in power production immense quantities of waste heat must be disposed of. At present, the power industry must dispose of 1.3 watts of heat for every watt of generating capacity at a coal-fired steam generating station and about 2 watts of heat per generated watt at a nuclear power plant.[1] Water, an excellent coolant because of its high specific heat, is used at most generating locations to carry off the excess heat. Once the heat is transferred to the water what do you do with the heated water?

A common practice has been to draw water from a body of water, use the water as a coolant for the power plant, then return the heated liquid back to the body of water from which it came. This method is now in an unfavorable light due to the concern over thermal pollution of bodies of water. As a result, power station designers look towards cooling towers as a solution.

In principle, cooling towers take the hot water used as a coolant by the generating station and by means of passing a draft of air through the water, cool the water enough to allow recir-

---

[1] Stockman, John, <u>Cooling Tower Study</u>, IITRI-C6187, IIT Research Institute, Chicago, 1970, p.2.

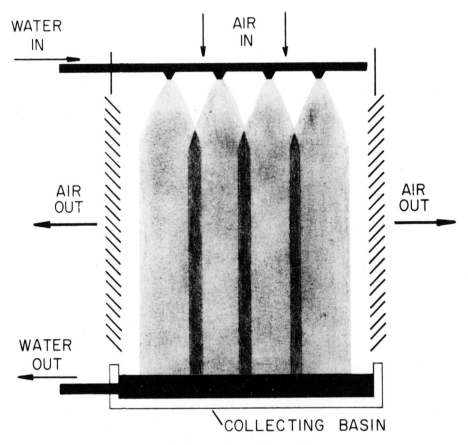

Figure 1. Schematic of operation. Heated water enters at top, passes through a draft of water where it is cooled, and is collected and recycled.

culation of the liquid through the power plant. See figure 1. Temperature drops of from 10 to 40 degrees Fahrenheit are possible for water passing through a cooling tower, with most of the waste heat being passed off into the atmosphere by evaporation. Exact details, of course, depend upon the type and number of units employed. Enormous amounts of heat and water vapor are released into the atmosphere from cooling towers and since heat and water vapor are prime determinants of meteorological conditions, the hot, moist plumes emitted can modify local weather conditions as the following example shows.

In Philadelphia, Pennsylvania the Atlantic Refining Company operates a refining facility which is skirted by the Schuykill Expressway. About 400 feet from the expressway is an eight-cell cooling tower designed to remove 5 million BTU's of heat from 24,000 gallons of water each minute. This is accomplished by sucking up 2.4 million cubic feet of air - enough air to fill 3000 average homes - through the water each minute. On December 7, 1959, weather conditions coupled with the plume to produce a fog which shrouded the expressway - a hazardous condition on the already treacherous freeway.

Analysis of the situation revealed that fog would appear if winds down - drafting in a westerly direction (towards the expressway) at six miles per hour or more plus low atmospheric temperatures around 30 - 40 degrees Fahrenheit and humidities above 60 percent occured or if low level inversion conditions with low temperatures and high humidity existed. These conditions do not occur often but the threat they pose warranted concern on the part of Atlantic.

Fortunately, company engineers found a solution to their unique but dangerous problem. They discovered that heating the plume prevented reoccurence of the fog.[2]

Clearly, short term effects are possible from cooling towers but what about long-term effects in areas where these towers are concentrated? Growth predictions show 32-40 natural draft cooling towers, identifiable by their huge, hyperbolic shape, being built in the United States, with 22 being built in Pennsylvania alone. Presently, the largest number of these towers, four, can be found at the Keystone generating plant in Western Pennsylvania and they were the focus of a study by John Stockman of the Illinois Institute of Technology.[3]

The towers at the Keystone site are natural-draft cooling towers that operate on a chimney principle as opposed to mechanical-draft cross flow towers which rely on a fan to induce the draft. Figure 2 details the two tower types. In the natural draft tower the height of the structure and the difference in

---

[2] Hall, William A., "Elimination of Cooling Tower Fog from a Highway", Journal of the Air Pollution Control Association, Vol. 12, No. 8, August 1962 pp. 379-383.

[3] Stockman, p. IV

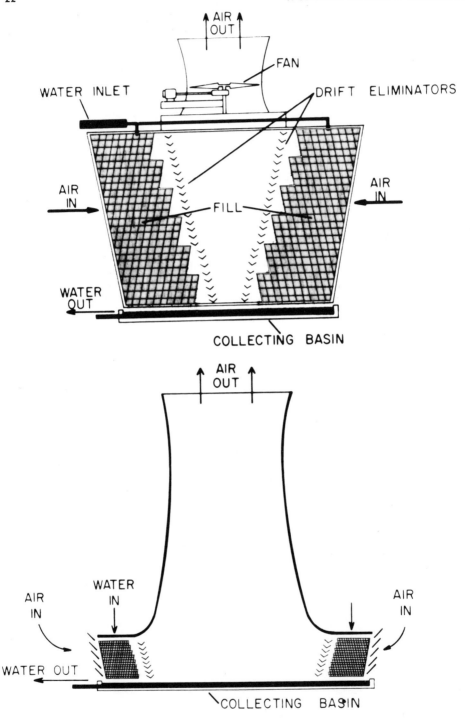

Figure 2 (Top) An inside look at a mechanical draft cooling tower. (Bottom) A cross-sectional view of a natural draft cooling tower.

(a)
A battery of mechanical draft cooling towers.

(b)

The Keystone generating station under normal conditions.

Figure 3

(c)

An early morning shot of the Keystone location. A temperature inversion and turbulence in the environment mixed the water vapor from two of the towers with small patches of ground fog to produce a layer of fog covering the entire area around and downwind of the station.

(d)

Plumes from the four Keystone towers produced this cloud formation which extended downstream for miles. The plumes penetrated a surface inversion layer and stopped at an elevated inversion (5500 to 8000 feet) where the air was close to saturation.

Figure 3

density between the warm, moist air inside the tower and the colder, dense air outside induce a strong draft which passes through the heated cooling water from the generating station. These behemoths stretch 325 feet into the air and discharge 10,000 gallons of water and 120 million BTU's of heat into the atmosphere each minute when the power plant is operating at 80-86 percent of its rated capacity of 1800 megawatts.

The Stockman study revealed that under "normal" atmospheric conditions the plume was visible up to an altitude of 200 meters before dissipating. When the temperature was in the range of 25-30 degree Fahrenheit and relative humidity approached 80% the plume was visible for thousands of meters. Even when the plume was visible for short distances aerial measurements of humidity could trace the path of the plume for up to 11,000 meters. On occasion, the plume was witnessed to evaporate and then recondense at higher altitudes downwind. Figure 3 shows some tower effects.

The results of the study proved to be inconclusive. No adverse weather effects could be directly linked with the towers at the Keystone station although precipitation enhancement during July 1969 appeared to be linked with tower operations. Cloud initiation was infrequent but the plume was observed to frequently merge with local stratus cloud cover.

Cooling towers have the potential to produce fog and drizzle, initiate clouds, and enhance rainfall. The potential hazards of resultant fogs or other inadvertant weather modifications can be lessened through research, locating towers away from highways, and by thorough examination of a chosen site's local meteorological conditions. The use of cooling towers is essential in order to preserve water supplies and prevent thermal pollution of water but not if the cost is modification of the weather.

REFERENCES

1. H. R. BEYERS, General Meteorology, McGraw-Hill, New York, 1959.
2. W. A. HALL, "Elimination of Cooling Tower Fog from a Highway," vol. 12, no. 8, August 1962, pp. 379-83.
3. C. L. HOSLER, "Wet Cooling Tower Plume Behavior," AIChE Cooling Tower Symposia, Houston, Texas, March, 1971.
4. J. STOCKMAN, Cooling Tower Study, IITRI C6187, Illinois Institute of Technology, Chicago, 1970.
5. R. D. WOODSON, "Cooling Towers," Scientific American, May 1971, pp. 70-78.

MAN-MADE ALTERATIONS OF THE NEAR EARTH SPACE ENVIRONMENT

Wilmot N. Hess

*NOAA Environmental Research Laboratories
Boulder, Colorado*

This paper will deal with questions of "pollution" in space - not in the atmosphere but at altitudes of hundreds or thousands of miles. There is a question about what pollution in space means. What is the pollution? It is particles - not fly ash or SO_2 molecules - but rather high energy electrons and protons in the Van Allen radiation belt.

In what sense are these particles pollution? If you are an astronaut on a space flight, they are pollution. You would like it if they were not there. They constitute a source of radiation that will limit the kinds of flights you can make. All manned circumterrestial flights so far have been at altitudes of less than 300 miles. At this altitude we are below the Van Allen belt. Lunar flights go through the Van Allen belt quite rapidly, but still the astronauts pick up about 1 rad dose of radiation in the Van Allen belt. Probably prolonged manned space flight in the heart of the Van Allen belt is not feasible now. With reasonably heavy shielding the dose rate could be about 1 rad/hour which is quite high.

So, the Van Allen belt represents pollution to manned space flight (and unmanned flights using sensitive optical instruments are also bothered.) However, in fairness I should say that many people do not consider the Van Allen belt pollution but rather lots of fun to study as a geophysical phenomena. In engineering they say, "One man's signal is another man's noise."

The aurora or northern lights could also be considered "pollution." The aurora is caused by energetic electrons coming down into the atmosphere. The aurora is also lots of fun to study, but it produces interference to communications. The ionosphere is disturbed in and around aurorae and frequently radio circuits in the auroral region don't work during aurorae.

I have tried to make the case that various kinds of energetic particles represent a special kind of pollution in space. What can we do about it? In the last few years people have been able to carry out several experiments to articicially modify the space environment. We can now produce artificial aurorae. We can change the population of the Van Allen radiation belt. We can artifically modify the ionosphere from the ground and we have other ideas about artificial experiments for the future. Let me discuss these several experiments and then return to the question of pollution.

ARTIFICIAL AURORAE

The first artificial aurorae were made by nuclear bombs exploded at high altitudes (1). The following table lists these explosions that have made aurorae. The aurorae are made by energetic electrons mostly from decay of fission products. These electrons follow magnetic field lines and enter the atmosphere below the explosion site and at the conjugate point and create aurorae.

More recently we have been able to make aurorae by using accelerators (2). A beam of $\frac{1}{2}$ amp of 10 kev electrons sent out from an Aerobee rocket at $\sim$ 250 km produced an aurora in January 1969. This is not a bright aurora - in fact, it had to be photographed by a special TV system to detect it at all. Brighter aurorae will be made by accelerators being developed now in the U.S. and in the USSR. These accelerator experiments are carried

TABLE 1

Explosion	Locale	Date	Yield	Altitude, km	L of burst
Argus I	South Atlantic	August 27, 1958	1 kt	~200	1.7
Argus II	South Atlantic	August 30, 1958	1 kt	~250	2.1
Argus III	South Atlantic	September 6, 1958	1 kt	~500	2.0
Starfish	Johnson Island Pacific Ocean	July 9, 1962	1.4 Mt	400	1.12
U.S.S.R. [1]	Siberia	October 22, 1962	Several hundred kt	?	1.9
U.S.S.R.	Siberia	October 28, 1962	Submegaton	?	2.0
U.S.S.R.	Siberia	November 1, 1962	Megaton	?	1.8

out to (a) study the propagation of beams in space and plasma instabilities that might disrupt the beam and (b) develop a system capable of sending beams long distances in space in order to map magnetic fields in space and also to study electric fields. An electron beam might be used to solve an interesting space science problem. Does the magnetic field line from one magnetic pole go out into the geomagnetic tail and then connect to the other pole or not? This magnetic field line might connect to the sun. By trying to observe at one pole the auroral ray produced by the accelerator at the other pole, we can possibly answer this question.

Another technique has been developed to make large scale optical displays in space. Scientists at the University of Alaska have developed a technique for putting out a beam of Barium ions by using shaped charges of explosives. Such a beam of Ba^+ ions launched from a rocket above Hawaii were photographed to display an entire magnetic field line. This technique can clearly map magnetic fields in space also.

IONOSPHERIC MODIFICATION

When an auroral beam enters the atmosphere from space, it modifies the ionosphere. The energetic electrons collide with atmospheric molecules and produce lots of low energy electrons. This results in a bump on the ionosphere – a local region of higher electron density and of higher electron temperature. These effects have been measured above natural aurorae. They probably occurred in connection with our accelerator produced aurorae but our attempts to measure it were inconclusive. These changes in the ionosphere have a substantial change in the manner in which radio waves propagate through this region. Communication black outs sometimes occur in connection with aurora.

Just last year, scientists in Boulder, Colorado, were able for the first time to modify the ionosphere from the ground below. A very high power radar in Colorado, putting out 2 megawatts at 5 to 10 megacycles/second beamed upwards, produced substantial alteration of the ionosphere in a few seconds (3). Figure 1 shows the onset of "spread F" which is a classic form of ionospheric disturbance which occurs naturally quite frequently. In this case, spread F was produced by the radar beam in a few seconds. A region of the ionosphere about 100 km wide and maybe 10 km thick was heated and expanded and the electron density decreased. Energetic electrons moving down field lines from the modified region at about 300 km altitude enter the dense atmosphere and produce air glow which has been measured by optical equipment on the ground.

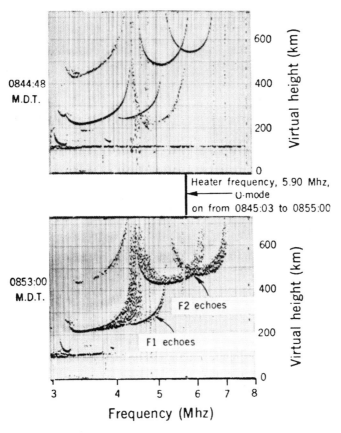

Figure 1. Ionograms (24 July 1970) showing the development of daytime spread F during excitation by the high-power ground-based transmitter.

PRODUCING ELECTROMAGNETIC WAVES

The first evidence that man could produce EM waves in space was found accidentally in connection with the signals sent out by the 18 kHz U.S. Navy transmitter at Jim Creek, Washington (4). When dashes were sent (150 msec long) they sometimes produced signals called "riser" -- a form of VLF wave seen naturally and here produced artificially about 100 msec after the signal started. Interestingly, the shorter dots from Jim Creek do not produce VLF waves. It is suspected that the riser is due to the presence of energetic electrons near the earth's equator at high altitudes that can cyclotron resonate with the transmitted EM wave as it passes through the equator.

More recently, people have tried to produce waves in space deliberately. Helliwell has used a VLF transmitter in Antarctica called "Longwire" to generate VLF waves. So far, only weak signals have been detected at long distances from this transmitter but probably this technique will work and produce magnetospheric waves.

Whistlers, which are circularly polarized VLF waves that propagate along magnetic field lines, have been produced by high altitude nuclear explosions (5). These waves have the interesting property that they can cyclotron resonate with energetic electrons in the magnetosphere very efficiently and can change their orbits substantially (Figure 2). Dungey (6) showed that the principal effect here was to change the direction of motion of the particle rather than its energy. This can result in removing electrons trapped in the Van Allen radiation belt. Very probably the rapid loss of electrons from the outer portion of the Van Allen belt is due to whistlers produced by lightning strokes scattering the trapped electrons down into the earth's atmosphere.

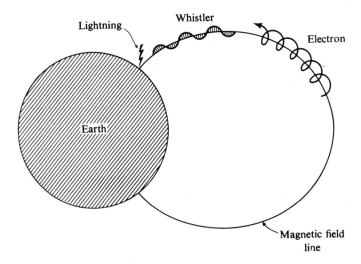

Figure 2. A whistler, a circularly polarized wave generated by lightning, propagates along a field line and can resonantly interact with a charged particle moving along the line.

It would be interesting to be able to produce whistlers from the ground easily because then we could probably control the population in the Van Allen radiation belt overhead quite readily.

Recently it has been found that the VLF waves, called risers, are associated with the occurance of X rays measured on high altitude balloons (7). Probably the X rays result from the precipitation of energetic electrons into the atmosphere producing bremstrahlung. Very likely the riser VLF wave interacted with the electrons to cause them to precipitate.

Waves have also been produced by accelerator-produced particle beams in space. Directed energetic particle beams should be unstable and tend to emit EM waves. An accelerator capable of emitting 70 ma of 40 kev electrons was flown on a rocket in August 1970. The rocket nose cone was ejected forward from the rest of the payload and carried a sensitive radio receiver. This receiver detected (a) plasma waves, (b) whistlers, and (c) other waves of uncertain origin (8).

INJECTION OF THERMAL PLASMA

A very promising new idea has been put forward to reduce particle populations in space. In 1966 Kennel and Petcheck (9), developed a theory to explain several features of the natural Van Allen belt electrons. They showed that this electron population was self regulating. If the population gets too big a plasma instability starts that produces whistler waves. These waves interact with the high energy electrons and scatter them into the atmosphere reducing the population. This theory agrees quite well with experimental data on maximum observed energetic electron populations and is generally accepted by most scientists.

Brice (10) suggested that one feature of the theory of Kennel and Petcheck should allow artificial control of the high energy electron population. The amplitude A of the whistler waves that scatters out the high energy electrons depends on the density of _thermal_ electrons n_e in the region where the waves are generated. The reason for this is that these thermal electrons are the medium in which the waves are formed. The high energy electron population P depends inversely on the whistler wave amplitude. That is

$$P \propto 1/A \propto 1/n_e$$

This shows that if we increase the _thermal_ electron density n_e we will decrease the _high energy_ electron population P. This says, if a thermal plasma is injected in space, it will allow higher amplitude whistler waves to occur and will scatter out lots of high energy electrons. This population control device has not been tested yet - it probably will be in a few years (11) - but the idea is so well based theoretically that it would be surprising if it did not work.

OTHER IDEAS

There are other interesting possibilities that have been suggested for modifying space.

The largest and probably most important process that might be used would be an atomic bomb in space. The Starfish nuclear explosion of July 1962 produced a large trapped population of energetic electrons from fission product decay in space (1). This transient population has only recently disappeared. The decay following these bomb-produced electrons and also the electrons from the three USSR high altitude nuclear explosions of October and November 1962, has provided the best data on the lifetimes of trapped electrons in the Van Allen belt. Starfish also suddenly moved large fluxes of trapped protons to different positions and allowed us to watch this particle population return to its natural state. Lots of interesting space physics has been done following these bomb effects. However, no more high altitude nuclear bombs will be set off by the U.S. or USSR or U.K., so this process of space modification is not available to us anymore - and properly so.

Another possibility suggested by Petchek (12) was to deposit a large mass of ionized matter into the outer regions of the radiation belt. This region of space is generally supposed to be in motion with material being convected inwards towards the earth. This convective motion may produce a $\vec{U} \times \vec{B}$ electric field and this electric force may accelerate the energetic electrons which make aurora. If a mass of ions, probably measured in tens of kilograms, is placed in the convecting region, it should slow down the convection and turn off the aurora maybe for several days. This idea is untested and is based on uncertain theory.

SATELLITE SWEEPING

Particles in the Van Allen radiation belt bounce from the hemisphere and slowly drift around the earth as shown in Figure 3. It might be possible to absorb or sweep out some of these energetic particles by putting a large satellite into the region of the belt (13). We can calculate how long it would take for the particles

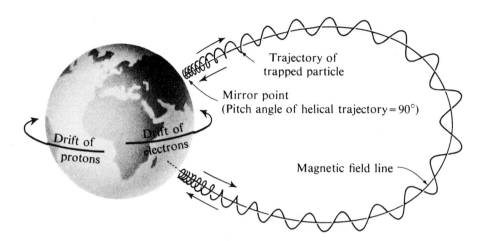

Figure 3. Motion of a charged particle in a dipole field. For low-energy particles the motion can be broken up into three components, as shown here: (1) spiraling along field lines, (2) bouncing back and forth along field lines from one hemisphere to the other, and (3) drifting in longitude around the Earth with electrons drifting East and protons West.

to be lost by striking the satellite. Let's assume that we have a satellite of radius ρ moving on a circle of radius R concentric to the drift path of a group of particles of bounce period τ.

The probability p_1 of absorption per particle motion past the satellite is

$$p_1 = \frac{2\rho}{2\pi R}$$

so the probability p_2 of absorption per second is

$$p_2 = \frac{p_1 2}{\tau} = \frac{2\rho}{\pi R \tau}$$

so the mean lifetime, T, for these particles is

$$T = \frac{1}{P_2} = \frac{\pi R \tau}{2\rho}$$

If we consider relativistic particles at R = 10,000 km which have bounce period τ = .15 sec and we take an artificial satellite of ρ = 1 km we get a mean lifetime of

$$T = \frac{\pi(10,000)(.15)}{2} = 2400 \text{ sec} = 40 \text{ min}$$

This shows that in a day the satellite will have swept out a slot 1 km wide in the radiation belt.

If we want to sweep out a broad swath of the radiation belt of width W we should put the satellite into an eccentric orbit in which case we have approximately

$$P_1 = \frac{\pi \rho^2}{2\pi RW}$$

and the mean lifetime is

$$T = \frac{RW\tau}{\rho^2}$$

If we take W = 1000 km for the conditions above we find

$$T = \frac{(10,000)(1000)(.15)}{(1)^2} = 1.5 \times 10^6 \simeq 2 \text{ weeks}$$

This says that if we put up a thick-skinned balloon-satellite of ρ = 1 km we could cut a pretty good sized hole in the inner portion of the Van Allen radiation belt in one year.

It has been suggested that this sweeping process is going on naturally at Jupiter (14). Five of the Jovian moons are inside the intense radiation belt that is known to exist near Jupiter. It seems probable that these moons are sweeping up energetic electrons and protons in the Jupiter radiation belt and that there are nearly empty slots there. If these slots are not found when the Pioneer spacecraft fly past Jupiter next year, it will tell us something very interesting about the particle motion in the Jovian radiation belt.

CONCLUSIONS

I have indicated that the following kinds of artificial modifications of the space environment are possible:

(a) launch a particle beam from a satellite or rocket to create an artificial aurora or to create plasma waves, whistlers and other waves,

(b) launch electromagnetic waves from transmitters on the surface of the earth to disturb trapped particles in the Van Allen radiation belt,

(c) place a large artificial satellite in orbit near the earth to absorb trapped energetic particles from the Van Allen belt,

(d) inject a thermal plasma at high altitudes near the equator to create a plasma instability and dump lots of trapped radiation belt particles into the ionosphere.

One or more of these techniques could be used to modify the particle environment in space. Probably the most promising technique here is the injection of thermal plasma to produce whistlers and precipitate electrons or the use of sweeping satellites. These techniques and others show good promise of being able to reduce the energetic particle populations in the Van Allen radiation belt to low enough levels to allow use of this position of space by manned flight or sensitive instruments. In this sense, we have a potential engineering solution to a space pollution problem.

REFERENCES

(1) W. N. Hess, The Radiation Belt and Magnetosphere. Blaisdell Publication Co., Waltham, Mass., pp. 155, (1968).

(2) W. N. Hess, M. C. Trichel, T. N. Davis, W. C. Beggs, G. E. Kraft, E. Stassinopoulus, and E. J. R. Maier, Artificial Aurora Experiment: Experiment and Principal Results.

T. N. Davis, T. J. Hallinan, G. D. Mead, J. M. Mead, M. C. Trichel, and W. N. Hess, Artificial Aurora Experiment: Ground-Based Optical Observations. J. Geophys. Res., $\underline{76}$, No. 25, (1971).

(3) W. F. Utlaut and R. Cohen, Modifying the Ionosphere with Intense Radio Waves. Science, $\underline{174}$, pp. 245, (1971).

(4) R. A. Helliwell, J. Katsufrakis, M. Trimpi, and N. Brice, Artificially Stimulated Very-Low-Frequency Radiations from the Ionosphere. J. Geophys. Res., $\underline{69}$, pp. 2391, (1964).

(5) G. M. Allcock, C. K. Branigan, J. C. Mountjob, and R. A. Helliwell, Whistlers and Other Very-Low-Frequency Phenomena Associated with the High Altitude Nuclear Explosion on July 9, 1962. J. Geophys. Res., $\underline{68}$, pp. 735, (1963).

(6) J. W. Dungey, Resonant Effect of Plasma Waves on Charged Particles in a Magnetic Field. J. Fluid Mech., $\underline{15}$, pp. 74, (1963).

and also

Loss of Van Allen Electrons Due to Whistlers. Planetary and Space Sci., $\underline{11}$, pp. 591 (1963).

(7) T. J. Rosenberg, R. A. Helliwell, and J. P. Katsufrakis, Electron Precipitation Associated with Discrete Very-Low-Frequency Emissions. J. Geophys. Res., $\underline{76}$, pp. 8445, (1971).

(8) D. G. Cartwright and P. J. Kellogg, Radiation from Artificial Electron Beam in the Ionosphere. Nature Physical Science, $\underline{231}$, pp. 11, (1971).

(9) C. F. Kennel and H. E. Petchek, Limit on Stably Trapped Particle Fluxes. J. Geophys. Res., $\underline{71}$, pp. 1, (1966).

(10) N. Brice, Artificial Enhancement of Energetic Particle Precipitation Through Cold Plasma Injection: A Technique for Seeding Substorms? J. Geophys. Res., $\underline{75}$, pp. 4890, (1970).

(11) D. J. Williams, ARC 1: A Pilot Program of Ionospheric Modification Through Artificial Stimulation of Mangetospheric Trapped Particle Precipitation. NOAA TN ERL SEL-19, September 1971.

(12) H. E. Petcheck, Modification of the Magnetospheric Convection Pattern by Mass Injection in Report of Ad Hoc Committee on Environment Modification Experiments in Space, University of Maryland, Institute for Fluid Dynamics & Applied Mathematics report to NASA, (1968).

(13) S. F. Singer, Artificial Modification of the Earth's Radiation Belt. Advances in Astronautical Sciences, $\underline{4}$, pp. 335, Plenum Press, (1959).

(14) W. N. Hess and G. D. Mead, The Jupiter Radiation Belt. Published in the Proceedings of the Irkutsk Conference on Solar Terrestial Physics, (1971).

ORBITAL POLLUTION CONTROL

Ervin Y. Rodin

Dept. of Applied Mathematics and Computer Science Washington University, St. Louis, Missouri 63130

ABSTRACT

This paper proposes a method to enhance the well known heat island effect over large cities. This effect causes warm air to circulate as a closed system, under a low inversion layer, over a city, thereby causing the pollutants to remain essentially entrapped, with a constantly increasing concentration. It is therefore suggested that, as a temporary measure (for the next 10-15 years) the application of additional heat content to a small portion of the interior of such a circulating regime might cause a chimney effect, with a resultant breakthrough in the inversion layer: followed either by a complete breakdown of this layer and the subsequent alleviation of heavy concentration doses by natural mixing; or by a ventilating of the city's airshed through the artificial "chimney". It is further proposed that the additional heat necessary to generate this corrective measure should come from a source, the operation of which a.) does not cause additional pollution and b.) can easily be utilized over continental expanses. Specifically, the end results of simplified calculations are presented which show that this problem may have a feasible solution by using a very large dirigible mirror, in stationary orbit. This mirror, since it is dirigible, could be directed at various cities during different hours of the day, and thus might represent an economically attractive proposition, for the attainment of this alleviative treatment.

1. INTRODUCTION

The fundamental parameter in the movement of pollutants in the air by the atmosphere is the wind; its speed and direction. These in turn are interrelated with the horizontal and vertical temperature gradients (also referred to as lapse rate, defined as the increase of temperature with height) in the atmosphere, both of large and of small scale. Usually, the temperature decreases upward in the atmosphere up to 10 kilometers or so. However, there are layers of limited vertical extent in which the temperatures increase with height. Such layers are commonly referred to as "inversion layers". Very stable inversion layers (warm air on top of cold) resist any kind of mixing and vertical motion of air flow. In an unstable situation (cold air over warm air) convection currents are set up as the warm air below attempts to rise. Pollutants are strongly affected by vertical motions of these air masses. Inversion layers, to a great extent, limit the height to which the pollutants may rise causing them to spread out horizontally. In an unstable situation, the pollutants will rise until a stable layer is encountered. In general, the greater the wind speed, the greater is the turbulence and the more rapid and complete is the dispersion of contaminants in the atmosphere.

The topography of an area may be extremely important to the general pollution problem; however, this is not necessarily always the case. It is well documented[1] that the pollution episodes in London have little to do with local topography. More data are required to define this effect properly.

It is by now very well known[2] that thermal inversions are a constant threat for the buildup of pollutants. Thus, a method is required to cause vertical motion of air masses and large eddies in order to facilitate mixing of pollutants in relatively "clean air". In some cities having large industrial and residential centers, huge amounts of heat energy are released, thereby contributing to the convection current. The height to which this convection current will rise depends primarily upon the pressure gradient of the atmosphere in the region and to a lesser extent on the local insulation rate, the amount of industrial activity, the humidity of the air, the amount of cloud cover and other factors. The rising convection current due to these causes is commonly referred to as the "heat island effect".[3]

2. A PROPOSED CONTROL METHOD

A method will now be presented to enhance and strengthen the heat island effect. The principal advantages of the method are the following:
- a) technologically and economically feasible;
- b) produces no additional pollution during its own functioning;

c) readily adaptable to a wide variety of local topographical conditions;
d) the same unit can be used for several locations.

The central idea is to put a large parabolic reflector into stationary earth orbit (about 22,340 miles above the earth's surface). The purpose of this reflecting "dish" is to intensify the sun's radiation at a preselected location on the earth's surface. This spot on the earth's surface might be specially treated to absorb as much of the radiant solar energy as possible. As the reflected energy is transmitted through the earth's atmosphere, some will be absorbed. This is due to the water vapor and air borne solid particles always present above the earth. Of the amount of energy striking the earth, some will be reflected back into the atmosphere. This in turn will be absorbed partially by the atmosphere lying close to the earth's surface and the rest will be reflected back again towards the earth. The result of solar radiation striking the earth is shown in Figure 1 below.

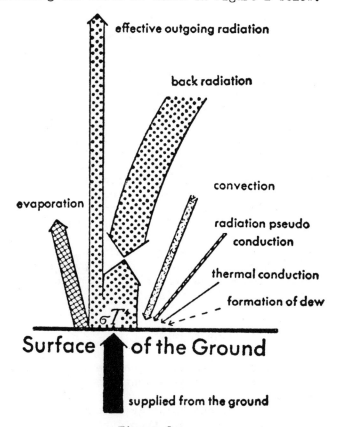

Figure 1
The transformation of solar radiation striking the Earth
(from Reference (2))

Thus, of the total energy that is collected at the reflector, only a certain percentage will be absorbed into the earth. The exact numbers are functions of certain parameters of the atmosphere and are ascertained by experimentation. We will discuss these parameters in a later section.

The energy which is absorbed by the earth will go towards heating the air immediately above the spot on the earth and will cause a rapid decrease in air density. This sudden drop in air pressure creates a 'partial vacuum' allowing the out-lying air to flow in rapidly to take up the difference in air pressure, thus causing local winds. This interpretation helps to explain, in part, the natural laws that will cause the artificially induced winds by the heat-island effect.

To calculate the average wind velocity induced by the heat-island effect necessitates the use of a relatively simple heat model. Thus, one should try to obtain for it a tractable mathematical model. The mathematical model requires some very gross assumptions which may or may not be valid. As usual, experimental data are necessary to validate it. A truly accurate model would have to take into account very subtle effects concerning local topography of the city under consideration, the location of mountains or rivers nearby, the local atmospheric conditions etc. The consideration of such effects would defeat the essential purpose of this research, which is the question of whether this idea is feasible or not in some general setting. We may conclude from the results of our mathematical model that if in some "idealized sense" enough local winds may be produced to reduce effectively the concentration of pollutants in the air, then it is a reasonable assumption to conclude that these same results are to be expected to occur in a city which does not deviate too far from the idealized model. Again, it is necessary to reiterate that further experimental research is necessary to state more accurately whether the model is adequate or not.

3. PHYSICAL ASSUMPTIONS FOR A MATHEMATICAL MODEL

It is instructive to envision the plume created from the reflected solar energy absorbed into the earth's surface, to be completely contained within the walls of a huge cylinder whose base is situated parallel to, and at some fixed altitude over the absorbing spot on the earth (see Figure 2, next page). This absorbing spot on the earth, which may be specially treated[6] so that there is little reflected energy, is located approximately at the focal point of the parabolic reflector. (This requires designing a reflector with automatic precision pointing devices utilizing an adjustable equatorial mounting and a photoelectric device for automatic solar tracking.)

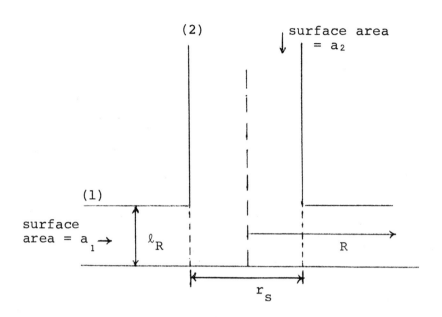

Figure 2

The "fumigating chimney"

We assume the incoming air flow, caused by the rapid pressure drop, to enter horizontally and up to a fixed thickness or altitude or possibly to a fixed small angle to the horizontal (See Figure 2). The incoming air will not break through the plume above the fixed altitude. Further, we assume that the flow velocity of the incoming air has a constant value from the earth's surface to the fixed altitude at any fixed distance from the axis of the cylinder.

At this point, we should mention one of the important assumptions of this model. If air is thought of as an ideal fluid, it is experimentally known that if the pressure would be suddenly decreased within a cylindrical volume in the fluid, the surrounding fluid would enter the volume up to the total height of the fluid. We might observe a velocity profile of the fluid for this situation (see Figure 3, next page). The velocity of the incoming fluid

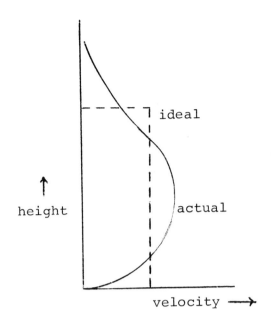

Figure 3

Velocity Profile

immediately above the bottom of the container of fluid would be zero. The velocity would sharply increase at points rising above the bottom, attain some maximum value and would slowly decrease to some value as the top of the container is reached. We assume for our model, however, that the velocity profile is of a shape indicated by the dashed lines in Figure 3. Hence, we tacitly assume that the incoming air flow due to the heating of the ground will not take place above some known altitude and that below this altitude the flow will have an unknown but constant velocity at any fixed distance out from the axis of the cylinder. This altitude in our heat model will correspond to the altitude of the inversion layer over a city. Thus, we see that this oversimplification not only reduces the complexity of a mathematical model but more importantly lends itself to a more accurate physical representation.

We also assume that the normalized pressure drop is equal to the normalized temperature drop from the interior to the exterior of our artificial cylinder. This is a reasonable assumption because air can be treated as an ideal gas and the total volume of the cylinder can be considered constant.

Using all of these assumptions we may write equations describing the heat transfer up through the cylinder. We then equate the total volumetric air flow up the cylinder to the total volumetric air flow into the bottom side of the cylinder, arriving in this manner at a formula relating average velocity of the incoming air as a function of some atmospheric parameters and of the radii of the reflecting dish and of the receiving spot on the earth. The average velocity will also be a function of the height of the inversion layer.

4. MATHEMATICAL DEVELOPMENT

4.1 Notation

We begin by defining some of the symbols to be used in the sequel,

$P \triangleq \dfrac{F}{a}$ = pressure $\left(\dfrac{lb_f}{ft^2}\right)$, where $lb_f \triangleq lb_m \dfrac{ft}{sec^2}$

ρ = density $\left(\dfrac{lb_m}{ft^3}\right)$

v = velocity (ft/sec)

a = surface area (ft^2)

$J = 778 \left(\dfrac{lb_f\text{-}ft}{sec} \bigg/ \dfrac{BTU}{sec}\right)$

h = enthalpy $\left(\dfrac{BTU}{lb_m}\right)$

C_p = specific heat at constant pressure $\left(\dfrac{BTU}{lb_m\text{-}{}^0R}\right)$

${}^0R$ = temperature (Rankine) ${}^0R = {}^0F + 460$

q = heat input (BTU/sec)

g_c = gravitational constant = 32.174 $\dfrac{ft}{sec^2}$

g = local gravity $\dfrac{\text{ft}}{\text{sec}^2}$

4.2 Energy Balance

Consider the system as indicated in Figure 2. Writing an energy balance at points (1) and (2), one obtains

(total energy)$_1$ + energy added

= (total energy)$_2$ + energy lost + work done.

Since there is no work done in the system, the last term is zero. At either point (1) or (2), the total energy is given by the sum of three terms: Pressure energy, velocity energy (kinetic), and enthalpy. At point (1) (for example)

$$\text{pressure energy} = P_1 \, a_1 \, v_1$$

$$\text{velocity energy (K.E.)} = \tfrac{1}{2}(\rho_1 a_1 v_1)\, v_1^2 \, g_1/g_c$$

$$\text{enthalpy} = h_1 \rho_1 a_1 v_1$$

Therefore, the total energy is given by

$$\Sigma = \left\{ \frac{1}{J}\left[\frac{P_1}{\rho_1} + \frac{1}{2}\frac{g_1}{g_c}v_1^2\right] + h_1 \right\} \rho_1 a_1 v_1 \quad (\text{BTU/sec})$$

Thus we may write:

$$\left\{\frac{1}{J}\left[\frac{P_1}{\rho_1} + \frac{1}{2}\frac{g_1}{g_c}v_1^2\right] + h_1\right\}\rho_1 a_1 v_1 + q_{in} =$$

$$\left\{\frac{1}{J}\left[\frac{P_2}{\rho_2} + \frac{1}{2}\frac{g_2}{g_c}v_2^2\right] + h_2\right\}\rho_2 a_2 v_2 + q_{lost}$$

4.3 Mass Balance, Simplification and Solutions

Since no mass should be lost through the idealized system, the mass balance equation is given by:

$$\rho_1 a_1 v_1 = \rho_2 a_2 v_2$$

or, the mass flow into the system (in $\frac{lb_m}{sec}$) must equal the mass flow out of the system.

Solving for v_2, one obtains

$$v_2 = \frac{\rho_1 a_1}{\rho_2 a_2} v_1$$

and using this equation in the energy balance equation, one obtains

$$\left\{\frac{1}{J}\left[\frac{P_1}{\rho_1} + \frac{1}{2}\frac{g_1}{g_c} v_1^2\right] + h_1\right\} \rho_1 a_1 v_1 + q_{in} =$$

$$\left\{\frac{1}{J}\left[\frac{P_2}{\rho_2} + \frac{1}{2}\frac{g_2}{g_c}\left(\frac{\rho_1 a_1}{\rho_2 a_2} v_1\right)^2\right] + h_2\right\} \rho_1 a_1 v_1 + q_{lost}.$$

If we define

$$K_1 = \frac{1}{J}\frac{P_1}{\rho_1} + h_1$$

$$K_2 = \frac{1}{J}\frac{P_2}{\rho_2} + h_2$$

$$\gamma = \frac{\rho_1 a_1}{\rho_2 a_2}$$

$$\alpha_1 = \frac{g_1}{g_c}, \quad \alpha_2 = \frac{g_2}{g_c},$$

then we can write our equation as a cubic in v_1:

$$v_1^3 + \frac{2J(K_1-K_2)}{(\alpha_1 - \alpha_2\gamma^2)} v_1 + \frac{2J(q_{in}-q_{lost})}{\rho_1 a_1 (\alpha_1 - \alpha_2 \gamma^2)} = 0$$

In order to solve the cubic equation

$$v_1^3 + Av_1 + B = 0$$

where $A = \dfrac{2J(K_1-K_2)}{(\alpha_1-\alpha_2\gamma^2)} \quad B = \dfrac{2J(q_{in}-q_{lost})}{\rho_1 a_1(\alpha_1-\alpha_2\gamma^2)}$

define

$$\mu = \left[-\frac{B}{2} + \left(\frac{B^2}{4} + \frac{A^3}{27}\right)^{1/2} \right]^{1/3}$$

$$\lambda = \left[-\frac{B^2}{2} - \left(\frac{B^2}{4} + \frac{A^3}{27}\right)^{1/2} \right]^{1/3}$$

Then, as is well known, the three solutions are given by

$$v_1^{(1)} = \mu + \lambda \ ; \ v_1^{(2),(3)} = -\left(\frac{\mu+\lambda}{2}\right) \pm \left(\frac{\mu-\lambda}{2}\right)\sqrt{-3}$$

The first of these solutions may be rewritten as

$$v_1^{(1)} = \left(-\frac{B}{2}\right)^{1/3} \left\{ \left[1 + \left(1 + \frac{4}{27}\frac{A^3}{B^2}\right)^{1/2}\right]^{1/3} + \left[1 - \left(1 + \frac{4}{27}\frac{A^3}{B^2}\right)^{1/2}\right]^{1/3} \right\}$$

or, with the abbreviations

$$X = \frac{4}{27}\frac{A^3}{B^2} \ , \ Y = -\frac{B}{2}$$

$$v_1^{(1)} = Y^{1/3} \left\{ \left[1 + (1+X)^{1/2}\right]^{1/3} + \left[1 - (1+X)^{1/2}\right]^{1/3} \right\}$$

Consider the term under the square root sign

$$1 + X = 1 + \frac{8}{27} \frac{J(K_1-K_2)^3(\rho_1 a_1)^2}{(q_{in}-q_{lost})^2(\alpha_1-\alpha_2\gamma^2)}$$

If this term is strictly positive, then the solutions of the cubic equation will be one real root and two conjugate imaginary roots. If the term is identically zero, there will be three real roots. If the term is strictly negative, there will be three real and unequal roots. The one real root in the first case is given by

$$v_1^{(1)} = \mu + \lambda$$

as indicated previously. While we shall not go into details of this, it can be shown easily that this is the only root having physical significance. It can also be shown that under the condition of zero thermal inversion

$$K_1 - K_2 \approx 0$$

so we may approximate our cubic equation by

$$v_1^3 = -B$$

This equation will give us the magnitude of the velocity for various values of the free parameters, for the case of zero thermal inversion. Later, we shall see how the effect of non-zero thermal inversion decreases the magnitude of the solution.

We must determine values for q_{in}, q_{lost}, ρ_1, α_1 and $1-\gamma^2$ (we will assume $\alpha_1 \approx \alpha_2 \approx 1$).

The amount of solar energy striking the earth's upper atmosphere has been experimentally determined to be 442 BTU per hour per square foot.[4] All of this energy will strike the reflector and we will assume the "effective surface area" is equal to the true surface area, i.e. curvature of the reflecting parabolic dish will be neglected. Some of this reflected solar energy will be lost through transmission to the earth's surface. Cloud cover, water vapor and pollutants will act to absorb some energy, and some energy will be reflected back into the atmosphere from the "spot" located on the earth. It would be possible to treat this "spot" specially with lampblack or another highly absorbing material so as to make re-radiation almost negligible.[6] Thus, the only remaining item to contend with is absorption during transmission through the atmosphere.

Due to these effects, we may write

$$q_{in} = K \frac{442}{3600} \pi r_r^2 \quad \frac{BTU}{ft^2\text{-sec}}$$

where r_r is the reflector radius and K is an absorptivity parameter. Ideally the value of K would be $\approx .9$ to 1.0.

The determination of q_{lost} is extremely difficult. This term includes heat losses through the plume. Among factors which affect this term are: gradient temperatures, surface areas, wind velocities, viscosity of air and other factors which will require experimental work to determine their full effects. However, let us write

$$q_{lost} = \beta \, q_{in}$$

where $0 < \beta < 1$. Hence

$$q_{in} - q_{lost} = (1-\beta) K \frac{442}{3600} \pi r_r^2$$

An important observation to be made here, is that the velocity is proportional to the cube root of $(1-\beta)K$, hence for values close to 1, this term will not have a dominant influence. We assume this coefficient to be 0.9. Again, experimental results will indicate whether we are reasonable in our assumptions.

Using the definitions of α_1 and γ, and assuming a nominal value of air density, ρ_1, to be 0.075 lbm/ft^3, and since $J = 778 \frac{\text{lbf-ft}}{\text{BTU}}$; we may write

$$B = \frac{2(778)(.9)\frac{(442)}{3600}(\pi r_r^2)}{(0.075)(2\pi R l_R)(1-\gamma^2)} \, .$$

Further, since

$$2Rl_R = \gamma r_s^2$$

where r_s is the spot radius

$$B = \frac{2292(\frac{r_r}{r_s})^2}{\gamma(1-\gamma^2)}$$

and

$$v_1^{(1)} = \sqrt[3]{-B}$$

If $v_1^{(1)}$ is to be a positive valued velocity, we must require $\gamma > 1$ or

$$2\pi R l_R > \pi r_s^2 \, .$$

Practically, this is a "funneling" effect; the air will be compressed as it proceeds from point 1 to point 2. (see Figure 2).

In the following typical computer drawn Figures 4 and 5, $v_1^{(1)}$ is graphed as a function of the reflector and spot radii for various values of R, the distance (in miles) from the center of the spot. We use two values of l_R, 200 and 400 feet, corresponding to nominal inversion altitudes.

In the bottom portion of each of the graphs, there are tables of the percentage decrease in wind velocity for various inversion temperatures (in degrees, F) and for various reflector radii. This "decrease factor" is determined from the following considerations.

Returning to the original equation for velocity given by

$$v_1^{(1)} = Y^{1/3}\left\{\left[1+(1+X)^{1/2}\right]^{1/3} + \left[1-(1+X)^{1/2}\right]^{1/3}\right\}$$

or

$$v_1^{(1)} = Y^{1/3}\left\{2^{1/3}\left[\frac{1+(1+X)^{1/2}}{2}\right]^{1/3} + 2^{1/3}\left[\frac{1-(1+X)^{1/2}}{2}\right]^{1/3}\right\}$$

and thus

$$v_1^{(1)} = 3\sqrt{-B}\left\{\left[\frac{1+(1+X)^{1/2}}{2}\right]^{1/3} + \left[\frac{1-(1+X)^{1/2}}{2}\right]^{1/3}\right\}.$$

For the condition $X = 0$ (no thermal inversion) the above equation yields the family of curves shown in the graphs. For the condition $X \neq 0$, the factor

$$1 - \left\{\left[\frac{1+(1+X)^{1/2}}{2}\right]^{1/3} + \left[\frac{1-(1+X)^{1/2}}{2}\right]^{1/3}\right\}$$

in percentage, is calculated for various values of T_1-T_2 and for the same values of spot and reflector radii as used for zero temperature inversion. The quantity T_1-T_2 is related to K_1-K_2 as follows:

$$K_1-K_2 = (\frac{1}{J}r + C_p)(T_1-T_2)$$

5. SUMMARY AND CONCLUSIONS

This report has been concerned with the possibility of using concentrated solar energy to create sufficient wind velocity to draw out heavy pollutants from below a thermal inversion layer. An extremely simple model was used to ascertain the feasibility of such a system. It was shown that under certain conditions, sufficient wind can be created. The problem becomes one of determining whether the simple model truly represents "physical reality"; that is, are the solutions of the wind velocity model comparable to wind velocities of a true "physical" system. This question may be answered only after sufficient experimentation.

It is the opinion of the author that a first step has been made. What is required is more experimental evidence to determine the feasibility of such a local pollution control method as advocated in this report. The results of the model indicate that this may be possible.

It is relevant to point out here some evidence to support this claim. It is well known that a glider pilot flying along a two-lane highway during the summer, experiences climb rates which in many cases may be as large as 500 to 1000 feet per minute. Is it possible that if this wind velocity were harnessed and controlled to a fixed spot on the earth's surface, it would be sufficient to disperse heavy suspended pollutants?

Another fact has been reported in Science News[5] recently, that indicates that increased rainfall has been recorded over large sections of the plains area in the United States; notably Kansas and Nebraska. An explanation of this occurrence, advocated by some observers, is that increased irrigation causes a change in the earth's absorptivity (a function of surface color and texture) by higher crop yields. Higher concentrations of solar energy are absorbed into the earth's surface thereby effecting changes in the local climate conditions. It is obvious that there are many local conditions effecting climate control other than surface color, but the point to be made here is that climate control or wind control might be possible on a large scale.

The "heat island effect", discussed earlier in this report, is known to occur naturally. It should be possible to enhance this effect by concentrated solar energy by the method advocated in this report. However, experimental evidence would be necessary to support this claim. Finally, recent laboratory research by a team at the General Electric Research and Development Center at Schenectady, New York[7], aimed at developing cooling towers for multi unit electric power stations, indicates that if such towers are built as indicated in the picture below, such a circular arrangement would "drill holes" in inversion layers, in the manner described in the present paper. Thus, the only difference between the G.E. scheme and the one proposed here is the power source; which in the case of the former will be a one-location unit, while in our case one unit might be sufficient for an entire continental expanse.

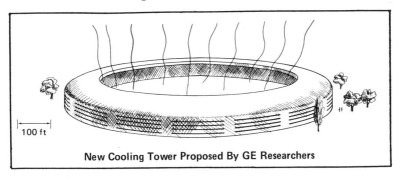

New Cooling Tower Proposed By GE Researchers

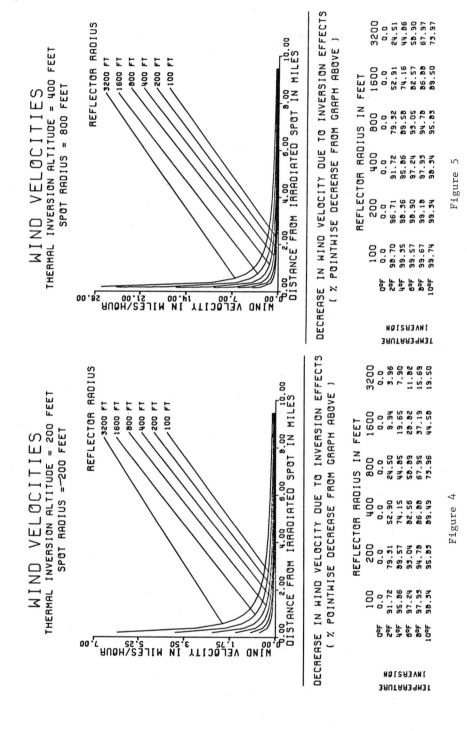

Figure 4

Figure 5

6. REFERENCES

1. Wexler, H., "The Role of Meteorology in Air Pollution," <u>Air Pollution</u>, published by Columbia Univ. Press, New York, for World Health Organization, 1961.

2. Arnold, G.R., "Local Inversions, Air Currents, and Smoke Pollution in Cahokia Bottom", D.Sc. dissertation, Washington University, St. Louis, Missouri, 1964.

3. Ibid., page 34.

4. Kreith, F., <u>Radiation Heat Transfer for Spacecraft and Solar Power Plant Design</u>, International Textbook Company, 1962.

5. Frazier, K., "Irrigation and Climate", <u>Science News</u>, Vol. 96, No. 26.

6. Hammond, A.L., "Solar Energy: A Feasible Source of Power?", <u>Science</u> Vol. 172, 197.

7. "Smog Blanket's Days Are Numbered", <u>Electromechanical Design</u>, May 1971.

GLOBAL AIR POLLUTION MONITORING FROM SATELLITES: PROSPECTS AND PROBLEMS

George Ohring, Joseph Otterman & Joachim Joseph

Department of Environmental Sciences, Tel Aviv University, Tel Aviv, Israel

1. INTRODUCTION

On the basis of the scale of the phenomenon being monitored, air pollution monitoring problems can be divided into two general categories: 1) local monitoring and 2) regional and global monitoring. Problems associated with horizontal scales less than ~100 km and vertical scales less than ~1 km - such as monitoring of single sources, neighborhoods, towns, cities, and metropolitan areas - fall into the first category. Problems associated with horizontal scales greater than ~100 km and vertical scales greater than ~1 km fall into the second category. A satellite observing system based upon remote sensing is particularly suited for problems of the second category.

The global problems of concern are the long term trends in CO_2 and aerosols, and the global cycle and trends of SO_2 and CO (SCEP, 1970; SMIC, 1971; and Robinson and Robbins, 1969). Also of global concern is the potential pollution of the stratosphere by SST's and the long term trends of the important stratospheric constituents O_3 and H_2O (SCEP, 1970; Mastenbrook, 1971; and Kohmyr et al, 1971). Regional problems of concern are the delineation of pollution episodes of large areal extent (e.g. Fensterstack, 1968) and the dilution and removal of pollutants downwind 100's and 1000's of kilometers from regional sources, about which relatively little is known. Many of these problems are reviewed in NASA SP-285 (1971) and COSPAR (1972).

2. OBSERVATIONAL REQUIREMENTS

2.1 Introduction

The observational requirements for a satellite observing system are primarily dependent upon the total amounts, distributions, and variability of the pollutants, and on the particular problem being addressed. Regardless of the problem, the satellite remote sensing system must have the capability of measuring pollutants at the low concentrations in which they are present in the atmosphere. The next two sections provide information on the observed characteristics of pollutant distributions.

2.2 Gases

We may divide the gases of concern into two groups: 1) those gases that control the atmospheric radiation budget, and, hence, the climate, and 2) all the other pollutant gases of environmental concern. Because of their importance to meteorology, the gases of the first group - H_2O, O_3, and CO_2 - have been measured on a global basis and much more is known about their concentrations, distributions, and variability than of the second group. As indicated above, man's activities may be causing changes in the atmospheric CO_2 concentrations and in the stratospheric H_2O and O_3 concentrations. Information on the observational characteristics of this group is presented in Table 1, and information on the second group is presented in Table 2. In these tables, the concentrations listed are average concentrations. The concentrations of H_2O and O_3 have significant meteorological variations. For the pollutant gases of Table 2, concentrations in towns and cities can easily be more than an order of magnitude greater than those shown. The total amounts listed in the tables represent the total amount of the gas in a vertical column of atmosphere and are useful for examining the feasibility of remote sensing from a satellite. Also shown in the tables are residence times, which are a measure of the life time of a gas molecule in a particular layer of the atmosphere and depend upon removal processes. Generally speaking, the smaller the residence time, the greater the spatial and temporal variability of the gas.

Several general conclusions can be reached concerning satellite remote sensing requirements from these tables. Because CO_2 is uniformly distributed in the atmosphere, only spot observations are required to monitor its long term trend. The tables indicate that the average concentrations of most pollutant gases are extremely low.

Table 1 OBSERVATIONAL CHARACTERISTICS OF GASES INFLUENCING ATMOSPHERIC RADIATION BUDGET

	CO_2	H_2O	O_3
Average concentrations (ppm)			
Troposphere	320	1000	0.01
Stratosphere	320	3	5
Total amount	2.5 m STP	0.5-5cm Prec. Water	0.2-0.4cm STP
Vertical Distribution	Constant concentration	Concentration decreases to tropopause with average scale height of 2 km	Almost constant concentration in troposphere; increase to peak at ~30 km
Residence times			
Troposphere	4 yr	10 d	1 mo
Stratosphere	2 yr	2 yr	2 yr

Table 2 OBSERVATIONAL CHARACTERISTICS OF POLLUTANT GASES

Gas	Average surface concentration (ppm)	Total amount (cm STP)	Vertical distribution	Residence time
CO	0.1	0.15	Uniformly mixed in troposphere	70-1000d
NO_2	0.002			~5d
NH_3	0.01	< 0.013		~5d
SO_2	0.01		Concentrated in 1st km; 0.001 in stratosphere	
N_2O	0.4	0.5	Nearly constant concentration	~4 yr
CH_4	1.4	1.2	Constant to 20km; decrease to 0.2ppm at 50km	~100 yr
H_2S	0.002			~2d

For most pollutant gases, surface concentrations in towns and cities can be an order of magnitude greater than those listed.

Sensitivity to concentrations less than 1ppm and total amounts less than 1 cm STP will be required for most pollutant gases. Aside from being able to detect pollutant gases in the low concentrations and total amounts given above, the measurement must be of a certain accuracy, depending upon the particular problem. Estimates of required accuracy for many of the important problems are contained in NASA SP-285 (1971). For most pollutant gases, accuracy requirements are in the 1-100 ppb range.

2.3 Aerosols

Atmospheric aerosols may be divided into three main tropospheric types: continental, marine, and background and several characteristic of the stratosphere and upper atmosphere. This division of course does not differentiate between natural and man-made aerosols. Such a discrimination is usually very difficult in nature and especially so in the case of the remote detection of aerosols from a satellite.

We shall concern ourselves here with a short description of the tropospheric and stratospheric types. A general property of the tropospheric aerosol is that the number or mass-concentration decays with a characteristic dimension or scale height of 1-2 km. (Elterman, 1964, Joseph, 1967, Manes, 1971, Woodcock, 1953). Another property is that the individual aerosol particles, if not composed of soluble substances dissolved in a droplet of water, have highly irregular shapes.

If one assumes each particle to have a determinable equivalent radius or diameter one can define a distribution of radii or size spectrum for any given aerosol. The three main tropospheric types have different size distributions (see, for example, SMIC, 1971). Characteristically, the maximum of the number concentration distribution is in the 10^{-2} to 10^{-1} μ radius interval. The total number concentration is extremely variable, ranging from ~700 cm^{-3} in "background" air to over $10^5 cm^{-3}$ in cities.

The stratospheric size distributions may be quite different from that in the troposphere. The mean size of the particles there is of the order of 1 micron, the total number density of the order of .1 cm^{-3} (NASA SP 285, 1971) and again they are often quite irregular or floccular in shape.

Most types of aerosols have hygroscopic components, leading to a dependence of the size distribution on relative humidity. It should be most strongly emphasized that the above properties are average ones. The temporal and spatial variability of aerosols is very large and not very much is known about it.

The parameters most relevant to their remote detection from a satellite are their vertical, zenith or nadir, optical depths for scattering and for absorption. Assuming the particles to be spherical and to have a certain complex index of refraction, $m = m' - im''$ which may be a function of height, one can define these quantities at any wavelength λ for the whole atmosphere:

$$t(\lambda)^{scatt., abs.} = \int_0^\infty dh \int_{r_1}^{r_2} Q_{Mie}[\frac{2\pi r}{\lambda}, m(h)] \pi r^2 \frac{dn}{dr}(r,h) = \quad (1)$$

$$= \int_0^\infty dh \, \beta_\lambda(h)^{scatt., abs.} \quad (2)$$

where $Q_{Mie}^{scatt.}$ or Q_{Mie}^{abs} are the theoretical scattering and absorption efficiencies, $\beta_\lambda^{scatt.}(h)$ and $\beta_\lambda^{abs}(h)$ are the volume scattering and absorption coefficients at height h. The optical depth is an integral over all heights in the atmosphere and all relevant sizes, $r_1 \leq r \leq r_2$.

The optical depth usually exhibits avarriation with wave length in the form of a power-law
$\lambda^{-\alpha}$, where $-2 \leq \alpha \leq 2$ (NASA SP-285, 1971; SMIC, 1971).

The range of variation of typical aerosol optical depths at say .5 micron, is between .01 and .25 both locally in time, and globally. Sometimes values of the order of one are reached in highly polluted areas and in smogs, forest fires, desert and savannah dust storms etc. The ratio between absorption and scattering is uncertain, but may be of the order of one, or even higher, especially in polluted areas.

An increase of the average global aerosol total optical depth in the visible part of the solar spectrum from .050 to .075 has been variously documented (SMIC, 1971) over the last 40 years. This would mean that the long term relative and absolute accuracy and stability of any monitoring of aerosols should be of the order of .001. This holds also if the attempt at detection is for the purpose of determining sources, dispersions and sinks of local or regional pollution with shorter time-scales (less than a year).

3. THEORY OF OBSERVATION

3.1 Gases

Remote sensing of gaseous pollutants is conceivable if the

pollutants interact with either the shortwave solar radiation or the
longwave terrestrial radiation fields. For most practical purposes,
this interaction is in the form of line or band absorption of the
radiation by the gases. Figure 1 shows the blackbody curves of the
solar radiation and the terrestrial radiation, absorption by the
atmosphere, and the locations of some of the gaseous absorption bands.
When one looks down at the Earth at wavelengths below ~4μ, one sees
solar radiation reflected by the surface and the atmosphere.
Observations of the intensity of the surface reflected solar radiation
in an absorption band can provide information on the amount of gas
causing the absorption. However, one must isolate the effects of the
gaseous absorption from effects due to surface reflectivity and
scattering by molecules and particles in the atmosphere. If one looks
down at the atmosphere at wavelengths beyond 4μ, one sees essentially
the terrestrial emission spectrum. At these wavelengths, the
observed intensities in an absorption band are a function of both the
vertical distribution of the absorbing gas and vertical distribution
of temperature. Thus, in the terrestrial emission spectrum, one must
isolate the effects of the absorption from the effects of the
temperature in order to derive information on the amount and
distribution of the absorbing gas.

There are basically two types of remote observations that can be
performed from a satellite: vertical (nadir) observations (instrument
points down toward Earth) and limb observations (instrument points
towards planetary limb or sun). In each type one can look at
terrestrial radiation or solar radiation. Figure 2 illustrates the
geometry associated with these observations. We shall outline the
theory for inverting observations of terrestrial radiation taken in
the vertical mode. Discussions of the other modes of remote sensing
may be found in Twomey and Howell (1963), Heath et al (1970), Gille
and House (1972) and NASA SP-285 (1971).

The observed radiance at wavenumber in an absorption band of a
pollutant gas can be written as

$$N(\nu) = B(\nu, p_t) + \int_{B(\nu, p_t)}^{B(\nu, p_s)} \tau(\nu, p) dB(\nu, p) \qquad (3)$$

where p is pressure, $\tau(\nu,p)$ is the transmission from the pressure
level p to the top of the atmosphere, B is the Planck function, p_s is
the surface pressure, and p_t is the pressure at which the transmission
is unity (effective top of atmosphere). Equation (3) indicates that
the observed radiance depends upon the temperature profile (since the
Planck function depends upon temperature) and upon the distribution of
the absorbing gas with altitude (since τ depends upon the gas
distribution).

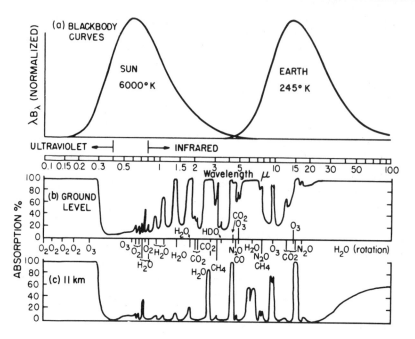

Fig. 1 (a) Blackbody emission spectra for 6000 K and 245 K, being approximate emission spectra of sun and earth; (b) and (c) Atmospheric absorption spectra for solar beam reaching the ground and 11 km, respectively (After Frisken, 1971).

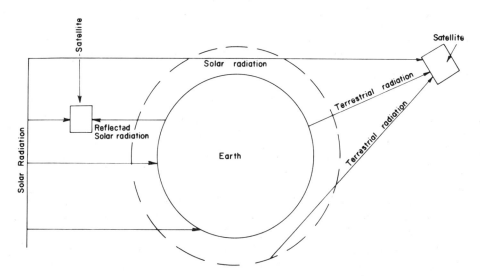

Fig. 2 Geometry of vertical (nadir) and limb viewing for observing solar and terrestrial radiation from a satellite.

Thus, to derive the gas distribution, we must know the temperature distribution. This can be obtained from radiosonde observations or from inversion of satellite observations in the 15μ CO_2 band, as is currently being done with the NIMBUS weather satellites. From the observed spectrum, $N(\nu)$, of the gaseous absorption band, one can obtain the distribution of τ with altitude by means of an iterative or statistical inversion scheme (Smith, 1970; Conrath, 1969). From the distribution of τ, the gas concentration profile can be derived. Equation (1) indicates that if the earth's surface and the atmosphere are at the same temperature, then the observed emission is independent of the amount and distribution of the absorbing gas and dependent only on the temperature. Thus, to obtain information on the gas distribution using vertical infrared emission observations requires the presence of temperature differences between the surface and the atmosphere. Furthermore, to obtain information on τ at all heights requires that the transmittance gradients $(d\tau/dp)$ of the different wavenumbers have their peak values at a wide range of heights. This requirement is met only for strong absorption bands.

The question arises, in the light of the low concentrations in which they are found in the atmosphere, whether the gases of concern can indeed be detected remotely. For those gases influencing the radiation budget - CO_2, H_2O, and O_3 - there is no problem in detection, and in fact, satellite experiments have already been flown for the purpose of measuring O_3 and H_2O. The only problem with these gases might be one of accuracy. The problem with the other gases might be one of detectability. We have estimated for some of the pollutant gases the minimum concentration required for detection from space, assuming that vertical viewing of terrestrial infrared radiation is the remote sensing technique. It is further assumed that each pollutant gas is located in the 1st km of the atmosphere at its average concentration, and that detection is possible if the gas produces an absorption of 5%. These estimates - and they must be considered as rough estimates only - are shown in Table 3. They indicate that detectable concentrations of CO, NH_3, and SO_2 are higher than the observed concentrations, but of the same order of magnitude. Thus, detectability of near surface concentrations of these pollutants from space using infrared sensing appears difficult. The detectable concentrations for N_2O and CH_4 are more than an order of magnitude less than the observed concentrations. Thus, detection of near surface concentrations of these gases from space in the infrared appears possible in theory.

The discussion in this section has been on passive techniques. Active techniques - such as lasers and radars - can also be considered. Hanst (1970) discusses the possibilities of lasers for remote detection of pollution. In general, however, the application of active techniques

Table 3

REMOTE INFRARED DETECTION OF POLLUTANT GASES NEAR THE EARTH'S SURFACE

Gas	Average Concentration (ppm)	Location of IR[1] Band Centers (μ)	Detectable[2] Concentration (ppm)	Transmittance Model Used or Source of Estimate of Detectable Concentration
CO	0.1	2.3, 4.6	0.5	Burch and Williams (1962a)
NH_3	0.01	10.5	0.05	Gille and Lee (1969)
SO_2	0.01	7.3, 8.7, 18.5	0.03	Estimate by Hanst (1970)
N_2O	0.4	7.78	0.03	Burch and Williams (1962b)
CH_4	1.4	3.3, 7.7	0.03	McClatchey et al (1970)

(1) Partial list. Other IR bands also available.

(2) Based on assumption that gas is located in 1st km of atmosphere and that 5% absorption is required for detection.

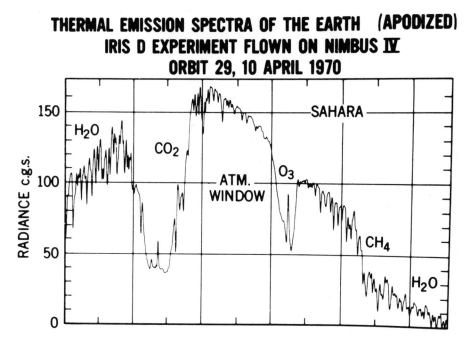

Fig. 3. NIMBUS 4 IRIS Spectrum taken over the Sahara desert showing atmospheric absorption bands (After Hanel and Conrath, 1970).

3.2 Aerosols

The way in which the remote sensing of aerosols depends on its optical depth may be brought out in the following extremely simplified manner. Let us assume an unpolarised beam of radiation I_o normally incident upon the atmosphere from below and observed from a satellite. Let the optical depth of any fraction of the atmosphere between the heights o and h be

$$t(h,\lambda) = \int_o^h \{\beta^{scatt.}(\lambda,h) + \beta^{abs}(\lambda,h)\} \, dh \quad (4)$$

Then the emergent specific intensity is given by

$$I(\infty,\lambda) = I_o(\lambda)e^{-t(\infty,\lambda)} + \int_o^\infty S(t(h),\lambda)e^{-[t(\infty,\lambda)-t(h,\lambda)]} \, dh \quad (5)$$

The function $S(t(h))$ contains two parts in general. The first is the radiation scattered from other directions into the zenith direction and the second any true emission by the atmosphere. Only in the case when function $S(t(h))$, the so-called source function, is negligible compared to the directly transmitted intensity, $I_o(\lambda)e[-t(\infty,\lambda)]$ does the solution reduce to the commonly used Lambert-Bouguer-Beer's Law. It is thus usually necessary to know also the laws governing the monochromatic scattering of the incident radiation by the spatially inhomogeneous turbid atmosphere, if one at all wants to achieve the necessary accuracies. The techniques for the extraction of $t^{scatt}(h)$ and $t^{abs}(h)$ from the spectral data are based on mathematical inversion methods, with all their accompanying difficulties. They are further complicated by the fact that formal scattering laws are available only for particles of simple shapes. The scalar scattering laws become matrices when polarisation is measured (Chandrasekhar, 1950). Nevertheless, measurements of the polarisation field are advantageous for several reasons. The polarisation field of a turbid atmosphere is very different from that of a clear one. Secondly, measurement of the polarisation is experimentally more accurate even though slightly more complicated, because one considers ratios of the intensities.

Several observational techniques are available for remotely sensing aerosols from satellites ranging from simple photography with optical filters to radiometric or spectrometric photo-electric techniques. Measurements may be made of the albedo, the polarisation and infra-red thermal emission fields of the Earth-atmosphere system or of solar and stellar radiation as influenced by the Earth-atmosphere system. All these can be made as a function of wavelength, geographic coordinates and solar or stellar zenith angle.

All these systems are passive. In the future, it may be possible to use satellite-borne lasers to monitor the radiation scattered back from the Earth's surface and atmosphere and infer the distribution in height of the aerosol, at least up to 15 kms (NASA SP-285, 1971) with relatively high resolution.

For detection and monitoring of stratospheric aerosols, limb-scanning satellite techniques are especially advantageous. These are not influenced by radiation from the lower troposphere and will not only detect the aerosol but also give its distribution with height in the stratosphere. In this type of measurement, one observes the sun or a star while these are being occulted by the atmosphere or the solar aureole at small scattering angles. Twilight measurements are also feasible using the solar beam as a slowly sinking or rising searchlight and observing the radiation scattered from the beam. Another advantage of limb type measurements is that the difference in wavelength dependence of aerosol and molecular scattering may be more fully utilised. This is because observations in both ultra-violet and infra-red parts of the spectrum may be made in addition to those possible in the nadir-direction.

In summary, it may be said that it is feasible at present to detect remotely the presence and determine the total optical depth of aerosols in the atmosphere, especially when its optical depth is at least of the order of that of the molecular atmosphere and the reflectance of the underlying surface is relatively low.

3.3 Potential Problems

There are several problems that may hinder pollution observations from space. Clouds may mask an area from being viewed. However, Glaser et al (1972) have shown that this problem is not serious. From an analysis of satellite observed cloudiness [Sherr et al (1968)], Glaser et al show that the probability of making a successful remote observation per 12 hour period per 200 km square for most regions of the Northern Hemisphere is greater than 0.5.

The required temperature contrast between surface and atmosphere for a successful determination of gas pollutants using the vertical terrestrial radiation mode is potentially a serious problem for the detection of pollutants concentrated near the surface.

The fact that the important radiative gases of the atmosphere – CO_2, H_2O, and O_3 – have wide absorption bands in most regions of the spectrum and the fact that the absorption bands of the pollutant gases overlap one another create observational problems. The latter problem may be alleviated by measurements of high spectral resolution.

4. RESULTS TO DATE AND SUMMARY

Although, to this date, no satellite experiment has been flown specifically for the remote observation of air pollution, a number of interesting results have been obtained from other satellite and manned spacecraft experiments. In addition, some results have been obtained from high altitude balloon flights designed to test remote sensing concepts. In this section we review some of these results.

4.1 Gases

The Infrared Interferometer Spectometer (IRIS) experiment, flown on the NIMBUS 3 and 4 satellites, was designed primarily for meteorological observations, i.e., for vertical soundings of atmospheric temperatures, humidities, and ozone concentrations (Conrath et al, 1970). Figure 3 shows a spectrum obtained with the NIMBUS 4 IRIS instrument over the Sahara desert (Hanel and Conrath, 1970). The spectrum covers the range 400-1500cm^{-1} (7 - 25µ) with a resolution of 2.8cm^{-1}. The absorptions due to H_2O, CO_2, and O_3 are clearly seen. Also indicated is the absorption due to the 7.7µ band of CH_4. Conrath et al (1970) searched a NIMBUS 3 IRIS spectrum, taken under clear sky conditions over the Gulf of Mexico, for absorptions due to N_2O, NO_2, SO_2, H_2S, and HNO_3. The search was unsuccessful, probably due to a number of factors, including masking by the H_2O, CO_2, O_3, and CH_4 bands, lack of contrast for low altitude pollutants, the low spectral resolution (~5cm^{-1}), and, possibly, low values of pollutant gases over the Gulf of Mexico. Vertical distributions of ozone (Prabhakara et al, 1970) and water vapor (Conrath et al 1970) have been estimated from such IRIS spectra. Water vapor profiles have also been obtained from the Satellite Infrared Radiometer Spectrometer (SIRS) experiment on the NIMBUS 4 satellite (Smith, 1970). Total O_3 amounts have been obtained from the Backscatter Ultraviolet experiment on NIMBUS 4 (Mateer et al, 1971), which observes Rayleigh backscattered solar ultraviolet radiation.

Barringer et al (1970) have performed a balloon experiment to measure SO_2 and NO_2 remotely using a correlation spectrometer. The absorption bands used were the 0.3 to 0.32µ UV band for SO_2 and the 0.4 to 0.44µ visible band for NO_2. The balloon floated along a trajectory at 35km, above most of the ozone layer (thus representing a good test for satellite sensing) and observed the Chicago area. They present a comparison of the total amounts of SO_2 observed by the balloon and surface concentrations obtained by direct sampling. Although it is difficult to compare quantitatively the balloon and direct determinations, qualitatively the results indicate that high values of total amount are generally associated with high values of surface concentration. NO_2 was also successfully observed, but no comparison is given with ground based observations. This same group has also used aircraft to perform similar observations of industrialized areas (Barringer et al, 1968; Barringer, 1968).

4.2 Aerosols

Determinations of the presence of aerosols have been carried out in several cases.

The most prevalant data are those derived from satellite photographs of dust storms or plumes (Ing, 1972).

Also photographs are available of the planetary limb taken by astronauts and showing aerosol layers.

4.3 Summary

The following observations appear feasible from space:

1) Vertical profiles of H_2O and O_3
2) Total amounts of most pollutant gases and total aerosol optical depth.
3) Stratospheric minor constituents and pollutants.

Observations of vertical profiles of pollutants in the lower troposphere appear extremely difficult.

Further research is required on the absorption and scattering properties of pollutants; the distributions, cycles, and variabilities of pollutants; possibilities of microwave and active sensing techniques; instrumentation for remote observation; and mathematical inversion techniques - especially for aerosols.

REFERENCES

Barringer, A., 1968: Chemical analysis by remote sensing. 23 Annual

ISA Instrumentation Automation Conference.

Barringer, A., J. Davies, and A. Moffat, 1970: The problems and potential in monitoring pollution from satellites - AIAA Paper 70-305, 15 pp.

Barringer, A., B. Newbury, and A. Moffat, 1968: Surveillance of pollution form air borne and space platforms. Fifth Remote Sensing Symposium, Ann Arbor, Michigan.

Burch, D., and D. Williams, 1962a: Total absorptance of carbon monoxide and methane in the infrared. App. Optics, 1, 587-594.

Burch, D., and Williams, 1962b: Total absorptance by nitrous oxide in the infrared. App. Optics, 1. 473-482.

Chandrasekhar S., 1950: Radiative Transfer, Clarendon Press, Oxford

Connes, P., J. Connes, W. Benedict, and L. Kaplan, 1967: Traces of HCL and HF in the atmosphere of Venus. Ap. J., 147, 1230-1237.

Connes, P., J. Connes, L. Kaplan, and W. Benedict, 1968: Carbon monoxide in the Venus atmosphere. Ap. J., 173-743.

Conrath, B., 1969: Statistical estimation of tropospheric temperatures and relative humidities from remote radiometric measurements. NASA GSFC X-622-69-14, 43 pp.

Conrath B.R. Hanel, V. Kunde, and C. Prabhakara, 1970: The infrared interferometer experiment on Nimbus 3. J.G.R., 5831-5857.

COSPAR, 1972: Application of space techniques to some environmental problems. Preliminary observing system considerations for monitoring some important climate parameters. (To be published by COSPAR working Group 6).

Elterman, L.R., R. Wexler and D.T. Chang, 1969: Features of tropospheric and stratospheric dust. App. Optics, 8, 893-903.

Fensterstock, J., 1968: Thanksgiving 1966, Air pollution episode in the eastern United States. National Air Pollution Control Administration Publication No. AP-45, 45 pp.

Frisken, W., 1971: Extended industrial revolution and climatic change. EOS, 52, 500-508.

Gille, J. and T. Lee, 1969: The spectrum and transmission of ammonia under Jovian conditions. J. Atmos. Sci., 26, 932-940.

Glaser, A., D. Hunt, J. Sparkman, and R. Danson, 1972: Probabilities for successful operational satellite soundings under normal global cloudiness. AMS 2nd symposium on meteorological observations and instrumentation, 7 pp.

Hanel, R., and B. Conrath, 1970: Thermal emission spectra of earth and atmosphere obtained from the Nimbus 4 Michelson interferometer experiment. Nature, 143-145.

Hanst, P., 1970: Infrared spectroscopy and infrared lasers in air pollution research and monitoring. App. Spect., 24, 161-174.

Heath, D., A. Krueger, and C. Mateer, 1970: The backscatter ultraviolet spectrometer (BUV) experiment. In NIMBUS IV User's Guide, Goddard Space Flight Center, 135-148.

Ing., J., 1972: A dust storm over central China April 1969. Weather, 27, 136-145.

Joseph, J.H. 1967: Detection of noctilucent clouds in the twilight layer from satellites. J.G.R., 72, 15, 4020-4025.

Joseph, J.H. and A. Manes, 1971: Seasonal and secular variations of atmospheric turbidity at Jerusalem. J.A.M. 10, 3, 453-502.

Joseph, J.H., A. Manes and D. Ashbell, 1972: Desert aerosols transported by Khamsinic depressions and their climatic effects, (Submitted for publication).

Komhyr W., E. Barrett, G. Slocum, and H. Weickmann, 1971: Atmospheric total ozone increases during the 1960's. Nature, 232, 390-391.

Ludwig, C.B., R. Bartle, and M. Griggs, 1969: Study of air pollutant detection by remote sensors. NASA CR-1380.

Manes A., 1971: Atmospheric turbidity over Jerusalem, M. Sc. Thesis, Dept. of Environmental Sciences, Tel-Aviv University. pp 81.

Mastenbrook, H. 1971: The variability of water vapour in the stratosphere. J. Atmos. Sci. 28, 1495-1501.

McClatchey, R., et al., 1970: Optical properties of the atmosphere. Environmental Research Papers No. 331, Air Force Cambridge Research

Laboratories, 85 pp.

Mateer, C., D. Heath, and A. Krueger, 1971: Estimation of total ozone from satellite measurements of backscattered ultraviolet earth radiance. J. Atmos. Sci., 28, 1307-1311.

NASA SP-285, 1971: Remote Measurements of Pollution. NASA SP-285, 253 pp.

Prabhakara, C., B. Conrath, R. Hanel, and E. Williamson, 1970: Remote sensing of atmospheric ozone using the 9.6 band. J. Atmos. Sci., 27, 689-697.

Robinson, E. and R. Robbins, 1969: Sources, Abundance and Fate of Gaseous Atmospheric Pollutants Supplement. Stanford Res. Inst. Report.

SCEP, 1970: Study of critical environmental problems. Man's impact on the global environment. MIT Press, Cambridge, Mass.

Sekera, Z. 1957: Light scattering in the atmosphere and polarisation of light. J. Opt. Soc. America, 47, 484-490.

Sherr P., A. Glaser, J. Barnes, and J. Willard, 1968: Worldwide cloud cover distributions for use in computer simulations. Allied Research Associates.

SMIC, 1971: Inadvertent climate modification. Report of the study of man's impact on climate. MIT Press, Cambridge, Mass., 308 pp.

Smith, W., 1970: Iterative solution of the radiative transfer equation for the temperature and absorbing gas profile of an atmosphere. App. Optics, 9, 1993-1999.

Twomey. S., and H. Howell, 1963: A discussion of indirect sounding methods with special reference to the deduction of vertical ozone distribution from light scattering observations. Mon. Wea. Rev., 91, 659-664.

Woodcock, A.H., 1953: Salt nuclei in marine air as a function of altitude and wind force..J. of Met., 10, 362-371.

Yamamoto, G. Tanaka M. and K. Arao, 1968: Hemispherical distribution of turbidity coefficients as estimated from direct solar radiation measurements. J. Met. Soc. Japan, 40, 287-300.

ENVIRONMENTAL OBSERVATIONS FROM SPACE - THE PROMISE AND THE CHALLENGE

E. S. Schaller

General Electric Company - Space Division

Environmental pollution is a problem of world-wide concern and study. Its affects on a local scale are readily apparent, and the monitoring and control of localized pollutants is, perhaps, best accomplished at or near the source. Like disease, however, most forms of pollution ignore our political and national boundaries and take their toll on a far broader, even global, scale. Neither the extent nor the impact of such pollution on our environment is fully appreciated or understood. Certainly our concern is justified, and our willingness to act has been demonstrated. In order to make intelligent, enlightened decisions on our own behalf and that of our decendants, we must gain a deeper understanding of the processes at work in our environment. To accomplish this, we must first collect and analyze data which describe and define these processes. Much of this data can be collected locally by in situ measurement, or remotely using instrumented aircraft. Where the phenomena occur over large areas or are even global in extent, we must seek a new viewpoint, and here lies the promise of environmental observation from space.

Earth observation satellites have been orbiting the globe for a number of years now, engaged in both experimental and operational missions for a variety of applications. The Nimbus Meteorological Satellite series, first flown in the early 1960's, has demonstrated the ability of a spacecraft to provide reliable, repetitive, global coverage of meteorological and environmental phenomena. The true value of this vehicle lies in its capacity to support a large number and variety of sensor payloads. Nimbus 1, launched in 1964 carried only two sensors, an Advanced Vidicon Camera System for daylight cloud cover imagery, and a High Resolution Infrared Radiometer for night coverage. Subsequent advances in sensor technology have provided Nimbus with new instruments with which to aid in understanding the complex relationships between air pollution and climate. Air pollution, we know, influences the radiative processes of the atmosphere, which, in turn, affects short term weather. In the long term, gaseous and particulate pollution affect the earth's heat budget. The Medium Resolution Infrared Radiometer (MRIR), flown on Nimbus 2 (launched May 1966) and Nimbus 3 (launched April 1969), measured the heat balance of the entire 200 million square mile area of the earth for the first time on a global basis each day the instrument was operated. Utilizing data from two-week periods in each of April, July, and October of 1969 and January of 1970, it was possible to perform a complete global heat budget study of the earth for all four seasons for the first time. Nimbus 4, launched in April of 1970, carries nine individual sensor subsystems whose measurements include the spatial distribution of atmospheric ozone, cirrus cloud content and contamination, and the amount of infrared radiation leaving the atmosphere in the CO_2 and water vapor bands.

The outstanding success of the Nimbus Program is measured in terms of the total achievement of its goals and the discovery of new applications for its data. These, in turn, have spearheaded new applications and requirements for earth observation missions.

The world will soon witness the launch of the first Earth Resources Technology Satellite, ERTS A, developed for the National Aeronautics and Space Administration by the General Electric Company. This vehicle is the first of a series specifically designed to develop and demonstrate new techniques for monitoring and managing the earth's resources and environment. In its broadest sense, the ERTS Program consists of the ERTS Observatory, a unique ERTS Ground Data Handling System, and an international "family" of over 200 ERTS investigators. For its part, the ERTS Observatory will acquire high resolution multi-spectral images of the earth's surface on a repetitive, global basis. To accomplish this, the observatory will carry two different sensor systems: a three-camera Return Beam Vidicon (RBV) system, and a four-channel Multi-Spectral Scanner (MSS) system. The Return Beam Vidicon System consists of three high resolution television cameras, each fitted with appropriate filters such that they operate in the red, green, and infrared bands of the spectrum, as follows:

Camera	Spectral Band (micrometers)
1	.475 - .575
2	.580 - .680
3	.698 - .830

All three cameras, each sensing a different spectral region, are focused on the same earth area of 100 x 100 nautical miles. The three cameras are shuttered simultaneously and the images are stored on the photo sensitive tube faces. Each image is then scanned by an electron beam to produce a video signal for transmission to the ground. Reshuttering of the cameras occurs automatically every 25 seconds. This results in a contiguous series of images along the spacecraft's direction of travel.

The Multispectral Scanner System, or MSS, gathers data by imaging the earth scene below in several spectral bands simultaneously, through the same optical system. This virtually eliminates the problems of later image registration on the ground. The MSS is a line scanning device which uses an oscillating mirror to scan normal to the spacecraft direction of flight. Spacecraft motion provides the along-track scan. For ERTS A, the MSS is a 4-band scanner operating in the solar-reflected spectral band region of 0.5 to 1.1 micrometers wave length. Optical energy is sensed by an array of detectors simultaneously in the four spectral bands. The detector outputs are sampled, encoded, and formatted for transmission in a continuous 15 megabit per second PCM data stream. Like the RBV System, the scanner ground swath is 100 nautical miles wide, but is continuous in the direction of travel. During ground processing of the MSS data in the ERTS Ground Data Handling System, the continuous strip imagery is converted into framed images with an area approximately equal to that of RBV.

ERTS A also carries a completely different type of payload from those I have already described. This is the Data Collection System, or DCS. DCS operates in conjunction with small, remote, automatic data collection platforms which are instrumented and deployed on the ground by individual investigators. Each DCS ground platform can collect data from as many as eight sensors, sampling such local environmental conditions as water quality, air quality, stream flow, temperature, and many others. The DCS platforms are designed for remote, unattended deployment. The data acquired by the platform is transmitted to the Data Collection System on-board the spacecraft and immediately relayed to one of the ERTS ground receiving stations. Data from any platform is available to investigators in less than 24 hours from the time the sensor measurements are relayed by the spacecraft.

In order that the data collected by ERTS have maximum utility to the investigators, the data must be collected systematically, repetively, and under nearly constant observation conditions. For this reason, the observatory will operate in a circular, sun-synchronous, near-polar orbit, at an altitude of 494 nautical miles. It will circle the earth every 103 minutes, completing

14 orbits each day, and will have viewed the entire earth every 18 days. This orbit was selected and will be controlled in order that the spacecraft ground trace repeats itself at the same local time every 18 days, within 10 nautical miles. The data acquired by the spacecraft then will permit quantitative measurements of earth-surface characteristics on a spectral, spatial, and temporal basis.

The ERTS Ground Data Handling System (GDHS) has been specifically designed to control the operations of the observatory, acquire and process its data, and provide the highest quality output products for use by the investigators. The ERTS Ground Data Handling System is located at the NASA Goddard Space Flight Center in Greenbelt, Maryland, and consists of the ERTS Operations Control Center and the NASA Data Processing Facility.

The Operations Control Center is the focal point for all ERTS mission activities. It provides control of the spacecraft and payload orbital operations necessary to satisfy the mission and flight objectives. The Control Center operates 24 hours each day, seven days a week, and its activities are geared to the operations timeline dictated by the 103 minute spacecraft orbit and the network coverage capabilities. The primary receiving stations at Fairbanks, Alaska; Goldstone, California; and the Goddard Space Flight Center provide contact with the spacecraft 12 or 13 of its 14 orbits each day.

The ERTS NASA Data Processing Facility (NDPF) is responsible for the processing of ERTS payload data and its dissemination to the user community. Payload data from the spacecraft is recorded in analog form on ground recorders, and input to the NDPF for conversion to film images. The initial video-to-film conversion, known as Bulk Processing, is accomplished using an Electron Beam Recorder which produces a 55 millimeter black and white image on 70 millimeter film. During the conversion process, initial radiometric and geometric corrections are made and annotation, image location, and grey-scale reference information are added. These 70 millimeter images are then processed photographically, inspected for quality, enlarged if required, and distributed to the users. Selected images may undergo Precision Processing to remove additional geometric errors. Precision Processing also performs precise image location and orthographic projection of the corrected image relative to the Universal Transverse Mercator (UTM) grid system. Further processing of ERTS image data is also available from the NDPF and provides the investigator with a digital computer compatible tape of the payload data for computer processing and analysis.

What I have given you in these past few minutes, describing the ERTS Program, is only the briefest glimpse into a very large

and complex program which represents a major first step in a new
application of space technology for the benefit of man. Let me
spend just a minute more with a few facts which I hope will put
the Earth Resources Technology Satellite Program in its proper
perspective. ERTS is a research and development program designed
to demonstrate the successful merger of space and remote sensor
technology. Yet, even as an experimental effort, a great deal
of operational thinking has gone into its design and development.
The data collection capability of the observatory and its ground
system are enormous. The observatory can collect some 1300 scenes
each day. Each scene is acquired in seven distinct spectral bands,
three from the RBV System and four from MSS. To the Ground Data
Handling System, this amounts to some nine thousand separate
master images to be generated each day, not including the data
from the Data Collection System platforms which is processed. To
fulfill its mission, the ERTS System must produce this data such
that the images faithfully portray the radiometric character of
the earth's surface, and display the observed patterns with geometric
accuracy. To satisfy its users, the system must make the
data available in a variety of formats and in a timely fashion.
In terms of format, the system will produce both 70 millimeter
and 9.5 inch images in both black and white and color, and in
both negative and positive transparencies and prints. The ERTS
System is capable of producing 300,000 data products each week.
In terms of timeliness, the data will normally be available
within two weeks of the time of observation.

Interest in the ERTS data is widespread. In addition to the
proposals for ERTS experiments received from within the United
States, more than 85 proposals have been received from governments,
universities, and industry in 27 other countries. Our host for
this conference, the University of Tel Aviv, is a participating
ERTS investigator. The role of the investigator is an integral
and indispensable part of the ERTS Program. His experimentation
with, and analysis of the ERTS data is the principal route to
developing and demonstrating the utility of data acquired by
satellite systems of this type for use in resource and environmental
observations of the earth. The results of these investigations
will provide much of the basis for defining the operational
earth observation systems of the future.

From what we know today, based on the accomplishments of the
past decade, the promise of these future systems is far more
reality than dream. ERTS, for its part, will survey more than
42 million square kilometers each week, and will repetitively
cover the same scene 20 times each year. The continued emphasis
on sensor technology is resulting in the development of new
devices specifically designed for space and aircraft monitoring
of critical environmental parameters such as gaseous and particulate
air pollutants. These microwave sensors provide the

capability for collecting data both day and night and are unhindered by cloud cover. Systems like the ERTS Data Collection System, with its remote, ground-based platforms, provide an effective method of communication with in situ monitoring devices. A low cost DCS satellite network could accommodate literally thousands of such platforms, and relay measurements from each to regional ground stations every two hours.

The very nature of the earth satellite uniquely qualifies it to provide a global view. Just as we climb the mountain to better see the valley, the satellite platform affords the better view of many features of the earth and our environment and it will play a major role in the Environmental Observation System of the future. One of our major challenges lies in properly defining this role within the context of our local, national, regional and world-wide environmental data requirements. We are in the process of adapting a new technology to the growing problem of man's survival and the quality of life. First, through programs like ERTS and others, we must understand those elements of the problem which can be addressed from the vantage point of space. I am sure we all agree that the scope of the problem is broader than that to which the satellite can contribute. It is essential then that we recognize the capabilities and limitations of remote sensing from space. From our past successes and the current trends of technology we can make some forecasts for the future of space systems, of remote sensing, of data processing, and of information extraction. We must also recognize that space observation is but one method of collecting environmental data. The Environmental Observation System of the future will depend upon the complementary interaction of space systems with other data collection methods, such as the ground observer, the aircraft, and the in situ sensor which may relay its data via a satellite system. Considering the fact that here, as elsewhere in our technological age, our ability to collect data often far outweighs our ability to process, analyze, and utilize that data, we must recognize the magnitude of the data processing effort required to extract meaningful information from repetitive, global observations.

Finally, the technologists, at least, must recognize the difficulty in taking action on the information which is produced. To some extent, this results from the skepticism with which the information is received. In large measure, however, it comes about because the actions are so heavily dependent upon political, economic, social, and other exogenous factors; some too difficult to contemplate.

In summary, let me first say that the global nature of our environmental problem, and the global capability of surveys from space are an intriguing match. We stand now on the very threshold of determining their compatibility.

The concept of environmental observation from space carries with it perhaps the most exciting prospect of space application for the benefit of man that we have yet imagined. While not the full answer to the problem, such observations may well contribute significantly to reversing the trend of environmental deterioration by helping us understand the processes at work and take the necessary action. In addition, such observations should underscore the need for international cooperation in terms of both environmental monitoring, and management. To be effective, the Environmental Observation System of the future - near future - must include a variety of data gathering devices, coupled with an international network of communications, processing, and analysis centers established on all levels -- local, national, regional, and international. Satellite and aircraft observations would be collected at the most effective locations and the data processed there, or transmitted to other analysis centers. The results of the data analysis would be disseminated routinely throughout the coverage area. Information relating to a critical or dangerous situation could be relayed on an emergency basis for immediate action.

To a large extent the technology required for such a system exists today. Detailed consideration has already been given to a direct readout ground terminal and processing system for ERTS, which would provide ERTS imagery and DCS data at world-wide locations. Many other elements of the system are also available today, including the communications capability. Nor is the concept new. On the contrary, our world-wide weather system functions in much the manner described.

If the system I've described closely parallels our weather reporting system, let me hope there is one major difference. We always say we can't do anything about the weather - let us not say that about our environment.

A MODEL FOR FLUID MECHANICAL STUDIES OF AIR POLLUTION

C.Y. Liu AND K.G. Agopian
University of California at Los Angeles

ABSTRACT

A chemical model designed for fluid mechanical studies of pollutants distribution in the atmosphere is presented. The 7-step kinetic mechanism consisted of different rate constants and stoichiometric coefficients for various hydrocarbons is verified against several sets of experiments. The kinetic mechanism is adopted to predict pollutants time history in "static" and "convective" conditions simulating containments distribution in an urban (Los Angeles) basin. The model is also used to construct simple scaling laws and to examine the efficiency of various automobile engines in reducing pollutant levels.

INTRODUCTION

The time history of pollutants distribution in atmosphere is governed by convection, diffusion and chemical reactions. Fluid mechanical studies of air pollution generally aim at the rapid prediction of pollution levels for a given weather condition; the appraisal on the effect of new sources; and the evaluation of control mechanisms. To accomplish these objectives, one must solve the pollutant concentration equations, in which information regarding convection, diffusion and chemical production or depletion are needed. Advective current is normally prescribed through correlation of meteorological data rather than solution to the fluid dynamical equations, as the topology of an urban basin becomes too difficult to be incorporated analytically or even numerically. Molecular diffusion and turbulent diffusion are

often represented by a single coefficient obtained through
measurements as a function of the mean wind speed, though molecular diffusion is practically negligible in its overall contribution. The important chemical processes in the lower atmosphere
requires some appropriate description. Though the production of
photochemical smog remains to be fully understood, in its description chemical processes numbering from three to ninety are considered important. In recent years, several reviews of atmospheric chemistry has become available (Leighton,[1] Altshuller and
Bufalini,[2] Haagen-Smit and Wayne,[3] etc.) and a number of simplified kinetic models are proposed (Eschenroeder,[4] Friedlander and
Seinfeld,[5] Westberg and Cohen,[6] Behar,[7] Wayne and Earnest,[8] Hecht
and Seinfeld[9] etc.). Resolution of the information given by the
model is generally proportional to the complexity of the model.
Discussion of all existing models is neither necessary nor feasible, the following four are among the frequently cited.

(1) Model of Eschenroeder (1969).[4] Seven chemical species participate in eight chemical reactions. Two different sets of
reaction rates and the associated stoichiometric coefficients are
obtained from the experimental data of Tuesday[10] and Altshuller[11]
for trans-2-butene and propylene. The original model was refined
in later publication by including more reactions.[12]

(2) Model of Behar (1970).[7] Six chemical species in six steps of
reactions are considered. One set of reaction rates is obtained
from fitting the data of Korth[5] using actual engine exhaust, which
is a complex mixture of various hydrocarbons.*

(3) Model of Westberg and Cohen (1970).[6] Seventy-one reactions
with a very large number of chemical species are taken into
consideration. One set of reaction rates is used for propylene.
Comparison is made with Altshuller's experiment,[11] also with
Wilson's experiment[14] to illustrate the effect of carbon monoxide.

(4) Model of Hecht and Seinfeld (1971).[9] A detailed 80-step
mechanism for propylene-NO_x-air system and a simplified 15-step
mechanism for the description of photochemical smog formation are
proposed. The 15-step model consists of adjustable rate constants
and stoichiometric coefficients dependent on the particular
hydrocarbon and the initial reactant ratio.

Many other models, such as the 15-step mechanism of Calvert[15]
and the 31-step mechanism of Wayne et. al.,[8] are developed with
the same objectives. All mentioned are obtained by some educated

---

*Major components of various hydrocarbon mixtures are ethane,
ethylene, acetylene, propylene, n-butene, benzene, toluene,
m- and p-xylene.

guess and some curve-fitting of experimental data. Our computation has shown that the application of one model to any experiment other than the one on which the model is based, would yield completely incorrect predictions. The search of a chemical model with wider applicability and broader verification is conducted. The present study, however, does not pretend to offer an ultimate answer to the kinetic modeling of pollutants; continuous effort by various investigators would hopefully lead to a correct and practical formulation. As far as fluid mechanical studies are concerned, the overriding consideration lies in the simplicity of the model and the variety of hydrocarbons included. Similar requirements must be posed for kinetic models pertained to describe the thermal and photochemical reactions in the engine exhaust of high speed aircrafts.

MODEL

The final objective in analytical modeling is to provide information for the chemical production term R_i in the species continuity equation for the concentration of the ith species, $C_i (i = 1,2,...n)$.

$$\frac{\partial C_i}{\partial t} + \nabla \cdot (\vec{U} \, C_i) = \nabla^2 (\kappa C_i) + S_i(\vec{x}, t) + R_i(C_1, C_2, ...C_n, t)$$

where

$\vec{U}$ is the advective velocity,

S_i is the given source or sink distribution for the ith species,

κ is the scalar diffusivity including molecular and turbulent diffusions,

$\vec{x}$ is the spatial variable,

t is time.

Aerodynamicists are well accustomed to fluid mechanical problems involving chemical processes, numerical integration of the diffusion equation including advection has been investigated. On the other hand, treatments of the ordinary different equations for chemical reactions have also been thoroughly studied. Perhaps the missing link in solving the species continuity equation is to provide an appropriate description of the photochemical processes. In view of the needs in fluid mechanical studies, the chemical model should involve a minimum number of reactions; it should include a single average hydrocarbon with which an average free-radical is associated; and it should also be identifiable with key chemical reactions known in the atmosphere. These objectives and conditions practically dictate the description of chemistry in air pollution. Our specific goals in constructing such a model are listed in their order of importance: (1) prediction of ozone production, (2) description of the peaking of NO_2 and the leveling off of NO, (3) estimation of hydrocarbon consumption, and (4) estimation of PAN (peroxyacetyl nitrate) and aldehyde production. The present model consists of seven reactions with constant reaction rates and stoichiometric coefficients as given in Tables 1 and 2, dependent on the hydrocarbon used.

$$NO_2 + h\nu \xrightarrow{k_1} NO + O \qquad (1)$$

$$O + O_2 + M \xrightarrow{k_2} O_3 + M \qquad (2)$$

$$O_3 + NO \xrightarrow{k_3} NO_2 + O_2 \qquad (3)$$

$$O + HC + O_2 \xrightarrow{k_4} a_{41} RO_x + a_{42} \text{ Ald.} \qquad (4)$$

$$O_3 + HC \xrightarrow{k_5} a_{51} RO_x + a_{52} \text{ Ald.} \qquad (5)$$

$$RO_x + NO \xrightarrow{k_6} NO_2 + RO_{x-1} \qquad (6)$$

$$RO_x + NO_2 \xrightarrow{k_7} a_{71} \text{ PAN} \qquad (7)$$

TABLE 1. REACTION RATES

Hydrocarbon	k_1	k_2	k_3	k_4	k_5	k_6	k_7
Propylene	.4	$.302 \times 10^7$	.22	497	$.18 \times 10^{-2}$	.5	.01
Trans-2-Butene	.4	$.302 \times 10^7$	.22	1300	$.65 \times 10^{-3}$	.5	.01
Engine Exhaust	.4	$.302 \times 10^7$	.22	1700	$.15 \times 10^{-4}$	1.5	.01
Isobutylene	.4	$.302 \times 10^7$	.22	575	$.10 \times 10^{-4}$	2.0	.01

All concentrations are expressed in pphm. Hence reaction rates k_1, k_2 are given in min.$^{-1}$; k_3, k_4, k_5, k_6 and k_7 are in units of $(\text{pphm})^{-1} (\text{min})^{-1}$. Temperature dependency of the reaction rates has not been taken into consideration.

TABLE 2. STOICHIOMETRIC COEFFICIENTS

Hydrocarbon	a_{41}	a_{42}	a_{51}	a_{52}	a_{71}
Propylene	1.0	1.0	1.0	.1	0.67
Trans-2-Butene	2.5	2.5	2.5	.12	.67
Engine Exhaust	2.5	2.0	2.0	1.0	.67
Isobutylene	2.0	2.0	2.0	.75	.50

The present model falls into the so-called "lumped parameter" family, in that several chemical processes are represented by an average reaction. The representation may be schematical and the entire sequence of fast reactions yields a combined reaction rate, the resulting reactions are not necessary to be stoichiometrically balanced. The significant primary processes in the atmosphere are the photolysis of NO_2 as described by reaction (1), and the rapid equilibrium achieved between NO_2, NO and O_3 as given by reactions (2) and (3). Since NO_2 absorbs radiation over a broad spectrum of sunlight including the violet and the ultraviolet rays; the reaction rate k_1 is an average value for the entire spectrum. The conversion of NO into NO_2 and subsequently NO_2 into peroxyacetyl nitrates (PAN) are represented by reactions (6) and (7), in which the involvement of NO_2 in terminating the free radical chain has been lumped (in the form of k_6 and k_7); and the depletion of PAN by photolytic process is included (through coefficient a_{71}). Reaction (6) is important in providing a real mechanism converting NO into NO_2 without the depletion of ozone. In the sequence of events leading to reactions (6) and (7), intermediate products bearing oxygen atoms are rapidly formed and destroyed, they are ignored in the present considerations as practically all of them are very short-lived. Reaction (4) indicates the attack of the "average" hydrocarbon by oxygen atoms and molecules resulting in the formation of some stable aldehydes. Attack of the hydrocarbon by ozones must be treated separately through reactions (5), as the reaction takes place in atmosphere in much slower manner. The effects of aerosol and particulates have not been included in the present formulation; the role that carbon oxides play in the reaction of contaminants has also been excluded. Smog containing a material portion of SO_2 would hence be poorly represented by the present model. The model is similar to that formulated by Eschenroeder[4] and Behar,[7] except that different representative reactions are emphasized, and, in the present work, a matrix of reaction rates and stoichiometric coefficients are obtained for a variety of hydrocarbons. For the first three primary reactions, the reaction rates are the same as that measured for each individual process, while the stoichiometric coefficients for reactions (4)-(7) are lumped as close to order of unity as possible.

It is not surprising that the system of ordinary differential equations, corresponding to reactions (1)-(7), is stiff, separate studies are being conducted to obtain the eigenvalues of the associated matrix for all values of t. The behavior of the solution near equilibrium, (namely, $t \to \infty$), is nevertheless quite indicative. In an example that the final value of $[O_3]$ and $[NO]$ are 1 pphm at 0.2 pphm respectively, the six eigenvalues of the system at large time are -0.14, -0.186, -0.229, -0.67, -0.8575, and -2.9×10^6. Since all eigenvalues are negative and real, pollutants at large time undoubtedly approach a stable equilibrium. The span between the largest and the smallest eigenvalues indicates the source of stiffness for the set of ordinary differential equations. Numerical treatments of the stiff system are well studied and thoroughly documented. However, the production and depletion of $[O]$ and $[RO_x]$ are so fast that they are practically in equilibrium in the time scale of our interest. The assumption of local equilibrium leads to two algebraic equations for $[O]$ and $[RO_x]$.

$$[O] = \frac{k_1 [NO_2]}{k_2 + k_4 [HC]}$$

and

$$[RO_x] = \frac{k_1 a_{41} [O][HC] + k_5 a_{51} [O_3][HC]}{k_6 [NO] + k_7 [NO_2]}$$

Similar assumptions regarding the steady-state of $[O]$ and $[RO_x]$ were introduced by Eschenroeder and Behar; also by Stephens in the description of atmospheric oxidents.[16] Leighton first recognized that the atomic oxygen concentration can be calculated as the steady-state between the formation by NO_2 photolysis and the disappearance by reaction to form ozone. In atmosphere containing a small fraction of HC and NO_2 all atomic oxygen indeed reacts in this fashion.

Available smog-chamber data show substantial divergence quantitatively and qualitatively; chamber configuration, initial concentrations and the type of hydrocarbon used are not very well controlled. Verification between experiments is not probable, if not impossible; the needs for controlled chamber experiments are indeed greater than ever. Nevertheless, five sets of measurements performed over a period of ten years in different chambers are used to verify the present model. Measurements by Tuesday[10] using trans-2-butene with a reaction time of about 20 minutes are compared with the present calculation, and also with the model of Eschenroeder (Figure 1). The peaking of NO_2 and the depletion of

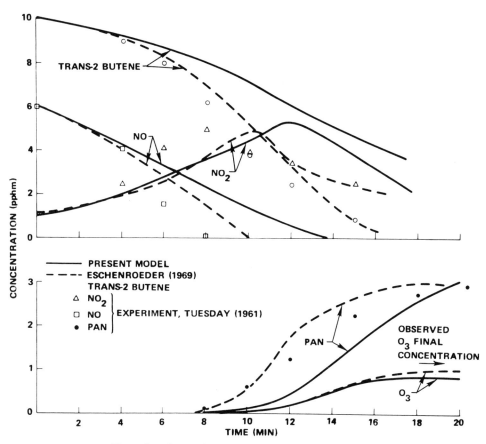

Figure 1. Comparison with Experiment by Tuesday.

NO are correctly predicted, the production of PAN and ozone are also in the right range, while the consumption of trans-2-butene according to the present model seems slow. Altshuller's measurement[11] using propylene is compared against the present model and Eschenroeder's calculation in Figure 2; the time-history of HC and PAN seem acceptable by comparison, yet the final ozone level in the 180-minute experiment shows a much lower value than that given by the present model as well as that predicted by Westberg and Cohen's 71-step computation. The present model and the 81-step mechanism of Hecht and Seinfeld designed specifically for propylene-NO_x-air system are verified against the measurements performed at the Gulf Research Company* in Figure 3. Again, the consumption of propylene and the conversion of NO into NO_2 are correctly predicted, but the level of ozone production according

---

*Data cited in Ref. 9.

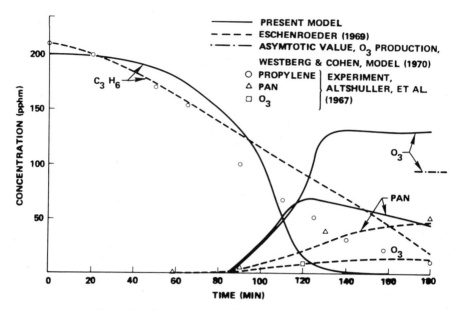

Figure 2. Comparison with Experiment of Altshuller et al.

to the present model seems too high. The series of experiments performed by Korth using automobile exhaust is exhibited in Figure 4 along with the computation by the present model and Behar's. Four cases labeled "C", "E", "I", "N" are selected; they represent initial hydrocarbon concentrations ranging from 75 pphm to 200 pphm and initial nitric oxides from 25 pphm to almost 200 pphm. Cases "C", "E", and "I", show acceptable prediction for the final values of pollutants at the end of the 4-hour period, but case "N" illustrates the worst comparison in the entire series of measurements. The present model, as well as Behar's, fails to describe the extreme situation when the initial nitric oxides overwhelms the initial hydrocarbon, fortunately it represents a fairly uncommon situation in atmospheric condition. The experiment of Westberg, Cohen and Wilson[14] is illustrated in Figure 5 along with the present calculation. All results shown illustrate the range of applicability of the proposed model. Perhaps the comparison is considered acceptable in view of the small number of chemical reactions selected and the invariance of rate constants and stoichiometric coefficients.

APPLICATIONS

Several simple applications of the model are performed before embarking on a large-scale computational effort in the integration

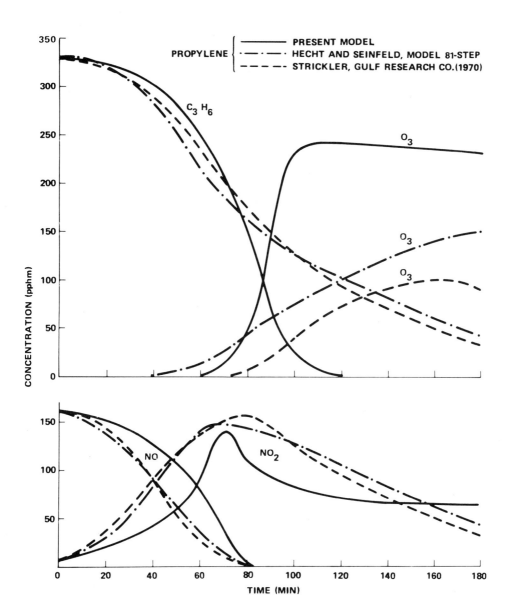

Figure 3. Comparison with Experiment Performed by Strickler at Gulf Research Company.

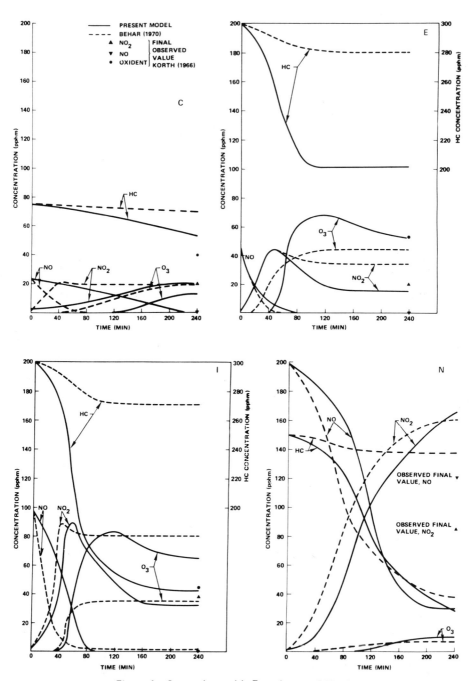

Figure 4. Comparison with Experiment of Korth.

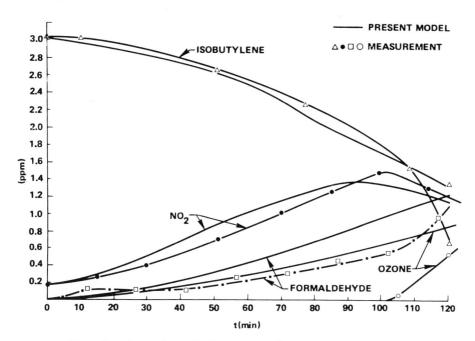

Figure 5. Comparison with Experiment of Westerberg, Cohen and Wilson.

of the system of the species continuity equations. First, we consider the atmosphere in the Los Angeles basin a homogeneous mixture free of advective and diffusive currents and completely absent of source or sink. Two limiting cases are studied: the "static" case in which the atmosphere behaves identically to that in a static smog chamber, and the "convective" case in which a constant residence time for the mixture is assumed. The latter case is equivalent to a continuous supply of premixed atmosphere in its initial conditions, and a continuous evacuation of pollutants in their chamber condition. An average residence time of 300 minutes is assumed, which corresponds roughly to a uniform wind speed of 7 mph swept over the basin of 1250 square miles. The "convective" case could be viewed as stream tube computation with a constant advective velocity. The source of hydrocarbon is the automobile exhaust, while the source of oxides of nitrogen is 72% from automobile and the remaining from industrial and power plants. Hence the average value of rate constants $k_4 = 1400$, $k_5 = 0.2 \times 10^{-4}$ are used. The time step for the integration of Eqs. (1-7) is chosen so small that the simple forward differencing and the Runge-Kutta second order routine would practically yield the same numerical result. The integration routine is then examined against the Gear's Method for stiff systems, no substantial discrepency in the numerical results has been observed.

TABLE 3

Time of day	Percentage of daily traffic per hour	Input Concentration (pphm/min.)		
		[HC]	[NO]	[NO_2]
0000-0600	1.45	0.036	0.012	0.0006
0600-1600	6.0	0.15	0.05	0.0025
1600-1900	8.0	0.20	0.067	0.0033
1900-2400	1.45	0.036	0.012	0.0006

Initial conditions for both cases are obtained from the time distribution of traffic for the Los Angeles basin and its daily amount of pollutants emitted. Approximately, 2600 tons of ozone and 950 tons of nitric oxides are emitted per day,[17] while the percentage of daily traffic per hour for the entire 24-day is also known as given in Table 3. The average molecular weight of 40 gm/mole is assumed for the highly reactive hydrocarbons, the volume of basin is estimated from the area of 1250 square miles and an inversion height of 1500 ft. Hence the hourly emission rate of hydrocarbon is known. The initial amount of oxides of nitrogen is separated into 95% of NO and 5% of NO_2. The resulting input conditions for both the "static" and the "convective" cases, simulating a smoggy summer day in Los Angeles, as shown in Table 3 above.

Numerical results are then compared with two sets of data obtained in 1969 (Scott Laboratory)[18] and 1971 (Air Pollution Control District)[19] respectively; they do not necessarily correspond to the same initial conditions as the traffic pattern for the day and the location of observation vary. Comparisons of ozone production, hydrocarbon depletion, and nitric oxide consumption for a 500-minute day starting at 6:00 A.M. are illustrated in Figure 6. In view of the uncertainities in the input conditions, the qualitative agreement between computational results and measured data is indeed surprising, particularly for [HC] and [NO_x]. Results from the "static" and the "convective" cases, in principle, bound the observed data, as illustrated by the [HC] and [NO_x] variation. The magnitude of the peak values for ozone, which is of prime interest in pollution control, seems correctly predicted. Results of the "static" model is closer to observed data in the morning when the effects of advection is negligible; and the "convective" model appears to work better in the afternoon. In spite of many crude approximations made, the distributions of pollutants in the atmosphere seems conclusively to be predominated by chemical processes. Some models, in which superposition of linear solutions are permitted (such as the Gaussian plume model), are hence suitable only for the descriptions of nonreactive pollutants.

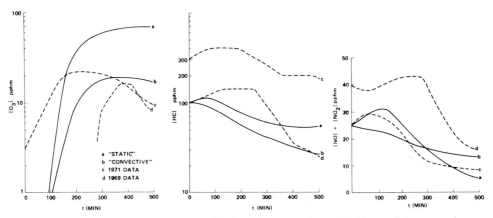

Figure 6. Comparison of Theoretically Obtained Concentrations with Measured Concentrations for $[O_3]$, [HC] and [NO] + $[NO_2]$ in Central Los Angeles Basin.

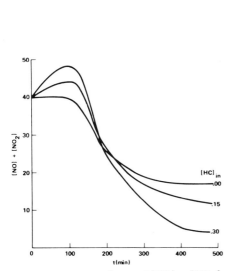

Figure 7. Dependency of [NO] + $[NO_2]$ on the Initial Value of [HC].

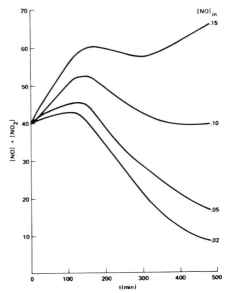

Figure 8. Effect of $[NO]_{in}$ on the Level of Oxides of Nitrogen.

The validity of the model is also examined by formulation of simple scaling laws through variations of some key parameters. The first three reactions which describe the NO_2-air photochemistry, are relatively better understood, so are the associated reaction rates. Reaction rates k_1 and k_2 are insensitive to temperature variation, but the dependency of k_3 on temperature T follows that

$$k_3 = \frac{C_1}{T} \exp(-C_2/T)$$

with C_1 and C_2 being known constants.[12] The effect of daily temperature variation with respect to some pollutant (say, ozone) level can then be studied through the variation of the rate constant k_3. Temperature on an average summer day increases from 65°F at 6 A.M. to about 95°F at 2 P.M., computation shows that the variation of species concentrations becomes pronounced only after 9 A.M. (or $t \geq 200$ minutes). The aforementioned temperature increase, corresponding to a reduction of k_3 from 0.40 to 0.22, affects the peak of $[O_3]$ and the final level of [NO] most. (See Table 4.) Both elements in turn affect human comfort to a large extent.

The levels of ozone, hydrocarbon, PAN and aldehydes are found directly proportional to the hydrocarbon input level, whereas oxides of nitrogen are inversely proportioned. The qualitative behavior, shown in Figure 7, agrees well with atmospheric measurements made during the early period of hydrocarbon emission controls when increase in the level of nitrogen oxide was observed. An other important input parameter is the initial value of NO_x. Computation shows that the input level of oxides of nitrogen reach a critical point at about 0.10 pphm, beyond this value the free radicals of hydrogen RO_x cannot consume enough NO_2 at a rate that would keep the nitrogen oxide decreasing, hence resulting in an increase in oxides of nitrogen (Figure 8). Among the residuals of all pollutants left overnight, an excessive amount of nitric oxide would be most damaging to the smog condition next day.

TABLE 4. TEMPERATURE EFFECT ON POLLUTANT CONCENTRATIONS

k_3	$[O_3]_{max}$	$[HC]_{max}$	[HC] level	[NO] level	$[NO_2]_{max}$	$\left(\frac{[NO]+[NO_2]}{[NO_2]}\right)_{max}$	$[PAN]_{max}$
.22	99	103.8	30.7	.23	41.7	31.7	31.7
.30	83	103.7	36.2	.13	41.8	35.0	35.0
.40	69	103.6	41.0	.07	41.9	37.4	37.4

Reaction rate k_3 is in $(pphm)^{-1}$. All concentrations are given in pphm.

TABLE 5. FINAL CONCENTRATION OF POLLUTANTS FOR DIFFERENT ENGINES

		STATIC MODEL		CONVECTIVE MODEL	
	Initial $[NO_x]$	40	25	40	25
$[O_3]$	Piston	99	91	16	13
	Steam	87	78	11.3	8.2
	Stirling	72	70	7.6	5.4
	Turbine	76	72	8.3	5.8
$[HC]_{level}$	Piston	30.7↑	47.4↑	23.2	25
	Steam	16.5	33.5	11.8	13.7
	Stirling	4.4	13.9	3.6	4.9
	Turbine	6.5	17.3	5.5	6.9
$[NO]_{min}$	Piston	.23	.1	1.57↑	1.60↑
	Steam	.22	.03	.9	.86
	Stirling	.59	.26	1.95↑	1.83↑
	Turbine	.50	.21	1.77↑	1.68↑
$[NO_2]_{max}$	Piston	41.7	23.6	19.9	13.8
	Steam	35.8	22.4	14.7	8.7
	Stirling	36.8	23.4	13.9	8.1
	Turbine	36.7	23.4	14.1	8.3

(Arrow ↑ indicates increase in production of particular species after going through a minimum. All concentrations are given in pphm.)

The kinetic model can also be used to evaluate different types of automobile engines based on the level of different pollutants. Alternates to the conventional piston engines are gas turbine, steam and Stirling engines; they all emit hydrocarbons and oxides of nitrogen in different proportions. Computation are performed according to the following figures of estimated emissions, as obtained by Caretto et al.[20]

	HC(gr/mile)	NO_x(gr/mile)
Piston (1971 California Limit)	2.2	4.0
Steam Engine	.7	.4
Stirling Engine	.07	1.0
Turbine (Rover) Engine	.2	1.0

Computations were made by replacing the entire Los Angeles Basin car population by, in turn, the steam, Stirling and turbine engines. We obtain the numerical results as given in Table 5 above.

At a given initial condition, the Stirling engine and turbine excel over others in practically all categories. Motor vehicles using steam as a working medium, according to the present calculation, does not seem successful in the reduction of key pollutants probably due to the inefficient fuel combustion process which discharges about a third of the hydrocarbons of the conventional engine into atmosphere.

CONCLUSIONS

A simple model suitable for fluid mechanical studies of pollutants distribution in the atmosphere is constructed, and it is verified against several sets of experimental data. The model is shown capable of predicting the behavior of key pollutants. Two limiting cases equivalent to a "static" and a "convective" smog chamber, are examined; their results set the upper and lower bounds of the pollutants levels. The "static" model gives good agreement for hydrocarbons and oxides of nitrogen levels but overestimates the ozone level. The "convection" version gives good agreement for the ozone level, while underestimating the hydrocarbon and oxides of nitrogen levels. Applications of the model to the formulation of simple scaling laws and the evaluation of vehicle engines are also performed. The model at its present stage of development is perhaps too simple to describe all relevant chemical spieces in the lower atmosphere, yet it identifies the importance of the problem as being a chemical one. The next task should be the verification of the model for the prediction of pollutants distribution including all transport processes.

ACKNOWLEDGMENT

The authors are grateful to Professor J.D. Cole for suggesting the study and his careful guidance throughout the course of the investigation.

REFERENCES

1. Leighton, P.A., (1961) <u>Photochemistry of Air Pollution</u>, Academic Press, New York, N.Y.

2. Altshuller, A.P., and Bufalini, J.J., (1971) "Photochemical Aspects of Air Pollution: A Review," Environ. Sci. and Tech., Vol. 5, p. 39.

3. Haagen-Smit, A.J., Wayne, L.G., (1967) in Chapter 6, Vol. I, <u>Air Pollution</u> edited by Stern, A.C., Academic Press, New York, N.Y.

4. Eshenroeder, A.Q., (1969) "Validation of Simplified Kinetics for Photochemical Smog Modeling," IMR-1096, General Research Corp., Santa Barbara, California.

5. Friedlander, S.K., Seinfeld, J.H., (1969) "A Dynamic Model of Photochemical Smog," Environ. Sci. and Tech., Vol. 3, p. 1175.

6. Westberg, K., Cohen, N., (1970) "The Chemical Kinetics of Photochemical Smog as Analyzed by Computer," AIAA Paper 70-753, AIAA 3rd Fluid and Plasma Dynamics Conference, Los Angeles, California.

7. Behar, J., (1970) "Simulation Model of Air Pollution Photochemistry," Project Clean Air, University of California, Vol. 4.

8. Wayne, L., Danchick, R., Weisburd, M., Kokin, A., Stein, A., (1970) "Modeling Photochemical Smog on a Computer for Decision-Making," 70-18, 63rd Air Poll. Cont. Assoc., St. Louis, Mo.

9. Hecht, T.A., and Seinfeld, J.H., (1972) "Development and Validation of a Generalized Mechanism for Photochemical Smog," Environ. Sci. and Tech., Vol. 6, p. 47.

10. Tuesday, C.S., (1961) "The Atmospheric Photo-oxidation of Trans-Butene-2 and Nitric Oxide," GMR-332, General Motors Research Lab.

11. Altshuller, A.P., Kopcznski, S.L., Lonneman, W.A., Becker, F.L., Slater, R., (1967) "Chemical Aspects of the Photo-oxidation of the Propylene-Nitrogen Oxide System," Environ. Sci. and Tech., Vol. 1, p. 889.

12. Eschenroeder, A.Q., Martinez, J.R., (1971) "Further Development of the Photochemical Smog Model for the Los Angeles Basin," CR-1-191, General Research Corp., Santa Barbara, California.

13. Korth, M.W., (1966) "Effects of the Ratio of Hydrocarbon to Oxides of Nitrogen in Irradiated Auto Exhaust," Public Health Publications, 999-AP-20.

14. Westberg, K., Cohen, N., Wilson, K.W., (1970) "Carbon Monoxide: Its Role in Photochemical Smog Formation," ATR-71 (8107)-1, Aerospace Corp., El Segundo, California.

15. Calvert, S., (1969) "A Simulation Model for Photochemical Smog," California Air Environment, Vol. 1, p. 1.

16. Stephens, E.R., (1966) "The Rule of Oxygen Atoms in the Atmospheric Reaction of Olefins with Nitric Oxide," Air & Water Poll. Int. J., Vol. 10, p. 793.

17. Teague, R.M., (1964) "Los Angeles Traffic Pattern Survey," Vehicle Emission I, SAE Progress in Technology Series, Vol. 6, p. 102.

18. Scott Research Laboratory, (1969) "Atmospheric Reaction Studies in the Los Angeles Basin," Final Report on Phase I, Vols. I and II.

19. Air Pollution Control District, County of Los Angeles, (1971) Profile of Air Pollution Control, (pamphlet).

20. Caretto, L.S., McElroy, M.W., Nelson, J.L., and Venturium, P.D., (1970) "Automobile Engine Development Task Force Assessment," Project Clean Air Task Force Assessment, Vol. 1.

A SYSTEMATIC METHOD FOR EVALUATING THE POTENTIAL ENVIRONMENTAL IMPACT OF NEW INDUSTRIAL CHEMICALS

E. J. Sowinski and I. H. Suffet

Environmental Engineering and Sciences Program
Drexel University, Philadelphia, Pennsylvania

Introduction

　　The increasing pervasiveness of toxic substances in the environment--including metallic compounds and synthetic organic compounds--poses a particular concern for future environmental quality. The nature of specific environmental problems, involving for example heavy metals as the organometallic methyl-mercury and organics as polychlorinated biphenyls (PCB's) indicates the need for systematic methods aimed at resolving the potential environmental impact of new industrial chemicals prior to full-scale industrial use (1,2). Current environmental legislation reflects this trend and the setting of environmental quality standards in the future can be expected to include this concept (3,4).

　　This report concerns a general systematic approach to aid in the prediction of the potential environmental impact of new industrial chemicals based on chemical fate guidelines. A discussion of specific analytical approaches for evaluating the chemical fate of two exemplary chemical types used in the electronics industry, namely, boron hydrides and organoantimony compounds, are included. Control aspects based on air pollution potential are discussed.

Background

　　In the past, the total concentration of individual elements rather than their specific chemical forms has been commonly used as a guide in setting standards for industrial air environments (Threshold Limit Values) and ambient air and water environments.

TABLE 1

CRITERIA FOR STANDARDS -
NEW INDUSTRIAL CHEMICALS

I. Toxicological

    A. Acute and subacute levels
    B. Chronic levels
    C. Nuisance levels

II. Aesthetic

III. Analytical Detectability

    A. Sensitivity
    B. Positive identification

IV. Fate in the Environment

    A. Chemical
    B. Physical
    C. Biological

TABLE 2

LOWER LIMITS OF DETECTABILITY
FOR BORON HYDRIDES

	Nanograms B_2H_6	Nanograms $B_{10}H_{14}$
FPD	1.00	0.71
Electron Capture	0.10	0.10
Coulometry	1.00	1.20

However, as knowledge of environmental chemistry and toxicology has improved, it has become increasingly evident that such limits have little meaning without characterization of specific chemical species which exist in the environment.

A set of general criteria which relate to the degree which specific chemicals must be controlled are described in Table I. These criteria have been applied in the past mainly in the water pollution area as a guide for controlling new chemicals, particularly pesticides (5). Some of these criteria have found general acceptance in defining standards for industrial chemicals. These include primarily toxicological, aesthetic and analytical criteria (Table I; items I, II and III) (6). However, an important dimension involving the environmental fate (Table I item IV) of specific chemical species appears to be commonly overlooked when the potential environmental impact of a new industrial chemical is considered.

The environmental fate of parent industrial chemicals may be subdivided into three basic areas: 1) Chemical Fate, 2) Physical Fate, and 3) Biological Fate. As defined here, chemical fate relates to chemical changes of parent compounds in manufacturing processes and in the environment, physical fate involves the distribution of specific chemicals in the environment, and biological fate involves chemical changes in receptor organisms.

The importance of chemical fate in predicting the potential impact of a new industrial chemical is depicted by a pollutant that exists as a unique manufacturing byproduct and/or by a pollutant which undergoes chemical changes in the environment. Such changes can result in secondary pollutants of greater or lesser environmental hazard than the original parent chemical. Environmental controls may be instituted for the parent compound which are unsatisfactory for byproducts although satisfactory for the parent. As a result a particular need exists for determining the chemical fate of parent industrial chemicals to prevent pollution from a process byproduct.

General Approach for Determining Chemical Fate

An important aspect in predicting potential environmental impact of new industrial chemicals relates to the question of whether or not specific by-products, once formed in a manufacturing process, are stable throughout the manufacturing process and in the environment. Three basic steps are involved when this question concerning the chemical fate of a particular compound is evaluated. First, the literature concerning the chemistry and analysis of the parent compound must be reviewed. The extent to which byproducts may differ from parent compounds must be particularly considered with attention given to all possible reactions, stabilities, and rates of decomposition at manufacturing process levels and at potential environmental levels. Important factors to be considered include the chemical properties of the parent compound, its method of use and the effect of environmental stresses on the parent compound. A pitfall exists in this area due to potential differences of stability and reactivity at trace concentrations as compared to high concentrations. For example, it has recently been shown that a particular compound (dichloroacetylene) which is unstable at high airborne concentrations (>200 ppm) is stable at low concentrations and exhibits toxic effects (7).

Secondly, sampling and analytical methods for qualitative and quantitative evaluation of the parent compound and its byproducts must be established. This measurement phase should follow some general guidelines:

1) The isolated sample should represent qualitatively and quantitatively, the ratio of parent and its breakdown products in the process and in the environment;

2) No changes in chemical integrity of the sample should be produced during the analytical procedure;

3) The analysis time should be minimal for subsequent use as an early warning system;

4) The method should be as free from interferences as possible.

In this area confirmatory analysis is required for positive identification of byproducts. For example, chromatographic retention parameters are not sufficient for positive identification of byproducts as the possibility of mistaken identification of PCB's as chlorinated hydrocarbon pesticides has been described (8). Consequently confirmatory techniques employing spectroscopy to measure molecular properties are required besides conventional chromatographic detection. These may include Infrared, Nuclear Magnetic Resonance and Mass Spectroscopy.

Finally, manufacturing byproducts must be identified and their stability and reactivity in the environment determined. For this purpose, pilot processes may be monitored and/or laboratory modeling procedures can be employed. The use of modeling procedures for screening the biological fate of new chemicals has recently received attention in the area of pesticides. For example a model ecosystem has been developed to study the biodegradability of pesticides (9). Also, an analogous procedure for modeling the physical fate (dispersion) of chemicals has been outlined in the area of air pollution (10). Figure I summarizes the chemical fate concept in the form of a template in order to facilitate its incorporation into a preventative environmental control program for industry. The factors influencing chemical fate and the steps to be taken in a preventative control program are outlined around the basic chemical fate concept. This overall concept for controlling new industrial chemicals represents a preventative technique for pollution control rather than a curative technique. Prevention of pollution at the source should be the guiding practice for industrial environmental control.

Chemical Fate Applied to the Control of Exemplary Industrial Chemicals

Chemicals from the metal hydride and organometallic classes, as used in the electronics industry, have been selected as models for chemical fate evaluations. A laboratory modeling system for determining chemical fate of these compounds requires a gas handling system which is capable of modeling a variety of manufacturing situations. A system which has been developed includes features for mixing reactant gases to simulate manufacturing situations, collecting byproducts in a static retention system and a gas sampling system for injection of byproducts into a gas chromatograph.

Emphasis must be placed on the fact that any single modeling procedure may not be applicable to all planned manufacturing processes in any given industry. However, the modeling of planned manufacturing processes in some manner is desirable in order to predict potentially harmful byproducts. Preferably this should be accomplished in early periods of industrial research and later in stepwise coordination with further programs of research and development. In this manner if harmful byproducts are detected, substitute manufacturing procedures can be substantiated or more adequate controls can be designed than would otherwise be the case.

An example of the importance of chemical fate as it relates to the control of an industrial chemical can be found in the case of boron hydrides and organoantimony compounds. These compounds are of concern as boron was recently named as one of fourteen specific agents requiring control in the United States Clean Air Amendments of 1970. Antimony, as a heavy metal on the other hand, fits into the same class of toxic pollutants as mercury and lead.

The most common of the boranes encountered industrially are diborane (B_2H_6 - gas Threshold Limit Value (TLV) = 0.1 ppm), pentaborane (B_5H_9 - liquid TLV = .005 ppm) and decaborane ($B_{10}H_{14}$ - solid TLV = 0.050 ppm). The family of boron hydrides has widespread industrial applications as catalysts in synthetic reactions, as parent compounds for the synthesis of carborane polymers and as

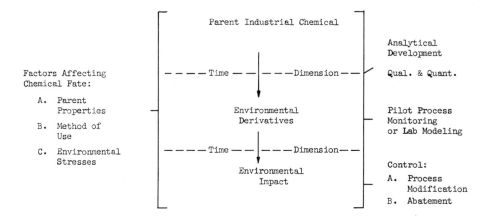

FIGURE I

CHEMICAL FATE TEMPLATE

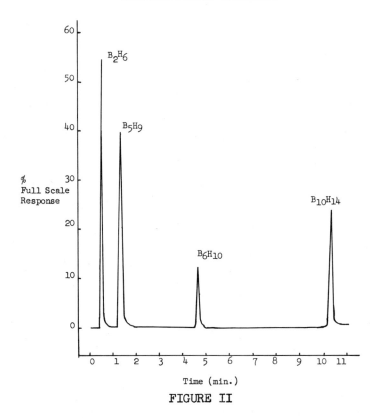

FIGURE II

TEMPERATURE PROGRAMMED CHROMATOGRAM OF B_2H_6 PYROLYSIS BY-PRODUCTS

1.28×10^{-8} Amp Full Scale
6% OV-17: 80/100 Chrom. W
40°C - 180°C

dopant sources for electronic semiconductor manufacturing.

Notable in the chemistry of diborane is its tendency to form higher borane byproducts, including pentaborane and decaborane, when subjected to elevated temperatures (11, 12). Diborane tends to be unstable to oxidation and hydrolysis; however, the higher hydride derivatives of diborane tend to exhibit greater stability (13). Since diborane is commonly subjected to elevated temperatures in its industrial applications, concern exists for the identification and fate of the more stable hydrides which may form as byproducts.

In the case of organoantimony compounds, triethylstibine and trimethylstibine are of health related significance in the manufacture of semiconductor materials. Also, antimony and its compounds are industrial hazards in the rubber industry, the printers trade, and the storage battery industry. Stibine (SbH_3), antimony hydride, is a particularly serious hazard which has a TLV of 0.10 ppm in air.

Methods of analysis for boron and its compounds are available; however, they lack the required selectivity and sensitivity for determining trace atmospheric concentrations related to air pollution and for determinations near the industrial air standards (TLV). For example, gas chromatographic separation has been reported for several boron hydride moieties; however, the detection methods were not specific for boron and only high concentrations have been separated (14, 15).

Sensitive methods for determining trace quantities of antimony and its compounds are not common. The most specific and sensitive methods available for the analysis of antimony have been wet chemical photometric methods using Rhodamine B or 3, 4, 7-trihydroxyflavone as the reagent (16, 17).

The lack of sensitive and specific analytical methods for evaluating the airborne fate of specific boron hydride and organoantimony compounds and the toxicity of these compounds at low levels has led to an investigation of gas chromatographic isolation combined with sensitive and selective gas chromatographic detection systems at ppb levels.

The detection systems which have been investigated are flame photometric detection, electron capture detection and microcoulometry. The principles of operation for these three detectors have been amply developed (18, 19, 20). Nonetheless they have not been combined with gas chromatography for evaluating the chemical fate of specific metal hydride and organometallic compounds.

In a recent report by the authors, a Melpar Flame Photometric detector (FPD) used with a 546 nm narrow bandpass filter in combination with gas chromatographic isolation has been defined as a selective and sensitive detector for volatile boron hydride compounds (21). Unfortunately antimony flame emission occurs in the UV region where narrow bandpass filters with high percentage transmittance are not available. However for limited applications triethylstibine has been found to respond to the FPD using a filter with a maximum transmittance of 10% at a wavelength of 268 nm and a half band width of 10 nm.

Table 2 outlines experimentally determined detection limits for all three detectors as applied to the boron hydrides. The electron capture detector is the most sensitive. The FPD is next in sensitivity and it does offer selectively for boron. Most importantly, however, is the fact that at these lower limits of detectability direct determinations can be made in the range of the industrial air standards for specific boron hydrides. For example 1.25 ng of decaborane vapor in 5 cc of air sampled with a gas syringe can allow for direct quantitation at the TLV of 0.050 ppm. Also, the detection of byproducts generated at potential environmental levels from a modeling system can be accomplished.

Figure II shows an application of the FPD-GC combination to the analysis of boron hydride byproducts from a laboratory scale pyrolysis reactor. Byproducts appear in a pyrolysis temperature range of 220°C to 260°C and they have been determined to be stable in the presence of oxygen at the ppm concentration range. A mass spectrometric confirmatory analysis has identified the byproducts as B_5H_9, B_6H_{10} and $B_{10}H_{14}$. With this information obtained from a model or pilot process it is now possible to design an actual manufacturing process which will avoid harmful byproducts. (i.e. Manufacturing processes which operate outside of the critical temperature boundaries preclude the formation of undesirable byproducts at the source.) As a result potentially costly abatement procedures can be avoided.

Summary

It should be recognized that research and development is required to provide preventative techniques for the control of new technology chemicals. For evaluating the potential environmental impact of new industrial chemicals it is necessary to develop selective and sensitive analytical methods for measuring byproducts from laboratory modeling procedures or pilot processes. This involves the concept of chemical fate.

Specific studies have demonstrated that analytical schemes can be developed for process byproducts at the industrial air standards and manufacturing situations can be modeled for volatile boron containing compounds.

This overall project, although patterned for the control of byproducts from speciality metal hydride and organometallic chemicals, can provide a useful model for establishing preventative pollution control techniques in many industries.

Acknowledgement

This research was made possible by a grant from the Western Electric Company to Drexel University.

References

1. Langley, D. G., Paper presented at the 162nd American Chemical Society National Meeting, Washington, D.C., September 15, 1971.

2. Gustafson, C. G., Environmental Science and Technology $\underline{10}$, 814-819, 1970.

3. United States Clean Air Amendments of 1970: P.L. 91-604 December 31, 1970.

4. Toxic Substances Control Act of 1971: H. R. 5276, H. R. 5390, H. R. 6316, and S. 1478.

5. Suffet, I. H., Ph.D. Thesis; Rutgers University, 1969.

6. Stokinger, H. E., American Industrial Hygiene Association Journal $\underline{24}$, 469-474, 1963.

7. Saalfield, F. E., Williams, F. W., Saunders, R. A., American Laboratory pp 8-16, July, 1971.

8. Veith, G. D., and Lee, G. F., Water Research $\underline{4}$, 265-269, 1970.

9. Metcalf, R. L., Sangha, G. K., and Kapoor, I. P., Environmental Science and Technology $\underline{5}$, 709-713, 1971.

10. Zimmer, C. E., and Larsen, R. I., Air Pollution Control Association Annual Meeting, Toronto; June, 1965.

11. Cotten, F. A., and Wilkenson, G., "Advanced Inorganic Chemistry" InterScience, New York 1962.

12. Schechter, W. H., Jackson, C. B., and Adams, R. M., "Boron Hydrides and Related Compounds", Callery Chemical Company May, 1954.

References (Cont'd)

13. Carabine, M. D., and Norrish, E. R. S., Proc. Roy. Soc. 296, 1, 1966.

14. Kaufman, J. J., Todd, J. E., and Koski, W. S., Anal. Chem. 29, 1033, 1957.

15. Borer, K., Littlewood, A. B., and Phillips, C. S., J. Inorg. Nucl. Chem. 15, 316, 1960.

16. Portmann, J. E., and Riley, J. P., Anal. Chem. Acta. 35, 35, 1966.

17. Filer, T. D., Anal. Chem. 43, 725, 1971.

18. Brody, S. S., and Chaney, J. E., J. Gas. Chromatog. 2, 42, 1966.

19. Lovelock, J. E., Anal. Chem. 35, 474, 1963.

20. Challacombe, J. A., and McNulty, J. A., Residue Reviews 5, 57, 1964.

21. Sowinski, E. J., and Suffet, I. H., J. Chrom. Sci. 9, 632, 1971.

EFFECT OF ADDITIVES ON BOILER CLEANLINESS AND PARTICULATE EMISSIONS

Ira Kukin
Apollo Chemical Corp.

INTRODUCTION

The use of concentrated metal-containing, additives in oil-fired furnaces is common practice today. This is particularly true with boilers having high superheat and reheat temperatures and high heat release. Such additives improve boiler cleanliness and reduce corrosion of the firesides and air heaters. They also are effective for reducing stack emissions. The plumes from the stacks, particularly when they result from hydrocarbons or sulfuric acid (from SO_3) are improved.

Fuel additives in general should provide for the following:

1) Boiler cleanliness.
2) High temperature vanadium corrosion protection.
3) Prevention of loss of operating capacity by maintaining design steam temperatures.
4) Cold end (air heater) corrosion protection.
5) Reduction of stack emissions from: (a) hydrocarbon particulates and (b) SO_3.
6) Improvement in the handling characteristics of the ash in the flue gas in oil-fired boilers equipped with precipitators and stack collectors.

Generally, the significant fuel additives in use today contain the following:

1) MgO (with or without small amounts of aluminum oxide or hydrate).
2) Manganese.
3) MgO with manganese.

The application of additives to solve particular boiler requirements hopefully has evolved from an art to almost a science. Utility managers today will no longer consider seriously any fireside additive that does not contain a high metallic concentration. The choice of additive, however, depends upon the needs of the particular boiler, as well as environmental requirements as detailed in this study.

THE HEAVY DUTY ADDITIVES: MgO vs Mn

Historical Role of MgO as a Fuel Oil Additive With High Sulfur and High Vanadium Fuels

The addition of MgO based products to the fuel oil or furnace will raise the fusion point of the fuel oil ash from an initial 950-1050° F. to approximately 1350-1450° F. at a Mg:V weight ratio of 1:1 to 1.5:1. A weight ratio of 1.5:1 is equal to a 3:1 molar ratio of Mg:V. This corresponds to a dilution rate with an MgO slurry containing 50% MgO of approximately 1:2000 with a fuel oil having a vanadium content of 200 parts per million of vanadium. The fuel oil ash, consisting generally of magnesium vanadates, magnesium sodium vanadates and magnesium sodium vanadyl sulfates, is less adherent to the superheat surfaces.

Depending upon the ratio of Mg:V used, the ash will range in texture from a soft but voluminous powder at high Mg:V ratios to a somewhat brittle "popcorn" configuration or even to a dense, layered, amorphous coating as the ratio of Mg:V is lowered. The "treated" slags are generally easier to remove by manual or air lancing; also, they are more water soluble so that they make for easier water or steam lancing, or preferably high pressure "steam blasting" for maximum cleaning. The treated slags also have a higher pH, by 1 to 2 pH units.

There are several disadvantages, however, in the use of MgO-based products. The deposits are highly susceptible to rapid bridging of the superheater tubes because of the increased boiler ash loading, as shown in Fig. 1 which compares a furnace treated with MgO compared to one treated with manganese.

Low Sulfur-Low Vanadium Fuels: Trend to Manganese

With fuels of 1% sulfur, or less, and a vanadium content of less than 100 parts per million, high temperature superheat fouling and corrosion of tube supports are less likely to occur than with fuels of 2.0% sulfur with 200 to 300 or higher parts per million of vanadium. With the lower sulfur fuels, it has been possible to eliminate completely any interim boiler cleanings between annual outages and frequently also to eliminate even annual boiler cleanings by the use of manganese at low treatment rates.

With the alternate use of magnesium oxide, however, periodic outages at bimonthly intervals frequently are required to remove the magnesium-containing ash deposits that build up in the superheater and reheat areas, and particularly in the air heaters.

With today's low ash containing fuels, the use of MgO is simply not justified. However, manganese additives can result in a realizable net gain in boiler efficiency. Manganese-containing products are available commercially in economical, concentrated, liquid slurries. They can be stored and metered into the fuel inlet system of power station boilers. One leading furnace catalyst containing 25.5% manganese is offered to the utilities industry at economical costs by the Apollo Chemical Corp., as Apollo MC-7. The treating costs with manganese are competitive and generally lower than comparable costs with MgO slurries.

Superheater section of boiler using 2.3% sulfur fuel treated with MgO. Note fouling and bridging.

Superheater section of boiler using 2.3% sulfur fuel treated with manganese.

Fig. 1: Comparison of MgO and manganese with high sulfur fuel.

SPECIFIC BOILER PROBLEMS SOLVED BY ADDITIVES

Loss of Steam Temperature When Converting From Coal to Oil Firing

<u>Use of MgO, or a combination of MgO with manganese.</u>
Many boilers now firing fuel oil were originally designed for coal firing and many of these boilers suffer from too much furnace heat absorption when fired on oil. This results in low final steam temperatures. The boiler will not reach design superheat temperatures. The use of MgO based additives to foul the furnace will often raise final steam temperatures. Unfortunately, the MgO also fouls the convection passes. This results in frequent forced outages for boiler cleaning. When furnace coating is not required, the use of MgO will result in an additional requirement to use the furnace wall blowers and/or attemperating water.

To summarize, the use of MgO added to the fuel oil to raise superheat has the following disadvantages:

1) Fouls the superheaters and air heaters, which shows up as an increase in draft differential across the superheaters and air heaters.
2) Increases requirements for attemperating water to spray the superheater and/or reheaters to obtain required steam pressures to the turbines.
3) Increases stack emissions (Fig. 2).

The best solution for the boiler that suffers a decrease in final steam temperature after a coal to oil conversion is furnace modification. The use of chrome ore on studded tubes with gunniting of the furnace walls reduces furnace radiant heat absorption. This then results in an increase in final steam temperature, making it possible to reach design conditions with higher megawatt capability.

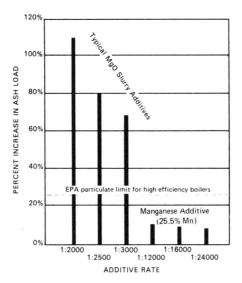

Fig. 2: Percent increase of ash load in fuel oil-fired furnaces by MgO additives compared to use of manganese

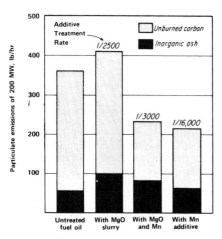

Fig. 3: Percent increase of ash

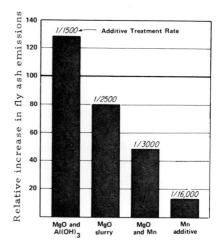

Fig. 4: Typical emissions from a 200 Mw steam generator

If it is not possible to modify the furnace of a boiler that suffers loss of temperature after a fuel conversion, a magnesium/manganese based additive approach gives adequate furnace fouling with a minimum of convection pass fouling. The magnesium still coats the furnace walls by impaction but the manganese reduces deposits in the convection passes. This probably results from the different types of ash modification obtained from manganese plus its combustion catalytic activity.

One such commercial product, containing a stabilized slurry of magnesium oxide and manganese, at a Mg:Mn ratio of 6.5:1, is offered to the utilities industry as Apollo UC-65 by the Apollo Chemical Corp. It has been found to be a good compromise in increasing superheat temperatures with minimum boiler fouling and minimum increase in stack emissions. This combination of magnesium oxide with manganese will frequently reduce the total amount of stack emissions. The manganese reduces the carbon emissions by 50 to 80%. With the use of MgO-type slurries, there is an increased ash input of the order of 100-150%.

A comparison of the total emissions in a utility station boiler is shown in Figs. 3 and 4. These results show that the use of manganese by itself, or in combination with MgO, reduces combustible particulate emissions, as well as grain loading.

High Temperature Corrosion From Vanadium, Sodium and Sulfur

Fuel oils containing sulfur, vanadium, nickel and sodium can cause high temperature corrosion. This corrosion results from the fuel oil ash forming a liquid phase which coats the boiler tubes allowing liquid phase corrosion. These compounds all have melting points in the range of 1000-1300° F.; they will be molten at the metal temperatures of a modern utility boiler and form tenacious slag deposits.

To prevent high temperature corrosion and slagging, it is necessary to raise the melting point of the troublesome ash constituents. This prevents the formation of the corrosive liquid phase. It results in a friable ash that normally responds to soot blowing. An alternate solution is to burn premium fuels containing low vanadium, sodium, nickel and sulfur concentrations-i.e., a total ash content of 0.01 to 0.02%, or less.

Manganese-based additives will raise the fusion point of the fuel oil ash. Where only a moderate increase in the ash fusion point is required, manganese treatment has the advantage because much less additive is required. The lower treatment rate possible with manganese results in improved boiler cleanliness. This minimizes and often eliminates forced outages for boiler cleaning. Field experience has shown that manganese is superior to magnesium in maintaining boiler cleanliness, particularly with fuels of low ash content.

Unlike MgO, manganese acts as a combustion improver and provides the potential for lower excess air firing.

Cold End Corrosion and Air Heater Fouling

<u>Sulfur trioxide (acidic stack fallout)</u>. Cold end corrosion is the result of SO_3 condensation, as sulfuric acid (H_2SO_4), on the air heater and gas duct surfaces. When the gas dew point is reached, a very sticky mist of sulfuric acid is formed. This is extremely corrosive at the temperature conditions present in this boiler region.

Acidic stack fallout will also occur with fuels of lower sulfur content in the range of 2-3% and in many cases even with fuels having a sulfur content of $1\% \pm 0.5$. The most troublesome acid fallout problems occur under the following conditions:

(1) A high excess air is required in order to obtain steam temperatures.

(2) Coal-fired furnace converted to oil firing-generally such a furnace will require a high excess air to help maintain steam temperatures. Also, the exit duct work in these furnaces, whether or not the utility continues to use the precipitators that were present in coal firing, mechanical or electrostatic, often present individual problems. In these systems, there generally occurs a high SO_3 concentration in the flue gas. A tendency exists for the SO_3 to precipitate out with a rise in dewpoint temperature. The surfaces of the duct work cool rapidly because of the longer path lengths that must be traversed before the flue gases enter the atmosphere. Rapid corrosion of the air heaters, the fans and the duct work, will occur.

It is virtually impossible to experience low temperature corrosion without cold end fouling. The sulfuric acid mist traps much of the particulate and fly ash carbon passing through the dewpoint area. Normally the cold end deposits are high in carbon; this absorbs SO_3 and intensifies corrosive attack.

Use of excessive amounts of magnesium oxide slurries can actually accentuate the fouling and corrosion problem. The MgO resulting from use of the MgO additive slurries is dead burned as it passes through the boiler's flame zone. This significantly reduces the reactivity of the MgO. This less reactive MgO then contributes to air heater fouling.

An alternate solution is to raise the unit's exit gas temperature. However, each 50° F. increase in exit gas temperature costs 1% in fuel economy. With today's fuels costing in excess of $4.00 per barrel, it is usually more economical to control cold end problems with an additive program.

Generally, when one obtains a reduction of SO_3 of 25 to 40% with the use of fairly substantial quantities of MgO, it results from the MgO coating the iron tubes in the furnace so that the iron surfaces are no longer exposed. Otherwise, the iron acts as a catalyst to increase the percentage conversion of sulfur to sulfur trioxide;

$$S + O_2 \longrightarrow SO_2$$

$$SO_2 + \tfrac{1}{2}O_2 \longrightarrow SO_3$$ (Reaction catalyzed by iron and/or vanadium; extent of reaction controlled by high excess air.)

Theoretically, at least, magnesium oxide would be expected to react with sulfur trioxide to form magnesium sulfate:

$$MgO + SO_3 \longrightarrow MgSO_4$$

$$40 \text{ lbs.} + 80 \text{ lbs.} = 120 \text{ lbs.}$$

Therefore, if the magnesium were not consumed by the reaction with the vanadium in the fuel oil, and if the magnesium oxide reacted 100%, it should be possible to completely remove SO_3 in this manner. This does not occur! Field experiences show that a stoichiometric reaction does not take place. Therefore, complete SO_3 removal with MgO is not practical. The decomposition temperature for $MgSO_4$ is 1550° F. and this probably explains why the free SO_3 is released again.

$$MgSO_4 \xrightarrow{1550° F.} \underset{\text{(solid)}}{MgO} + \underset{\text{(gas)}}{SO_3}$$

<u>Air heater corrosion protection</u>. Both manganese and MgO slurries, at proper feed rates, can provide air heater corrosion protection. Both products can retard SO_3 formation, generally by reducing the catalytic effect of iron and vanadium. Reduced air heater carbon deposits and fouling results with manganese, thus providing improved basket protection.

It is only when massive amounts of MgO are used that SO_3 corrosion is arrested, but at the expense of blockage of the air heaters. Successive layers of MgO and $MgSO_4$ build up on the air heaters. As a consequence, air heaters must be water-washed at frequent outages to remove the insulating deposits.

<u>Air heater cleanliness</u>. Regarding air heater cleanliness, manganese is far superior to MgO slurries. Not only does the lower dosage rate with manganese reduce the unit ash loading, but the combustion catalytic activity of the manganese reduces the amount of unburned carbon and fuel oil residue by as much as 50-70%. Although this carbon residue represents only 0.2 to 0.4% of the fuel oil, it acts as a binder for air heater fouling and as an absorbing agent for acidic SO_3. Most boilers treated with manganese have required approximately one-third the air heater washing frequency that they required when treated with MgO based products.

NEW PROCESS TO PREVENT COLD END CORROSION AND STACK FALLOUT PROBLEMS

Generally, the application of MgO to reduce SO_3 is restricted to boilers that operate with excess air of 3% or less, but certainly no more than 5% excess air. Where it is not possible to restrict the amount of excess air to below 3 to 5%, then the use of MgO or dolomite added to the fuel oil or furnace can actually increase the SO_3 content in the flue gas[1]. A preferred method to reduce low temperature corrosion and SO_3 emissions is the use of a supplementary cold end neutralizing additive.

Wilkinson and Clarke[2] have reported on work carried out in England at the Marshwood generating station with boilers firing fuel oil in which back end injection with ammonia was attempted for reducing SO_3 in the flue gas. It was found that the addition of ammonia at 20 to 33 lbs./hour was required to control SO_3 attack in a boiler producing 550,000 lbs. steam per hour at 915° F. superheat temperature with a fuel containing 4.2% sulfur. At this feed rate there was an increase of 3" of draft loss in the air heater within eleven days. Further, the resultant deposits were coke-like mixtures of fused ammonium bisulfate and carbon, which were still acidic.

The economic losses to a utility because of corrosion of the air heaters can be substantial. This is particularly severe when a boiler is load limited because of insufficient air. This occurs when the air heaters are plugged with iron sulfate deposits or have become corroded.

In addition, utility managers report a worsening of the stack plume as the air heaters become plugged or inoperative. Frequent outages to water wash the air heaters then become necessary.

One leading process that utilizes back end feed of a neutralizing complex, either as a solid or liquid injection, is offered to the utility industry by the Apollo Chemical Corp. It already is in use in over 50 large power boilers, both with fuel oil, as well as with coal. Worldwide patents have been applied for detailing this process which overcomes the problem where it occurs -i.e.-in the flue gas. This is far superior to previous attempts to use MgO slurries added to the fuel oil as a method of decreasing the content of SO_3 in the air heaters and in the stacks emitting to the atmosphere.

Test work has been carried out in Japan[3], as well as the United States and the Caribbeans to clarify under what conditions the use of MgO slurries will reduce SO_3 in the flue gas. It is reported that with certain furnace designs, particularly where high excess air must be maintained, the use of MgO fuel additive slurries has been found to actually increase the amount of SO_3 in the flue gases.

Summarizing, the preferred method to eliminate cold end corrosion in oil fired furnaces is the use of a manganese, carbon-destroying, catalyst in the fuel oil. This often allows for lower excess air firing, modifying of the vanadium ash, coating of the furnace iron tubes thus preventing the catalytic action of iron and vanadium in converting SO_2 to SO_3 and significantly reducing the flue gas carbon grain loading.

Then, if air heater corrosion and fouling persists, or if the stack plume indicates SO_3 (acid) fallout as evidenced by a deep blue attached or detached plume, a supplemental flue gas neutralizing treatment can be used, one that is injected into the economizer outlet. A dramatic demonstration of the use of cold end, supplemental neutralizing feed with a leading commercially available product, Coaltrol MH-2, offered to the utility industry by the Apollo Chemical Corp. is shown in a study made at the Union Electric Co. at its Sioux plant outside St. Louis.

In this coal fired unit of 500 Mw capacity producing 3,290,000 lbs. per hour of 1000° F. steam at supercritical pressures, there was a persistent buildup of deposits of iron sulfate in the tubular air heaters in the bypass section. This bypass section, accounting for 10% of the exit flue gas, required forced outages at almost bi-monthly intervals in order to water wash and manually remove the deposits in the tubular air heaters.

Attempts to improve this condition by raising the steam temperatures by installing heating coils around the tubular air heaters were tried but were unsuccessful or otherwise economically prohibitive. The injection of ammonia into the air heater resulted in tenacious acid deposit buildup.

One of the twin tubular air heaters then was treated by tail-end injection of a specially processed magnesia-containing complex of 15-30 micron size[4]. This product was found to be over 90% active for removing SO_3 within a residence time of 4 seconds.

As shown in Fig. 9, this method resulted in a dramatic improvement in the air heaters. No cleaning was required during the 12 week test trial. The untreated section had to be water washed at monthly intervals.

UNTREATED   TREATED (12 weeks)

Fig. 5: Reduction of air heater fouling by back-end neutralizing additive. Union Electric Company's (Sioux Plant) coal-fired B & W Universal boiler (500 Mw; 3,290,000 lb./hr. of 1000° F steam at supercritical pressure).

POOR HANDLING CHARACTERISTICS OF FUEL OIL ASH IN OIL FIRED BOILERS EQUIPPED WITH ELECTROSTATIC PRECIPITATORS

Further, flue gas supplemental feed, has proven superior to the use of fuel oil treated with MgO with regard to precipitator performance where electrostatic precipitators are in use in oil-fired furnaces. The problem of cracking of the porcelain insulators resulting in shorting out of the emitting wires can often be reduced by supplementary air cooling of the insulators, but this will not entirely eliminate the problem.

This problem, it is thought, originates because of the accumulation of carbon and acidic flue gas by-products on the insulators. When subjected to the high temperatures, it destroys the insulation material.

This tail end neutralization has been found useful not only for improving the overall efficiency of the precipitators, but also it often permits the ash hoppers to be operated without interruption. The additive dries up the fuel ash in the back end, by virtue of the acid neutralizing properties of the injected additive. The ash no longer tends to clump, but rather flows freely. Indeed, this tail end feed will often eliminate the necessity to revert to the more expensive method of water sluicing of the ash in order to remove it from the hopper collectors.

STACK PLUME VISIBILITY AND PARTICULATE EMISSIONS

Manganese Reduces Total Stack Emissions

The use of manganese will result in significantly less particulate stack emissions[5, 6, 7, 8] as much as 80% less, as shown in Table 3.

Table 3

REDUCTION OF CARBON PARTICULATES BY MANGANESE

Test Period	Additive Slurry of:	Parts per Million Mn	Flue Gas Particulates- Average Concentration (Mg/SCF)	Average Percent Carbon (in particulates)
IA	None		0.17	72
IB	Manganese	45	0.04	31
IIA	None		0.14	58
IIB	Manganese	45	0.03	32

In this situation carried out at an oil company in the United States, the fuel consisted of heavy asphaltic bottoms. This pitch was burned in a package boiler producing 200,000 lbs. of steam per hour at 900° F. superheat temperatures of 900 psi. Measurement of the flue gas particulates showed a decrease from 0.17 to

0.04 Mg/SCF, with a reduction of the carbon content of the emitting particulates of from 72 to 31%.

Refinery Carbon Plume Problems

Plume in petrochemical plants and refinery waste heat boilers burning refinery pitches, waste oils and slops. The use of manganese can have a significant effect in reducing particulate emissions when burning waste oils and bottoms in petrochemical and refinery heaters. In one example, a Canadian refinery was burning a petrochemical waste pitch in a boiler. Because of the appearance of the stack, which exceeded a Ringelmann 2 value, the refinery could burn only 100 barrels per day of the pitch, rather than the 800 barrels produced daily and for which the heater originally was designed. Otherwise, the smoke number exceeded the local regulations that prohibited the stack having a Ringelmann value in excess of 1. This meant that the refinery had to burn natural gas which was extremely limited in availability.

It was economically prohibitive to modify the furnace to meet the regulations. A quick solution was found to the use of the manganese oxide slurry (25% by weight of manganese) at a feed rate of 1 gallon/75 barrels of pitch. The required feed rate of manganese was determined with an in-line Bailey meter. From a visual point of view, the manganese kept the Ringelmann number below 1, permitting the refinery to burn the total amount of pitch, 800 barrels/day, for which the furnace was designed.

Case History: Burning of waste pitch from ethylene cracking process. A chemical refinery on the Gulf Coast of the U.S. had a severe smoking problem, since resultant plume opacity was not meeting local air pollution regulations. Three C.E. boilers, each rated for 460,000 lbs. steam production per hour, burned a mixture of natural gas and the pitch produced from the cracking of middle distillate fuel used as feed stock to produce ethylene. The pitch had a low sulfur content (below 0.2%). Unfortunately, the stacks showed a noticeable carbon emissions plume having a Ringelmann 2+ value.

Manganese, in the form of a manganese oxide slurry containing 25.5% Mn, was fed into one of the three units. The initial feed rate was 1 gallon per 4,000 gallons of oil. The response was immediate and dramatic. The plume was reduced to essentially zero visibility. This was noted both by visible observation and recorded on a Shell Oil Co. light transmission meter which monitors stack plume.

The additive rate was reduced gradually to 1 gallon to 8,000 gallons while still maintaining stack plume improvement. The exact feed rate required depended upon the amount of pitch burned and the excess air used. This relationship is not linear, but is easily controlled by the operators, resulting in minimum treatment costs.

The use of the additive then was extended to the other two boilers. All three units are being treated at the above rate and maintain a stack plume reading below a Ringelmann 1. This dramatic response with the manganese was accomplished with a refinery pitch, one of the most troublesome fuels to burn.

SUMMARY

1) High temperature slagging and corrosion in oil-fired boilers is reduced ideally with the use of a slurry containing both magnesium oxide and manganese in combination.

2) Manganese additives reduce internal boiler fouling of superheaters and reheaters at low ash input ratios.

3) Fuels with low vanadium contents should be treated with manganese, rather than magnesium oxide.

4) High sulfur fuels but with low vanadium contents are preferably treated with tail end chemical injection of an active neutralizing agent to remove SO_3.

5) Overall decrease in stack emission ideally is obtained with a dual additive application. This consists of the addition of manganese added to the fuel oil and a neutralizing agent added to the economizer outlet.

6) Manganese additives were found to eliminate stack smoke, particularly in refineries burning pitch or polymerized (waste) bottoms.

7) Corrosion and fouling from SO_3 in coal fired units can be eliminated by aspirating an activated neutralizing agent in powder form into the economizer outlet of the furnace.

REFERENCES

1) W. T. Reid, ASME paper #69-PWR-5

2) T. J. Wilkinson & D. G. Clarke, J. Inst. of Fuel, XXXII (2), 61-72, (1959)

3) M. Aramaki et al, Tech. Rev. Mitsubishi Heavy Industries Ltd., (5) (1969)

4) Providence Evening Bulletin, October 6, 1970, pp. 1, 11 (for reprint, write to Apollo Chemical Corp., Clifton, N. J. 07014)

5) A. Belyea, Power, 110 (11), 59-60, (1966)

6) V. J. Cotz, Power, 114 (2), (1970)

7) I. Kukin, APCA-NES Spring Technical Session, April 15, 1971 (for reprint, write to Apollo Chemical Corp., Clifton, N. J. 07014)

8) R. J. Bender, Power, 116 (1), 42, (1972)

HIGH TEMPERATURE DUST CONTROL--APPLICATION OF A COMBINATION PARTICLE COLLECTOR AND HEAT EXCHANGER

H. Gregor Rigo, James J. Stukel

Department of Mechanical Engineering, University of Illinois, Urbana-Champaign, Illinois 61801

ABSTRACT

High temperature convective heat exchange surfaces are damaged by dust erosion and their function is impaired by dust deposition.

A novel heat exchange surface is presented in this paper which minimizes deposition on, and erosion of, boiler tube passes in fossil fuel fired units. This is accomplished by removing particulate phases from an air stream. A mathematical model is used to analyze this modified boiler tube. This demonstrably conservative predictor indicates that essentially all the material entrained in the gas stream entering a ten tube bank of the particulate precipitating heat transfer surface can be removed. Additional investigations are presented which demonstrate that there is no measurable difference between the heat transfer effectiveness of a conventional boiler tube and the "flow separator and heat exchanger."

INTRODUCTION

On December 23, 1971, the United States Federal Government passed its New Source Emission Standards. These regulations require that fossil fuel fired boilers emit less than 0.1 pounds of particulate matter per million BTU's (0.18 g/NM^3). In addition, incinerators are allowed only 0.08 grains per standard cubic foot (0.18 g/NM^3) corrected to 12% CO_2.

If it is assumed that some form of particle collection mechanism must be employed, it would appear to be advantageous to remove the aerosol before the products of combustion enter the power generating or recovery zones of a stationary combustor. For example, it is well known that the leading edges of boiler

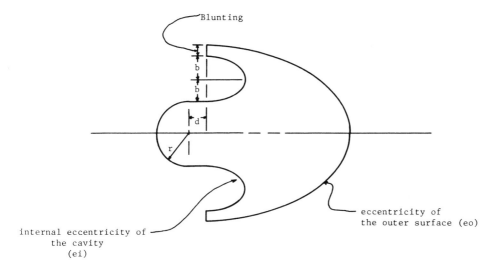

Figure 1. Tube cross section with dimensions for mathematical generation

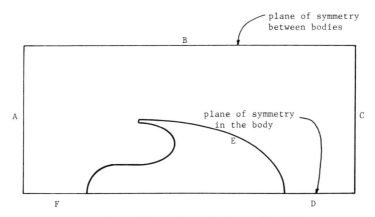

Figure 2. Identification of flow field boundary condition zones

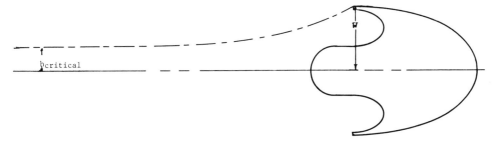

Figure 3. Collector with limiting particle trajectory

tubes in coal fired boilers are eroded by abrasion. In addition, slag bridging is common in large power recovery incinerators. Certainly, removing the ash from the combustion product stream between the primary combustion chamber and the heat transfer zones would be desirable.

With this general consideration in mind, several co-current requirements and goals should be delineated. It is imperative that any potential collector be cooled. Its structural integrity must be maintained. Since heat transfer is a boiler's function, this requirement is probably beneficial. Second, the collector should be self-cleaning. While soot blowing is practiced at existing coal fired boilers, reducing its frequency and intensity are two more goals for this device. Third, increases in overall power consumption for a facility should be minimal. Fourth, conversion of an existing facility to utilize the device should not involve major system modification.

THEORY

The specific formulation studied in this paper is shown in Figure 1.* It can be generated in five sections. The leading nose is a quarter circle, and the nose is extended parallel to the plane of symmetry a distance "d". The internal cavity is specified by an opening width and an eccentricity. The wing tip is blunted and the following portion is defined by a continuity consideration and an external eccentricity.

Since this geometric configuration is both multi-valued (in the vicinity of the cavity), and irregular - piecewise continuous, computerized numerical solutions are performed. It is assumed that the gas cloud approaching the collector is unaffected by the presence of the solid phase; that the particles are uniformly or iso-disperse; and that the gas properties are constant. If it is further assumed that the momentum thickness (boundary layer) is thin when compared to a characteristic collector dimension, the fluid dynamics problem is reduced to the evaluation of the potential flow field. In two dimensions, the governing equations for the flow field are continuity ($\nabla \cdot u = 0$) and conservation of momentum

$$\frac{Du}{Dt} = \frac{1}{\rho}\nabla P + \nu \nabla^2 u$$

If the stream function, Ψ, is defined in terms of the velocity components

$$u = \frac{\partial \Psi}{\partial y} \; ; \; v = -\frac{\partial \Psi}{\partial x}$$

then, continuity is identically satisfied for all Ψ. Upon substitution of the stream function into the momentum equation, and utilization of the potential flow requirements delineated above, the momentum equation becomes: $\nabla^2 \Psi = 0$.

---

* U. S. Patent Pending.

The appropriate boundary conditions are uniform streaming at infinity, and no normal flow on the planes of symmetry or on the collector surface. Written in terms of the stream function, the boundary conditions become (refer to Figure 2 for letter locations:

at the ends of the flow field (A,C) $u = \frac{\partial \Psi}{\partial y} = U_\infty$

along (B) $v = \frac{-\partial \Psi}{\partial x} = 0 \to \Psi = $ constant

along (D,F) $v = \frac{-\partial \Psi}{\partial x} = 0 \to \Psi = 0$

along (E) no normal flow $\to \Psi = 0$

The governing equation can be written in terms of finite differences. Employing Taylors series approximations for the second derivatives, the Laplacian can be written in Cartesian coordinates and its value defined in terms of adjacent functional values.

$$\nabla^2 \Psi = 0 = \Psi(x+\delta,y) + \Psi(x,y+\delta) + \Psi(x-\delta,y) + \Psi(x,y-\delta) - 4\Psi(x,y)$$

It should be noted that the irregularity of the collector necessitated the handling of irregular stars [1]. Several methods for solving the field of stream function values exist. The approach found to be fastest [2] involved an interative relaxation of all residuals larger than a floating minimum. In order to assure accuracy in the procedure, the residuals were recalculated frequently during the process.

After defining the flow field, the fractional collection efficiency for the obstruction is determined. If it is assumed that regardless of the collector Reynolds number and proximity of the obstruction the particle is surrounded by a low Reynolds number fluid, then a non-dimensionalized force balance shows for spherical particles that:

$$\overline{\Psi} \frac{d^2 x}{dt^2} + \frac{dx}{dt} - u_f(x) = 0$$

$$\overline{\Psi} = \frac{2}{9} \frac{\rho}{\rho_p} \frac{u_\infty D_c}{\nu} \frac{a^2}{D_c^2}$$

The collection effectiveness is determined by using particle marching techniques. A particle is launched toward the collector at infinity and at a specified distance from the collector plane of symmetry (h). Its position is tracked stepwise in time by noting that the force equation defines the particle acceleration and then utilizing Newton's Laws of Motion which define velocity and position given acceleration and time. If it is assumed that particle stream lines do not cross, the efficiency is defined to be: $\eta = h_c/W$. This is a direct consequence of the concept of the limiting particle trajectory (see Figure 3). All material launched within h_c of the collector centerline will be collected. All material launched beyond h_c will miss the obstruction.

The single tube collection efficiency can be extrapolated to include tube banks within the context of the expressed limitations if filtration theory is applied:

$$\eta = 1 - \frac{n}{11}(1 - BR\,\eta_j) \qquad j = 1$$

The results of the computer collection efficiency study can be expressed in terms of the scaling parameter $\overline{\Psi}$ and the characteristic collector diameter D_c. If

$$D_c = \frac{(r+d)}{(r+2b)}(2r+2b+d) \tag{1}$$

a single fractional efficiency plot results (see Figure 4). Subsequent work generated the following empirical equation for the particle precipitating heat transfer surface collection effectiveness:

$$\eta(\overline{\Psi}) = 0.5\,(1+\tanh(C_1\overline{\Psi} + C_2/\overline{\Psi})) \tag{2}$$

$C_1 = 0.6314; \qquad C_2 = 0.1411$

Since the existing work on collection efficiencies for cylinders does not include fluid field compression effects resulting from the presence of neighboring cylinders, the collection efficiency for a cylinder must also be obtained. The coefficients to fit Equation 2 for a cylinder are: $c_1=0.7211$; $c_2=0.1313$.

The heat transfer effectiveness of the particle precipitating heat transfer surface is obtained through an application of the equivalent wedge flow similarity solution. The model assumes that the heat transfer rate at any point on an arbitrary cylinder is equivalent to the heat transferred at a similar point on an equivalent wedge [3,4]. The technique involves the determination of the wedge angle and a solution to the boundary layer and energy equations for wedge flow.

The wedge angle, m, has been shown to be [3]:

$$m = -\,(xdv/dz)/U_\infty \tag{3}$$

Since z is the developed chord length to the point in question, dU/dz is the tangential acceleration and U the tangential velocity at that point in the potential flow field; m can be determined directly from the potential flow field.

The appropriate momentum and energy conservation equations in the boundary layer are:

$$\rho u \frac{\partial u}{\partial x} + \rho v \frac{\partial u}{\partial y} = \frac{\partial}{\partial y}\left(\mu\frac{\partial u}{\partial y}\right) - \frac{\partial p}{\partial x} \tag{4}$$

$$\rho c_p \left(u \frac{\partial T}{\partial x} + v \frac{\partial T}{\partial y}\right) = \frac{\partial}{\partial y}\left(k\frac{\partial T}{\partial y}\right) + \mu\left(\frac{\partial u}{\partial y}\right)^2 + u \frac{\partial P}{\partial y} \tag{5}$$

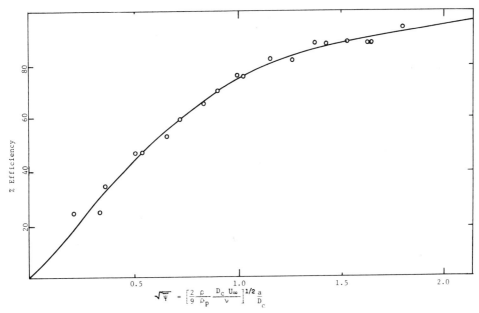

Figure 4. Generalized fractional efficiency curve

$$\sqrt{\overline{\Psi}} = \left[\frac{2}{9} \frac{\rho}{\rho_p} \frac{D_c\ U_\infty}{\nu}\right]^{1/2} \frac{a}{D_c}$$

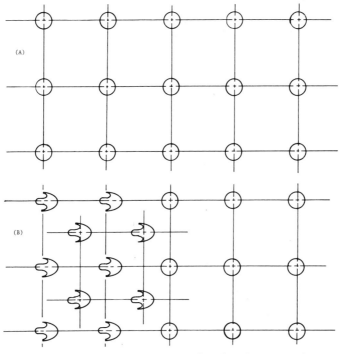

Figure 5. Original boiler tube arrangement for boiler simulation (A) and boiler tube arrangement in the modified boiler (B)

Continuity is:

$$\frac{\partial(\rho u)}{\partial x} + \frac{\partial(\rho v)}{\partial y} = 0 \qquad (6)$$

The boundary conditions are:

@ $y=0$; $u=0$, $v=0$, $T=T_w$

$$\text{as } U \to U_\infty; \quad \frac{\partial U}{\partial y} \to 0, \ldots, \frac{\partial^n U}{\partial y^n} \to 0 \qquad (7)$$
$$\text{as } T \to T_\infty; \quad \frac{\partial T}{\partial y} \to 0, \ldots, \frac{\partial^n U}{\partial y^n} \to 0$$

After defining a dimensionless temperature, utilizing a similarity variable which employs the stream function, and assuming that the gas properties are isotropic but temperature dependent, then the Equations 4, 5, 6 subject to 7 can be solved simultaneously as total differential equations.

A Runge-Kutta digital computer program was executed to determine the relation between wedge angle and surface temperature gradient. Noting that:

$$N_u = \frac{\sqrt{Re_z}}{T_w - T_\infty} \frac{dT}{dy}$$

Assuming that three boundary layers were present on the collector - from the nose to the cavity opening, from the wing tip to the far end, across the trapped vortex in the cavity - then an average heat transfer coefficient ($\bar{h}$) can be obtained by integrating the local heat transfer coefficients over all the surface.

$\bar{h}$ is of the order of 10 Btu/hr-ft of tube-°F for boiler conditions for the particle precipitating heat transfer surface. While $\bar{h}$ varied with geometry, it was found to be within 9% of the value predicted for an equivalent round tube (based on $2D_c$) using Hilpert's model [5].

By way of example, a vertical tube package boiler will be considered. If gas is directed through a package boiler from an incinerator, it will pass through the radiative section usually used for a burner, turn 180°, and approach the convective banks. These tube banks normally consist of 80 two-inch tubes arranged in line (see Figure 5A). For the modeled comparison, it will be assumed that the first ten rows are replaced with particle precipitating heat transfer surfaces arranged in a "stagger bank" (see Figure 5B).

The analysis is straightforward and utilizes empirical gas relations, the work of Hilpert, and the continuity laws. As heat is removed from the gas stream, the gas temperature must decrease due to conservation of energy. Continuity of mass says that the gas velocity must also decrease. These two effects can be incorporated by writing Ψ in terms of temperature.

The gas properties become:

$$\nu = 3.76 \times 10^{-9} T^{1.7}$$
$$k = 6.205 T^{0.85}$$
$$C_p = 6.6 \times 10^{-2} T^{0.19}$$
$$\rho = 39.8\, T^{-1.0}$$
$$\dot{m} = 1.432 \times 10^4 U_\infty T^{-1}$$
$$\overline{\Psi} = 3.475 \times 10^{-8} \frac{\dot{m}a}{D_c} T^{0.3}$$

Since the temperature in a power recovery boiler installed on an incinerator would probably range between 1600°F (the ash fusion point) and 600°F, for constant collector and particle size, Ψ can vary significantly. If it is assumed that each tube row removes heat as a plane sink and that mixing past each row is perfect, then using Hilpert's equation, the temperature of the gas after passing between rows can be shown to be:

$$T_2 = \frac{6.62 \times 10^{-2} T_1^{1.19} \dot{m} - BP(4.66 \times 10^3) T_1^{0.046} (\dot{m} D_c)^{618} (T_1 - T_w)}{6.6 \times 10^{-2} \dot{m}}$$

Combining the above equation with Equation 1 for evaluation:

$$\eta = 1. - \prod_{j=1}^{n} (1 - \eta(\overline{\Psi}(T_i - 1)))$$

For the particle distribution shown in Figure 6, and allowing an equivalent mass accumulation on the tubes, the increase in time between soot blowing is determined to be:

$$\frac{t}{t_E} = \frac{\sum_{i=1}^{n} f(i) \eta_E(i)}{\sum_{j=1}^{n} \{f(j)[1 + \eta_{PPHTS}(j)] \eta_E(j)\}} = 4.1$$

APPLICATION

Assuming that the power recovery train analyzed is a packaged oil boiler with the first ten rows modified, the boiler's cost would increase $10,000 over a conventional package unit. This cost increase is offset for a new facility by a simplification of any proposed tail end air pollution control equipment. A cost savings of order $45,000 can be realized.

A thirty-minute soot blowing period every eight hours with four thirty pound per minute lances consumes approximately 1.1% of a boiler's steam load average over an eight-hour period. This can be resolved into an average increase in steam rate of .84% if the soot blowing frequency is changed to once every 24 hours as the model predicts for the package unit under consideration. For a 75-ton per day incinerator plant, this yields a net increase of 125 pounds per hour of salable steam. At current

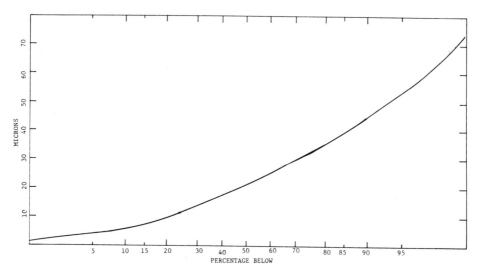

FIGURE 6. Incinerator fly ash distribution

stream value for medium size industrial facilities, the production value of the boiler plant is increased $1,200 annually. This savings is sufficient to offset the capital cost increase in eight years (6% interest).

CONCLUSION

A novel form of a high temperature dust control apparatus was proposed, analyzed, and theoretically applied. On a new plant, its application results in an immediate cost savings and for an existing facility, the capital increase is offset by an operating cost decrease.

REFERENCES

1. Allen, N. de G., <u>Relaxation Methods</u>, McGraw-Hill Book Company, New York, 1954.

2. Rigo, H. G., <u>Evaluation of a Particulate Precipitating Heat Exchanger</u>, unpublished Master's Thesis, University of Illinois, 1971.

3. Brown and Levingood, NACA Tech. Note 2800, September 1952.

4. Eckert and Levingood, NACA Tech. Report 1118, 1953.

5. Hilpert, R., <u>Warmeabgabe von geheizen Drahten und Rahren</u>, Forsch. Gebiete Ingenieuriv," 4 (1933), 220.

NOMENCLATURE

Operators

D	–	material derivative
d	–	total derivative
∂	–	partial derivative
$\nabla \cdot$	–	gradient $\left(\frac{\partial}{\partial x} + \frac{\partial}{\partial y} + \frac{\partial}{\partial z}\right)$
∇^2	–	Laplacian $\left(\frac{\partial^2}{\partial x^2} + \frac{\partial^2}{\partial y^2} + \frac{\partial^2}{\partial z^2}\right)$
$\prod_{j=1}^{n} F$	–	PI $(F_1 \cdot F_2 \cdot F_3 \ldots \ldots F_n)$
$\sum_{j=1}^{n}$	–	summation

Symbols

a	–	particle diameter
BP	–	bypass ratio
Cp	–	heat capacity (Btu/hr-°F-ft^3)
Dc	–	characteristic dimension
h	–	launch distance of aerosol from centerline or convective heat transfer coefficient
M	–	wedge angle
m	–	mass flux
P	–	pressure
T	–	temperature (°R)
t	–	time
U	–	velocity along direction of free stream
v	–	velocity normal to the free stream
w	–	actual width of collector
x	–	Cartesian coordinate parallel to free stream
y	–	Cartesian coordinate normal to free stream
z	–	developed chord length to a point on the collector
ρ	–	density (#/ft^3)
η	–	efficiency
ν	–	kinematic viscosity
μ	–	dynamic viscosity
δ	–	infatisimal displacement
ψ	–	stream function
$\overline{\psi}$	–	inertial deposition scaling function

Dimensionless Groups

Nu	–	Nusselt number (hx/k)
Re	–	Reynolds number (ux/ν)
f	–	normalized fractional size distribution

Subscripts and Superscripts

—	- average value
p	- particle
w	- evaluated at the wall
E	- evaluated for unmodified boiler
PPHTS	- particle precipitating heat transfer surface
C	- critical value
∞	- evaluated at infinity
1	- previous value
2	- current value

SOLUTIONS TO PROBLEMS OF SOIL POLLUTION BY AGRICULTURAL CHEMICALS

I. J. Graham-Bryce
Rothamsted Experimental Station, Harpenden, Herts, U. K.

INTRODUCTION

Pollution is an imprecise concept that is difficult to define satisfactorily. Rather than add to the many definitions that have been attempted, I shall discuss methods of improving contaminated soils and of avoiding such contamination, taking the point of view that ideally agricultural chemicals should not persist in biologically active forms for longer than necessary for their intended effects and should be free from undesirable side effects. Although this discussion is limited to soils, it must be emphasised that soil cannot be considered in isolation, and a process that simply passes a persistent chemical on to another part of the environment is not an allowable solution to the problem of soil pollution. Failure to consider the fate of a chemical once it has left the immediate area where it was applied has contributed to many of the most difficult problems in the past.

INCIDENCE OF SOIL POLLUTION

In modern agriculture, a wide range of agricultural chemicals is used on a very large scale and, in relation to the amounts used, there have been remarkably few serious problems. Before discussing methods of improvement, I shall consider briefly the incidence of pollution by inorganic and organic crop-protection chemicals and by fertilisers. Problems of salinity resulting from irrigation will be excluded.

Crop-Protection Chemicals

With present knowledge it is possible to describe in some detail the circumstances in which biocides are likely to cause soil pollution (1). Apart from localised contamination arising from accidental spillage, the chemicals that may cause serious problems tend to have a broad spectrum of activity and to be cheap so that they are used widely in large amounts. They must also be persistent which implies that their physical and chemical properties must be within a fairly well-defined range. To restrict losses by leaching and evaporation they must be adsorbed by soil particles and be relatively involatile, although not to the extent that they are inactive biologically in soil. To avoid degradation they should be resistant, even under the extreme conditions of the adsorbed state, to chemical processes such as oxidation, reduction and hydrolysis that occur in soils.

Several inorganic pesticides have many of these properties. For example, copper is a non-specific toxicant which is not detoxified by degradation and can accumulate progressively in soils where copper fungicides are used, particularly in vineyards (2) and orchards (3,4) which are sprayed repeatedly. In extreme cases the activity of soil fauna is so severely disrupted that organic matter accumulates as a mat at the soil surface. The use of copper is decreasing with the advent of modern synthetic fungicides which are usually more specific, more reactive, and more easily degraded, but large quantities of copper fungicides will continue to be used for some time against fungous diseases such as apple scab, potato late blight, cocoa black pod and grape downy mildew. Very persistent and toxic residues of heavy metals were also left by other older inorganic pesticides, now largely replaced, such as lead and calcium arsenate. There are extreme examples of old orchard soils containing several hundred p.p.m. of such residues and remaining unproductive for long periods, either because the residues damage plants or lead to unacceptable amounts in subsequent crops. Kearney *et al* (5) discussed the risks of similar accumulation in soil from modern arsenical herbicides.

It has been suggested that mercury fungicides used as seed dressings may contaminate soil with a similar toxic and persistent heavy metal. However organomercury compounds can be lost from soil as vapour, either in their original form or after degradation to mercury itself (6). Also, the amounts used are small. In Britain, a typical 1% phenyl mercury acetate cereal seed dressing is applied at a rate equivalent to only 4-5g/ha. Amounts entering the environment from agricultural sources are probably minor compared with those from industrial sources and with natural amounts (7).

Of the chemicals possessing the properties which may lead to pollution, the outstanding examples are, of course, the organochlorine insecticides which are by far the most common contaminants in soils (8), particularly orchard soils and soils used for growing vegetables where large amounts of insecticide are applied. Organochlorine insecticides have the added disadvantage that they can accumulate in the fat deposits of animals and so may be concentrated as they pass along food chains, which start with soil animals such as earthworms feeding on organic matter in soil. There is little evidence that such persistent residues directly affect soil fertility. However they may act as a reservoir for redistribution to other parts of the environment and they may intensify later pest problems by inducing resistance in pests or by killing predators of pests (such as certain staphylinid and carabid beetles) more readily than their prey (such as cabbage root fly) (9). By altering the composition of the soil fauna they may also affect the rate at which organic matter decomposes. In a recent extreme illustration of this effect, isobenzan* applied at 2·24kg/ha in New Zealand so drastically affected soil fauna that plant debris accumulated as a mat above the soil surface and plant growth was retarded (10). These effects, which recall those caused by heavy metal contamination, were still apparent after four years.

For insect control, persistence was originally considered desirable to give prolonged protection from a single application. For herbicides used in agriculture however, long persistence is undesirable because of the possibility of harming subsequent crops, and most herbicides are too short-lived to cause pollution of soils. However, for the selective elimination of weeds in a growing crop, or for total weed control, a degree of persistence is required. Some of the longer-lived herbicides such as picloram, some triazines and substituted ureas, can persist for longer than one year and lead to losses, particulary when a sensitive crop succeeds a resistant one. Kearney et al (5) compared the relative persistence of some of the relevant chemicals. It can be argued that many such examples of persistence do not really constitute pollution, but should be regarded more properly as constraints to the cropping sequence which a farmer can adopt in his rotation, probably no more severe than those dictated by pest and fertility considerations, and that they can be avoided by ensuring that sensitive crops are not grown in the year after the persistent herbicide is applied.

* Names of crop protection chemicals used in this paper are taken from British Standard 1831:1969, Recommended Common Names for Pesticides (British Standards Institution) which gives their chemical names and structures in full.

Fertilisers

The types of problem associated with fertilisers differ from those associated with biocides. Unlike biocides, fertilisers are applied to increase amounts of plant nutrients which are already present in soil. Large quantities of fertilisers must be applied to soil to maintain fertility and get the yields required in modern intensive farming. For optimum yields, a correct balance of nutrients is necessary and occasionally so much of a specific nutrient is added that yields begin to fall. Occasionally also, trace element deficiencies may be induced by fertiliser use, although these may often be cured by regulating soil pH. Hypomagnesaemia in sheep is another example of the type of problem that can be aggravated by using fertilisers, in this case by applying potassium and nitrogen injudiciously to certain pasture soils (11). However most such problems can be avoided by good management and by applying existing knowledge.

Enriching soil may bring other problems not properly described as soil pollution. For example, it is often possible to grow enough grass per acre to support numbers of stock which the same area of land cannot accommodate in other ways: disposal of sewage may become a major problem or the danger of damaging soil structure by poaching may increase, particularly in wet regions. Also, increasing the nutrients in soil by any method probably means more occur in drainage water or water running off agricultural land. Losses of nutrients from agricultural land in Britain have been discussed by Cooke and Williams (12) and by Tomlinson (13).

Occasionally novel fertilisers or related materials intended to improve soil conditions can have damaging effects, which can be more justifiably considered as soil pollution. For example pulverised fuel ash added to sandy soils to improve water retention was found to contain toxic concentrations of boron, which affected susceptible cereal crops for several years (14). Sewage sludge, used as a soil amendment in some areas, often contains large amounts of heavy metals which cause lasting contamination. The effects can often be unexpected, for example, some boron-containing fertilisers have been found to cause highly undesirable accumulation of fluoride in leaves (15).

Except for a few local examples such as those quoted, the problems associated with fertilisers largely arise from increasing fertility, especially the nitrogen content of soils as discussed earlier. They can mostly be avoided or reduced by proper management, and will not be considered further here.

DECONTAMINATION OF POLLUTED SOILS

The processes whereby the effects of an undesirable contaminant present in soil can be diminished are few. In principle, they amount to accelerating degradation by microbial, chemical or photochemical reactions, to reducing the biological availability in soil, or to transferring the chemical to another part of the environment. However, such transfer is satisfactory only when it does not cause undesirable effects elsewhere and if it results in faster degradation, as for example would happen with a pesticide that is degraded photochemically in the atmosphere. The means by which soils can be decontaminated may be discussed under the headings of cultivation and related methods, the addition of adsorbents designed to decrease availability, and finally amendments to increase rates of degradation.

Cultivation and Related Techniques

Generally chemicals are lost more readily from soils when applied to the soil surface, as with sprays, than when incorporated into the soil as are some chemicals applied to control soil-borne pests (16). This is because of the greater opportunity for evaporation and photochemical degradation at the soil surface. Whereas cultivation should therefore be avoided for removal of a surface-applied chemical, repeated cultivation can hasten the removal of an incorporated chemical by bringing fresh portions to the surface (17).

It is difficult to calculate from first principles the amount of pesticide that will evaporate under complex conditions in the field, but Hartley showed (18) that the potential capacity for evaporation may be estimated from a knowledge of the rate at which water is lost under similar conditions, using the relationship that rate of evaporation is proportional to vapour pressure $\times \sqrt{M}$, where M is the molecular weight. Water losses in the field have been measured under many conditions, for example 14 tonnes/ha/day can easily evaporate from moist soil during the summer in Britain into air at about 50% relative humidity. Table 1, which includes some of Hartley's results, shows the calculated amounts of some relevant crop protection chemicals which would evaporate from an inert surface under similar climatic conditions.

Table 1

Predicted Losses by Evaporation from 1ha of Inert Surface

Chemical	Vapour pressure mm.Hg at 20°C	Molecular Weight	Predicted rate of loss
Mercury	1.2×10^{-3}	201	6.45 kg/day
γ-BHC	9.4×10^{-6}	291	2.02 kg/month
HEOD	2.6×10^{-6}	381	0.06 kg/month
DDT	1.7×10^{-7}	355	1.35 kg/year
Simazine	6.1×10^{-9}	202	0.04 kg/year

In soil, vapour pressures are decreased because chemicals are dissolved in soil water and adsorbed on soil surfaces. Provided the adsorption isotherm is linear and partition between solution and vapour reasonably independent of concentration, vapour pressure should be approximately proportional to the concentration of pesticide in the soil until the solution and adsorbed phase become saturated and the vapour pressure reaches that of the pure chemical. As an example, Spencer et al (19) found that the vapour density of HEOD in soil increased linearly until the concentration reached 25 p.p.m. when it was the same as that of the pure chemical.

It is sometimes suggested that evaporation of chemicals should increase with increasing soil moisture content because the chemical is displaced from the adsorbent surfaces by water molecules. However, even air-dry soils usually contain a few percent of water, considerably more than the amounts of pesticide usually present, so that this factor would be expected to exert measurable effects only in the uppermost layers of soil when very dry. This is supported by the measurements of Spencer et al (19), who found that moisture content had little effect on the vapour pressure of HEOD until the soil contained less water than that required for a monolayer on the particle surfaces. The vapour pressure then decreased considerably, implying that there would be very little evaporation under these conditions in the field, although associated rises in temperature could offset this to some extent.

Apart from these very dry conditions, the rate of evaporation, of an incorporated pesticide is more usually limited by the rate at which it can be transported to the surface. By maintaining the soil moist, this transport can be accomplished largely by the "wick action" of water moving to the surface and bringing dissolved chemical with it. When the moisture content falls below that required to maintain a capillary network, however, water movement is very much slower and transport depends much more on molecular diffusion of the pesticide itself. Except for volatile chemicals, molecular diffusion coefficients in soil are of the order of 10^{-7} cm^2/s (20,21) so that this process is slow and it is under these conditions that cultivation may be particularly valuable in bringing fresh chemical to the surface more rapidly.

Bringing a chemical to the soil surface may increase the rate of photochemical degradation in addition to increasing the rate of evaporation. Photochemical degradation of some chemicals may be enhanced by adsorption on soil, for example the herbicide paraquat is photochemically stable to sunlight in solution but adsorption shifts the wavelength for maximum absorption into the range of normal sunlight thus increasing photochemical degradation (22).

For relatively immobile residual herbicides such as the triazines, cultivation may make unwanted activity less by the opposite process: that of redistributing the chemical from the surface more uniformly through the soil and thereby diluting it to an ineffective concentration. Although this may decrease availability, it probably does little to accelerate disappearance from the soil as a whole.

Cropping with plants that accumulate pesticides has been suggested as another method for decontaminating soils. For example, uptake by resistant plants such as Johnson grass, maize and sorghum accounted for about 25% of the atrazine applied to soil in pot experiments by Sikka and Davis (23). The general usefulness of this method has not been established however and clearly it would require caution where it resulted in more of a persistent chemical being present in food eaten directly by stock or people.

Heavy irrigation has also been suggested as a means of accelerating the removal of contaminants from soil. To estimate the potential of such leaching, soil may be considered as a crude chromatographic column on which the chemical is adsorbed. By analogy with chromatography, a band of chemical will be leached downwards as an increasingly diffuse bell-shaped pulse whose peak concentration moves at f times the rate of movement of the solvent (here water) where f is the fraction of the chemical that is not adsorbed and therefore freely mobile in the soil solution. For a chemical that is distributed between the solid particles and the solution in the ratio 20:1 as a result of adsorption (a ratio less than many measured in practice), f would be approximately 0·01 at 20% soil moisture content. It would therefore take 20cm of rain, resulting in a net 100cm of "solvent" movement in soil at 20% moisture content, to move the band of pesticide 1cm. For many of the most important persistent chemicals that are strongly adsorbed, therefore, the quantities of water required for effective leaching would be enormous and could have undesirable effects by removing other materials.

Furthermore although this chromatographic analogy can be used to

give a rough guide to the effects of leaching, recent experiments
with the 4-hydroxypyridine herbicide, haloxydine, emphasise that it
may be a considerable over-simplification (24). Instead of moving it
as a band, leaching by rainfall of haloxydine applied to the soil
surface produces a very diffuse distribution with a more or less
uniform concentration throughout the 15-20cm profiles examined,
although there is no detectable adsorption in the laboratory. In
some soils, the maximum concentration remains in the top horizon
even after more than 30cm of rain have fallen over a period of
several months. The results may be explained on the hypothesis that
the herbicide solution, which is applied during dry weather, is
drawn into fine pores in the structural aggregates from which it is
only released back to the main channels of water flow with difficulty.
Such behaviour would explain similar results obtained with picloram
(25, 26); direct observations of movement of systemic fungicides in
soil have also shown related effects (27). If this hypothesis is
correct, it is possible that chemical which had been held in an
unavailable form in the fine pores could be released unexpectedly
when the structural units were disturbed on cultivation and possibly
damage a subsequent crop, although retention is probably in micro-
structural units, too small to be affected by cultivation.

Adding Materials to Reduce Biological Availability

As adsorption by soil affects the activity of soil-applied
chemicals, it is perhaps not surprising that adding adsorbent
materials has often been suggested as a means of making persistent
chemicals less harmful. The technique has been considered
particularly for residues of persistent herbicides (28, 29), although
it has also been suggested for organochlorine insecticides (30).
The most popular adsorbent for this purpose is activated carbon.
There is no doubt that adding activated carbon can decrease the
activity of chemicals in soil, although substantial quantities may
be required. For example to achieve reasonable protection from
triazine herbicides, a ratio of activated carbon to herbicide of
several hundred may be needed (28). The exact figure depends on
many factors such as soil type, the crops to be grown and the type
of activated carbon used.

The popularity of this approach is surprising because the
chemical is not lost from the soil, but merely stored in a less
available form. Adding large quantities of adsorbent modifies soil
properties in other ways and means that larger quantities of chemical
will subsequently be required, thus working against the generally
accepted policy of trying to reduce rates of application. Also the
method is not cheap: inactivating 1kg/ha of triazine herbicide could
easily cost £25/ha.

In practice the most promising outlets for adsorbents seem to be as localised applications around individual crop plants or as root dips to protect seedlings which can then be grown safely in herbicide-treated soil. Although the use of adsorbents seems unlikely to become generally accepted as a routine treatment for large areas of arable soil, it could possibly become established in agricultural practice as a specialised technique to be used occasionally as, for example, when unusually large quantities of herbicide are required for non-recurrent control of a resistant weed.

In addition to active carbon, some other organic materials, which are discussed below in connection with increasing the rates of degradation, probably exert their influence partly by adsorption, and various other adsorbents such as clays have occasionally been suggested.

Additives Intended to Accelerate Degradation

The possibility of enhancing the frequently-demonstrated ability of microorganisms to degrade pesticides in soil has often been considered. In general, direct inoculation of soil with alien degrading organisms has not had much success because the introduced species do not survive well in competition with the natural population. The prospects for stimulating microbial degradation indirectly appear to be better. In suitable cases, rates of biocide degradation parallel the levels of general microbial activity in soil, as indicated by such parameters as soil organic matter decomposition (31) and may therefore be stimulated by adding energy-rich materials such as sugars or various composts (28) or by raising the temperature or maintaining the soil moist.

Manipulation of microbial processes in other ways has also been suggested. For example, Guenzi and Beard (32) found that DDT was degraded to DDE faster in anaerobic than aerobic conditions, especially in the presence of alfalfa residues. The possibility of exploiting this by flooding polluted soils was suggested but this again is a drastic solution that could have other undesirable effects. There is also much accumulated evidence that successive applications of some pesticides persist for progressively shorter periods in soil because of adaptation and selection of degrading microorganisms. Conceivably this could be put to practical use in suitable situations.

Stimulating chemical degradation of pesticides has received less attention. However on the basis of known detoxification pathways in resistant plants such as maize, Castelfranco and Deutsh (33) suggested that adding nucleophilic reagents such as calcium polysulphide to soil should accelerate decomposition of simazine. This type of approach does not seem to have been exploited in practice, but has clearly not been exhausted as indicated by more recent reports (34) of attempts to produce "self-destructing" DDT containing catalysts that promote degradation. Any successful exploitation of this type of approach will have to overcome the problems associated with the very large bulk of soil in the field and its considerable capacity for buffering the activity of added reactive chemicals.

ALTERNATIVES TO PERSISTENT CHEMICALS

Many of the methods for lessening pollution discussed so far are drastic and expensive. They would have significant and often unknown effects on soil properties, in addition to their intended effects. Most would be justifiable only in extreme situations where emergency measures were required. Replacement of persistent chemicals by alternative methods of pest control or by less persistent and more specific chemicals is a much better approach where possible. Much effort is currently being directed towards developing techniques such as biological control of pests by predators, sterile male release and the use of pathogens for pest control (35). While some progress has been made and these methods may hold hope for the future, there is little prospect that they will generally replace chemical control. They will not be discussed in detail here but it is perhaps worth emphasising that the possibilities of "polluting" the environment by injudiciously introducing predators that upset unsuspected ecological balances, or by introducing exotic pathogens, are at least as great as those arising from the use of chemicals, and new methods must be very carefully considered before they are attempted in practice.

In many cases persistent chemicals have already been replaced by less persistent ones and to that extent many of the problems discussed so far are largely historical. As is well known, there has been a general tendency to replace organochlorine insecticides by less persistent organophosphorus and carbamate insecticides. Although on balance this is clearly an improvement, it should be stressed that many of these substances have a much greater acute toxicity to mammals and are more dangerous in the short term. Also, it cannot be assumed that the effects of a very active chemical last only for as long as the parent compound can be detected chemically. Biologically active degradation products may be formed, and ecological effects can persist for long periods after the chemical has disappeared (36). For the control of soil-inhabiting pests such as wireworms, wheat bulb fly larvae and vegetable root fly larvae, some measure of persistence is often essential and replacement of organochlorine insecticides for this purpose is particularly difficult. Even if a chemical is intrinsically sufficiently toxic to the pest, to be a successful replacement it must achieve a delicate balance between adequate persistence for control under the varied conditions in the field and sufficiently rapid breakdown to avoid pollution. Although some of the more persistent organo-phosphates and carbamates approach this ideal (37), they tend to be less reliable than the organochlorines, and for controlling some pests, such as carrot fly in Britain, no really satisfactory alternatives have been found. The greater cost of newer chemicals is also a major disadvantage, particularly in developing countries.

As yet the factors influencing activity and persistence are not well enough understood for suitable chemical structures to be predicted theoretically. When no practical alternative to a persistent chemical for the control of a pest exists, the best approach to avoiding pollution is to improve the efficiency of its use, so that less is applied. This is also desirable economically, and could sometimes enable alternative chemicals to be used that are now excluded because they are too expensive.

IMPROVING UTILISATION OF AGRICULTURAL CHEMICALS

A large proportion of the chemical applied to foliage eventually finds its way down to the soil. To reduce contamination of soil, it is therefore important to apply less above the ground, as well as directly to the soil. In theory very substantial improvements should be possible. Even under favourable conditions, such as when systemic chemicals are applied as seed dressings to be taken up by the developing plants, the amounts taken up are only 1 or 2% of what is applied (27).

Placing the chemical in the critical region of the soil rather than broadcasting it uniformly is a well established method of reducing rates of soil-applied pesticides. The possible economies were discussed by Price Jones (38) and are illustrated in Table 2. There are many other examples involving row and band treatments.

Table 2

Influence of Placement on Effective Application Rate (g/ha) (38)

Pest	Insecticide	Method of Application		Seed Treatment
		Broadcast	combine-drilled	
wireworms	γ-BHC	840	280-420	56-84
wheat bulb fly	HHDN	–	1120-2240	140

It cannot always be assumed however that such a reduction in rate is automatically paralleled by a reduction in harmful effects.

As Price Jones emphasised, seed treatment can be more harmful as it puts the chemical in a very sensitive part of the ecosystem. The harm caused by seed dressed with organochlorine insecticides to seed-eating birds is a well-publicised illustration of this.

Method of application can also influence persistence. Granular formulations of insecticides usually persist longer than emulsions, and losses are fastest from dusts and wettable powders. (16,39) With granules different rates of release can be obtained by varying the composition of the carrier and additives, and in theory matching the shape of the dose/time pulse released by the granule to the behaviour and susceptibility of the pest should lead to more efficient use of the pesticide, but difficulties arise in practice because of variations in weather. Also, the rate a chemical moves within the soil may limit the rate at which it is redistributed from the source, rather than the properties of the granule, unless the chemical is very mobile, which is unlikely for persistent compounds. However this is a subject where more research is needed, particularly as recent microencapsulated formulations may extend the possibilities of regulating release rates and thereby allow smaller amounts to be applied initially but still ensure that enough chemical is present at a later stage.

Application to the foliage is probably at least as inefficient as application to the soil. A large proportion of the chemical is rapidly lost from the target site, particularly under tropical conditions. Phillips (40) showed that 95% of the HEOD applied to cotton leaves is lost in 24 hours at $40^{\circ}C$ and a windspeed of 60 metres/min. For adequate pest control, it is therefore necessary to apply large amounts which compensate for these losses. Much of the resulting contamination of soil and the general environment could be avoided if persistence on the target area could be controlled and less used. The possibilities of using microencapsulation formulations for this purpose are currently being investigated at Rothamsted Experimental Station. Using specially prepared leaking capsules, it has been possible to decrease the amounts of DDT lost over 35 days under warm conditions ($28^{\circ}C$ day/$20^{\circ}C$ night, 55-60% R.H. and 10 hours day light) from 90% with a standard wettable powder to about 20%. (40)

The possibility of exploiting insect behaviour provides one of the most fascinating approaches to using less pesticide. To some extent many existing practices exploit behaviour, for example in the control of soil pests discussed earlier, the reduction in rate achieved by using seed treatments relies on attraction of the pest towards the seed where the pesticide is concentrated. More specific attempts to exploit behaviour have been reviewed by Cherrett and Lewis (41). The most spectacular results can be achieved with social insects such as ants which, because of aggregation into nests, are often difficult to control by general spraying or even direct application of insecticides. They are particularly suitable for control by poison baits because they carry food back to the nest where it is shared with other members of the colony. The control of the imported fire ant in the southern U.S.A. provides a well-known example of the reductions in rate which can be achieved using baits. Originally, control by spraying large areas with HEOD or heptachlor at 2·24kg/ha from the air was attempted. This failed to control the pest but caused secondary pest outbreaks and damage to wild life. After much further research and development, excellent control was achieved by scattering from the air baits consisting of corncob grit with the insecticide kepone and soyabean oil as an attractant. By exploiting the foraging behaviour of the ant to accumulate the slow-acting toxicant, it has been possible to reduce the rate of active ingredient in these baits to only 10g/ha. Research to extend this type of approach to other social insects such as leaf cutting ants is currently underway; the use of microencapsulated formulations which give delayed release may have further possibilities here, particularly in combination with attractants.

While social insects provide extreme examples of the possibilities of exploiting behaviour, there is little doubt that utilisation of chemicals could be improved in other cases by a more thorough understanding of pest behaviour in relation to pesticide distribution patterns from different methods of application. At sufficiently small rates, many pollution problems would disappear and ultimately this kind of approach offers more realistic hope of a long term solution to problems of soil pollution than the search for methods to remove unwanted contaminants resulting from applying unnecessarily large amounts.

ACKNOWLEDGEMENTS

I am grateful for helpful discussions with several colleagues, particularly Dr J F Newman, Dr D Price Jones and Dr T E Tomlinson at Jealott's Hill Research Station and Dr C Potter at Rothamsted Experimental Station.

REFERENCES

1. Graham-Bryce, I J and Briggs G G, R I C Reviews, 1970, 2, 87.
2. Taschenberg, E F, Mack, G L and Gambrell, F L, J. agric. Fd Chem 1961, 9, 207.
3. Hirst, J M, LeRiche, H H and Bascombe, C L, Plant Path., 1961, 10, 105.
4. Raw, F, Rep. Rothamsted exp. sta., 1961, p146 and 1962, p158.
5. Kearney, P C, Nash, R G, and Isensee, A R in Chemical Fallout: Current Research on Persistent Pesticides ed. M W Miller and G C Berg, Chap. 3, pp54-67, Springfield, Ill : Charles C

Thomas, 1969.
6. Kimura, Y and Miller, V L, J. agric. Fd Chem. 1964, 12, 253.
7. Newman, J F, The Ecological Effects of Agricultural Chemicals: to be published.
8. Edwards, C A, Persistent Pesticides in the Environment, CRC Monoscience Series, London, Butterworths, 1970.
9. Wright, D W, Rep. Nat. Veg. Res. Sta., Wellesbourne, 1963, 13, 50.
10. Kelsey, J M and Arlidge, E Z, N Z J. Agric. Res. 1968, 11, 245.
11. Wolton, K M, N.A.A.S. Quart. Rev., 1963, No. 59, 122.
12. Cooke, G W and Williams, R J B, Water Treatment and Examination, 1970, 19, 253.
13. Tomlinson, T E, Outlook on Agriculture 1972, 6, 272.
14. Walker, W A, A.D.A.S. Gleadthorpe Experimental Husbandry Farm Report 1971, p30.
15. Bovay, E, Fluoride, 1969, 2, 222.
16. Edwards, C A, Residue Reviews, 1966, 13, 83.
17. Lichtenstein, E P and Schulz, K R, J. econ. Ent. 1961, 54, 517.
18. Hartley, G S, in Pesticidal Formulations Research, Physical and Colloidal Chemical Aspects A.C.S. Advances in Chemistry Series No. 86 p115.
19. Spencer, W F, Cliath M M and Farmer, W J, Soil Sci. Soc. Amer. Proc. 1969, 33, 509.
20. Graham-Bryce, I J, J. Sci. Fd Agric., 1969, 20, 489.
21. Ehlers, W, Farmer, W J, Spencer, W F and Letey, J. J Soil Sci. Soc. Amer Proc. 1969, 33, 505.
22. Slade, P, Nature, 1965, 207, 515.
23. Sikka, H C and Davis, D E, Weeds, 1966, 14, 289.
24. Riley, D., Results to be published.
25. Scirfes, C J, Burnside O C and McCarty, M K, Weed Science 1969, 17, 486.
26. Herr D E, Stroube, E W and Ray D A, Weeds, 1966, 14, 248.
27. Graham-Bryce, I J and Coutts, J, Proc 6th Br. Insectic. Fungic. Conf. 1971, Vol. II, p419.
28. LeBaron, H M, Residue Reviews, 1970, 32, 311.
29. Foy, C L and Bingham, S W, Residue Reviews, 1969, 29, 105.
30. Lichtenstein, E P, Fuhremann, T W and Schulz, K R, J. agric. Fd Chem., 1968 16, 348.
31. McCormick, L L and Hiltbold, A E, Weeds, 1966, 14, 77.
32. Guenzi, W D and Beard, W E, Soil Sci. Soc. Amer. Proc., 1968, 32, 522.
33. Castelfranco, P and Deutsch, D B, Weeds, 1962, 10, 244.
34. Anon, Chem. and Eng. News, 1970, 48, 12.
35. Price Jones, D and Solomon, M, Editors, Biology in Pest and Disease Control Blackwell Scientific Publications, Oxford, 1972 in press.
36. Barrett, G W, Ecology, 1968, 49, 1019.
37. Griffiths, D C, Rep. Rothamsted exp. Stn. for 1967, p332.
38. Price Jones, D in Biology in Pest and Disease Control ed.

Price Jones, D and Solomon, M, Blackwell Scientific Publications, Oxford, 1972 in press.
39. Schulz, K R, Lichtenstein, E P, Liang, T T and Fuhremann, T W J. econ. Ent. 1970, 63, 432.
40. Phillips, F T, Results to be published.
41. Cherrett, J M and Lewis, T in Biology in Pest and Disease Control, ed. Price Jones, D and Solomon, M, Blackwell Scientific Publications, Oxford, 1972 in press.

CONTROL OF AIR POLLUTION AFFECTING OR CAUSED BY AGRICULTURE

John T. Middleton[1] and Ellis F. Darley[2]

[1]*U.S. Environmental Protection Agency, Washington, D.C.* [2]*University of California, Riverside, California, USA*

Air pollution is a serious environmental quality problem in the United States and many countries of the world. It is a problem of concern to both the developed and developing nations. The importance of the problem is shown by the increasing attention given by national and international organizations.

Air pollution in the United States results from the emissions of five principal pollutants from four major categories. About half of the air pollutants come from our transportation system, chiefly the internal combustion engine. Fuel burning in stationary sources accounts for about one-fourth of the total while industrial processes account for about one-sixth and incineration and other waste disposal methods account for another one-sixth.

The pollutants from these sources interfere with visibility, damage property, irritate the senses, have adverse effects on public health and welfare, and damage vegetation and animals. The annual cost of direct observable air pollution damage to field, flower, fruit and nut crops, and vegetables in the U.S. has been estimated to amount to about $80 million, of which nearly $26 million occurred in California (1,2). This loss to selected crops is somewhat less than 1/2 percent of the total annual cost of pollution from damage to health, materials, and residential property (3).

POLLUTANTS AND THEIR EFFECTS

The air pollutants responsible for damage may be either particulate or gaseous in nature (4). The solid particles released into the atmosphere are sometimes the cause of soiling fruits and vegetables, damage to the tissue of the exposed leaves and fruits, and growth reduction, as well as adding a toxic burden to forage crops used as feedstuffs for livestock. Liquid particles, such as acid aerosols and toxic mists, are sometimes responsible for leaf spotting. The greatest amount of damage to animals and vegetation is usually caused by gaseous air contaminants which directly injure plants and indirectly injure animals by the toxic effects produced after the animal has consumed contaminated forage and food supplements.

A number of particulate materials, such as carbon and cement dusts, occur in the atmosphere and form undesirable deposits on agricultural produce. Injury to leaves and fruits of many plants may result from acid aerosols originating from the combustion of sulfur-containing fuels, smelting of ores and chemical manufacture. Sulfuric and other acid aerosols produce small to large spots on the upper exposed surface of leaves. Although injury usually occurs during periods of fog when the droplets become large enough to settle out and the leaf readily wets, acid aerosol damage also occurs any time the gas effluent dew point permits droplet formation.

There appears to be little doubt that naturally deposited dust from certain cement plants is responsible for injury to leaves of deciduous and coniferous species and occasionally for death of the latter. The finer particles of certain cement-kiln dusts from electrostatic precipitators interfere with carbon dioxide exchange and may cause considerable leaf injury.

Fluorides in particulate form are less damaging to vegetation than gaseous fluorides. Fluoride may be absorbed from depositions of soluble fluoride on leaf surfaces. However, the amount absorbed is small in relation to that entering the plant in gaseous form. The fluoride from particulates apparently has great difficulty penetrating the leaf tissue in a physiologically active form. Fluorides absorbed or deposited on plants may be detrimental to animal health. Fluorosis in animals has been reported due to the ingestion of vegetation covered with particulates containing fluorides.

The principal gaseous air pollutants causing damage to vegetation are ethene, fluorides, nitrogen dioxide, ozone, peroxyacetyl nitrate, and sulfur dioxide. Other gaseous pollutants, such as carbon monoxide, mercaptans, and other odor-producing compounds do not cause damage to plants but are either toxic or offensive to man. Crop damage consists of both injury to leaves, flowers and fruits, and their premature loss, as well as reduced plant growth and lowered yields through altered biochemical and physiological processes.

The most characteristic symptoms of the plant damaging pollutants are: <u>Ethene</u> - distortion and drying of flowers and curling of plant leaves. <u>Fluorides</u> - lesions confined to the leaf margins of broad-leaved species are generally tan or dark brown in color. Affected tissues may become brittle and fall from the unaffected portion of the leaf. The leaf tips of conifers become yellow or dark brown. <u>Nitrogen dioxide</u> - no pathological color change in leaves, but exposed plants are retarded in growth. <u>Ozone</u> - a stippling, mottling, or bleaching of the upper leaf surface. Color of the stipple marking may vary from light tan or ivory to reddish or almost black. <u>Peroxyacetyl nitrate</u> - a silver, bronze, or otherwise metallic sheen on the under leaf surface of affected plants. <u>Sulfur dioxide</u> - injury occurs between the veins and the affected tissue is bleached to a light tan or may become brown. The veins characteristically remain green. Injury on conifer needles is similar to that caused by fluorides.

The photochemical oxidants, ozone and peroxyacetyl nitrate, together with the hydrocarbon ethene, account for about 91 percent of crop losses based on damage costs. Sulfur oxides account for 5 percent, fluorides for 3 percent and nitrogen oxides about 1 percent of the damage.

POLLUTION SOURCES

Development of a control strategy for abatement of damage to agriculture requires, among other things, an inventory of emission sources and an estimate of the amounts emitted (5). The annual emissions in 1969 for the 5 major pollutants in the United States are presented in Table 1. Transportation accounts for most of the emissions of carbon monoxide, hydrocarbons, and nitrogen oxides and contributes insignificant amounts of particulate matter and sulfur oxides; the latter two contaminants come largely from fuel combustion, industrial processes, and solid waste disposal. Of the transportation sources, the motor vehicle accounts for about 87 percent of the carbon monoxide emissions, 86 percent of the hydrocarbons, and 79 percent of the nitrogen oxides.

Table 1. Nationwide Emissions for Selected Categories, 1969
(million of tons per year)

Source Category	Particulates	Sulfur Oxides	Carbon Monoxide	Hydro Carbons	Nitrogen Oxides
Transportation	0.8	1.1	111.5	19.8	11.2
Fuel Combustion	7.2	24.4	1.8	0.9	10.0
Industrial Processes	14.4	7.5	12.0	5.5	0.2
Solid Waste Disposal	12.8	0.4	26.1	11.2	2.4
All Sources	35.2	33.4	151.4	37.4	23.8

A comparison of emissions from motor vehicles with those from incineration of domestic, industrial, and municipal wastes, and those from the open burning of agricultural crop residues and forests is shown in Table 2.

Table 2. Nationwide Emissions for Motor Vehicles, Incineration, and Open Burning of Agricultural Refuse and Forests, 1969
(million of tons per year)

	Particulates	Sulfur Oxides	Carbon Monoxide	Hydro Carbons	Nitrogen Oxides
Motor Vehicles	0.4	0.3	97.8	17.1	8.7
Incineration	1.4	0.2	7.9	2.0	0.4
Agricultural Refuse	2.4	N*	8.3	1.7	0.3
Forests	8.8	N	9.4	2.9	1.6

*N=negligible

The emissions from forests shown in Table 2 were estimated from the destruction of 5.7 million acres of forests by unprescribed fires and the intentional burning of an additional 2.9 million acres. The emissions from agricultural refuse were estimated from the open burning of nearly 280 million tons of materials, such as crop residues, brush, weeds, grass, and other vegetation. The burning of farm refuse and forest materials produces about 28 times as much particulate matter as do motor vehicles, produces negligible amounts of sulfur oxides, and accounts for about 18 percent of the nitrogen oxides emitted by motor vehicles.

Particulate matter from agricultural burning contributes significantly to reduced visibility, especially during periods of air stagnation. Hydrocarbons and nitrogen oxides from open burning participate in the formation of plant damaging ozone and peroxyacetyl nitrate (6). The importance of agricultural sources of hydrocarbons and nitrogen oxides in the photochemical process depends upon the amounts released relative to those from motor vehicles in the same air basin.

A comparison of the hydrocarbons from motor vehicles and agricultural sources has been made for three air basins in California and for the State as a whole (7). The Central Valley, comprising the Sacramento and San Joaquin Air Basins, represents the agricultural heart of California. The burning of agricultural refuse contributes about 7 percent of the hydrocarbons as compared to motor vehicles. The San Francisco Bay Area Basin is an important agricultural region in the midst of several metropolitan centers, in which the hydrocarbon contribution from open burning is about 0.4 percent of that from motor vehicles. The South Coast Air Basin is a highly industrialized and urban area wherein the remnants of agriculture surround the metropolitan center of Los Angeles. There, agricultural waste burning contributes a mere 0.03 percent of the hydrocarbons as compared with motor vehicles. The contribution of hydrocarbons from all 11 air basins comprising the entire State is about 1 percent of that coming from motor vehicles.

A further comparison may be made between the motor vehicle and the burning of that portion of agricultural refuse comprising orchard prunings, grain straw, and native brush, and noting the amounts of olefin and saturate hydrocarbon percentages produced per ton of fuel consumed. Olefins are considered to be fairly reactive photochemically in producing ozone and peroxyacetyl nitrate, while the saturates are much less so. For total hydrocarbons, burning the identified agricultural fuels yields about 10 percent of that of burning an equal weight of gasoline in uncontrolled motor vehicles. A comparison by classes of compounds shows agricultural fuels to produce 26 percent of the ethene, 13 percent of all olefins including ethene, and 7 percent of the saturates emitted from uncontrolled motor vehicles.

Carbon monoxide does not adversely affect agriculture but it does adversely affect man. An analysis of carbon monoxide emissions for the same 3 principal basins and statewide shows the percent contributions compared to the motor vehicle to be 10 percent for the Central Valley, 0.8 percent for the San Francisco Bay Area, only 0.04 percent for the South Coast Basin, and 0.2 percent statewide.

AGRIBUSINESS SOURCES

Pollutants in addition to those from burning agricultural wastes and forests come from a number of agribusinesses and industries producing goods from agricultural products. In 1969 nationwide, cotton ginning emitted 50,872 tons of particulate matter while grain handling and milling produced 1,750,000 tons, phosphate fertilizer manufacture released 18,597 tons, and kraft paper from wood pulp yielded 378,000 tons. In addition to particulate matter, phosphate fertilizer production emitted 3,306 tons of fluorides, while kraft paper production accounted for emissions of 966,000 tons of carbon monoxide and 78,000 tons of sulfur oxides.

POLLUTION CONTROL

The abatement and control of air pollutants affecting agriculture is needed for improved crop production and to reduce the cost of air pollution damage. Control of motor vehicle created air pollution can be achieved by application of performance standards limiting emissions to levels providing acceptable air quality (8,9,10). Motor vehicle emissions can be reduced through the introduction and use of advanced automotive power systems, limiting emissions by improved combustion with cleaner fuels, and application of control devices.

Motor vehicle pollution control was first begun in California in 1960 (11). Federal controls were initiated in 1967, essentially by adoption of California's standards for crankcase and exhaust emissions. National crankcase and exhaust standards have been tightened since 1967, limits have been set for fuel evaporation, and more exacting test conditions have been set for compliance by 1972 and subsequent model year motor vehicles. Currently effective standards provide for complete control of crankcase emissions, a 94 percent reduction in fuel evaporation losses, 69 percent reduction in carbon monixide, and 80 percent reduction in hydrocarbons over uncontrolled vehicles.

Emission standards for 1975 and subsequent model years require a 90 percent reduction in hydrocarbon and carbon monoxide emissions over those allowed from 1970 vehicles and by 1976, a 90 percent reduction in nitrogen oxides emissions from those measured from 1971 model vehicles. These emission standards are designed to provide air of acceptable quality for motor vehicle created pollutants by mid-1980.

The control of fuel combustion and industrial processes through the use of best available control technology and the application of performance standards (12) will effect reduction of particulate matter, sulfur oxides, and acid mists requisite for desired air quality.

Particulate matter, including solid fluoride material, from a number of existing industrial sources has been satisfactorily controlled for many years through the use of cyclones, electrostatic precipitators and fabric bag houses; extended application of this technology will be required if acceptable air quality for particulates is to be achieved.

Limitation of sulfur in fuels through use of naturally occurring low sulfur coal, or sulfur-cleaned high sulfur coals, or inherently low sulfur content oil, or desulfurized oil, and the application of stack gas cleaning techniques or their combination, are accepted practices for control of sulfur oxides emissions. The application of new source performance standards for stationary sources of pollution assures that adequate control systems will be employed wherever such sulfur emitting industries are located.

Gaseous fluorides are collected both by wet scrubbers and conversion to solid particulate fluoride which then is removed by conventional particulate collectors. Additional reductions can be made by limiting the amounts of fluoride containing materials in the production process.

Reduction of pollution from agricultural and forest practices can be accomplished by methods which avoid burning or employ the most efficient burning techniques. A common practice in California, particularly with many row and vine crops, is complete incorporation of crop residues into the soil. The same techniques may be recommended for heavier woody materials where incorporation does not enhance disease and insect development or adversely alter soil properties.

Whereas the amounts of generated wastes burned vary from 2 to 90 percent, the average burned in California is about 27 percent; the remaining 73 percent is presently incorporated into the soil.

For those wood crop residues which cannot be easily incorporated through normal cultivation practices, there are a number of mechanical systems commercially available to reduce volume and size of residues thereby permitting their ready incorporation into the soil.

Established meteorological criteria under which agricultural burning is permitted in California at elevations below 4000 feet assure adequate dispersion of pollutants and thus minimize surface pollution (13). As an example of implementation of this system, a burn day is announced for the San Joaquin Air Basin when all of the following criteria are satisfied: (a) near the time of day when the surface temperature is at a minimum, the 3000-foot temperature is not warmer than the surface temperature by more than 13° F.; (b) the expected 3000-foot temperature is colder than the expected surface temperature by at least 11° F. for 4 hours; (c) the expected daytime wind speed at 3000 feet is at least 5 miles per hour.

Individual air pollution control agencies may also add criteria on how dry material should be, time of day for ignition and wind direction. For example, the Sacramento agency will not permit burning when the wind is from the north, even though the basin criteria may be met because of expected hazardous conditions often associated with such winds.

Remote and sparsely populated areas at altitudes between 2000 and 4000 feet above sea level may apply for exceptions to the agricultural burning regulations. Typical requests for this exception come from those concerned with improvement of pasture and range lands for which there are currently no alternative methods.

Present forest prescribed burning practices are directed toward pollutant reduction through use of most efficient burning techniques. These include burning under optimum conditions of fuel moisture, atmospheric humidity, wind speed and direction, ignition methods, and ground slope of burn area to assure that the fire will remain contained. These criteria also apply generally to slash disposal following timber harvest. Additionally, where prescribed burning is done under established trees for the purpose of reducing fire hazard, eliminating undesirable species, and game and recreation considerations, the piled material can be burned more efficiently if the material is dry, is ignited from the top, and additional material is thrown on the pile only after burning is well under way.

In some areas of the country, depending on the angle of the slope and nature of the soil, forest debris is smashed and worked into the soil with beneficial effects. Whole tree harvest is being tried in less accessible forest areas wherein the entire tree is air-lifted to a roadhead so that not only the lumber manufacturer but all other wood users have a ready access to the tree. This results in better timber utilization with minimal generation of waste.

The importance of improved resource utilization is steadily becoming recognized and has led to greater recycling of agricultural and forest wastes into feeds, paperboard, packing materials and building materials, soil conditioners and chemicals (14). Best raw product utilization and minimum burning at maximum efficiency with least possible emissions should contribute materially to restoring air of good quality with concomitant reduction in air pollution damage.

BIBLIOGRAPHY

1. Barrett, Larry B., and Thomas E. Waddell. The Cost of Air Pollution Damages - a Status Report. U.S. Environmental Protection Agency Report 56pp., 1971.

2. Millecan, Arthur A. A Survey and Assessment of Air Pollution Damage to California Vegetation in 1970. California Department Agriculture Report 48pp., 1971.

3. Anonymous. Environmental Quality. Council on Environmental Quality 2nd Annual Report 360pp., 1971.

4. Darley, E.F., Carl W. Nichols, and John T. Middleton. Identification of Air Pollution Damage to Agricultural Crops. California Department Agriculture Bulletin 55(1): 11-19, 1966.

5. Middleton, John T. Planning Against Air Pollution. American Scientist 59:188-194, 1971.

6. Darley, E.F., F.R. Burleson, E.H. Mateer, John T. Middleton, and V.P. Osterli. Contribution of Burning of Agricultural Wastes to Photochemical Air Pollution. Journal of Air Pollution Control Association 16:685-690, 1966.

7. Anonymous. Air Quality and Emissions - 1963-1970. California Air Resources Board Report 31pp., 1971.

8. Anonymous. Control of Air Pollution from New Motor Vehicles and New Motor Vehicle Engines. Federal Register 35(219): 17288-17313, 1970.

9. Ibid. Federal Register 36(128): 12652-12663, 1971.

10. Anonymous. National Primary and Secondary Ambient Air Quality Standards. Federal Register 36(84): 8186-8201, 1971.

11. Middleton, John T. and Diana Clarkson. The California Motor Vehicle Pollution Control Program. Traffic Quarterly 15(2): 306-317, 1961.

12. Anonymous. Standards of Performance for New Stationary Sources. Federal Register 36(247): 24876-24895, 1971.

13. Darley, E.F. Burning Agricultural Wastes and Effects on Agriculture. Proc. of Symposium on Agriculture and the Environment in the Southern San Joaquin Valley. Univ. of Calif. Extension Service, Bakersfield (In Press).

14. Anonymous. Solid Waste Disposal and Management in California. Univ. of Calif., Div. Agric. Sciences Research Task Group Report (In Press).

THE CREATION AND CONTROL OF PEST SITUATIONS THROUGH ENGINEERING: PAST, PRESENT, AND FUTURE

Kyle R. Barbehenn

Center for the Biology of Natural Systems,
Washington University, St. Louis, Missouri

It's a fair observation that most Americans have a rather poor sense of history and a visit to the countries surrounding the Mediterranean Sea makes one pause to reflect on the past. Aquaducts, viaducts, pyramids and other massive structures bear witness to the fact that ancient civilizations had good engineers. Barren hills, deserts on the march, and landlocked seaports buried under many meters of alluvium suggest that the ecologists of those early days didn't have much say about how the landscape should be managed. Certainly the ancient engineers had failures as well as successes and what role their short-term successes and long-term failures played in the rise and fall of civilizations is a matter for historians to determine. Engineers, however, are only parts of socities and engineering is but one of many processes that influence long-term stability of civilizations. We cannot understand the whole by looking at a fraction of the parts but we may start by determining how some of the more important parts are interrelated in their structure and function. The purpose of this symposium is to examine some of the relationships between engineering and pest situations.

The usual concept of a pest has its origins in human well-being and personal comfort. One quickly thinks of rats and mice, of cockroaches and flies, and of the weedy plants that infest agricultural land, lawns, and lakes. Ther term "pest," however, was originally applied to a fatal epidemic disease--bubonic plague or the "black death." Thus, at first glance, there is a wide variety of organisms that one might consider controlling as pests and several engineering skills have been applied to this end. Operationally, however, we must be concerned with pest <u>situations</u> and the particular pest species' population must be considered in relation to the rest of the environment.

Extending this line of reasoning, a pest situation is one that degrades the quality of the human environment, either in the short or the long run. To a considerable extent, pests are pests because of the way we have managed (or mismanaged) the environment. If we define engineering broadly as any physical or chemical alteration of the environment whose goal is the betterment of human welfare (however narrowly viewed), then it is proper to ask how engineering can either create or control pest situations.

At first glance our topic might seem very narrow and conceptually simple but, in fact, it intrudes on virtually all aspects of man's endeavors and aspirations, directly or indirectly. This can be appreciated by organizing the problem according to various sectors of human concern and by classes of organisms that may produce pest situations (Table 1). Certainly the impact of pests on public health and on agriculture, forestry, and fisheries is well known and those of you who may be concerned with housing, transportation, and other public works have encountered pest situations of various sorts. Of less economic concern but of increasing personal concern as our standard of living increases are those pests that offend our dignities and interfere with our use of leisure time. In each of these sectors we can list a wide array of organisms that are involved in pest situations and note how engineering may be involved with causes and cures. For the sake of brevity, but to illustrate the point, I have listed some examples from the public health sector.

For practical purposes, we can start with viruses and, with little difficulty, list representatives of every major class of plants and animals. Some pest situations are extremely serious. Some are relatively trivial and were included to illustrate the diversity of the problem and not because more important examples are unavailable. Whether ragweed is important depends on what part of the world you live in and whether you happen to be allergic to the pollen. Wasp stings cause more fatalities in the U.S.A. than do bites from copperheads, rattlesnakes, cottonmouths and coral snakes. The role of engineering in controlling public health problems may be as well known as elementary sanitation and, in causing problems, may be as unpredictable as providing roosting sites for birds.

We will take a closer look at some public health problems but just comment on pest situations in other sectors. All organisms which are judged to be desirable, like man, are subject to diseases, predation, and competition from less desirable organisms. Other panel members will discuss some of these relationships. Fortunate-

Table 1. An Outline of Pest Situations by Sector of Human Concern and Classification of Organisms, with Notes on How Engineering May be Involved with Causes and Cures in One Sector.

	Sectors of Human Concern			
Class of Organisms	Public Health	Food & Fiber Production & Protection	Housing, & Public Works	Esthetics & Recreation
Virus	"Common cold": climate control of buildings			
Bacteria	Typhoid: sewage disposal and water treatment			
Fungi	Histoplasma: may be associated with pigeon and starling roosts in cities			
Algae	Blue-greens: toxins suspected of causing fish-poisoning in humans by entering food chains in disturbed marine areas			
Higher Plants	Ragweed: pollen is allergenic, plants are favored by disturbed ground			
Protozoa	Endamoeba histolytica (dysentery): sewage disposal and water treatment			
Coelenterates	"Sea-wasps": local control includes barriers such as nets			
Helminths	Hookworm: fecal contamination of soil			
Molluscs	Snails: Intermediate hosts for many parasites; irrigation and drainage ditches			
Arthropods	Wasps: nest sites are influenced by building design			
Vertebrates	Sharks: may congregate around marine garbage dumps			

ly, viruses do not attack non-living materials but the attacks of fungi, termites, and marine borers on wooden structures and the rapid growth of fouling organisms such as barnacles on ships, for instance, are too well known to dwell on here. We cannot fill in all cells of the table and the role of engineering is not of prime importance in many pest situations. I will conclude this section by noting that the excellent job that engineers have done in removing waste products from cities would be appreciated more if they didn't pour the material into water needed for recreation. The Tubifex worms and blue-green algae of Rock Creek Park in Washington, D.C. may provide some interest for a few old-fashioned biologists but the presence of higher forms of aquatic life would certainly have more general appeal.

THE ROLE OF ENGINEERING IN SOME PUBLIC HEALTH PROBLEMS

In most of the developed world, the role of engineers in controlling many communicable diseases has been outstanding. This is simply due to the fact that diseases transmitted by fecal contamination can be controlled by flushing a toilet, washing hands, and purifying water. Affluent societies demand and receive such protection although the system isn't everywhere perfect and having a typhoid shot and drinking tea may provide added insurance in some modern cities. Needless to say, the fact that most less-developed areas are plagued by enteric diseases is largely a fault of our economic system and cultural patterns. Where relatively simple physical and chemical techniques can be used to provide lasting solutions to health problems, engineers, given the financial resources, can do a good job. Not all problems are simple however. They may be created as unanticipated side effects of engineering operations, they may involve complex ecological relationships with other organisms, and they may involve patterns of human behavior. Three examples will be given to show how public health pest situations require a comprehensive approach to planning the human ecosystem.

Sleeping Sickness

A good system of roads is deemed essential to the economic development of any region since it facilitates the distribution of people and goods and opens new areas for the exploitation of raw materials. The lack of a good means of transportation, in fact, is a major reason for the present survival of many forests in underdeveloped countries. That the people and goods transported over roads can carry diseases and vectors to new areas is obvious, but who would imagine that a well-constructed road could otherwise influence the incidence of a disease?

Hughes and Hunter (1972) review many cases involving the relationship between roads and trypanasomiasis (sleeping sickness) caused by the protozoan Trypanasoma gambiense in Africa. The parasite is transmitted from man to man, or from certain reservoir species such as pigs and cattle to man, by the tsetse fly (Glossina palpalis). The fly breeds in forest edges and scrubby vegetation, especially along streams. Not only do infected people carry the disease organisms into areas previously free, but moving vehicles are attractive to the tsetse fly and carry the vector into tsetse-free areas. The highest incidence of the disease in some regions is directly associated with highways that facilitate man-fly contacts. Over and above the obvious problem of transporting the disease and its vector, particular hazards for transmission are produced wherever roads cross streams. This provides a natural situation for travelers to stop and refresh themselves and, while trees and shrubs along the banks may provide some welcome shade and help to stabilize the stream bank, they also increase the probability that the tsetse fly will be present in abundance.

Malaria as a Socioeconomic Disease

Throughout the history of mankind, malaria has been our greatest scourge. One measure of its impact can be seen in the incidence of genetic forms of resistance in man, such as the sickle-cell trait. Individuals who are heterozygous for the sickle-cell trait are resistant to malaria while the homozygous individuals succumb to anemia--a vicious evolutionary trade-off. No history of man would be complete without a consideration of malaria and the role of engineering figures prominently in some past solutions to the problem.

Malaria is caused by various species of the protozoan genus Plasmodium, the most important of which is P. falciparum. Several species of Anopheles mosquitoes transmit the parasite from man to man. The mosquitoes must breed in water--hence, the most obvious role of engineering in relation to the epidemiology of malaria is in the management of surface water.

The early historical relationships between malaria and past civilizations must necessarily be a matter for some speculation but some pieces of the probable picture can be reconstructed from our present knowledge of the disease and historical records (Garcia 1972). Malaria contributed to the destruction of the medieval civilization of Ninfa near the Pontine marshes of Rome. Apparently the debilitating role of malaria was enhanced by deteriorating socioeconomic conditions imposed by feudalism. Social instability was disastrous to an agricultural system whose methods held the vector population in check. A similar situation probably occurred in the ancient civilization of Ceylon as repeated invasions of Tamils disrupted the highly organized agricultural system. The history of malaria in more recently "developed countries" is better documented and it seems clear that in the United States, for instance, malaria disappeared progressively with a variety of factors associated with improved living conditions. In much of the U.S.A., malaria became a rare disease before the development and use of DDT (Chandler 1944).

Malaria remains a serious problem in many underdeveloped tropical and subtropical countries although the immediate situation has certainly improved in many areas through the efforts of the World Health Organization. The first real demonstration that a serious outbreak of malaria could be controlled however, occurred about 70 years ago--long before modern pesticides and drug therapy came into being (Garcia 1972). This feat was largely an engineering operation around the coastal towns of Klang and Port Swettenham, Malaya, and consisted of clearing jungles and controlling surface waters to prevent the breeding of the local anopheline vector. It must be emphasized that the ecology of the particular species of mosquito is a very important consideration. The right way to manage the habitat to control one species might produce opposite results for another species. In the case of <u>An. maculatus</u> (a major vector in Malaya) its preference for sunlit seepage areas and side pockets of streams as places to breed makes subsurface drains and stream channelization appropriate tools for control. Removing the canopy of vegetation that shades a potential breeding site is precisely the wrong thing to do. Thanks to the research efforts of medical emtomologists, the epidemiology of malaria in Malaya is well understood and the disease has been effectively controlled in cities, towns, and agricultural estates. The major problem now is in rural areas where the financial cost of effective, long-term control is excessive. Here, as elsewhere, malaria is mainly a disease of socially disadvantaged people.

A happy alternative to the expense of applying the joint skills of engineers and ecologists appeared with the discovery of DDT, other new-generation insecticides, and therapeutic drugs. The hope that malaria could be eradicated from wide regions of the globe by breaking the transmission cycle seemed highly realistic. That the high hopes have not been realized in many areas of the world might be discussed briefly as an example of how simple solutions may have, at best, variable consequences to complex problems.

The presence of malaria requires the parasite, a susceptible human population, and a vector population. The cycle of the disease can be broken by reducing any of the links to such a low level that the probability of transmission becomes nil. <u>If</u> this can be achieved solely by the use of chemicals, then the war is won. However, no malaria mosquito, to my knowledge, has been exterminated more than locally and temporarily by the use of chemicals alone and in fact, genetic resistance to DDT and other pesticides has developed in approximately 14 species of anopheline mosquitoes around the world (WHO 1971a). Even if the cycle appears to be broken and resistance is not a problem, the fact that the vector population persists means that the disease can recover by either reintroduction or from local foci that remain undiscovered. In effect, an area is "hooked" on chemical control and the consequence of a lapse in control is well illustrated by events in India and Ceylon during 1967 and 1968 (Garcia 1972). In Ceylon alone, about two million new cases developed and such situations may be exacerbated by the fact that residents of endemic areas lose their partial immunity after an area has been free from the disease fro some time, a fact known for 30 years (Chandler 1944). In addition, the malaria parasite has become resistant to drugs, especially in Vietnam.

The main argument for using chemical control instead of habitat management and for using DDT in preference to certain other pesticides is the relative cost. It now seems clear that a program that is not successful in eradicating malaria is a failure in the long run and efforts have been made in recent years to develop alternatives which are ecologically more sound (WHO 1971b). Malaria will remain a disease of the poor as long as the cost/benefit relationships are calculated in dollars, rupes, and pesos and as long as programs are based on outside intervention with artificial methods which are not harmonized with the cultural patterns of the people and with the rest of the ecosystem.

Bilharziasis--the Man-made Disease

With the level of success thus far achieved with malaria, schistosomiasis (bilharziasis) has become the most important disease of man, especially in the tropics. There are several forms of the disease caused by tiny blood flukes but our discussion will be limited to <u>Schistosoma mansoni</u> (an intestinal form) and <u>S. haematobium</u> (the urinary form) in Africa. Man is the primary host and certain species of aquatic snails serve as intermediate hosts. Eggs are shed in either the urine or feces or man and human infections are acquired by contact with water that is infested by free-swimming larval forms that have left the snails. The irony is that engineering schemes to provide irrigation water and hence improve man's standard of living in many areas have increased the problem enormously, hence, Weir (1959) has termed bilharziasis a "man-made" disease.

Conceptually, the reasons for the problem are quite simple. The essential kinds of snails can live in various aquatic situations but become dormant during dry periods. In many areas where rainfall is seasonal and the water quickly leaves the land, snails are absent or rare and schistosomiasis is not present. A dependable water supply is a boon to agriculture and, with irrigation, two or more crops can be grown each year instead of a single seasonal crop. Dams are built to harness the rushing water, irrigation ditches are dug to distribute the water on the land, vegetation grows along the water ways, snail populations expand their numbers, distribution, and seasons of activity, humans from areas where schistosomes are endemic move into the new crop land, and the incidence of schistosomiasis mushrooms (Van der Schalie 1972, Hughes and Hunter 1972).

The control of schistosomiasis is a difficult complex problem. One obvious engineering solution is to prevent human urine and feces from entering the water system but constructing simple sanitary facilities such as pit privies is difficult in low lying areas with high water tables. Encouraging people to associate with such foul-smelling situations that are foreign to their culture is a major problem. The worms are long-lived and may shed their eggs for several years. Treating a population with drugs would be an expensive possibility but the ideal drug remains to be discovered. Killing worms that inhabit blood vessels without killing the host is a real pharmacological challenge. Despite some side effects, hycanthone (HC) and related drugs recently showed much promise as a cure. The promise may be short-lived. Hamsters and mice infected with schistosomes were treated with hycanthone. After six to twelve months, the surviving parasites resumed production of viable eggs that gave rise to schistosomes resistant to HC and related drugs (Rogers and Bueding 1971). In addition, HC and related drugs were found to be mutagenic in _Salmonella_ (Hartman et. al. 1971) and the researchers expressed considerable alarm over the possible adverse consequences of massive field trials on people.

The main hope in controlling bilharziasis is by controlling the snails. Thus far (WHO 1967), molluscicides have been somewhat successful locally but extending treatments to vast areas for long periods of time seems to be unfeasible. The last snail is very hard to kill and individuals are very prolific. Mechanical control may be feasible with proper management of irrigation ditches but new machines must be designed (Van der Schalie 1972). Biological control, especially by introducing competing species of snails, shows considerable promise (WHO 1967) but much more research on control methodology is needed. Barring some miraculous breakthrough, bilharziasis will remain a man-made disease for a long time.

While I may have over-simplified the picture in some cases, these examples show the results of several interactions. Engineers modify the physical environment, disease vectors respond by changing their abundance and distribution, and the behavior of humans determines how the new values for the variables effect changes in the patterns of disease. The uniqueness of individuals perhaps make the human element the most unpredictable factor in the ecosystem. Engineers can now send a man to the moon but they can't keep him from tripping over a wire. However optimistic we once may have been, it is clear that our technological genius provides no quick and sure solution to many problems that involve living organisms.

The roads, the dams, and the irrigation ditches are now facts and some of their long-term consequences remain to be seen. In various situations, the side-effects of increased human disease could have been judged as either probable, possible, or unknown. Ecologists, like engineers, need both facts and understanding to make predictions and basic studies of many pest species have been badly neglected with the advent of modern pesticides. In the long-run, the undesirable side-effects of certain environmental alterations may outweigh whatever benefits accrue to society. Who can judge the relative merits of trade-offs involving food versus disease? Such decisions are not up to engineers and ecologists but we both have a responsibility to society to reveal the foreseeable costs of engineering projects and to evaluate the relative merits of alternative solutions to problems. The fledgling Environmental Protection Agency of the U.S.A. is a step in the right direction and the current U.N. meeting in Stockholm may provide even broader guidelines for assessing the environmental impacts of man-made alterations. It is now fashionable to point to the need for inter-disciplinary approaches to education, research, and problems of economic and social development. Here, I can only ask you to consider the following papers and to draw your own conclusions regarding the role of engineering in the cause and cure of pest situations.

References

Chandler, A.C. 1944. Introduction to parasitology (7th Ed.). John Wiley & Sons. New York. 716 p.

Garcia, R. 1972. Ecological considerations in malaria eradication. Environment 14: (in press).

Hartman, P.E., K. Levine, Z. Hartman & H. Berger. 1971. Hycanthone: a frameshift mutagen. Science 172: 1058-1059.

Hughes, C.C. & J.M. Hunter. 1972. The role of technological development in promoting disease in Africa. p. 69-101 In: Farvar, M.T. & J.P. Milton (Eds.). The careless technology. Natural History Press. New York.

Rodgers, S.H. & E. Bueding. 1971. Hycanthone resistance: development in Schistosoma mansoni. Science 172: 1057-1058.

Van der Schalie, H. 1972. World Health Organization Project Egypt 10: a case history of a schistosomiasis control project. p. 116-136 In: Farvar, M.T. & J.P. Milton (Eds.). The careless technology. Natural History Press. New York.

Weir, J. 1959. Bilharziasis as a man-made disease. WHO Chronicle 13:19
W.H.O. 1967. Epidemiology and control of schistosomiasis. WHO Tech. Rept. Ser. No. 372. Geneva. 35 p.
W.H.O. 1971a. Insecticide resistance--the problem and its solution. WHO Chronicle 25: 215-218.
W.H.O. 1971b. Alternative methods of vector control. WHO Chronicle 25: 230-235.

THE NEO-TECHNOLOGICAL LANDSCAPE DEGRADATION AND ITS ECOLOGICAL RESTORATION

Zev Naveh

Faculty of Agricultural Engineering, Technion - Israel Institute of Technology, Haifa, Israel

INTRODUCTION

As stated by Dr. Kyle R. Barbehenn, a pest situation is one that degrades the quality of the human environment. Such pest situations are created in the industrial society through neo-technological denudation and destruction of the natural landscape with rapidly increasing speed and extent. As these pest situations do not have immediate discernible effects on human health and, as most of their damages cannot be evaluated in market priced cost/benefit analyses, they are largely ignored by those who create them.

The object of this paper is to show that they are a no lesser threat to the quality of our environment and to human survival than air and water pollution and that, therefore, they should be recognized, prevented and controlled without delay.

NATURAL ECOSYSTEMS AND ECOLOGICAL STABILITY

The landscape from the ecological point of view can be defined as our visual perception of the functional and structural integration of the lithosphere, hydrosphere and atmosphere with the biosphere into mosaics of natural and cultural ecosystems and their man-made artifacts.

Natural ecosystems can be considered as the basic, life units of nature, in which plants, animals and micro-organisms are functioning as a whole together with their life space - the physical environment. This is manifested chiefly by the interception of radiation energy through photosynthetic activity of green plants and its conversion into chemical and potential energy, while flowing from the producing plants through trophic food chains of animal herbivores, carnivores and decomposers and by interchange and circulation of material between the biotic and abiotic components of these natural systems. On these two processes, namely energy flow and material cycling, depends the proper function of the biosphere as a whole and our life included.

Open systems, mature and stable natural ecosystems, such as forest shrub-and grasslands, lakes and swamps, maintain a dynamic equilibrium through complex, auto-regulative or "homeostatic" control mechanisms and feedback loops. This steady state is reached in the ecosystems by feeding on negative entropy ("neg-entropy") gaining thereby in order, diversity and storing information.[1,2,3]

The vegetation canopy of terrestrial ecosystems is the main basis for biological productivity and cybernetically controlled stability and "the basic agent for the ordering that is neg-entropy and creation".[4] It is thereby also the main living component of open, natural landscapes and their chief stabilizing agent.

During the long history of man's intervention with the natural landscape, he has modified and destroyed the natural vegetation cover and its underlying soil mantle by burning, cutting, grazing, lumbering, cultivating, flooding, draining, irrigating, drilling, mining, asphalting, constructing, etc. Thereby he has degraded these natural ecosystems into less diverse and stable ones or converted them into more simplified and unstable agricultural and cultural ecosystems. In this way, man not only reduced the natural basis of biological productivity and diversity and interrupted vital biological recycling processes. He has also removed the negative feedback controls, inbuilt in the vegetation. If these are not replaced by his own, man-made cultural and engineering controls, positive, disruptive feedbacks are released. This is especially so since man himself has become more and more subjected to the positive feedbacks of his increasing inputs of energy and information.[5]

In other words: By disturbing and destroying the compact vegetation layer, man has prevented its protective function as a reducer of destructive kinetic energy of moving water, wind, soil, sand and stones, as a reducer of radiation intensity and as an absorber and filter of polluted water, air and dust. Our man-made landscapes have become, therefore, more and more vulnerable to external disturbances, to degrading of order and to environmental injuries and pest situations, spreading like epidemic diseases.

The ruins of once flourishing civilizations which are now surrounded by man-made deserts and bad-lands in the drier and mountainous parts of the Old and New world, bear sad testimony to the final results of such uncontrolled pest situations. They have been caused by unfortunate combinations of vulnerable environments and predatory man-land relations. In these, man has become such an "oversuccessful" parasite that, in killing his host, he has also accomplished his own death.

During our present, neo-technological revolution, we are acquiring almost unlimited powers to raze off natural vegetation canopies and their "living sponge" in order to shape the face of the land according to our desires and needs. Today our bulldozers can move up to 200,000 tons of soil per day, our shovels can raise 185 cubic yards per bite and our trucks can haul 100 tons of soil in one load. Thus, the replacement of stable, healthy and attractive natural landscapes by unstable, unhealthy and ugly cultural ones is proceeding as an exponential function of spiralling demands and expectations of exploding populations and their technological skills.

Only in a few instances has modern man succeeded in creating a higher order through neg-entropy, diversity and information in his new, man-made landscapes. Amongst these cases are, e.g. the polder reclamation in the Netherlands and the arid-land reclamation in Israel. In the over-all balance, however, the scales are weighing heavily against him and his neo-technological landscapes are becoming as desolate and hostile as the man-made deserts left behind by above-mentioned, vanishing civilizations.

CAUSES AND EFFECTS OF TECHNOLOGICAL LANDSCAPE DEGRADATION

Space does not allow me to give a detailed account of the neo-technological impact on the landscape. Many references could be cited, but I would like to mention only some outstanding ones such as the Chicago University symposium on "Man's role in shaping the face of the earth"[5], the more recent Conservation Foundation conference on "The future environments of North America"[7], Dorst's monography "Avant que la nature meurt"[8], and the recent compilation by Detwyler on "Man's impact on environments"[9], Comprehensive charts of modern man's impact on the British countryside presented by Nicholson[10] and a condensed description of habitat degradations in North America by Fossberg.[11]

Most neo-technological landscape degradation processes and creation of environmental pest situations can be summarized under the following headings:

1. <u>Impoverishment of the natural and rural landscape</u> by intensive agri-industrial monocultures and their maintenance by high inputs of fossil fuel energy, chemical fertilizers, herbicides and pesticides.

2. <u>Disturbance and destruction of terrestrial and aquatic ecosystems</u> through large-scale virgin land deforestation and clearing, draining, inundating and irrigation development schemes in tropical and subtropical countries.

3. <u>Large-scale disruption and destruction of natural ecosystems</u> through urban sprawl and industrial expansion and dereliction highway, airport and other engineering constructions, surface mining, mass recreation and environmental pollution and military activities. The dangers of rapidly vanishing open landscapes and their deterioration into urban and industrial wasteland are especially great in densely populated and rapidly developing industrial countries and regions.

Unfortunately we are still very far from being able to express these interactions between man, landscape and environmental stability as differential equations and as multivariate correlation matrices between rate of changes and intensities of environmental, disruptive factors. But, like doctors using a group of symptoms occuring regularly together as a "<u>syndrome</u>" for the diagnosis and and therapy of diseases, we could also use indications of environmental disease and deterioration to define the stage of a pest situation, predict its further course and act accordingly. Such clearly distinguishable environmental syndromes are the <u>indications of soil instability</u>, man-induced, accelerated erosion, following increased run-off after vegetation disturbance and denudation and, if not controlled in time, lead to depletion of soil fertility, flooding, siltation and finally to total landscape dessication and collapse. [12][13]

In Israel, one of the early cradles of settled, agricultural civilizations, Naveh and Dan[14] have recently shown that man-induced soil-vegetation degradation cycles are manifested by erosive and aggradative processes which correspond roughly to major changes of land use patterns throughout the long history of Israel.

In the U.S., as an example of a highly industrialized country, erosion from run-off already amounts to 4 billion tons of sediments each year. Three quarters of these are derived from about 230 million acres of agricultural land which suffered from significant water erosion problems.

The conversion of natural or agricultural ecosystems and their "living sponges" into impervious, asphalt bedded or paved surfaces of cities and roads and into disturbed and denuded slopes and industrial wastelands, has greatly magnified this problem. Thus, after denudation of the vegetation cover and soil exposure for urbanization and road development, erosion has been increased several thousand times.[15][16] In the middle sixties there was in the U.S. already more than one mile of road for every square mile of land and each mile of new freeway occupied 25 acres. About 4 million acres have been disrupted and disturbed by surface mining with an annual increase of 150,000 acres.[17] In West Germany, each year, about 260,000 square km of open space - an area of the size of Munich - are lost and the built-up and degraded area has increased 50% since 1939.[18]

The physical and biological implications of increased sediment loads are manifold and have severe repercussions on the maintenance of stream channels, harbors, irrigation canals, ponds, reserves, fish yields, recreation etc. and on the transportation of absorbed nutrients, toxic and radioactive material and pathogens which impair water quality and endanger both human and aquatic life.

Closely related to the erosion syndrome are other, no less serious results of the agri-industrial environmental degradation, namely the destruction of biotic soil communities[19] and the disruption of nitrogen cycling[20], endangering not only future soil fertility but also the biological productivity of the biosphere.

Also in developing countries, neo-technological landscape degradation, followed by erosion and land destruction, has accelerated the already existing process of environmental deterioration. One of the main causes is the careless use of heavy machinery for wholesale clearing of tropical forests and for ploughing up large, slopy tracts of virgin land composed of friable and erodible tropical soils.[21] In these and in many other cases, recently documented in a special symposium[22], the neo-technological injury to the landscape has acted as an "ecological boomerang" to land and people and nullified much of the benefits of development projects. In many well-intended but ill-fated development projects, engineers and landplanners did not recognize the close interdependence of physical, biotic and sociological factors in these "primitive" ecological systems. Thus, also here, technological innovations, after removal of natural environmental negative feedbacks on land use, without substitution by new, effective socio-economical control mechanisms, resulted in positive, destructive feedbacks of land mismanagement, followed by land degradation. This was the case, for example, in East Africa where bushclearing and control of tsetsefly infected land and the development of bores and wells in water deficient savanna pastures has opened large areas for uncontrolled cultivation and grazing. Thereby, in a short period of misuse, these highly potential savanna ecosystems have been turned into man-made deserts.[23]

Other serious causes for landscape degradation are related to the damaging effects of phytotoxic air pollutants, such as SO_2, HF and oxidants on trees and other plants. Unfortunately the infamous case of total forest and land denudation by sulphur dioxide fumes in the copper basin of Tennessee followed by gully erosion and detrimental climatical changes[24], has been repeated by many others in North America and Europe. New incidents of severe injuries to forests and especially to pine trees over large areas, in addition to heavy losses of agricultural crops, are not only caused by concentrations of industrial complexes but also by single, ill-located industrial plants on mountain sites or valleys and, recently, also by the combined phytotoxic effect of urban sources of smog, especially in Southern California.[25,26,27,28]

Of direct, adverse impact on the quality of urban environments is the decline of trees and shrubbery in cities and along highways, caused by these pollutants and especially by automobile exhaust fumes.

Direct and indirect landscape degrading effects of highways are not restricted only to above-mentioned man-induced erosion, floods and to air pollution, but include also vegetative injury by the use of chemicals for melting of ice, and especially by salting, herbicides and other spray materials, accidental spills of chemicals and litter and debris which carry urban dereliction far across the open country.[15]

A new, serious threat to the most attractive and popular wildlands, parks, reserves and beaches serving as outdoor recreation, is their deterioration by uncontrolled overuse, trampling, despoiling and finally conversion into "outdoor recreation slums".

In addition to the impairment of ecological health and stability, landscape degradation has also other serious biological, economical and cultural implications. These are caused by the following:

1. The loss of natural variety in species and ecotypes and the destruction of their natural habitats as the only effective genepool for further biological evolution. Along with 280 mammals and 350 birds, about 20,000 flowering plants - one out of every ten - are now facing extinction.[29] They will soon no longer be available as a source for domestication and amelioration for human uses, as well as for future "non-economic" riches, not yet realized.

2. The loss of undisturbed natural habitats and well-balanced ecosystems as scientific "outdoor laboratories" and as integrated ecosystem studies. Such studies are vital tools in our future, more rational management of land and resources. With impending loss of these ecosystems around our urban and rural environments, we

might lose the last opportunity of understanding and learning how "nature functions" as a whole and of applying this knowledge to our man-modified and manipulated ecosystems.[30]

3. The loss of unspoiled natural wealth, solitude and beauty in remote wilderness areas as well as in more accessible, open rural landscapes, as an indispensible ingredient of human evolution, education and culture and as a "landscape therapeutic" countermeasure to the stress, anxieties and monotony of modern urban life.

We can best summarize the biological, social and psychic implications of neo-technological landscape degradation by citing the great urbanologist Mumford[31]:

"When we rally to preserve the remaining redwood forests or to protect the whooping crane, we are rallying to preserve ourselves, we are trying to keep in existence the organic variety, the whole span of natural resources, upon which our own further development will be based. If we surrender this variety too easily in one place, we shall lose it everywhere, and we shall find ourselves enclosed in a technological prison, without even the hope that sustains a prisoner in jail - that someday we may get out. Should organic variety disappear, there will be no 'out'."

THE CONTROL OF LANDSCAPE DEGRADATION AS PART OF THE ENVIRONMENTAL REVOLUTION

Neo-technological landscape degradation is part of the deep cultural crisis in modern man's relation with his environment. It cannot, therefore, be remedied only by scientific and technological means, but also requires far reaching changes in prevailing attitudes, values and aims of our society. These can be brought about only by the combined impact of education, public pressure and the fear of impending eco-catastrophies, as part of an over-all "environmental revolution".[10]

It requires the change from our ignorant and arrogant anthropocentric attitude, denouncing the evolutionary continuity of man in the natural world, into a more realistic and humble appreciation of the interdependence of man and his natural surroundings and his responsibilities to the organic world and its resources.

It requires a change from our present age of anxiety, insecurity and stress, after downfall from the illusionary "Paradise of the Impossible" brought about by eighteenth century rationalism and nineteenth century materialism, into a new, realistic confidence in modern man's potentials, as expressed so well by R. Ardley in his recent, challenging inquiry into the evolutionary sources of order and disorder:[32]

"When we renounce our hubris; when we see ourselves as a portion of something far older, far larger than we are; when we discover nature as our partner, not our slave, and laws, applying to us as applying to all: then we shall find our faith returning. We have rational faculties of enormous order. We have powers granted never before to living beings. But we shall free those powers to effect human solutions of justice and permanence only when we renounce our arrogance over nature and accept the philosophy of the possible."

It requires a change from our wasteful "cowboy economy" of maximum consumption and production of material values into a "spaceship earth economy" in which the final limits of our planet and the fearful consequences of our overstepping the maximum allowable technological burden are fully comprehended.[33]

It requires a change from the myoptic economic determinism of our narrow techno-economical subsystem into a more far-sighted ecological determinism of our Total Human Ecosystem as the highest integrative level of man-plus his total environment.[34]

It requires the realization that the dangerous hypertrophy of technology in all spheres of life and the bulldozer mentality of technocrates must be replaced by a new qualitative humanism and "eco-ethic" in which maximum human happiness and human values and the assurance of optimum quality of existence for all by control of quantity through quality, should be made major goals of the society.

The chief practical consequences of this environmental revolution can be summarized as follows:

1. Need for more balanced policies of population control and socio-economical development in densely populated countries with all its difficult and far-reaching political, religious and social implications.

2. Need for effective environmental conservation to avoid further global deterioration and contamination and to ensure the highest possible quality of the environment with all its socio-economical implications on higher costs of production and with highest priorities for development of non-polluting energy sources and on recirculation and re-use of waste materials.

3. Need for far-reaching changes in legislation and law enforcement for planning, zoning and decision making in land use,

for active cooperation of ecologists and ecological expertise in all these phases, strengthened by ecological research, ecosystem analyses and inventories and by education and training of technologists, engineers and specialists in ecological principles and practices.

ECOLOGICAL DETERMINISM IN LAND USE

A pre-requisite for effective control and ecological restoration of landscape degradation is the application of above mentioned ecological determinism in land use, based on comprehensive land development masterplans of our Total Human Ecosystem space. The needs for sustaining maximum natural variety in open and cultural landscapes could be reconciled with all other needs of the post industrial society. Further, disastrous landscape degradation could be prevented chiefly through:

1. Conservation of viable parts of all biological, scenic and scientific significant natural ecosystems as vital, protective environments.

2. Re-creation and restoration of maximum possible natural variety, order and stability in man-modified and disturbed habitats.

3. Optimization of landscape values for recreation and multiple use of open and agricultural landscapes.

4. Minimization of damage to natural variety and natural processes in the course of construction of highways, powerlines, industrial and urban development and utilization of natural resources.

Many promising beginnings in these directions have already been made. Thus, comprehensive approaches to landplanning on ecological basis were first applied in Australia[35] and are now used in many developing countries. The Netherlands were the first densely populated, industrialized country which succeeded in applying comprehensive landscape masterplans in large-scale reclamation and consolidation schemes, with fullest attention to the preservation and improvement of the open landscape.[36] Here also intensive ecological research is being carried out with new, advanced methods based on cybernetic theories of landscape dynamics and its interaction with human activities.[37]

New dynamic concepts of nature conservation are also being applied in Great Britain by the greatest, single body of practicing ecologists in the world, the Government Nature Conservancy, in close cooperation with universities. Here also much effort is being spent on the preservation of the British countryside, on fitting industrial plants, powerstations and motorways to the landscape and - with active cooperation of industry - on rehabilitating industrial derelict land.[38]

In this field of rehabilitation of industrial dereliction and especially surface mined waste land, much has also been achieved in Germany, especially in the Rhein-Ruhr region, and in industrial zones in the United States.[39]

In most countries, however, the application of ecological determinism by new methods of landscape evaluation, planning, zoning and utilization - as developed by McHarg in the U.S. and by other landscape architects and planners in North America and Europe, and especially in Germany[40] - is still very limited in scope and extent. Many of these landscape development schemes are handicapped by insufficient legislative and executive powers and by vested interests. However, at the same time, a new generation of ecological minded and trained land-planners and managers are now being educated all over the world and this, in itself, may become an important stimulus in further more enlightened land planning and development. Let us also hope that the efforts of international agencies, and especially those of IUCN (International Union for Conservation of Nature and Natural Resources) and the United Nations UNESCO and FAO will bear fruit and that such new, ambitious programs as "Man and the Biosphere" will have a real impact on world-wide land use and planning, as well as on other fields of environmental management.

NEW METHODS OF VEGETATION ENGINEERING AND "ECO-TECHNICS" AND THEIR IMPLEMENTATION IN ISRAEL

The development of ecological methods or "ecotechnics" of vegetation management and engineering is of greatest importance for the minimizing of technological injury to the landscape. The natural successional trends, after destruction of original vegetation cover, are utilized or simulated in the creation of stable and diverse plant communities in disturbed or destroyed habitats.[11] An important landmark in this respect is the ecological vegetation management of right of ways along highways, railroads, electric power transmissions and pipelines, developed by one of the outstanding American ecologists, F. Egler.[41]

Thanks to the development of new vegetation engineering methods, great advances have been achieved in the biological stabilization and revegation of denuded areas, cuts and fills of highway embankments and even of unaccessible, steep mountain slopes and ridges. Mixtures of suitable plants are sprayed together with synthetic and organic mulching materials and fertilizers by hydro-seeding machines and, on steeper slopes, this is combined with mechanical devices for slope stabilization and erosion control. Most of these methods have been developed and successfully applied in temperate climates with favorable growth conditions. In semi-arid countries, like Israel, where long dry and hot summer droughts alternate with torrential winter rains, the danger of land denudation and erosion are much greater and the natural process of revegetation is much slower. Therefore also ecological restoration is much more difficult. Its success depends on the selection of drought resistant and hardy plants and on suitable methods which will enable them to establish and maintain themselves in harsh conditions with minimum care and irrigation, while fulfilling their ecological, engineering and aesthetic functions.[42]

This has already been accomplished by the biological stabilization of moving sand around buildings and along highways in the densely populated Coastal Region[43] and by the biological protection and landscape reclamation of waterways and drainage canals in the Jordan Valley and in other intensive agricultural regions.[44] Special mechanical methods for slope preparation and for hydroseeding and stabilization of steep slopes have been developed[45,46] and after several years of selection of drought resistant grasses, legumes, shrubs and trees, these methods and plants are used in pilot rehabilitation landscape projects. In these a dynamic ecosystem approach is applied, differing from the European approach of phytosociological reconstruction of the potential natural vegetation with exclusion of exotic species, as well as from the American, pragmatic, horticultural approach of choosing pasture grasses and garden plants without reference to their surroundings. Its aim is to establish in these man-made habitats semi-natural, multistructured, diverse and stable communities of both indigenous and exotic plants, and to simulate a condensed natural succession by using fast growing "pioneer" grasses, legumes and dwarfshrubs together with slower growing but more persistant and taller shrubs and trees.[47] One of these first projects is being carried out in an abandoned limestone quarry near Haifa, as a joint venture between an artist and landscaper, a specialist in erosion control and vegetation engineering, and a plant specialist and ecologist. It is sponsored by the cement factory which used this quarry and it has already yielded early promising results.[48] It might well be that in such cooperative efforts between the creativeness and inspiration of artists, the technological skill and ingenuity of engineers, the scientific knowledge and experience of ecologists and the goodwill of industrialists, lies our hope for more stable, healthy and attractive environments of the post-industrial society.

BIBLIOGRAPHY

1. Odum E.P. 1971. Fundamentals of ecology, 3rd Ed., Saunders Co., Philadelphia.
2. Ashby W.R. 1956. An introduction to cybernetics, Chapman & Hall, London.
3. Cantlon J.E. 1969. The stability of natural populations and their sensitivity to technology, Brookhaven Symp. in Biology 22, Diversity and stability in ecological systems: 197-205.
4. McHarg J.L. 1971. Design with nature, Doubleday & Co., New York (paperback), p.53.
5. Odum H.T. 1971. Environment, power and society, Wiley & Son, New York.
6. Thomas W.L. Edit., 1956. Man's role in changing the face of the earth, University of Chicago Press, Chicago.
7. Darling F.F. & Milton J.P. Edit., 1966. Future environments of North America, The Natural History Press, New York.
8. Dorst J. 1965. Avant que la nature meurt, Delachaux et Niestle, Neuchatel.
9. Detwyler T.R. Edit., 1971. Man's impact on environment, McGraw Hill, New York.
10. Nicholson M. 1970. The environmental revolution, Hodder & Stroughton, London.
11. Fosberg F.R. 1966. Restoration of lost and degraded habitats, In: Future environments of North American: 503-515.
12. Riney T. 1963. A rapid field technique and its application in describing conservation status and trends in semi-pastoral areas, African Soils VIII/2.
13. McHarg J.L. 1966. Ecological determinism, In: Future environments of North America: 526-538.
14. Naveh Z. & Dan J. 1971. The human degradation of mediterranean landscapes in Israel, Intern. Symp. on Mediterranean Ecosystems, Valdivia, Chile 1971 (in press).
15. Scheidt E. 1971. Environmental effects of highways, In: Man's impact on the environment: 419-427.
16. Leopold L.B. 1971. The hydrological effect of urban land use, In: Man's impact on the environment: 205-216.
17. U.S. Dept. of Interior. 1967. Surface mining and our environment, A special report to the nation.
18. Buchwald K. 1970. Die Industriegesellschaft und die Landschaft, Naturschutz und Landschaft, 8:23-41.
19. Albrecht W.A. 1956. Physical, chemical and biochemical changes in the soil community, In: Man's role in changing the face of the earth: 648-676.
20. Commoner B. 1968. Threats to the integrity of the nitrogen cycle: Nitrogen compounds in soil, water, atmosphere and precipitation, In: S.F. Singer, Ed., Global effects of environmental pollution, Symp. Dallas, Texas, 1968: 70-94.
21. Philips J. 1961. Agriculture & Ecology, London.

22. Naveh Z. 1970. Ecology and development. 2nd World Congress of Engineers & Architects: Dialogue in Development. Tel Aviv, 1970.
23. Hursch C.R. 1948. Local climates in the Copper Basin of Tennessee as modified by the removal of vegetation, U.S.D.A. Cir. 774.
24. Treshow M. 1970. Environment and plant response. McGraw Hill Co., New York.
25. Hepting G.H. 1964. Damage to forests from air pollution. J. of Forestry, 62:630-634.
26. First European Congress on the influence of air pollution on plants and animals. 1969. Proceedings, Wageningen, 1969.
27. Berge H. 1970. Immisionsschaeden. In: die nichtparasitaeren Krankheiten, Edit. B. Rademacher, Paul Parey, Berlin.
28. Scurfield G. 1960. Air pollution and tree growth, Forestry Abst. 21:1-20.
29. Tinker J. 1971. One flower in ten faces extinction. New Scientist and Science Journal, 50:408-414.
30. Schultz A.M. 1967. The ecosystem as a conceptual tool in the management of natural resources. In: Natural Resources - Quality and Quantity, Ed. Ciriacy-Wantrup:139-161.
31. Mumford L. 1966. Closing statement in: The future environments of North America,:727.
32. Ardley R. 1970. The social contract,: Collins, London.
33. Boulding K.E. 1966. The economics of the coming spaceship earth, in: Environmental quality in a growing economy, Ed. H. Jarett, John Hopkins Press.
34. Egler F. 1971. The way of science, a philosophy of ecology for the layman, Hafner Publ., New York.
35. Christian C.S. 1958. The concept of land units and land systems, Proc. 9th Pacific Sci. Congr., Bankok, 1957, 20:74-80.
36. Benthem R.J. 1966. Erholung im laendlichen Raum, Der Landschaftsaufbau in den Niederlanden, Garten und Landschaft, 8:124-135.
37. Westhoff V. 1971. The dynamic structure of plant communities in relation to the objectives of conservation, Proc. 11th symposium Brit. Ecol. Soc, July 1970:3-14, Blackwell Scientific Publisher, Oxford.
38. Stamp D. 1969. Nature conservation in Britain, Collins, London.
39. Knabe W. 1965. Observations on worldwide efforts to reclaim industrial waste land. In 5th Symp. Brit. Ecol. Soc., 1964: 263-296, Blackwell Scientific Publ. Oxford.
40. Krysmanski R. 1970. Die Nuetzlichkeit der Landschaft, Bertelsmann Universitaetsverlag, Duesseldorf.
41. Egler F. 1968. Vegetation management (right of way) McGraw Hill Encyclopedia of Science & Technology, 14:287-288a, McGraw Hill, New York.

42. Naveh Z. 1969. Adaptation of drought resistant plants to vegetation engineering in Israel. 2nd Symp. on sedimentation problems,:166-177, Hebrew University, Jerusalem.
43. Tzuriel A.D. 1969. Sand dune stabilization research in Israel, Symp. on vegetation engineering, Tel Aviv, 1968 (Hebrew).
44. Naveh Z. & Elan S. 1970. Biological protection of waterways and drainage canals in mediterranean and semi-arid conditions, in: Hydrology of Drainage Canals, A. Kinori: 180-207, Elsevier, Amsterdam.
45. Kalvariski J. 1969. Preparation of slopes for planting and seeding, Symp. on vegetation engineering, Tel Aviv, 1968: 22-23 (Hebrew).
46. Morin J. 1969. Mechanized biological stabilization of slopes, Symp. on vegetation engineering, Tel Aviv, 1968:19-21 (Hebrew).
47. Naveh Z. 1970. A dynamic ecosystem approach to vegetation engineering in Israel, Proc. XVII, Intern. Hort. Congr. Tel Aviv, 1970:94.
48. Naveh Z., Morin J. & Danziger J. 1972. Progress report on the "Nesher" limestone quarry reclamation project, Technion, Haifa (Hebrew).

SOME ASPECTS OF THE ROLE OF ENGINEERING IN THE COMPETITION BETWEEN INSECTS AND MAN

D. SCHRÖDER

European Station, Commonwealth Institute of Biological Control, Delemont, Switzerland

I. HISTORICAL AND ECOLOGICAL BACKGROUND

For about two million years man lived as a hunter and gatherer, and his numbers have been regulated by negative feedback mechanisms, in much the same way as the numbers of other predators and herbivores: Man was a component species integrated into the ecosystem.
- With the domestication of plants and animals some 10,000 years ago, and the improvement of agricultural and other techniques thereafter, the position of man changed; he became the dominant species and altered with increasing efficiency the ecosystems according to his own needs and wellbeing. This development was accompanied by an initially slowly, but later rapidly increasing human population. It took an estimated two million years to reach the one-billion mark, but at present within only 15 years (from 1960 to 1975) one billion will be added to man's population: This rise in man's population is generating an enormous increase in the demand for food and other resources, and causes an increasing alteration of the biosphere.

Right from the beginning of an agricultural practice man has selected only a small number of plants and animals for domestication. Among the first crop plants we find wheat, barley, rice and maize (corn), and today two-thirds of the cultivated land is planted to cereals. In time, man has extended the area under cultivation to some three billion acres, about ten per cent of the earth's total land surface, and a considerably larger fraction of the land capable of supporting vegetation (Brown, 1971).

From an ecological view-point, cultivation is a process of degradation: Ecosystems which evolved over long periods and reached a certain level of self-organization and stability (e.g. prairies and forests) are simplified, both in their organization and in their species diversity. Under cultivation the flow of energy is amplified, and the increase in the ratio of primary production to biomass is accompanied by a decrease in the stability of the ecosystem, the yearly rhythms become more stressed (Margalef, 1968).

II. THE CREATION OF PEST SITUATIONS THROUGH AGRICULTURAL ENGINEERING

Although insects have always played an important role in the life of man as competitors for food and carriers of diseases, this competition became more and more prominent in the past two centuries, and engineering has directly or indirectly been involved in this development. Insect pest situations of great economic importance have been, and still are, created and / or aggravated mainly by three different processes:

1) <u>The mechanization and the general intensification of agriculture, with large areas being planted to a single crop plant only (monocultures)</u>. - In comparison with adjoining uncultivated land of comparable ecological characteristics, the insect fauna of agro-ecosystems is generally less diversified, both in quality and quantity. Depending on the level of organization of the pre-agricultural ecosystem a smaller or larger proportion of the original insect fauna is eliminated if subjected to cultivation. The number of species eliminated is directly proportional to the diversity of the pre-agricultural ecosystem.

Cultivation favours those insect species that have the ability to adapt to crop situations which allow the availability of an unnaturally high supply of food at certain periods of the year, and at varying sites within the cultivated area. Therefore pest species must be able to synchronize their life cycle with that of the crop plants; they have to be able to quickly increase in numbers, to disperse easily and to reconstruct their populations quickly when subjected to heavy losses (Margalef, 1968). In agro-ecosystems species with these opportunistic characteristics are favoured by a lasting increase in the supply of a limiting resource (the food plant), and at the same time they profit from the drop in species diversity which reduces the frequency of interactions and checks that previously prevented them from exploiting fully the resources of the ecosystem, and of building up beyond a certain population level. Outbreaks are characteristic of systems with low species diversity (Pimentel, 1961). Fortunately, only a relatively small fraction of the phytophagous insects has become pests of crop plants, and a still smaller number has apparently a potential to produce major outbreaks. However, the economic losses caused by the existing pest insects are considerable, and can locally reach catastrophic proportions.

2) The entry of insect species into previously uncolonized regions. - The development and expansion in world transportation and the increasing volume in world trade have resulted in the accidental introduction of insect species (pests as well as indifferent species) into regions where colonization had formerly been prevented by natural barriers, like oceans, deserts, mountains, etc. A great number of these introduced insect species developed to pest status in the area of introduction mainly for the following reasons: a) they were translocated without their natural enemies and diseases; b) they met with ecological conditions more favourable for a population build-up; and c) they came into contact with new host plants which do not exist in their area of natural distribution.

During the colonization of the Americas, of Australia, New Zealand, and other parts of the world, man has willingly or unconsciously assisted in the colonization of new regions by many insect species. For example, about one-fourth of the North American insect pest species have been accidentally introduced (Lindroth, 1957), including a number of the major insect pests in agriculture, like the European corn borer (Ostrinia nubulalis), the Oriental fruit moth (Grapholita molesta), the Hessian fly (Mayetiola destructor), the Japanese beetle (Popillia japonica), and the European wheat stem sawfly (Cephus pygmaeus), to mention some well-known examples.

Nowadays, with the modern transportation systems and the much shorter transit times favouring the survival of insects in transit, it becomes more difficult to prevent the undesirable introduction of insects. The Plant Quarantine Division of the U.S. Department of Agriculture reports 7,800 species of insects as intercepted between July 1, 1968 and June 30, 1969 (Girard, 1971).

3) The introduction and cultivation of crop plants outside their area of natural distribution. - During the period of discovery and exploration of new continents and the subsequent colonization, many crop plants and ornamentals, especially those of tropical and subtropical zones, have been introduced into new regions. Here the introduced plants were exposed to a variety of indigenous phytophagous insects and diseases, some of which adopted the introduced plants (unoccupied niches) as new host plants and became serious pests before an ecological homeostasis could develop. This has often been observed between phytophagous insects and their host plants in areas where they occurred together over long periods. - Sugar cane and cocoa may serve as examples for the adoption of

introduced crop plants by indigenous insects and diseases, which in time became serious pests.

Commercially cultivated sugar canes are varieties of the horticultural species Saccharum officinarum, or are hybrids derived predominantly from them. They seem to have originated in Melanesia. After introduction into new regions a number of insect species adopted the sugar cane as a host. This is demonstrated (a) by the regional character of sugar cane pests, e.g. the species of Diatraea (s.l.) are confined to the New World and 21 species are cane pests; (b) by the host-plant relationships of the pests, e.g. African species of Sesamia maintain populations on a variety of native grasses; and (c) by the presumed origin of the cultivated sugar canes (Pemberton & Williams, 1969). The process of adopting indigenous insects continues as is demonstrated by Numicia viridis (Tropiduchidae) in Natal (Dick, 1963), Yanga guttulata (Cicadidae) in Madagascar (Breniere, 1965) and Aeneolamia postica (Cercopidae) in Mexico (Iturbe & Ruano, 1963), which only recently became cane pests.

In 1879 cocoa was first introduced from South America into the mainland of West Africa, which became a major area of cocoa production some 20 years later. The early success of cocoa in West Africa would suggest that it took some time before local insects and diseases started to attack it. During the last 40 years, however, the cocoa production suffers heavy losses mainly from the attack of two polyphagous mirids (Sahlbergiella singularis and Distantiella theobroma), the lesions of which become infested by fungi (mostly Calonectria rigidiuscula), and a polyphagous mealybug (Planococcus njalensis), a vector of a virus capable of infecting cocoa (Swollen shoot disease - SSVD)(Clark et al., 1967). Five elements, three insect species, a fungus and a virus, formerly all independent and of no economic importance developed into a highly destructive complex of organisms through adaptation to the introduced cocoa.

III. THE CONTROL OF INSECT PEST THROUGH ENGINEERING

The occurrence of insect pests, and their increasing importance as competitors, forces man to react; but he finds himself in a dilemma: his policy of increasing the primary production is bound to create new, and to aggravate existing, insect pest situations. In other words, man helps the insect competitors to multiply, and afterwards he strives to reduce their numbers. During centuries of insect pest control man has developed and practiced a great number of control methods. From all the known methods, the use of insecticides, the so-called chemical control, has gained great economic importance, and is actually the most commonly applied method.

The modern use of insecticides commenced about a 100 years ago. In 1867, outbreaks of the Colorado potato beetle in the United States were controlled by the use of Paris green, containing arsenic, As_2O_3, and cupric oxide, CuO. During the following decades a number of additional inorganic compounds were applied as insecticides, like calcium arsenate and lead arsenate, but all the arsenates were later replaced by other types of insecticides mainly because of their high mammalian toxicity and their persistent residues.

A new area of insecticidal control began in Europe in 1939 with the discovery of the insecticidal value of DDT, a synthetic organic compound. The extremely successful use of DDT in all fields of insect control has initiated much research and resulted in the production and testing of thousands of synthetic organic chemicals. The majority of the currently used insecticides belong either to the group of chlorinated hydrocarbons, or to the group of the organophosphorus esters, the largest and most versatile group of pesticides in use at the present time. In the United States approximately 400 basic insecticides are presently registered, of which 90 per cent are synthetic organic compounds.

The use of these modern insecticides over the past 30 years has brought relief from insect competition in many pest situations and helped to increase crop-yields. Insecticides have proved to be highly effective, they can be used to rapidly control large insect populations with almost immediate effect, and they can be employed as needed. - However, experience has revealed a number of disadvantages the importance of which becomes more and more evident.
a) Pest insects develop resistance if they are exposed for long periods to the selective pressure of insecticides (the first case of resistance to DDT (Musca domestica) was discovered in 1946 in Sweden, and by 1968 some 230 species of insects and mites in various parts of the world have developed resistance to one or more groups of insecticides (FAO-Report, 1969)); b) broad spectrum pesticides are disrupting ecosystems with adverse effects on beneficial insects; c) residues remain on the treated product or area and persistent pesticides accumulate in food chains.

These and other intolerable effects of the long term use of certain groups of pesticides on a variety of ecosystems, and the biosphere as a whole, are causing great concern. It has been realized that the indiscriminate use of these chemicals has to be stopped, and the use of persistent insecticides has to be limited, or better completely avoided. The use of DDT and some other persistent products has already been banned in some countries, but large volumes are still being used, especially in less developed areas.

There is no doubt that insecticides will continue to be of great importance in insect pest control, but new types of pesticides are needed and these will have to be used more cautiously in order to prevent catastrophic consequences in the biosphere. Research is under way to produce selective or narrow-spectrum insecticides with rapid degradability, but their development is handicapped by factors involving toxicological and economic principles. Nevertheless, a solution will have to be found in order to solve these and other problems.

A significant proportion of the ill-effects of insecticides on ecosystems was, and still is, due to their indiscriminate routine application. The decision to use insecticides should (ideally) depend on an overall assessment of each pest situation. Information is necessary on the life history, the behavior and the population dynamics of the target species, in order to employ control measures with optimal effectiveness. If the decision to employ insecticides has been taken, preferably those products should be used that will cause the least disruption of the ecosystem. They should be applied against the most vulnerable life stages of the pest, and under the direction of employees trained in their proper application.

In order to reduce the extensive use of insecticides, growing attention is being paid to those methods of insect pest control which reduce the danger of increasing the instability of agro-ecosystems; which prevent the development of resistance; and which do not affect the health of man and life stock. Such control methods make use of biological, technical, cultural and even chemical means, but their effect is either exclusively on the pest species, or at least more specific than the presently applied insecticidical control. One type of these control methods rather tries to prevent pest situations, like cultural control and the selection of resistant hosts, whilst the other type is used to control pest situations, like microbial, autocidal and biotechnical controls, but both types may be used in either way.

The principle of cultural control is to make the environment less favourable for the pest insect. It is usually based on modifications in time and manner of performing operations in the production of a crop (U.S. Acad., 1969). One of the most widely used kinds of cultural control are crop rotations against insects with a restricted host range and a limited power of migration. The destruction of crop remains after harvesting, tillage to destroy susceptible stages of pest insects in the soil, the destruction of alternate hosts, etc., are further possibilities. Cultural control is a cheap control method which does not contaminate the environment, but it requires a good knowledge of the life history and habits of the pest, and timing is often critical.

The selecting and breeding of resistant plants, which is based on non-preference, anti-biosis, or tolerance of the plant to attack, is a method with great potentialities. The major advantage of the use of insect-resistant plants in insect pest control is that the effect on the pest is specific, cumulative, and persistent. The use of resistant plant varieties is a safe and low cost control, which makes it especially useful in less developed areas. The main limitations are the considerable time which is required to select resistant plants, and the difficulty to combine resistance with other desirable agricultural characteristics (de Wilde & Schoonhoven, 1969).

The use of large numbers of sterilized individuals of a pest insect to locally eradicate or control its population (autocidal control) is receiving much attention, after the successful eradication of the screw-worm fly, Cochliomyia hominivorax, from the island of Curacao and from the south-eastern portion of its range in the United States. This impressive success needed only eight weeks, but it was a result of 20 years of research, and of tremendous technical planning and organization. Large numbers of screw-worm puparia had been produced, and sterilized by gamma radiation, and were then released into the natural population (Baumhover, et al., 1955). - Since then several other pest insects have been or are being controlled, by what is now commonly named the "sterile-male technique", and research is being continued to improve this technique (use of cobalt-60 and chemosterilants) and the possibilities of its application in practical pest control. - Autocidal control has great potentialities when technical and financial difficulties can be met, and after a comprehensive knowledge of the life history, the habits and the population dynamics of the target insect species has been accumulated.

There are a number of biotechnical control methods which are being investigated, some of which may become important means of pest control, according to results so far obtained in small or large scale experiments. The use of chemical attractants in combination with insecticides in traps and baits, or with chemosterilants on baits, can be highly effective, as was demonstrated by the eradication of the Oriental fruit fly, Dacus dorsalis, from the Pacific island of Rota. - The potentialities of other biotechnical control methods, like the use of antimetabolites, feeding deterrents, and especially of juvenile hormone mimics (sesquiterpenses, farsenol, farsenic acid, etc.), are still under investigation.

Besides such biotechnical control methods that employ technical and chemical means, there is a further group of control methods which make use of biotic antagonists of the pest insects, like microorganisms (microbial control) or parasites and predators (biological

control). Microbial control of pest insects certainly has a great
potential, although most of the successes so far have been of
technological rather than operational nature. Only certain Bacteria,
e.g. Bacillus thuringensis, can be readily mass-produced, and
applied just like insecticidal sprays. Before the potential of
microbial control can be fully realized the problems of virulence,
high cost, safety and licensing will have to be solved. - Biological
control, the usefulness of which has often been disputed in spite
of a number of successes of great economic value (DeBach, 1965),
would seem to be the "ideal" method of control from the viewpoint of
an ecosystems approach of control, but the possibilities of
application in an intensified agriculture are admittedly restricted.

IV. DISCUSSION

Man has largely changed his environment, and today agro-ecosystems are extending over a large part of the land surface. The
increase in primary production has resulted in a decreasing
stability of the exploited ecosystems, and greatly assisted the
insect competitors to increase in numbers and to invade previously
unoccupied areas. Insect pest problems became more pressing, and
man had somehow to control them. But what is commonly termed
"control" of insect pests is in fact the temporary elimination or
the reduction in numbers of pest insects in locally defined areas.

Control workers so far have reduced the community-centered
problem of pest appreciation to the conceptional level of single-
species populations (Clark, et al., 1967). This unacceptable concept,
in combination with powerful broad-spectrum insecticides, had
serious consequences for the species diversity and stability of the
treated ecosystems, and often aggravated insect pest situations, or
even created new ones. Instead of keeping the pest insects under
control, a number of them developed resistance to certain groups of
insecticides, and the occurrence of outbreaks was generally
accelerated. This forced man to use greater volumes of more effective
pesticides at shorter intervals, and at increasing costs, but with
decreasing control effect and increasing disadvantages to the treated
ecosystems, the health of man, and the biosphere as a whole.

The negative consequences of the concept of single-species
populations and the use of broad-spectrum insecticides are now more
and more being understood. Research is under way to reach a better
appreciation of pest situations and to integrate insect pest control
into a systems-oriented management of the exploited ecosystems. It
will, however, take a long period of time before this higher quality
of the use of the supporting resourses of man will be acquired, if
it ever will be. In the meantime one has to try to integrate all
available control methods in such a way as to reduce the negative

effects of insect pest control. This requires the co-operation of ecologists, entomologists, control workers, and last not least, of the engineers. How fruitful their co-operation can be has been demonstrated in several successful large scale control programs, an example of which has been the eradication of the screw-worm fly. - The new concept of insect pest control has to orientate towards the regulation of pest situations. Man has to try to increase the stability of ecosystems and he has to exploit potentially favourable ecological relationships.

"The principle prerequisite, however, for the effective investigation and management of ecosystems is man's full appreciation of the fact that, being the dominating and only truly creative species, he must learn much more in the broadest possible way about the consequences of his unceasing but poorly-integrated efforts to improve his environment, and about how best to adjust his activities and population numbers for his own well-being. In other words, if man - a component species - is to advance far in the management of ecosystems, he must acquire more knowledge and understanding of how to manage his own kind." (Clark, et al., 1967).

REFERENCES

Baumhover, A.H., Graham, A.J., Bitter, B.A., Hopkins, D.E., New, W.D., Dudley, F.H. & Bushland, R.C., 1955: Screw-worm control through release of sterilized flies. - J. Econ. Entomol., 48: 462 - 466.

Brenière, J., 1965: Entomologie. Institut de Recherches Agronomiques à Madagascar. Document No. 38, Résultats d'activités 1964, Canne à Sucre: 22 - 25.

Brown, L.R., 1970: Human food production as a process in the biosphere. - Scientific American, 223: 161 - 170.

Clark, L.R., Geier, P.W., Hughes, R.D. & Morris, R.F., 1967: The ecology of insect populations. - Methuen & Co Ltd. London: 232 pp.

DeBach, P. (ed.), 1965: Biological control of insect pests and weeds. - 2nd. ed. Chapman and Hall Ltd. London: 844 pp.

Dick, J., 1963: Rep. Exp.Stn. S. Afr. Sug. Ass. 1962-1963: 55 - 57.

FAO - Report, 1969: Report of the fourth session of the FAO working party of experts on resistance of pests to pesticides held in Rome, Italy, 19 - 26 August 1968. - Meeting Rep., FAO (1969) No. PL: 1968/M/10: 45 pp.

Girard, D.H., 1971: List of intercepted plant pests. - 1969
U.S. Dept. Agric., Agric. Res. Service, ARS 82-6-4:
77 pp.

Iturbe, A.C. & Ruano, M.A., 1963: The sugar cane froghopper and its control in Mexico. - Proc. int. Soc. Sug. Cane Technol., $\underline{11}$: 650 - 657.

Lindroth, C.H., 1957: The faunal connections between Europe and North America. - Almquist & Wiksell., Stockholm: 344 pp.

Margalef, R., 1969: Perspectives in ecological theory. - The University Press, Chicago - London, 2nd ed.: 111 pp.

Pemberton, C.E. & Williams, J.R., 1969: Distribution, origins and spread of sugar cane insect pests. - In: Pests of sugar cane (J.R. Williams, et al., editors). Elsevier Publ. Comp. Amsterdam - London - New York: 1 - 9.

Pimentel, D., 1961: Species diversity and insect population outbreaks. - Ann. Ent. Soc. Am., $\underline{54}$: 76 - 86.

U.S. National Academic of Science, 1969: Insect-pest management and control. - Principles of plant and animal pest control, volume $\underline{3}$. - Publ. 1695, National Academy of Science: 508 pp.

de Wilde, J. & Schoonhoven, L.M., (ed.), 1969: Insect and host plant. - Proceedings of the International Symposium "Insect and Host Plant", Wageningen, the Netherlands, 2 - 5 June, 1969. North Holland Publ. Comp. Amsterdam - London: 810 pp.

SULFUR DIOXIDE REMOVAL FROM GASES
U.S.: LIME-LIMESTONE

I. A. Raben
Bechtel Corporation, San Francisco, California

INTRODUCTION

Opening Remarks

It is an honor and pleasure to speak to The Society of Engineering Science at its First International Meeting on Pollution: Engineering and Scientific Solutions. The time has come when we can no longer take our environment for granted. Clean air and pure water are important to the quality of our lives.

Among the pollutants identified, sulfur dioxide is a major contributor. My paper will present a U. S. status report on lime-limestone scrubbing for the removal of sulfur dioxide from gases.

I will discuss the chemistry involved, the main design problems, the status of pilot plant work, and the status of full-size installations.

In the United States, the most advanced SO_2 removal process is lime-limestone scrubbing. This is a nonregenerative process, producing a throwaway product, calcium sulfate. This product is essentially insoluble and, therefore, does not contribute to water pollution. When properly dewatered, it can be used as landfill.

CHEMISTRY OF LIME-LIMESTONE SCRUBBING OF SO_2

The chemistry of lime-limestone scrubbing is complicated because of the large number of species present in the system at equilibrium.

The three raw materials - gas, fly ash, and limestone - each contain several constituents that affect the chemical composition of the system.

For the power plant, the gas supplies SO_2, SO_3, CO_2, NO and NO_2; the ash contributes Na, K, Ca, Cl, Fe, Si, and others. Limestone gives Ca, Mg, and other constituents in minor proportions - Na and K.

The main reactions in the scrubbers are assumed to be: (1) absorption of SO_2; (2) hydrolysis to form H_2SO_3 acid; and (3) reaction of sulfite ion from H_2SO_3 with calcium ion from $CaCO_3$ or $Ca(OH)_2$.

These reactions are affected in several ways by other constituents in the system. Detailed studies of the system chemistry have been carried out by TVA (Ref. 1) and the Radian Corp. (Ref. 2). TVA studied the effect of supersaturation, ionic strength, and sulfite oxidation. Radian developed a computer program using 41 species and 28 equations to predict equilibrium compositions in the scrubber circuit.

The main equations in the scrubber can be written:

(1) $SO_2 + H_2O \rightarrow H_2SO_3$

(2) $H_2SO_3 \rightarrow HSO_3^- + H^+$

(3) $HSO_3^- \rightarrow H^+ + SO_3^=$

(4) $CaCO_3 \rightarrow Ca^{++} + CO_3^= + H^+ \rightarrow CaHCO_3^+ \rightarrow Ca^{++} + HCO_3^-$

(5) $Ca^{++} + SO_3 \rightarrow CaSO_3$

(6) $CaSO_3 + \frac{1}{2}O_2 \rightarrow CaSO_4$

If the system is assumed to be one of sulfurous acid formation followed by the reaction of acid with lime or limestone, then the following effects may influence the overall kinetic rate: (1) diffusion of SO_2 to and through the gas film at the liquid surface; (2) dissolution of SO_2; (3) hydration of SO_2 to H_2SO_3, H^+ and HSO_3^-; (4) dissociation of HSO_3 to form $SO_3^=$; (5) diffusion of H_2SO_3 and ions through the liquid film at the droplet surface and into the droplet interior; (6) hydration of CaO to $Ca(OH)_2$ when CaO is used; (7) dissolution of $Ca(OH)_2$ or $CaCO_3$; (8) reaction of $Ca(OH)_2$ or $CaCO_3$ with H^+ to give Ca^{++}; and (9) reaction of Ca^{++} with $SO_3^=$ to precipitate $CaSO_3$.

Available data indicate reactions in steps (3), (4), (8) and (9) are rapid. The controlling mechanisms are therefore either gas diffusion, liquid diffusion, CaO hydration, or dissolution of $CaCO_3$ or $Ca(OH)_2$. For the most used design case, introduction of $CaCO_3$ into the scrubber, gas phase mass transfer of SO_2, and $CaCO_3$ dissolution are the controlling steps. This case was studied by Boll (Ref. 3) in a three-stage scrubber and he found that these two steps were most critical.

Further studies are required to better define the chemistry and kinetics of this limestone scrubbing system. The EPA Test Facility test program will evaluate and study this subject.

Process System Design

The development of the lime-limestone scrubbing system has taken three process routes: (1) introduction of limestone into the scrubber circuit (Figure 1); (2) introduction of limestone into the boiler to produce CaO, followed by scrubbing of the flue gas; and (3) introduction of lime into the scrubber circuit.

The most used scheme is process route (1), introduction of limestone into the scrubber. This approach has the advantage of minimum effect on the power plant. This process approach can achieve high SO_2 removal and minimize scaling and plugging in a reliable fashion.

The disadvantage of this process is that limestone is less reactive than lime. To offset this limitation, a higher stoichiometric ratio of limestone to SO_2 must be used, more slurry must be recirculated (higher L/G), and a countercurrent scrubber with several stages is required.

The second process approach of introducing limestone into the boiler furnace produces a calcined limestone. The calcium oxide (CaO) produced enters the scrubber with the flue gas. This process has presented problems of the boiler fouling, overflowing, and inactivating the lime; and of increased scaling in the scrubber at the dry-wet interface. This approach is not applicable to sulfuric acid plants or to refineries or smelters.

Scrubbing efficiency can be increased in the third process by using lime in the scrubber circuit. This process has two disadvantages - the higher cost of lime over limestone, and increased scaling problems.

With this review of chemistry and kinetics of the lime-

limestone scrubber process and the basic process descriptions for lime-limestone scrubbing for SO_2 removal, the principal design considerations and problems will now be discussed (Ref. 4 and 5).

In designing a scrubber system, the following main design criteria must be considered: (1) scrubber type and size, system turndown; (2) gas reheat, mist eliminators; (3) instrumentation, system pressure drop; (4) scale control, corrosion, materials of construction; (5) waste disposal; (6) ash and sulfur content of fuel; and (7) percent SO_2 removal required.

In addition, the scrubber design involves additional variables, such as:

o   Liquid to gas ratio, pH control, number of stages

o   Percent solids in slurry, hold tank residence time

o   Stoichiometric ratio of limestone to SO_2, limestone particle size.

The major criteria for scrubber selection is its capability to remove both sulfur dioxide and particulates with high efficiency (SO_2 removal greater than 85 percent and particulate removal greater than 99 percent). Other factors considered are ability to handle slurries without plugging, cost, control and pressure drop.

The scrubber types that have been tested to date include: venturi, turbulent contact absorber (TCA), marble bed absorber, spray column, tray column, cross flow absorber and screen or grid absorber.

The venturi scrubber has been used when both particulate (fly ash) and sulfur dioxide must be removed. The venturi has good capability to remove fly ash down to 0.02 gr/SCF with pressure drops of 10 to 15 in. H_2O and liquid-to-gas ratios of 10 to 15 gpm per 1000 cu. ft. gas for typical dust loadings and particle size distributions from power plant stack gases. The venturi contains an adjustable throat area which permits control of pressure drop over a wide range of flow conditions. The venturi scrubber is limited in SO_2 removal to 40 to 50 percent with lime-limestone due to the short liquid residence time. It therefore must be used with an after-absorber to achieve 85 to 95 percent SO_2 removal.

The turbulent contact absorber (TCA) is a countercurrent multi-stage scrubber consisting of screens which both support and restrain the plastic spheres. The spheres move in a turbulent fashion providing good gas-liquid contact and scale removal. The number of stages generally runs between two and four for high SO_2

removal. The pressure drop per stage is approximately 2 to 2.5 in. H_2O.

The marble-bed absorber utilizes a 4 in. bed of packing of glass spheres (marbles) which are in slight vibratory motion. A turbulent layer of liquid and gas above the glass spheres increases mass transfer and particulate removal. This scrubber has mainly been used in the process where the limestone is added to the boiler for calcining and the flue gases are scrubbed to remove SO_2. Pressure drop is generally 4 to 6 in. H_2O.

The packed-bed absorber must use open packing to prevent plugging. Packed towers have been tested by Research Cottrell in a pilot plant with high SO_2 removal and no significant scaling with low sulfur coal. Pressure drops are low - 0.4 in. H_2O per foot of packing. Scale control is extremely important in this type of scrubber.

The spray column is a countercurrent type scrubber which is free of scaling and plugging. It has low pressure drop. The spray tower will require high liquid-to-gas ratios and several stages of sprays to achieve high SO_2 removal. It has been tested by TVA and Ontario Hydro. The spray column will be tested at the EPA Test Facility, Paducah, Kentucky.

The tray column offers high liquid hold-up and high SO_2 removal at relatively low pressure drop. The main disadvantage is scaling, which causes high pressure drop. High liquid-to-gas ratios $L/G = 40$) are required. In addition, undersprays are used to wash off soft scale. Solids in the slurry liquor are being tested over a range of 5 to 10 percent to determine their effect on scale control.

The cross flow absorber has a short gas path with the scrubber installed in a horizontal position. It has low pressure drop and has been tested with packing or sprays. It requires high L/G to obtain high SO_2 removal.

The screen or grid scrubber has been recently tested by TVA. It contains five to ten screens (7/8 in. openings). High liquid-to-gas ratios (L/G = 50) and a stoichiometric ratio of 1.5 give high SO_2 removal (85 percent). Low pressure drops were observed.

Scrubber Size and Turndown

In order to keep the investment as low as possible, scrubber sizes of 400,000 ft^3/min must be developed. This is required because the stack gas rate from a power boiler is so high. For a

1000 MW plant, the gas rate is 2,500,000 ft^3/min at 125 °F. This requires six parallel scrubbing trains to scrub the gases.

An additional design consideration is scrubber turndown due to changes in boiler operation. Some scrubbers can be turned down to 40 to 50 percent design and others require compartmentalization. Finally, one or more trains may be taken off line to maintain efficient operation.

Materials of Construction

Since the pH of the circulation slurry will vary between 5.0 and 6.0, stainless steel (316 SS L) or carbon steel lined with rubber or plastic coatings are used in the scrubber with good results based on limited operating experience.

System Pressure Drop

The total system pressure drop depends on whether both fly ash and SO_2 are removed in the same scrubber system. For removal of fly ash, pressure drops of 10 and 15 in. H_2O are required.

For SO_2 removal, typical pressure drops are 5 to 7 in. H_2O. This gives total system ΔP's of 20 to 27 in. H_2O for both sulfur dioxide and particulate removal.

If an electrostatic precipitator is used for particulate removal (ΔP = 1 in. H_2O) followed by an absorber for SO_2 removal, the system pressure drop could be reduced to 11 to 15 in. H_2O, depending on the scrubber chosen.

The best choice of systems will depend on ash and sulfur content of coal and means of waste disposal. Each design must be evaluated on a separate basis.

Mist Elimination

In order to obtain high SO_2 removal, high slurry circulation rates containing 5 to 15 percent solids are necessary. The high gas velocities cause mist formation, which must be eliminated to protect the fan and the reheater. Mist eliminators of the radial and chevron type are being used. The mist eliminators require washing to remove the soft deposits before plugging or excessive pressure drop results.

Gas Reheat

The gas temperature in a scrubber system is about 120 to 125 °F. In order to restore plume buoyancy and prevent condensation on the downstream, equipment reheat is required. This reheat can be accomplished by means of steam heated coils located in the gas duct or by burning to heat the flue gas by mixing with hot gases. The degree of reheat required has not been established, but the average reheat temperature is about 175 °F.

Waste Disposal

Limestone scrubbing produces about 1600 tons/day of calcium sulfite/calcium sulfate waste products, from a 1000 MW power plant. The fly ash amounts to 1000 tons/day, producing a total solid waste of 2600 tons/day. Since calcium sulfate is essentially insoluble, disposal in suitable areas is the best solution. The main problem is finding suitable sites. Limestone quarries appear to be the best locations at this time. Additional processing using a thickener, filter, or centrifuge may be necessary to dispose of solids when a pond does not exist at the plant site.

Scale Control

All of the above design considerations are influenced by the percentage of sulfur and ash in the coal and the percent of SO_2 removal desired. This is particularly true in the design of scrubbers. The liquid-to-gas ratio, which can vary from 20 to 50 gpm slurry per 1000 cu. ft/min of gas, effects SO_2 removal and control of scaling. The key to the scaling problem is the control of the supersaturation of $CaSO_3/CaSO_4$ in the liquor. This is accomplished by limiting the sulfur loading in the absorber, and providing sufficient residence time to break the supersaturation in a delay tank. The required delay time is not known. Experiments to date indicate delay times should be between 5 and 15 minutes. Control of sulfur loading for low sulfur coals is less difficult for the same L/G because the amount of SO_2 removal per pass is less (600 to 100 ppm). For high sulfur coals, the SO_2 removal is 2400 ppm to 400 ppm (or four times greater).

pH control is important for good limestone utilization - its effect on scaling, corrosion, and blinding. Laboratory and pilot plant studies indicate that outlet pH should not be less than 5.0 for good operation. The stoichiometric ratio of alkali to SO_2 is important for good SO_2 removal. Generally, it has been found that a stoichiometric ratio of 1.5 is required to obtain good operation.

The proper choice of limestone is important for good reactivity. Its grain structure should have a low degree of grain interlocking so that when ground, it will expose a large surface area. Particle size after grinding should be at least 90 percent through 200 mesh. The limestone should have a high calcium purity - 98 percent or better.

The percent of solids in the circulating liquor provides surface for solid precipitation and scale control in the absorber. Tests to date indicate that 5 to 15 percent solids in the liquor slurry produce the best overall results.

The number of stages required in an absorber have been found by tests to vary between 2 and 4, depending upon L/G, inlet SO_2 concentration, and percent of SO_2 removal desired.

PILOT PLANT STUDIES

In order to develop the necessary process technology for the design of large scale units, several extensive pilot plant programs are being carried out in the United States. The most extensive program utilizes the EPA Test Facility at Shawnee (TVA) power plant in Paducah, Kentucky. TVA has also conducted an extensive test program over the past several years, both in the laboratory and in pilot plants of 2000 CFM size. Other extensive pilot plant studies include the Mohave pilot plant in Nevada, the Detroit Edison pilot plant at the Rouge station in Michigan, and Cholla pilot plant in Arizona. Numerous pilot plant studies of short duration have been carried out to confirm design criteria for large scale installations for particulate and SO_2 removal.

Certain pilot plants generally process gases from combustion of high sulfur coal. These are:

o EPA Test Facility - 2.5 to 3.5 percent sulfur coal

o Detroit Edison Rouge Pilot Plant - 2.5 to 3.5 percent sulfur coal

o TVA Pilot Plant - 2.5 to 3.5 percent sulfur coal

Low sulfur pilot plants are

o Mohave Pilot Plant - 0.5 to 1 percent sulfur coal

o Cholla Pilot Plant - 0.5 to 1 percent sulfur coal

EPA Alkali Scrubbing Test Facility

The Office of Air Programs (OAP) of the Environmental Protection Agency (EPA) has sponsored a program to fully characterize wet limestone scrubbing for the removal of SO_2 and particulates from boiler flue gas.

The test facility consists of three parallel scrubber systems, each designed to treat 30,000 ACFM of flue gas from an existing coal-fired boiler at the TVA Shawnee Power Station, Paducah, Kentucky. Operation of the test facility began in March, 1972.

Bechtel Corporation, San Francisco, Ca., as the major contractor, designed the test facility, developed the test program, and will direct the test efforts. TVA constructed the facility and will operate the unit during the test program, which is scheduled for 30 months.

The test program objectives are to: (1) investigate and solve operating problems such as scaling, corrosion, and erosion; (2) investigate scrubber performance; (3) study solid disposal methods; (4) develop mathematical models to allow economic scale-up to full-size scrubbers; (5) optimize operating conditions for maximum SO_2 and particulate removal; and (6) perform long-term testing.

The scrubber systems were designed to be equivalent to a 10-MW power plant, so that confident extrapolation to full-size units could be obtained. The 30,000-CFM scrubber was judged to meet this requirement.

The scrubbers selected were: (1) venturi followed by after-absorber (Figure 1); turbulent contact absorber/TCA; and (3) marble-bed absorber (Hydro-Filter).

The facility can be operated under various alkali addition models. These include: limestone in the scrubber circuit, hydrate in the scrubber circuit, and limestone injection into the boiler furnace.

The scrubbers are either of stainless steel or rubber-lined carbon steel construction. All major piping pumps and tanks are lined with rubber or fiberglas reinforced polyester.

The basic test periods are:

(1) "Break-in" testing will evaluate operating problem and variable control and establish attractive operating configurations.

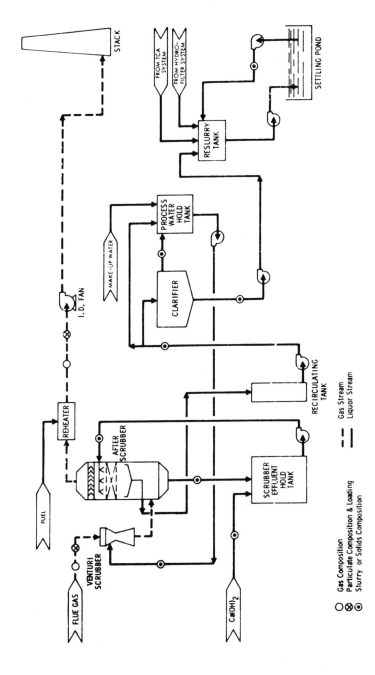

Figure 1. Typical Process Flow Diagram For Venturi System

(2) "Screening" testing will define the influence of process independent variables on dependent variables (SO_2 and particulate removal). Results of these tests will be used to perfect mathematic models for commercial scale-up.

(3) Primary testing will be used to optimize scrubber system performance and demonstrate process economics and reliability.

The level of variables for the break-in and screening experiments for limestone in the scrubber circuit are given in Table 1 for the venturi. (For further details see Ref. 6 and 14)

TVA Pilot Plant Program

TVA has carried out a wet scrubbing development program for the removal of SO_2 and particulates from boiler flue gases which are emitted from a coal-fired electrical power generating plant. The limestone throwaway process was selected because of its relative simplicity.

A limestone wet scrubbing pilot plant (2000 to 3500 CFM capacity) was operated by TVA for about one year (Jan. to Dec., 1971) for the purpose of evaluating scrubber types that appeared to be candidates for 550-MW Widows Creek installation. This pilot plant was installed on the Colbert Steam Plant located in Alabama. Flue gas was supplied from a 200-MW boiler.

The following scrubber types were evaluated in the pilot plant: (1) ventri-rod (venturi type) installed in a horizontal position followed by a Tellerette packed cross flow scrubber; (2) ventri-rod installed in bottom of spray tower; (3) turbulent contact absorber (TCA) containing three stages of mobile plastic spheres; and (4) multi-grid scrubber containing wire mesh and no packing (developed by TVA).

The scrubbers are designed for 3500 CFM of boiler stack gas containing 2500 to 3500 ppm SO_2. The following range of variables were tested:

(1) Stoichiometric ratio - 1.2 to 1.5

(2) Liquid/Gas ratio - 20 to 60

(3) Percent solids in slurry - 5 to 15

Pilot plant test data from a three-stage TCA scrubber

Table 1

LEVELS OF INDEPENDENT VARIABLES FOR EXPERIMENTS WITH LIMESTONE IN SCRUBBER CIRCUIT: VENTURI SYSTEM

	Controllable Factors	Levels for Break-In (Group I) Venturi & spray tower		Levels for Break-In (Group II) Venturi & Pall-ring		Levels for Factorial Experiments Venturi & spray tower Venturi & Pall-ring		Levels for Sensitivity Experiments Venturi & spray tower	
Primary Variables	Scrubber configuration					1 2	Venturi & spray tower Venturi & Pall-ring		
	Stoichiometric ratio	1 2 3 4	1.25 1.50 1.75 2.00	1 2	1.50 1.75	1 2	1.25 1.75		1.50
	Percent solids recirculated	1 2 3 4	4% 8 12 16	1 2	4% 10	1 2	8% 12		10%
	Gas flow rate	1 2	15,000 cfm 30,000	1 2	15,000 cfm 30,000	1 2	15,000 cfm 30,000		30,000 cfm
	Liquor flow rate-venturi		300 gpm		300 gpm	1 2	300 gpm 600 gpm		300 gpm
	Liquor flow rate-after-scrubber	1 2	300 gpm 600		300 gpm	1 2	300 gpm 600		600 gpm
	E.H.T residence time		60 min.		60 min.	1 2	15 min. 60		60 min.
	Flow configuration		14		14 15	1 2	14 15		14
Secondary Variables	Plug position*	1 2 3 4	40% open 60 80 100		60% open	1 2	40% open 80		60% open
	Solids handling configuration		Clarifier/pond		Clarifier/pond		Clarifier/pond	1 2	Clarifier/pond Clarifier/filter
	Limestone type		A		A		A	1 2	A B
	Limestone particle size		90% through 300 mesh		90% through 300 mesh	1 2	90% through 300 mesh 90% through 200 mesh		90% through 300 mesh
	Limestone injection point*		E.H.T.	1 2	E.H.T. P.W.H.T.		E.H.T.		E.H.T.
	Excess air		18%		18%		18%	1 2	18% 36
	Coal type		A		A		A	1 2	A B
Variables Held Fixed	Saturation sprays		On		On		On		On
	Cooling of inlet liquor		No		No		No		No
	Coagulant		No		No		No		No

* These variables have been considered "primary variables" within the statistical designs for Break-In testing.

indicate SO_2 removal efficiencies of 70 to 95 percent. Gas rates were varied from 1400 to 2700 CFM with system pressure drops of 6 to 24 in H_2O. Table 2 summarizes the test data.

The most recent pilot plant tests utilized a multi-grid scrubber. The grid type scrubber indicates that SO_2 removal may vary from 65 to 90 percent without spheres (as shown in Table 2).

Operating problems observed included erosion of the plastic spheres and high mist carryover. Erosion of the plastic spheres in the TCA absorber using 15-percent solids in the slurry was approximately 30 percent weight loss after 1000 operating hours. The mist eliminator (chevron type) accumulated solids during the tests. Analyses of the solids indicate the material to be calcium sulfite/sulfate rather than fly ash. Tests have begun to develop a more efficient eliminator.

On the basis of this evaluation, the multi-grid scrubber has been selected by TVA for further development and installation on a commercial size unit. The multi-grid scrubber has the advantage of being relatively nonplugging, with good SO_2 removal at moderate pressure loss (85 to 90 percent at 6 in. H_2O Δp and good turndown characteristics). A new pilot plant was designed to closely duplicate the preliminary design for the Widows Creek Unit. The objectives of the tests is to demonstrate a reliability of 70% or more SO_2 removal.

The test conditions are as follows: (1) Ca/SO_2 mole ratio - 1.5, percent solids in slurry - 14; (2) L/G to ventri-rod - 10, L/G to multi-grid - 50; (3) limestone type - longview; grind - 70% through 200 mesh; (4) gas velocity - 13 ft/sec, grids, number and spacing - 5 at 4 feet; (5) slurry retention time, min - 5; and (6) SO_2 inlet concentration, ppm - 2800, gas rate - 2100 ACFM.

Detroit Edison Power Plant

The Detroit Edison Co. decided to build a pilot scrubber installation because of a lack of demonstrated performance of a full-scale system and the availability of a variety of scrubber designs.

The pilot scrubbing installation was designed for 2500 ACFM of flue gas from an operating boiler at the Rouge plant.

(a) Percent of SO_2 removal, scaling problems, system turndown

(b) Levels of entrainment, mechanical problems

(c) Effectiveness of alkali such as lime or limestone.

Table 2

SCREENING TESTS USING TURBULENT CONTACT ABSORBER AND MULTIGRID SCRUBBERS

Test	Duration, hr.	Gas flow Flow rate,[a] acfm	Velocity, ft./sec.	Number of stages of spheres	ΔP across scrubber, in. H$_2$O	L/G Humidification	Scrubber	Scrubbing liquor pH F-11	F-12	F-13	Solids, %	SO$_2$ removal, %
P-44 (1)	6	2100	12.6	3	7.4	7	48	7.3	5.4	5.8	13	96
P-44 (2)	6	2100	12.6	3	8.5	7	40	7.4	5.5	5.8	14	98
P-44 (3)	4	2100	12.6	3	7.5	7	30	7.6	5.6	5.9	13	77
P-44 (4)	6	2300	13.7	3	8.2	7	24	7.5	5.6	5.9	16	74
P-44 (5)	6	2300	13.7	3	7.2	7	18	7.4	5.4	5.9	12	65
P-44 (6)	6	2300	13.7	3	5.4	7	18	7.3	5.5	5.9	13	63
P-44 (7)	6	2300	13.7	2	5.6	7	24	7.1	5.5	5.9	13	73
P-44 (8)	6	1800	10.8	2	4.7	8	30	7.1	5.7	6.0	17	78
P-44 (9)	6	1600	10.8	2	5.5	8	40	7.1	5.5	5.9	14	74
P-44 (10)	6	1800	10.8	2	5.7	8	50	7.6	5.5	5.8	11	74
P-44 (11)	6	1800	10.8	1	4.0	8	50	7.6	5.7	6.0	15	68
P-44 (12)	10	1800	10.8	1	3.1	8	40	7.3	5.7	6.0	15	74
P-44 (13)	3	1800	10.8	1	2.7	8	30	7.5	5.6	6.0	15	76
P-44 (14)	4	2300	13.7	1	4.4	7	24	7.5	5.8	6.0	16	76
P-44 (15)	4	2300	13.7	1	3.0	7	18	7.0	5.6	6.0	15	67
P-44 (16)	4	2300	13.7	0[b]	1.7	7	18	7.9	5.8	6.1	15	64
P-44 (17)	4	2300	13.7	0[b]	2.4	7	24	7.6	5.9	6.2	14	73
P-44 (18)	4	2300	13.7	0[b]	3.8	7	48	8.0	5.8	6.2	21	80
P-44 (19)	4	1800	10.8	0[b]	0.8	8	30	7.8	5.8	6.2	16	60
P-44 (20)	5	1800	10.8	0[b]	1.1	8	40	7.4	5.9	6.0	11	64
P-44 (21)	5	1600	10.8	0[b]	1.5	8	48	7.1	5.7	6.1	10	74
P-44 (22)	6	2300	13.7	0[b]	4.1	7	44	7.5	5.8	6.0	16	87
P-44 (23)	6	2100	12.6	0[c]	2.0	7	24	7.5	5.8	6.2	10	69
P-44 (24)	6	2100	12.6	0[c]	3.2	7	48	7.4	6.0	6.2	13	85
P-44 (25)	6	2500	14.9	0[c]	4.2	6	24	7.2	5.9	6.1	15	72
P-44 (26)	6	2500	14.9	0[c]	8.0	6	48	7.4	5.8	6.1	20	89

[a] At scrubber outlet conditions.
[b] Multigrid scrubber with five wire mesh grids.
[c] Multigrid scrubber with 10 wire mesh grids.

Scrubber types evaluated were: two venturi in series, venturi followed by sieve tray absorber and venturi followed by TCA absorber.

The results of the investigations conducted with the two venturi scrubbers indicate outlet dust loading of less than 0.02 gr/SCF can be attained. Using limestone, an SO_2 removal of 90 percent was obtained with 20 percent excess lime.

The second pilot plant configuration consists of a Lurgi venturi scrubber followed by a Peabody tray absorber. It is designed to treat both dirty and clean stack gases at 2500 ACFM. The SO_2 concentration in the flue gas averages 2500 ppm. The liquid-to-gas ratio to the Lurgi venturi is 20 to $\frac{1}{2}$, and to the absorber is 40 to 1. Stoichiometric ratio of CaO to SO_2 will be controlled at 1.5. Percent of solids in slurry will be tested at 5 to 10 percent.

Further testing will continue to supply data for design of full-size unit.

Mohave/Navajo Pilot Facility

This Mohave/Navajo SO_2 removal research program (Ref. 7) was initiated to test alkali absorption processes on a pilot plant basis using gas from a boiler burning low sulfur coal. Coal from the Black Mesa mine contains 0.3 to 0.8 percent S.

A flexible, dual loop system was designed and installed at Mohave Power Station. Two absorbers were in each loop. The four absorbers installed contain a single stage venturi, a turbulent contact absorber, a combined Lurgi impingment tray and low-drop, egg-crate packing.

Three absorbents, limestone slurry, lime slurry, and soda ash solution have been tested.

Gas flow rates can be varied from 500 to 4000 ACFM. Parametric variations of L/G, gas flow rate, and pH are made, while measuring SO_2 absorption efficiency.

Other Pilot Plant Studies

Other low sulfur pilot plants included the Cholla unit of Arizona Public Service, which consisted of a proprietary packed tower using limestone as the alkali. Extensive testing indicates good removal of SO_2 without scaling.

Several short test programs using venturi scrubbers were carried out to verify design criteria. These will not be described in detail. They generally consist of 2000 ACFM scrubbers with testing over a 4 to 6-week period.

Pilot plant programs will continue until enough data have been accumulated to design full-size units reliably.

Smelter Research For SO_2 Control

The Smelter Control Research Association (Ref. 8) was founded and incorporated as a nonprofit association by the eight primary copper producers of the United States for the specific purpose of developing processes for removal of SO_2 and particulates from smelter stack gas. Results of the Association's experimental programs will be published and made available to the public.

Sulfur dioxide produced at a copper smelter occurs at two concentration levels. The converters and roaster produce off gases concentrated in SO_2, averaging around 4 to 6 percent SO_2. The reverberatory furnace produces more dilute gas with SO_2 averaging 1 to 2 percent.

Technology is available for converting the concentrated stream to sulfuric acid. There are seven sulfuric acid plants operating today.

The Smelter Control Research Association reviewed 100 processes and selected wet limestone scrubbing since it is closest to commercial availability.

A pilot plant has been constructed and is in operation at the McGill, Nevada Smelter of the Kennicott Copper Corp. The pilot plant is rated at 4000 CFM, which is of sufficient size to permit confident extrapolation to commercial size (80,000 CFM). The pilot plant has been designed with a view of optimum versatility and will test wet lime and sodium sulfite scrubbing as well as limestone scrubbing. The plant employs a turbulent contact absorber in series with a venturi.

Limestone water slurry flows countercurrent to the gas stream from the turbulent contact absorber to the venturi.

The pilot plant has been in operation with limestone for approximately two months on a 24-hour-day basis. The plant is now in closed-loop operation.

It too early to present final conclusions with respect to the effectiveness of the system in smelter gas control.

Results to date indicate 70 percent removal from 0.7 percent SO_2 and 55 percent from 1 percent SO_2.

High calcium limestone has been used in all tests. Bench scale tests indicate all limestones tested were equally reactive.

Limestone usage has been varied from 1.2 to 2.0 times the stoichiometric rate. Both 90-percent, minus-200 mesh stone and 95-percent, minus-325 mesh stone have been tested in the pilot plant. Any difference in reaction rates could not be observed, although one might expect greater reactivity from the finer grid. No significant improvement in SO_2 removal has been noted.

The current efforts are directed to establishing the factors controlling SO_2 removal in order to optimize the process. Development of techniques for effective control of scale formation is also a major objective. Further research will continue.

STATUS OF FULL-SIZE INSTALLATIONS

Although the process technology for SO_2 removal has not been completely developed, the U. S. has decided to install full-size units. To date, there are 22 full-size scrubber installations either operating, under construction, or being engineered in the U. S. These installations fall into three categories:

(1) Particulate/SO_2 scrubbers treating gas from high sulfur coal
(2) Particulate/SO_2 scrubbing systems for low sulfur coal
(3) Scrubbers for particulate only. Table 3 summarizes these data.

The first full-size units installed were in 1968 and 1969 at Union Electric's St. Louis Station and Kansas Power & Light's Lawrence Station by Combustion Engineering. Both employed finely pulverized limestone injected into the combustion chamber to yield a highly reactive quick-lime. The lime was carried by the flue gas into the wet scrubber, where it neutralized the SO_2. Spent lime and calcium sulfate were powdered with the fly ash. Lack of prior design and operating experience led to a number of problems including scaling and corrosion.

Most of the scrubbing facilities now being built for high sulfur coal make use of ground limestone added directly to the scrubber liquids.

Table 3

Scheduled Operating Date	Unit Size, Mw	Utility and Station
SCRUBBER SYSTEMS FOR PARTICULATE ONLY		
September 1970	125	Pennsylvania Electric Holtwood
November 1971	166	Colorado Public Service Valmont #5
		Arizona Public Service Four Corners Plant:
December 1971	175	Unit #1
February 1972	175	Unit #2
March 1972	225	Unit #3
March 1972	360	Pacific Power & Light Dave Johnston #4
February 1973	4 units totaling 500	Duquesne Power & Light Elrama Station
PARTICULATE/SO_2 SCRUBBING SYSTEMS LOW-SULFUR COAL		
December 1972	115	Arizona Public Service Cholla Plant
December 1973	2 @ 125	Nevada Power Company Reid Gardner #1 and #2 (Sodium System)
January 1974	750	
January 1975	750	Salt River Project Navajo Power Station Units 1, 2, 3
January 1976	750	
1976-1977	2 @ 680 each	Northern States Power Co. Sherburne County, Minnesota
PARTICULATE/SO_2 SCRUBBER SYSTEMS HIGH-SULFUR COAL		
1968	140	Union Electric Meramec #2
1969	125	Kansas Power & Light Lawrence #4
November 1971	430	Kansas Power & Light Lawrence #5
February 1972	180	Commonwealth Edison Will County #1
March 1972	Pilot Plant	EPA - TVA Shawnee
August 1972	800	Kansas City Power & Light, La Cygne
1973	550	Tennessee Valley Authority Widows Creek #8
Late 1973	50% of 340	Detroit Edison St. Clair #6
February 1973	6 units totaling 400	Duquesne Power & Light Phillips Station
1973	50% of 195	Potomac Electric Power Dickerson #3

The full-size scrubbing systems presently operating are:

o   Commonwealth Edison, Will County No. 1 - 180 MW
o   Kansas Power & Light, Lawrence No. 5 - 430 MW
o   Pennsylvania Electric, Holtwood - 125 MW
o   Colorado Public Service, Valmont No. 5 - 166 MW
o   Arizona Public Service, Four Corners - 575 MW (total)
o   Pacific Power & Light, Dave Johnston - 360 MW

Commonwealth Edison, Will County No. 1

This scrubbing system (Ref. 9) is the first commercial-size limestone scrubber where the limestone is added to the liquor circuit. The Will County Unit is a Babcock & Wilcox system and consists of two trains, each designed for 375,000 CFM of flue gas containing 2800 ppm SO_2.

Each scrubber train consists of a venturi followed by a TCA absorber. Limestone is ground to 90 percent through 325 mesh in ball mills and pumped as slurry to the recirculation tank of the TCA absorber.

The fresh limestone slurry of 20-percent solids is added to the absorber recirculation tank where water dilutes the slurry to 8 percent solids. The spent or waste solids are taken off the venturi pump discharge line and flow to a pond. The water is then recycled to the scrubbing system.

The venturi recirculation tank provides a reaction time of 6 minutes; the absorber tank provides a 4-minute reaction time. Space is available to add two tanks to increase absorber hold-up time to 6 minutes if required.

The flow of slurry to each venturi is 5800 gal/min (L/G = 18.5) and to the absorber is 8750 gal/min (L/G = 28.1).

Flue gas passes from the boiler after precipitation and goes to the venturi. The gas pressure drop through the venturi is 9 inches water. The removal of fly ash is effected by the collision of particulates with small water droplets.

The gas flow from the venturi sump to the bottom of the absorber where the gas is contacted countercurrently with limestone slurry for removal of SO_2. Three stages of plastic spheres are provided in the absorber. These spheres provide surface for chemical reaction and act as cleaners to prevent scale buildup. The gas pressure drop through the absorber is 6 inches. Space for a third stage is available.

The flue gas leaving the absorber is reheated from 128 °F to 200 °F to give gas buoyancy and limit condensation in the fans, ducts, and existing stack. Steam reheater coils are used and foot blowers are provided to maintain tube cleanliness. The materials of construction are: venturi (carbon steel with Plasite 7122 and kaocrete, absorber (rubber-lined corten steel) and flues from absorber to reheater (corten steel with flakeline 103).

Pumps, recirculation tanks, valves, and all piping in contact with slurry above 6 inches are rubber-lined carbon steel. Piping less than 6 inches in diameter is 3162 stainless steel.

The cost of the 180 MW scrubbing unit was $8,000,000. Estimated operating costs are 20 to 26¢/10^6 BTU.

The Will County unit has now been operating approximately 1000 hours. Preliminary data indicate SO_2 removal to be 80-percent based on 2800 ppm SO_2 in, and 600 ppm in the outlet gas. Observations of the stack during this limited operating period indicates a clear stack was obtained.

There have been mechanical problems with the valves in the slurry system. These have been replaced. An expansion point of the recirculation pump failed and has been removed. Some scale deposits have been observed, but these were not serious and did not cause unit shut-down. The disposal of waste solids is still an important process consideration. Ponding at Will County is limited and work has begun to dewater the solids for removal to a more permanent site.

Kansas Power and Light, Lawrence No. 5

This is a Combustion Engineering system which involves injection of limestone into the boiler furnace where it is calcined. The CaO and flue gas then flow to 6 marble-bed scrubbers, in parallel, for removal of SO_2 and particulates. This unit went into operation in late 1971. The present state-of-design of this unit is based on the results obtained from the prototype unit installed on Unit No. 4 in 1969.

Based on very limited data, SO_2 removal has been reported to be in the 70-percent range.

Pennsylvania Electric's Holtwood Station

This is a retrofit design for particulate removal only. It is a retrofit system which utilizes a venturi scrubber. High removal of particulates has been reported.

Colorado Public Service Valmont Station

This is a retrofit design utilizing a Universal Oil Product TCA scrubber for removal of particulates to meet the new U. S. standards. Some mechanical problems were reported but have been solved. Outlet grain loadings in the 0.02 gr/SCF have been reported.

Arizona Public Service Company Wet Scrubber Installations

Arizona Public Service Company (Ref. 10) is installing three scrubber units at its Four Corners plant near Framington, New Mexico, with a generating capacity of 575 MW. The other installation is at the Cholla Power Plant in Arizona, with a rating of 115 MW. The Four Corners plant burns coal with an average ash of 22 percent and sulfur content averaging 0.7 percent. The Cholla Plant uses coal with ash averaging 8 percent and sulfur content of 0.5 percent. The wet scrubber concept of cleaning flue gases was selected because of concern for effective use of electrostatic precipitators with low sulfur coal and for SO_2 removal capabilities.

The Cholla installation is being designed by Research-Cottrell, Inc. and will use a high energy, flooded-disc venturi for particulate removal and an absorber tower following the venturi for SO_2 removal. The absorber will utilize a wetted film, fixed packing with limestone as the absorbent. The Cholla plant is scheduled for startup in December 1972. An extensive 5000 CFM pilot plant program has been conducted to optimize chemical tests.

TVA's Widows Creek SO_2 Removal System

The Widows Creek scrubbing system (Ref. 12) is the first commercial installation for Tennessee Valley Authority (TVA). TVA has carried out extensive pilot plant studies and selected the grid tower for Widows Creek (see pilot plant section of this paper). The scrubber will utilize limestone added to the scrubber liquor. There will be five trains of scrubbers in parallel with each scrubber having 5 to 8 grids on screens. Design conditions are: liquid-to-gas ratio of 50 to 1; stoichiometric ratio, 1.5; percent solids in slurry, 15 percent; slurry delay time, 5 minutes; limestone, 70 percent, 200 mesh.

The unit is scheduled for startup in December 1973. Further pilot plant tests will be conducted at Colbert Power Station (2100 ACFM) and at the EPA test facility at Shawnee Power Station (30,000 ACFM scrubber).

Pacific Power & Light Company's Dave Johnston Scrubber System

The Dave Johnston scrubber system was designed and constructed by Chemical Construction Company. It consists of three venturi trains for particulate removal from a 360-MW boiler which burns low sulfur coal. Operation began in March 1972. A clear stack appearance has been reported. Quantitative testing to determine outlet gas grain loadings will start in june 1972. SO_2 removal due to the alkalinity of fly ash has been reported to be 40 to 60 percent. No serious operating problems have been reported.

Detroit Edison Company's St. Clair Unit No. 6 Scrubber System

Detroit Edison Company (Ref. 11) will utilize the Lurgi/Peabody Scrubbing System at its St. Clair Station. This system will consist of two scrubber trains. Each train contains a Lurgi venturi and a Peabody tray absorber. Limestone will be used as the alkali and will be ground to 90 percent through 300 mesh. This unit is scheduled for operation in December 1973. The pilot plant at Rouge Station has been operated to confirm design variables and demonstrate reliability. The Lurgi venturi has been designed for a liquid-to-gas ratio of 20 to 1, and the absorber for an L/G of 40 to 1.

Economics

Information concerning economics of limestone scrubbing can be found in greater detail in Ref. 13.

SUMMARY AND CONCLUSIONS

This paper has described the increasing activity in process development for lime-limestone scrubbing for SO_2 removal from gases. Twenty-three full-size installations are operating or being engineered to provide important data for industry. The U. S. EPA Test Facility has started an extensive research and development program for lime-limestone scrubbing. The years 1972 and 1973 will provide critical data to meet our goal for clean air.

REFERENCES

1. TVA Special Reports on Pilot Plant Operations, Aug. 1971, Nov. 1971.

2. Lowell, P., et al., "A Theoretical Description of the Limestone Scrubbing Process," (June, 1971) Vo. 1. Report No. PB 193-029, Clearinghouse for Technical Information, Va., 22151.

3. Boll, R. A., "Mathematic Model for SO_2 Absorption by Limestone Slurry," Paper presented at Limestone Scrubbing Symposium, sponsored by Air Pollution Control Office, Perdido Bay, Pensacola, Fla., Mar. 1970.

4. Slack, A. V., "Sulfur Dioxide Removal from Stack Gases," Noyes Data Corp., Park Ridge, N. J. 07656, 1971.

5. Slack, A. V., et al., "Sulfur Oxide Removal From Waste Gases," Journal of Air Pollution Control Association, March 1972.

6. Epstein, M., et al., "Test Program for EPA Test Facility at the TVA Shawnee Power Station," paper presented at Second International Lime/Limestone Wet Scrubbing Symposium sponsored by Environmental Protection Agency, New Orleans, La., No. 1971.

7. Shapiro, J. L., et al., "Mohave/Navajo Pilot Facility for SO_2 Removal," Ibid.

8. Campbell, I., et al., "Sulfur Oxide Control at the Copper Smelter," Ibid.

9. Gifford, D. C., "Will County Unit 1, Limestone Wet Scrubber," Ibid.

10. Mundth, L., "Wet Scrubber Installations at Arizona Public Service Co. Power Plant," Ibid.

11. McCarthy, J. H., et al., "Detroit Edison Pilot Plant and Full Scale Development Program for ALKALI Scrubbing System," Ibid.

12. Kelso, T. M., et al., Part I & II "Removal of SO_2 from Stack Gases, by Scrubbing with Limestone Slurry," TVA Pilot Plant Tests, Ibid.

13. Sherwin, R. M., et al., Economics of Limestone Scrubbing, Ibid.

14. Epstein, M., et al., Mathematical Models for Pressure Drop, Particulate Removal and SO_2 Removal in Venturi, TCA and Hydro-Filter Scrubbers, Ibid.

SPECIAL PROBLEMS OF THE SMELTER INDUSTRY

G. Bridgstock

The Lummus Company Canada Limited, Toronto

Within the context of the subject of this symposium "Pollution: Engineering and Scientific Solutions", this paper gives no consideration to economics and their effect on the utilization of engineering and scientific solution which have been proposed for the smelter industry.

Other papers in this symposium have dealt primarily with processes available for the removal of Sulphur from power generating station combustion gases. In 1969 in the United States this was the largest single source of emission accounting for 73% of total sulphur oxide emissions as shown in Table 1 (1).

TABLE 1

Source	Emissions 10^6 ton/hr
Transportation	1.1
Fuel Combustion in Stationary Sources	
Coal	19.8
Fuel Oil	4.6
Industrial Processes	7.5
Solid Waste Disposal	0.2
Coal Refuse Banks	0.2
	33.4

Industrial processes, the second largest single source of emission, include the smelting industry and in particular the non-ferrous smelting industry which is the subject of this paper.

Though the smelting industry is a relatively small emitter of sulphur oxides, compared with the power generating industry, there are considerably fewer smelters than power generating stations, but each smelter is individually a larger single emitter of sulphur oxides. This has given rise in the United States and other countries to the establishment of different emission standards for smelters as compared with power generating stations and other industrial plants.

The United States Federal Government Standards applicable to all sulphur oxide emissions from any industry are ambient standards as shown in Table 2. (1)

TABLE 2

National Sulphur Oxides Air Quality Standards

Standard	Concentration ug./cm	ppm	Description
Primary	80	0.03	Annual arithmetic mean
	365	0.14	24 hr. maximum not to be exceeded more than once per year.
Secondary	60	0.02	Annual arithmetic mean
	260	0.1	24 hr. maximum not to be exceeded more than once per year.
	1300	0.5	3 hr. maximum not to be exceeded more than once per year.

Individual states have in some cases elected to reduce the ambient standards established by the Federal Government, and in other cases have established an emission standard in addition to the ambient standard. These emission standards (2) which vary from state to state, at present limit the emission of sulphur to from 5% - 10% of the total sulphur feed to the smelter with maximum emissions of between 5,000 and 10,000 lbs/hr.

Other countries worldwide have established or are preparing similar standards which in most instances are ambient or ambient related standards but in some cases are also emission standards.

Ambient standards can in general, and subject to local weather conditions in some special instances, be met by the installation of tall stacks which ensure wide dispersion of sulphur oxide in the

smelter gases before the gas reaches ground level. For example
International Nickel Company has built a stack 1250 feet high at
its plant in Sudbury. The use of tall stacks does not however enable
the smelter operator to meet an emission standard. It is this
emission standard which creates the special problem of the smelter
industry today.

The major source of the non-ferrous metals Copper, Zinc, Lead
and Nickel is at present, and has been for the past several decades,
ores containing mineral sulphides of these metals. Present technology
enables the economic recovery of sulphide minerals from ores containing as little as 0.5% copper in the form of concentrate which is
subsequently smelted for the recovery of the metal. Ores containing
as little as 0.2% copper (4 lbs. copper per ton of ore) are being
treated but are only economical because of the presence of other
metals such as Gold and Molybdenum which are recovered.

In the treatment of non-ferrous metal concentrates for the
recovery of refined metal it is necessary to separate the metal from
sulphur, iron and other gangue materials in the concentrate. To date
the majority of processes used to effect this separation have been
pyrometallurgical. In these high temperature operations sulphur is
oxidized to sulphur dioxide and removed as a gas, iron is oxidized
and along with the gangue removed as molten slag and the metal recovered initially in the molten state. Though the basic process for
sulphur removal is the same, the type of equipment used varies depending on the particular metal being recovered. Many non-ferrous metal
concentrates contain some or all of the precious metals Gold, Silver
and Platinum group. The smelting process provides a very efficient
means of recovering these elements in the molten product. Subsequent refining treatment to produce pure Copper, Nickel, Lead or
Zinc also enables the precious metals to be recovered.

The result of the use of differing process equipment is smelter
off-gases containing from 0.25% SO_2 up to in some special instances
13% SO_2. Smelting plants are in general located at or near mine
sites and close to the source of production of concentrate, their
feed material. This means that smelters are usually found in
locations remote from large centers of population.

The result therefore of the combination of remote location and
generally low strength gases has been to allow all SO_2 gases to pass
to atmosphere. When however particular location, gas composition
and available markets have prevailed the industry has developed and
used processes for the removal of SO_2 from off-gases, in the majority
of cases as sulphuric acid but in some instances as elemental sulphur.

The problems now facing the established non-ferrous smelter
industry and to a lesser degree new smelters stem from the basic

problem of restricting the quantity of sulphur emitted to the atmosphere. These problems fall into two interdependent categories:

1. How to remove the sulphur?

and,

2. What to do with sulphur after it has been removed?

The interdependence of these two problems can be illustrated by an example which faced the copper smelting industry in Wales in 1864 (3). At that time a novel type of furnace for roasting sulphide materials had been developed which enabled better control of combustion air to be made thus producing furnace gas containing a very small amount of oxygen. Sulphuric acid chambers were therefore erected to recover sulphur from the gas as sulphuric acid. This effectively reduced pollution from SO_2 but the operators were faced with the problem of how to dispose of or utilize the sulphuric acid they were producing.

Removal of Sulphur Oxides

Existing non-ferrous smelters have been designed primarily to produce one or more of the non-ferrous metals. This has resulted in plants which suffer from disadvantages such as air leakage into flues, batch operations etc. when it becomes necessary to remove sulphur products from gas streams.

Proposed new smelters do not suffer to the same degree from these problems as they are able to take advantage of more recent technological developments in smelting processes.

The proven and established sulphuric acid plant technology has existed for many years to enable sulphur oxides to be removed from smelter gases containing more than 3.5% (by vol.) SO_2. The end products of this process have been sulphuric acid and an off-gas essentially free of SO_2.

Alternatively in very specialized cases technology has been developed and used to enable SO_2 to be removed from smelter gases for the recovery of elemental sulphur or ammonium sulphate.

Treatment of smelter gases containing less than 3.5% SO_2 has, primarily for economic reasons, not been considered to the same degree in the past. A survey by Meisel (4) presents an extensive summary of 29 processes in varying stages of development for the recovery of sulphur from metallurgical flue gases which indicates the increasing attention that is being paid throughout the world to the problem of removal of SO_2 from these gases.

Alternative Treatment Processes

An alternative solution to the problem of sulphur oxide emissions is the development and use of processes other than the traditional pyrometallurgical processes. These processes which take place in solution would, by the elimination of the high temperature oxidation of sulphur, avoid the production of sulphur dioxide thereby eliminating the need to remove SO_2 from large volumes of gas containing low concentrations of SO_2.

One such process for the treatment of mixed nickel-copper sulphide concentrates was developed by an operating company in the early 1950's. This process, which leaches the concentrate in ammoniacal solution under pressure, produces refined nickel and converts sulphur in the concentrate to ammonium sulphate. This same company has developed other hydrometallurgical pressure leaching processes for copper and zinc concentrates which produce elemental sulphur as a by-product.

Other hydrometallurgical processes in varying stages of development, none of which have been developed to full scale operation use sulphuric acid, ferric chloride or chloride leaching solutions to recover the metal with sulphur as a by-product (5).

Present indications are that hydrometallurgical processes will not have the same degree of applicability to low grade concentrates that pyrometallurgical processes have. In addition they have a major disadvantage that they are unable to recover precious metals to the same degree as the smelting processes.

Disposal of Sulphur Products

Though technology exists or is being developed to treat smelter gases for the removal of SO_2 and production of sulphur containing compounds, this only solves part of the smelting industries problem.

As stated earlier, the majority of smelting plants are located at or very near to the mine site, generally in remote locations long distances from other industrial areas. The result is that, though sulphur could be recovered as sulphuric acid, potential consumers of this acid are so far away that freight costs for shipping the acid made the delivered cost of acid so high that the consumer can more cheaply produce acid from elemental sulphur.

Similarly markets for sulphur or liquid SO_2 are at such distances from the smelter sites that potential users are able to obtain their requirements more economically from other sources.

Other alternatives under consideration in the smelting industry involve the conversion of SO_2 to a throw away product such as calcium sulphate. The use of limestone scrubbing for removal of SO_2 has been dealt with in another paper in this session. Consideration is being given to the use of sulphuric acid plants with subsequent neutralization of the acid with limestone to produce calcium sulphate for smelters which cannot ship acid to markets. This however only replaces an atmospheric pollution problem with the problem of the disposal of large quantities of calcium sulphate, as for every ton of sulphur in smelter feed, approximately 5 tons of calcium sulphate are produced. For a copper smelter treating 1000 tons of concentrate per day, a site must be found near the smelter on which approximately 600,000 tons per year of solid calcium sulphate can be dumped.

In addition most limestone deposits providing raw material for either limestone scrubbing of gas or neutralization of sulphuric acid contain some magnesium carbonate which will be converted to magnesium sulphate. This soluble compound if allowed to drain into local water systems can have unpleasant reactions on local population and it therefore becomes necessary to treat these solutions for the precipitation of magnesium and sulphate as separate compounds.

The Copper Smelting Industry

As a practical example a block flow diagram shown in Fig. 1 illustrates a typical copper smelter operation using the long established pyrometallurgical process.

In this process sulphur in the concentrate is almost completely removed as SO_2 in two gas streams, the reverberatory furnace gas and the converter gas. Minor amounts of sulphur are contained in the solid dump slag with small amounts in gases from open launders, ladles and other equipment handling molten matte and slag. As an alternative to this process, concentrate is partially oxidized by roasting before being fed to the reverberatory furnace. The roaster off-gas will contain 8% SO_2 and this results in some reduction of sulphur in the reverberatory gas.

The reverberatory furnace is operated continuously and produces a continuous stream of low SO_2 content gas. The converting operation is a batch process with a consequent cyclic gas volume. In addition as iron and sulphur in the matte are progressively oxidized the SO_2 content of the gas stream varies. Treatment of these gas streams for the removal of SO_2 is therefore difficult; large volume low SO_2 content reverberatory gases cannot be readily treated and air leakage, the cyclic volume, varying SO_2 content of the converter gas render this gas stream difficult to treat.

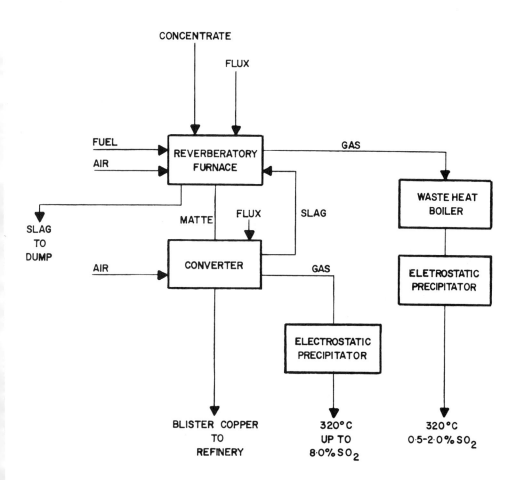

The problem of removal of sulphur from converter gas has now been essentially solved by modifications to converters and their associated hood and flue system and by the development of sulphuric acid plant technology. Many copper smelters throughout the world are now and have been for several years successfully treating converter gas for the removal of sulphur dioxide as sulphuric acid. The development of closed hoods to reduce air leakage and the recent development of the Syphon Converter have assisted considerably in producing converter gases more readily treatable in an acid plant.

Alternatives for the fuel fired reverberatory furnace have been developed to overcome the problem of removal of SO_2 from the large volume, low SO_2 content furnace gas. Flash Smelting developed in Finland in 1950 essentially combined in one furnace, the roasting and reverberatory furnace operations. This furnace, by utilizing the heat available from oxidation of part of the iron and sulphur in the concentrate, reduces the amount of fuel to be burnt to smelt the concentrate. This factor combined with other factors, such as reduced air leakage into the furnace and higher matte grade, results in the production of flash smelting furnace off-gas containing between 8 - 13% SO_2. Recent modifications using oxygen enriched combustion air have enabled the SO_2 content of off-gas to be increased to approximately 18%. These uniform volume gases containing high uniform SO_2 content can be readily treated in sulphuric acid plants or in plants to produce elemental sulphur.

Electric furnace smelting, either directly or combined with roasting, has provided another process as an alternative to the fuel fired reverberatory furnace. Pre-roasting of the concentrate when necessary to remove sulphur and provide control of matte grade produces a gas, uniform in volume and of strength suitable for treatment in an acid plant. The electric furnace, by use of electric power to provide the heat for smelting the charge and thereby eliminating the combustion products from fuel, produces a much smaller volume of gas with some increase in SO_2 concentration compared with the reverberatory furnace gas. This gas stream can be combined, if necessary with either converter gas or roaster for subsequent treatment in an acid plant.

Within the last five years continuous smelting processes have been developed to varying degrees by different operating companies in several countries. These continuous processes have aimed at combining the reverberatory smelting and converting operations in one piece of equipment. One advantage claimed for these continuous processes is the production of a single off-gas containing all the sulphur at a grade suitable for treatment in an acid plant. A full size production unit is now under construction for one of these processes.

In addition to the above developments which are essentially directed toward process equipment changes enabling elimination of low grade SO_2 reverberatory gas, other action is being taken to develop processes which will remove SO_2 from these gases. These processes range through alkali absorption processes, both regenerative and non-regenerative, organic absorption processes and solid absorption processes.

Conclusion

In conclusion technology has been developed to varying stages which will enable the non-ferrous industry to control and largely eliminate the emission of sulphur dioxide. Implementation of this technology is however subject to two other major problems, namely disposal of the sulphur products produced from sulphur dioxide and of equal and perhaps greater importance the cost of implementation of the removal of sulphur dioxide and disposal of sulphur product.

References

1. R.L. Duprey: "The Status of SO_x Emission Limitations" Chemical Engineering Progress Vol. 68 No. 2, February 1972.

2. Bureau of Mines Information Circular IC 8527 1971; Control of Sulphur Oxide Emissions in Copper, Lead and Zinc Smelting.

3. W.O. Alexander PhD. F.I.M.: "A Brief Review of the Development of the Copper, Zinc and Brass Industries in Great Britain from AD 1500 to 1900". REVIEW Vol. 1 No. 15, 1955 Published by Murex Limited.

4. G.M. Meisel: "Sulphur Recovery from Metallurgical Fluegas, Survey and Economics: Paper No. A72-86 A.I.M.E. Annual Meeting San Francisco 1972.

5. J. Dasher: "The Copper Smelter: Revolutionize or Replace". Paper No. A72-18 A.I.M.E. Annual Meeting San Francisco 1972.

SULFUR DIOXIDE REMOVAL FROM WASTE GASES: A STATUS REPORT UNITED STATES: RECOVERY PROCESSES

R. E. Harrington

Environmental Protection Agency Office of Research and Monitoring

The Environmental Protection Agency (EPA) is currently engaged in the massive job of reviewing from a nationwide viewpoint State implementation plans for control of pollutions to meet National Ambient Air Quality Standards by 1975. Although this task is not yet complete, the trends in technology requirements are clear. It is not suprising to find that flue gas desulfurization will be a key approach to the do-ability of SO_x control within the necessary time period and for the next decade.

What control process options then will the various sources likely have available over the next 2 years? Apparently several, although the level of maturity of candidate processes will vary substantially. One thing is clear, however. Although several processes will have progressed into demonstration, it is not likely that they will have been operated for a sufficient period to have clearly defined all operating problems and process economics. Several factors relating to each process will have varying levels of uncertainty associated with them. Among these will be the disposal of the process product. The best measure of the magnitude of the product disposal problem will undoubtedly have to await the test of time at the market place when the owners of control processes begin to compete for disposal sites and market outlets.

Most observers have classified flue gas desulfurization processes into throw-away and by-product recovery processes. This classification of throw-away implies a non-marketable product which will probably involve a disposal expense. A recovery process implies a marketable product, the sale of which will tend to offset production, handling and marketing costs. These are the connotations that will be used in this paper.

As with most things in the real economic-motivated world, however, definitions too are relative. Product value after all, is set by supply and demand. A product, especially a physically or chemically agressive product like SO_2 or H_2SO_4 can easily assume a negative value if there is no market or storage potential and its production is mandated. Sulfur and sulfuric acid may well find no market in many localities particularly if competing flue gas desulfurization products are superior and abundant or if the source is just too far from potential markets. Smelters located in the Western states of the United States which use only a small fraction of the acid they would potentially make if controlled to 90%, could be an example.

Conversely in many areas, the lime-limestone fly ash product of a lime/limestone scrubbing process which is normally agreed to be a throw-away process, may find useful application as a fill material, soil stabilizer, or building material. This product has been shown to have promise for these and other applications.

With this in mind, let us consider then the so-called recovery processes. If we start with the premise that flue gas desulfurization processes will for at least the next decade bear the brunt of the SO_x control load, the argument for a large number of process options is easily validated. One of the strongest arguments is based on the facts that present sulfur and sulfurous product markets can in fact absorb much of the products that could be produced by control processes provided they are of adequate quality and more competitive than current sources. Markets for individual sulfur products are not large however and will quickly saturate if even a fraction of the potential capacity of the control processes were directed their way. The two exceptions to this are the elemental sulfur market and the sulfuric acid market. Even with these materials the full impact of directing control process production into either could produce saturation when it is considered that there are many industrial consumers that could not arbitrarily convert to by-product acid or sulfur. The cost and practicality of transporting sulfur and sulfuric acid long distances are also major factors limiting the control process by-product markets. A wide variety of recovery processes, therefore, is an economic and practical necessity resulting from the original control premise.

The remainder of this paper will review several of the candidate recovery processes being developed. The listing is not exhaustive; indeed, new processes or process variations appear almost monthly. Most are in early stages of development such that no reliable economic or performance data are available. It will be several years before useful economic and technical comparisons between processes can be made. The order of their appearance in the following discussion has no significance; the depth of discussion for the most part relates to the availability of data and does not reflect any merit factor for the individual process.

Magnesium Oxide Scrubbing

The magnesium oxide (MgO) scrubbing process uses a slurry of magnesium oxide and/or carbonate to react with the SO_2 in flue gas to form sulfite and sulfate salts. Spent magnesium sulfite and sulfate salts are calcined to produce a concentrated stream of 10-15% SO_2 and regenerated MgO for reuse in the scrubber loop. Since the reactant is repeatedly recycled, it must be protected from contamination by fly ash. Therefore, fly ash must be efficiently removed from the flue gas prior to passing it into the MgO desulfurization process.

EPA and Boston Edison are cost sharing the development of a Chemico MgO scrubbing and regeneration process on a 125 MW oil-fired unit at Boston Edison's Mystic Station. This facility is currently undergoing start up and should be in continuous integrated operation over the next several months. It is still too early in the test program to assess the technical and economic merit of the system. Areas of potential problems include: Potential scaling and plugging problems, attainment of the 90% design efficiency, potential erosion/corrosion problems, effectiveness of the regeneration step, and overall system reliability. This demonstration represents the first time that the individual steps of scrubbing, centrifuging, and calcining have been operated on an integrated basis for the Chemico system.

In the Boston Edison development/demonstration system, particulate-free SO_2 flue gas from an oil-fired power plant enters a venturi absorber where it contacts a dense spray of magnesium oxide/carbonate/sulfite/sulfate scrubbing slurry. Fresh MgO slurry is added as a makeup to the venturi absorber. The scrubbed SO_2-free flue gas passes through a separator-demister section and exhausts to the stack. A bleed from the absorber is sent to a centrifuge for separation of solids; the mother liquor is returned to the absorber system. The wet cake from the centrifuge is dried in a direct-fired rotary kiln. Hot exhaust gases from the drier are sent to the stack where they provide reheat for the flue gas from the absorber.

The anhydrous crystals leaving the dryer are regenerated in a direct-fired calciner at about 1900°F to SO_2 and MgO. The MgO is recycled while the calciner off-gas containing about 15% SO_2 is oxidized to SO_3 and absorbed as H_2SO_4 in a conventional sulfuric acid plant.

In the EPA-Boston Edison demonstration, only the equipment for absorption, centrifuging, and drying will be located at the power plant. To avoid the capital cost of an acid plant as part of the current demonstration, the spent reactant is being shipped to the

Essex Chemical Plant at Rumford, Rhode Island, where it is calcined and used to produce 50 tons per day of 98% sulfuric acid and fresh MgO for reuse at Boston Edison. The SO_2 produced during regeneration will provide feed for this plant's total acid output.

A second full-scale magnesium oxide SO_2 removal facility is planned for Potomac Electric Power's Dickerson No. 3 unit. Approximately 100 MW of the 195 MW of this unit will be processed. Since this facility burns 3% sulfur, 8% ash coal, the scrubbing facility will consist of two venturis in series: the first scrubber will remove fly ash while the second will scrub with MgO slurry for SO_2 removal. The calcining and acid plant facilities located at the Essex Rumford Chemical Plant will be used in this demonstration also. This facility is scheduled for start up early in 1974.

Catalytic Oxidation (Cat-Ox)

The Cat-Ox process is a Monsanto developed version of the well-known contact sulfuric acid process. The process consists of passing the flue gas through a fixed catalyst bed where the SO_2 is oxidized with excess O_2 in the flue gas to SO_3. The SO_3 is then absorbed to form sulfuric acid.

Besides being designed for the dilute SO_2 concentrations found in flue gas, the Cat-Ox process differs from the conventional contact process in two ways: (1) the flue gas entering the process must either already be at the required 850°F reaction temperature or additional heat must be supplied. The heat of reaction of oxidizing SO_2 to SO_3 is insufficient to maintain the required temperature, (2) flue gas entering the system is not dried prior to entering the converter.

Monsanto has proposed two versions of the process: (1) the "integrated system" for use on new plants and (2) the "reheat system" for use on existing plants.

In the "integrated system", hot flue gas is passed directly from the boiler through a highly efficient electrostatic precipitator to remove at least 99.6% of the fly ash. The gas then flows through a vanadium pentoxide catalyst where the SO_2 oxidation occurs. Flue gas from the converter is cooled in an economizer followed by an air heater. Sulfuric acid is condensed in a direct contact packed-bed absorber producing about 80% sulfuric acid product.

The "reheat system" is similar to the "integrated system." However, since the flue gas enters the electrostatic precipitator at the typical temperature of 250° to 325°F this equipment need not be designed for high temperature service. High efficiency dust removal however is still required. The temperature of the entering flue gas is raised to reaction temperature by using the heat from the catalytic converter exhaust gases. Additional heat is supplied by the combustion of oil or gas.

This process is now being scaled-up on a 100 MW boiler at Wood River Power Plant of the Illinois Power Company. The $6.7 million cost of the demonstration will be shared by EPA and Illinois Power. Start up, planned for July 1972, will be followed by a one-year test program.

A primary advantage of the Cat-Ox process is that no raw materials, other than the flue gas and catalyst makeup, are required. The low concentration H_2SO_4 produced may be a serious disadvantage, depending on the marketing situation.

Wellman-Lord

In the Wellman-Lord process SO_2 is absorbed in a solution of sodium sulfite to form a concentrated liquor of sodium bisulfate. The spent scrubbing solution is regenerated by evaporation of water and decomposing the bisulfite in a crystallizer producing SO_2 and sodium sulfite crystals. The vapor from the crystallizer is cooled to condense the water and produce a high purity gaseous SO_2 which can be liquified or further processed to sulfur, or sulfuric acid. The sulfite solids are redissolved and recycled to the scrubber. Sulfate formed in the scrubber circuit cannot be regenerated and must be removed from the system.

This process was first successfully demonstrated at the Paulsboro, New Jersey plant of Olin Mathieson. It was installed in 1970 to control a sulfuric acid plant emitting 45,000 scfm of gas containing 4,000 to 6,000 ppm of SO_2. It has been operating successfully at better than 95% efficiency since start up. Two additional plants have been built and were put in operation in Japan in mid-1971. One is on an oil-fired boiler producing about 100,000 scfm of flue gas with 2,000 ppm SO_2. It was built by the Mitsubishi Chemical Machinery Company at the Chiba plant of the Japan Synthetic Rubber Company. The second is installed to control tail gases from a Claus furnace emitting about 35,000 scfm with 7,000 ppm SO_2. It was built by Sumitomo Chemical Engineering for the Negishi plant of Toa Nenryo. Both are reported operating successfully at over 90% removal efficiency for nearly a year. High levels of oxidation of sulfite to sulfate with the attendant need for high purge rates remain as a process problem to be solved.

As of March 1972, four more plants have been reported sold for industrial applications in the United States. One will go on a Claus unit operated by Standard Oil of California, another will control emissions from a kraft recovery boiler.

Sodium Citrate Process

The Sodium Citrate Process is being developed by the U. S.

Bureau of Mines, Salt Lake City. As presently conceived, the one molar mixture of sodium citrate prepared by reacting one and one-half moles of sodium hydroxide per mole of citric acid, is used as the reactant for scrubbing SO_2 containing effluent. The efficiency of removal of SO_2 from the effluent gas by the sodium citrate is quite sensitive to SO_2 concentration. As a result, the process appears to be much more attractive for use on high concentration SO_2 effluent streams such as smelter or refinery effluent gases.

The sodium citrate-sulfite is regenerated by reaction with hydrogen sulfite. An oxidation-reduction reaction occurs which produces elemental sulfur suspended in the sodium citrate liquor. The sulfur can be concentrated using mechanical techniques such as centrifugation or filtration. In the Bureau of Mines' version of the process, it is proposed that the ultimate separation be achieved by increasing the temperature of the sulfur-sodium citrate mixture to the melting temperature of sulfur and tapping-off the resulting molten sulfur make.

Work started on the process at the Bureau of Mines in late 1968. The development progressed from bench-scale tests through two larger continuous pilot units. In early 1971, pilot studies began on a continuous unit installed at Magma Copper Company's plant at San Manuel, Arizona. This pilot unit treats 400 cfm of stack gas from a reverberatory furnace. Considerable mechanical difficulty has been experienced but the pilot unit has established the technical feasibility of the process. A large demonstration unit will be built based on results obtained from the above pilot studies. Funds have been allocated for this purpose and negotiations are about complete for construction of this unit.

Ammonium Scrubbing

The use of ammonia scrubbing for SO_2 control has been perhaps one of the most studied processes in the world. Pilot units in the range of 10 MW have been operated to demonstrate the capability of the process to remove SO_2. The lack of a market for ammonia sulfate requires that the product be regenerated or used in another chemical process. Historically, the economics of the regeneration step has limited the process potential. Developments in Romania and Czechoslovakia in which the product is used in the production of fertilizer, opens a route for limited capacity for an ammonia scrubbing SO_2 control process.

An ammonium bisulfate scrubbing process is currently being developed by EPA and TVA. Although the process has never been operated in an integrated fashion on a commercial scale, most of its individual process operations have been proven to be technically feasible on large-scale industrial units. The ammonium bisulfate process removes SO_2 and forms ammonium sulfate. After condensing out the water, the pure stream of SO_2 can be further processed to sulfuric acid, elemental sulfur, or liquid SO_2. Ammonium sulfate is recovered from the acidified liquor by crystallization. The ammonium sulfate is decomposed by a thermal decomposition process to ammonium bisulfate and ammonia. The ammonium bisulfate is returned to the scrubber. To the extent that sulfites are oxidized to sulfates throughout the process, an excess of ammonium sulfate or bisulfate is produced. This excess may be either sold as a product or treated with lime for further ammonia recovery.

EPA and TVA are involved in a pilot plant development of all but the decomposition step of the ammonium bisulfate process. The pilot plant will process a 3,000 scfm side stream of flue gas from a coal-fired boiler burning 4% sulfur coal.

Sulfoxel-UOP

Universal Oil Products Corp. (UOP) has been developing a sulfur producing process for control of SO_2 from flue gas which has now been scaled to a large scale pilot unit. UOP is not ready to reveal the process description, other than the following: Flue gases are given a water pre-scrub using a venturi scrubber, to remove fly ash and sulfur trioxide. The flue gas is then scrubbed in an UOP TCA scrubber, using an unidentified liquid, to absorb the sulfur dioxide. The spent scrubber liquid is processed using a reducing gas to recover the absorbed sulfur oxides as molten sulfur and regenerate the scrubber liquid. Laboratory testing has been carried out by UOP. A pilot plant to handle 80,000 scfm of flue gases from a coal-fired boiler is in the start-up phase at Commonwealth Edison's statesline power plant at Hammond, Indiana.

Double Alkali

The concept of double alkali scrubbing is not new, yet because of its increased complexity has not matured rapidly. Its principal advantage comes from the use of a highly reactive soluable, non-plugging reactant in contact with the flue gas, thus eliminating many of the problems common to slurry or scaling processes. Sorption is followed by a subsequent chemically separate regeneration process. When used with lime as the regenerating material and the resultant calcium sulfur salts are discarded, the process is not a truly recovery process as defined earlier.

Several variations on the process are possible. One of the first, suggested by Dr. B. T. Kown of Rutgers University, called for the limestone to be injected into the boiler where it is calcined and absorbs sulfur trioxide. The flue gas is sequentially scrubbed with water, to remove fly ash and calcium solids, and then with a sodium hydroxide solution, to absorb the sulfur dioxide. The solids slurry effluent from the water scrubber is reacted with the spent liquor from the sodium hydroxide scrubber. The sodium sulfite reacts with the calcined limestone to produce solid calcium sulfite and a regenerated sodium salt.

A number of companies are working on the use of double-alkali process for specific applications. General Motors has pilot tested the double-alkali system using lime regeneration. Future plans include the installation of a full-scale industrial-size system at a General Motors plant. Zurn is presently designing a double-alkali unit for Caterpillar Tractor for installation on an industrial-size boiler in its Joliet, Illinois plant. The design basis for this recovery system is 90% removal of SO_2 when burning 2-1/2% sulfur coal.

Molten Carbonate Scrubbing

The molten carbonate scrubbing process was developed by Atomics International Division, North American Rockwell Corp., Canoga Park, California. In this process, sulfur oxide laden flue gas after fly ash removal, is scrubbed at 850°F with a molten eutectic mixture of lithium, sodium and potassium carbonates. The effluent scrubber liquid, containing the absorbed sulfur oxides as sulfites and sulfates, is reacted with coke at 1400°F to reduce the sulfites and sulfates to sulfides. The molt is then treated with steam and carbon dioxide at 850°F to produce hydrogen sulfide for feed to a Claus sulfur plant, and regenerate the alkaline carbonates for recycle. Laboratory development began in 1966 and is now completed. Most of the laboratory development work was done under sponsorship of EPA. The bench-scale scrubber consisted of a 4 inch ID spray chamber equipped with a single spray nozzle. Removal efficiencies greater than 95% were obtained at gas velocities up to 25 feet per second. Construction is to begin in the Summer of 1972 on a $4 million pilot plant at Con Edison's Arthur Kill generating station on Staten Island, New York. This pilot plant will accommodate about 10,000 kw of power plant capacity.

While the process has several attractive features such as high sulfur capacity and produces elemental sulfur as a product, there are also several known major engineering problems that must be solved for the process to work satisfactorily. The major problem area is that of materials of construction. Like other recovery processes, the molten carbonate process requires thorough pre-cleaning of flue gas which would require a high temperature precipitator. Concern for the effect of carry-over of molten salt into the boiler tubes could be a significant factor in limiting the process's acceptance in the industry.

Char Sorption Process

Westvaco is developing under EPA sponsorship a variation of the dry char sorption process. This process uses fluidized beds of activated carbon to remove SO_2 from waste gas streams. The char is regenerated to produce either elemental sulfur or concentrated sulfur dioxide as a by-product. The reductant is hydrogen sulfide. In one mode of regeneration conversion of the sorbed acid to sulfur dioxide is accomplished by reaction with the hydrogen sulfide in the regenerator/stripper. The hydrogen sulfide first homogeneously converts a proportion of the absorbed sulfuric acid to elemental sulfur at 300°F in the fluidized bed. Subsequently the carbon containing this sulfuric acid/sulfur mix is heated to 600°F to produce concentrated (90%) sulfur dioxide. The regenerated carbon is cooled and recycled for reuse.

In another mode of operation, the process option to produce elemental sulfur is anticipated. If the ratio of H_2S to H_2SO_4 is increased to three, elemental sulfur is the product and can be vaporized from the carbon, condensed and collected as product. This modification would have some similarity to the Sulfreen process where the hydrogen sulfide and SO_2 in the Claus tail gases are reacted together on a fixed bed of carbon to form sulfur which is volatilized with hot inert gases for collection.

This process is still in the early stages of development. Although most steps have been carried out individually on bench or pilot scale, the entire process has not yet been operated as an integrated system on any scale. The current EPA contract will result in the integration and operation of the process on an actual flue gas.

Potassium Formate-Consolidation Coal

Sulfur oxide laden flue gas (fly ash free) is scrubbed with 85% potassium formate solution at 200°F. The carbon dioxide in the scrubbed flue gas is removed by a hot potassium carbonate process, for use in the process, before the flue gas is vented. The spent potassium formate solution is processed in three steps to produce hydrogen sulfide, for feed to a Claus sulfur process, and regenerate the potassium formate solution for recycle. A sulfur dioxide removal efficiency of 90% is claimed.

Bench-scale testing was completed by the Research Division of Consolidation Coal Co. at Library, Pennsylvania prior to September 1970. In August 1971, the process was being tested as a system at a gas rate of 750 cfm. It is expected that the results of these tests should be available for evaluation late in 1972.

Metal Oxide Sorption

Efforts to develop a viable dry sorption process have in general proceeded much more slowly than wet processes. Dry metal oxide processes received extensive examination in the mid 1960's but were generally dropped because of the inability to develop a gas-solid contactor which would provide adequate mass transfer without excessive pressure drop or solid sorbent attrition. Two processes; metal oxide processes are still being actively developed however.

The Cupric Oxide Sorption developed by Shell International Research, The Hage (Netherlands) removes SO_2 from flue gas at 400°C using cupric oxide held by an alumina carrier. The cupric sulfate thus formed is then reduced with hydrogen to release SO_2 and regenerate the cupric oxide. Two towers packed with the absorbent are provided; the two alternating between absorptions and regeneration every hour. The SO_2 can be fed along with H_2S to a Claus plant for production of elemental sulfur or fed to a sulfuric acid plant.

Shell has made pilot plant tests of the process in the Netherlands. The pilot plant capacity was approximately 600 scfm. The world's first commercial plant is now under construction in Japan at the Yokkaichi refinery of Showa Yokkaichi Sekiyu Co. This plant, to be completed in March 1973, will treat approximately 70,000 scfm of flue gas from an oil-fired boiler. The recovered SO_2 will be fed to a Claus plant along with H_2S from elsewhere in the refinery to produce elemental sulfur. The developer's commercial position is not known.

In a similar process being developed by Esso Research & Engineering and Babcock & Wilcox, SO_2 laden stack gas is treated by contacting it with a unique but undisclosed dry sorbent that absorbs SO_2. According to B&W representatives, the sorbent is regenerated using a reducing gas to produce H_2S for sulfur recovery. (Earlier reports said the process produced an SO_2 containing gas stream that can be used for production of sulfuric acid.) The regenerated sorbent is then recycled to the absorber.

The developers report that pilot testing at a plant of Indiana & Michigan Electric Company has been underway since late 1969. The process is considered ready by the developers for demonstration on a 25 MW to 100 MW scale.

Stone and Webster/Ionics

The Stone and Webster/Ionics process while both interesting and potentially promising, is in the very early stages of development. Two technical and economic questions are key to the success of this process (1) the process is dependent on the ability of a membrane electrode to withstand the hostile environment of the flue gas scrubbing liquor and (2) the ability of the process to attain high electrical regeneration efficiency. The Stone and Webster/Ionics process is based on absorption of SO_2 with a sodium hydroxide. A solution of sodium sulfite and bisulfite is stripped of its SO_2 with a solution of sulfuric acid and sodium sulfate to form a sodium sulfate. The high concentration SO_2 may be dried and condensed to a liquid, converted to elemental sulfur, or fed directly to a sulfuric acid plant. The sodium sulfate solution resulting from acidification and stripping of the scrubber effluent is fed to an electrolytic cell which converts it to sodium hydroxide, which is recycled to the absorber, and to sulfuric acid, which is recycled to the stripper.

Experience with an actual flue gas was gained from the operation of a small pilot plant (200 scfm) which was operated at the Gannon Station of the Tampa Electric Company, Tampa, Florida, from July 1967 through January 1968. EPA sponsored a small study of oxidation rates for various scrubbers which was completed in 1970. Cell development work which would be necessary for scale up of the study has continued in the Ionics laboratories. EPA is currently negotiating to conduct a pilot scale study of this process. If successful, it is anticipated that a 3,000 scfm pilot plant will be built at a coal-fired power plant and that a cell development program aimed at production of a cell of the size required for a power plant module will be undertaken. This program would take 18-24 months and if successful, could be followed by design and construction of a 70 MW demonstration unit.

Summary

The problem of product disposal is among the most difficult to be solved by the owners of sulfur oxide control processes. The use of a variety of processes producing a variety of products will help solve this problem by providing a broader spectrum of sulfur products to be integrated into existing markets and permitting matches of products to appropriate markets.

This list of recovery processes is not exhaustive. Rather, it represents some of the most advanced processes being developed and offered to the user industries in the United States.

Most processes are in early stages of development and/or demonstration. Their selection and use over the next two to five years will not be without a measure of uncertainty as to performance, economics, and reliability. Yet the selection is wide and will certainly contribute to the assurance that the problems of SO_2 control can be solved.

SULPHUR DIOXIDE REMOVAL FROM WASTE GASES: A STATUS REPORT JAPAN

Jumpei Ando[1] and Shoichiro Hori[2]

[1]Faculty of Science and Engineering Chuo University, Tokyo; [2]Faculty of Hygienic Science Kitasato University, Tokyo

1. INTRODUCTION

Desulfurization of waste gases is assuming growing importance in Japan parallel to the mounting use of heavy fuel oil, which contains 2.5% sulfur on the average. Although many plants for hydrodesulfurization of heavy fuel oil--with a total capacity of nearly 500,000 barrels per stream day--are in operation and in 1971 by-produced 350,000 metric tons of elemental sulfur, approximately 5 million tons of sulfur dioxide was emitted by the combustion of heavy fuel oil in that year. In addition, sulfur dioxide in waste gases from chemical plants, oil refineries, smelteries, and coal-burning boilers presents some problems. The yearly average of sulfur dioxide concentration in the polluted districts of major cities ranged from 0.05 to 0.09 ppm in 1970.

Many processes for sulfur dioxide removal from waste gases have been developed and used recently on a commercial scale. A salient feature of the desulfurization efforts in Japan is that they are oriented towards processes that yield salable by-products. This is because Japan is subject to limitations in domestic supply of sulfur and its compounds as well as in land space available for disposal of useless by-products.

2. EMISSION AND REGULATION OF SO_2 IN JAPAN

2.1 SUPPLY OF ENERGY IN JAPAN

Along with the rapid industrial advance in Japan supply of

energy doubled in the past five years (Table 1); most of the increase was accounted for by imported oil. In 1970, 71% of the total energy supply depended upon petroleum while in 1965 the dependence was 58%.

Table 1 SUPPLY OF PRIMARY ENERGY (10^{12} kcal)[1,2]

		1965	1970	1975
Electric power	Hydraulic	143	187	212
	Atomic	0	11	102
Coal	Domestic	316	252	226
	Imported	136	392	627
Oil	Domestic	7	8	8
	Imported	960	2,189	3,425
Natural gas	Domestic	20	27	29
	Imported (LNG)	0	13	45
Other energy sources		29	19	17
Total		1,656	3,105	4,704

2.2 FUEL AND SULFUR

Most of petroleum is imported in the form of crude oil. About 85% of the crude oil comes from the Middle East and is rich in sulfur. The total amount of sulfur present in imported petroleum exceeded 3 million tons in 1970.

According to the Government's petroleum supply plan for 1971 through 1975, demand for oil is expected to reach approximately 230 million kiloliters in 1971 and 390 million kiloliters in 1975. Even though efforts have been made to import low-sulfur oil, the amount of sulfur seems to grow steadily.

In Japan, most of crude oil is treated by topping (atmospheric distillation). The residual oil from topping is known as "heavy oil" and is used for fuel. Approximately 90% of sulfur in crude remains in heavy oil. Some of heavy oil has a sulfur content as high as 4%.

In 1971, consumption of heavy oil amounted to approximately 130 million kiloliters. About one-fourth of the heavy oil was subjected to hydrodesulfurization giving nearly 350,000 tons of sulfur as by-product. Still 2.5 million tons of sulfur in heavy oil burned produced 5 million tons of SO_2, which constituted the chief source of SO_2 emission. About a third of heavy oil was burned in electric power plants and the rest in other plants and buildings.

Domestic production of coal is decreasing because coal is more expensive than oil. About a half of domestic coal has been burned in power plants. The coal mined in Japan contains about 1% sulfur on the average. Imported coal is used in the iron industry. The SO_2 problem is much less with coal than with heavy oil.

2.3 EMISSION STANDARD FOR SULFUR OXIDES AND NEED FOR DESULFURIZATION

The emission standard for sulfur oxides is given by the following equation:

$$q = k \times 10^{-3} He^2$$

q: amount of sulfur oxides, Nm^3/hr

k: a value between 2.92 and 26.3 assigned to each district

He: effective height of stack (meters)

The smallest k value 2.92 has been applied recently to plants to be built newly in districts that are already heavily polluted such as city areas of Tokyo and Osaka, and industrial areas such as Yokkaichi and Kawasaki. The sulfur content of fuel oil for new power plants in these districts is required to be about 0.5% or less.

It is difficult to reduce the sulfur content of heavy oil to below 1% by topped-crude hydrodesulfurization (direct desulfurization). By so-called indirect hydrodesulfurization, heavy oil is subjected to vacuum distillation and the vacuum gas oil is desulfurized to 0.2% sulfur, but the residual oil which contains much sulfur cannot be desulfurized. In such a situation, flue-gas desulfurization has become more and more important.

3. HISTORY OF WASTE-GAS DESULFURIZATION IN JAPAN

3.1 AMMONIA SCRUBBING

In 1931, there was built an ammonium sulfate plant based on a sulfite process with an annual capacity of 80,000 tons.[3] In this plant, sulfur dioxide gas produced by roasting pyrite was absorbed by aqueous ammonia to form an ammonium sulfite solution, which was oxidized with air to produce ammonium sulfate (Figure 1). Four absorbers were used in series. The third and the fourth absorbers in effect worked as desulfurizers for the waste gas from the second absorber. For the oxidation, "ceramic atomizers" were used. Each atomizer consisted of porous porcelain tubes through which compressed air was passed to form fine bubbles. The atomizer was so effective for oxidation that there was no need for an oxidation catalyst.

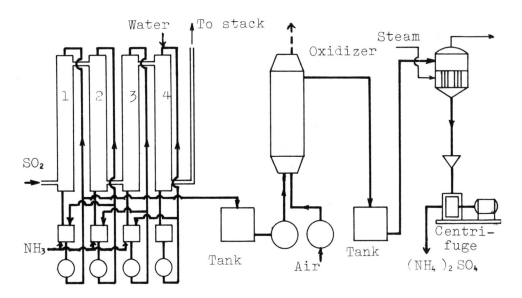

Figure 1 Ammonium sulfate plant by sulfite process (80,000ton/year, built in 1931)

This process was then applied to desulfurization of tail gas from several sulfuric acid plants including a plant of Degussa, Germany which was built in 1964. Sulfur oxides were recovered nearly completely with almost no fumes from the stack.

Another process of ammonia scrubbing was developed by Sumitomo Chemical in 1938. Tail gas from a sulfuric acid plant was absorbed by aqueous ammonia to form ammonium sulfite, to which sulfuric acid was added to produce ammonium sulfate and concentrated SO_2. The latter was returned to the sulfuric acid plant.

3.2 ZINC OXIDE PROCESS

Since 1952, Mitsui Mining and Smelting Co. has operated a desulfurization plant for tail gas 30,000 Nm^3/hr from sulfuric acid plants. SO_2 in the tail gas is absorbed with a zinc oxide slurry to form zinc sulfite. Sulfuric acid is added to the sulfite to produce zinc sulfate and concentrated SO_2. The latter is returned to the sulfuric acid plant.

3.3 LIME-GYPSUM PROCESS AND ROTARY ATOMIZER

Study of the wet-lime process was started by Japan Engineering Consulting Co. (JECO) in 1955 to recover SO_2 in waste gas with milk of lime to form calcium sulfite, which is oxidized to produce salable gypsum with a good quality. As the ceramic atomizer was not suitable for the oxidation because of the scaling problem involved, a rotary atomizer called "TOPCA" was developed.[4] Air is introduced into a rotating cylinder (Figure 2 (b)) or onto the surface of the cylinder (Figure 2 (a)). The air forms a thin film on the cylinder surface and then breaks into many small bubbles, 0.5 to 5mm in diameter (3mm on the average).

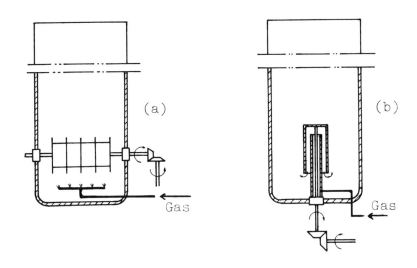

Figure 2 Two types of rotary atomizers

The bubble size increases when the amount of air exceeds a certain limit (critical volume) which is related to the size of the cylinder and to the speed of the rotating surface. A rotary atomizer with a cylinder 700mm in diameter and 1,400mm in length, rotating at 260 r.p.m., can treat up to 5,000 Nm^3/hr of air to generate bubbles between 0.5 and 5mm in diameter either in a solution or in a slurry and to produce about 7 tons/hr of gypsum. The atomizer is free from scaling, erosion and corrosion, because the cylinder is covered with a film of air (or other gas) during operation.

The lime-gypsum process has been developed by S. Hori with the assistance of K. Murakami and S. Nakagawa using the rotary atomizer. The first commercial plant by the lime-gypsum process was built in 1964 by Mitsubishi Heavy Industries and JECO at the Koyasu plant, Nippon Kokan K. to treat 60,000Nm^3/hr gas from a sulfuric acid plant. Three other commercial plants have been built recently by Mitsubishi and JECO.

3.4 OTHER WET PROCESSES

In the last few years many desulfurization plants have been built using various wet processes as the use of heavy oil increased remarkably resulting in a serious problem due to SO_2. Those processes will be described later.

3.5 DRY PROCESSES

Tests on limestone injection into boilers of power plants were carried out by the Central Research Institute of Electric Power Industry and Chubu Electric in 1965 and 1967. The removal of SO_2 was not satisfactory. Since 1966, the Agency of Industrial Science and Technology awarded contracts to Mitsubishi Heavy Industries (active manganese oxide process) and Hitachi Ltd. (activated carbon process) to build pilot plants and then prototype plants. Each prototype plant, built in 1968, was capable of treating 150,000Nm^3/hr flue gas from oil-fired power plants (50MW equivalent). Commercial plants by these processes have been completed recently as will be described later.

4. MAJOR PROCESSES AND THEIR CLASSIFICATION

At present, twenty-seven commercial plants for waste-gas desulfurization are in operation, thirteen plants are under construction, and several others are being planned. Many processes are used in these plants and also in pilot plants. These processes are classified by the types of reactions and products as shown in Table 2.

About a half of the commercial plants in operation use the "direct-recovery process" to produce sodium sulfite for paper mills. Oxidation should be minimized for the direct-recovery process, while it should be completed for the "oxidation process" by which gypsum or ammonium sulfate is produced. By the "regeneration process", the reaction products of SO_2 and absorbents are decomposed either at elevated temperatures or by reducing gas, to produce concentrated SO_2 and to regenerate the absorbent. By the "reduction process" SO_2 is finally reduced to H_2S or to S. A cheap reducing agent is required for the process. The "removal process", which does not produce any salable by-products, is not popular in Japan. The process may become popular when an oversupply of sulfur and its compounds occurs with an increasing number of plants using the various recovery processes noted above.

Table 2 CLASSIFICATION OF DESULFURIZATION PROCESSES AND MAJOR PLANTS

Type	Absorbent	By-product	Process developer	Unit capacity (1,000Nm³/hr)	Date of completion
Direct recovery (wet)	NaOH	Na_2SO_3	Kureha Chemical	$300^a)$, $300^a)$	1968
			Oji Paper	$184^a)$, $220^a)$, $130^b)$, $130^b)$	1966 to 1969
			Showa Denko	$150^a)$, $270^a)$, $80^c)$	1970 to 1971
			Bahco-Tsukishima	$220^b)$, $110^b)$, $45^b)$	1971
Oxidation (wet)	$Ca(OH)_2$	$CaSO_4 \cdot 2H_2O$	Mitsubishi-JECO	$60^c)$, $100^a)$, $52^d)$, $90^g)$	1964 to 1972
	NH_4OH	$(NH_4)_2SO_4$	Nippon Kokan K.	$150^d)$	1972
	$dil.H_2SO_4$	$CaSO_4 \cdot 2H_2O$	Chiyoda Chem.Eng.	$90^e)$, $33^e)$, $14^e)$, $158^a)$	1972
Oxidation (dry)	MnO_x	$(NH_4)_2SO_4$	Mitsubishi	$326^a)$	1972
	Charcoal	$CaSO_4 \cdot 2H_2O$	Hitachi Ltd.	$420^a)$	1972
Regeneration (wet)	Na_2SO_3	SO_2	Wellman-Lord (MKK)	$200^a)$	1971
			Wellman-Lord (SCE)	$60^e)$	1971
	MgO, MnO	SO_2	Grillo-Mitsui S.B.	pilot plant	1971
Regeneration (dry)	Charcoal	SO_2	Sumitomo S. M.	$175^a)$	1971
	CuO	SO_2	Shell	$120^a)$	1973
Reduction(dry)	Na_2CO_3	H_2S to S	NRIPR	pilot plant	1968
Removal(wet)	$Ca(OH)_2$	Waste $CaSO_3$	Chemico-Mitsui M.	$384^f)$	1972

a) Oil-burning boiler b) Kraft-recovery boiler c) Sulfuric acid plant
d) Sintering plant   e) Claus furnace   f) Coal-burning boiler   g) Smeltery

5. DIRECT-RECOVERY PROCESS

Many commercial plants have been built since 1966 to recover SO_2 to produce sodium sulfite for paper mills. Waste gas containing SO_2 is washed with a sodium sulfite solution to form a bisulfite solution which is neutralized with sodium hydroxide to produce a sodium sulfite solution (Figure 3). A part of the solution is recycled for SO_2 absorption and the rest is sent to paper mills. Solid sodium sulfite is also produced.

The types of absorbers used are packed towers for the Kureha process, vertical-cone type scrubbers for the Showa Denko process, Bahco scrubbers for the Bahco-Tsukishima process, and cyclonic and screen-type scrubbers for the Oji process. More than 90% of SO_2 is recovered. The process is simple and the investment cost is low. The demand for sodium sulfite, however, is limited. Overproduction of the sulfite has already occurred. Tests are being made on treating the sodium bisulfite solution with powdered limestone to recover sodium sulfite solution and to produce calcium sulfite, which is then oxidized to form gypsum.

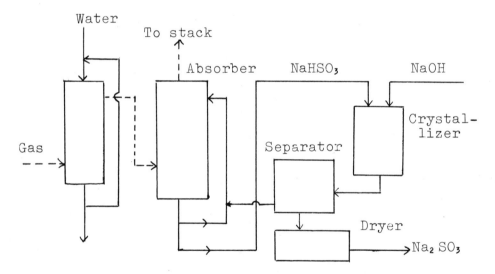

Figure 3   Kureha sodium sulfite process (300,000Nm³/hr)

6. OXIDATION PROCESS (WET)

6.1 MITSUBISHI-JECO LIME-GYPSUM PROCESS[6,7,8,9]

Waste gas is treated with milk of lime to produce calcium sulfite which is then oxidized by air to produce salable gypsum. In a commercial plant built in 1964 to treat tail gas from a sulfuric acid plant, a spray tower was used for desulfurization to minimize the scaling problem. Extensive tests have been made by Mitsubishi Heavy Industries to prevent scaling and to improve the scrubber, resulting in the establishment of many items of know-how for scale prevention.[6] In a new prototype plant designed to treat 100,000Nm^3/hr flue gas from an oil-fired boiler of Kansai Electric Power Co., two plastic-grid packed towers are used in series (Figure 4).

Prior to SO_2 removal by the packed towers, flue gas is washed with water for cooling to 60°C and for the removal of 80 to 90% of the dust in the gas. As the water becomes acidic and dissolves some components of the dust such as heavy metals, it is neutralized

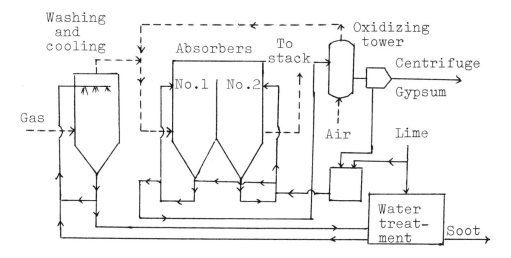

Figure 4 Mitsubishi-JECO lime-gypsum process
(100,000Nm^3/hr)

with milk of lime to precipitate the metal ions and is then filtered. The filtrate is recycled to the process. Sea water can be used for the washing.

The gas is then introduced into No. 1 absorber and then to No. 2 absorber to attain more than 90% desulfurization. Milk of lime is charged to the No. 2 absorber. The reaction product in the No. 2 absorber, a mixture of calcium sulfite and lime with a small amount of gypsum, is led to the No. 1 absorber. Calcium sulfite slurry at pH 4.5 is discharged from the No. 1 absorber and sent to an oxidation tower. In the tower the sulfite is converted into gypsum by oxidation with fine bubbles of air at 50 to 80°C under 5kg/cm^2 pressure. Very fine bubbles are produced by means of rotary atomizers invented by JECO which is very effective involving no operational problems.

Gypsum is centrifuged from liquor, the liquor and wash water are recycled to the cooling step of the gas. Gypsum thus obtained has more than 98% purity and good quality, which make it salable at ¥2,000/ton for use in cement and gypsum board.

6.2 CHIYODA DILUTE SULFURIC ACID PROCESS[7]

SO_2 in waste gas is absorbed in a packed tower with dilute sulfuric acid (2 to 4%) which contains a soluble iron catalyst and is saturated with oxygen. About 90% of SO_2 is absorbed and partly oxidized into sulfuric acid (Figure 5).

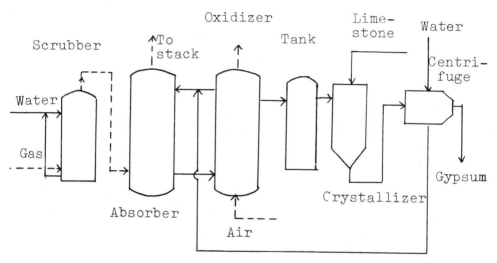

Figure 5 Chiyoda dilute sulfuric acid process
(158,000Nm³/hr plant under construction)

The product acid is led to the oxidizing tower into which air is bubbled from the bottom for oxidation. SO_2 is converted into sulfuric acid to increase the concentration of the acid. Most of the acid nearly saturated with oxygen is returned to the absorber. Part of the acid is treated with powdered limestone (minus 200 mesh) at 50 to 55°C to produce gypsum which is salable with a good quality. The gypsum is centrifuged from the mother liquor and washed with water. The mother liquor and wash water are sent to the absorber to maintain a proper concentration of the acid.

The process is simple and the plant is easy to operate, although a fairly big packed tower is required to ensure high recovery of SO_2. Four plants are under construction with this process(Table 2).

6.3 NIPPON KOKAN AMMONIA PROCESS[7]

This process is a combination of ammonia recovery from coke oven gas and SO_2 recovery from waste gas. Ammonia in coke oven gas is recovered by means of an ammonium bisulfite solution to form an ammonium sulfite solution, which is used to recover SO_2 in waste gas from an iron ore sintering plant. A prototype plant designed to treat 150,000Nm³/hr waste gas containing 400 to 1,000 ppm SO_2 is in operation at Keihin Works, Nippon Kokan K. (Figure 6).

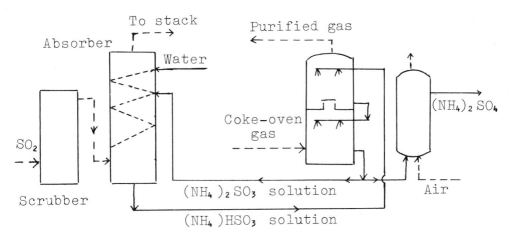

Figure 6 Nippon Kokan ammonia process (150,000Nm³/hr)

A spray tower is used for ammonia recovery. A Jinkoshi type absorber (a tower with screens on which liquid film is formed) is used for SO_2 absorption to ensure high recovery, viz. over 90%. A part of the ammonium sulfite solution is led into an oxidizer to be converted into ammonium sulfate by oxidation with air.

Tests also have been made on reacting the ammonium sulfite solution with calcium hydroxide to precipitate calcium sulfite and to recover ammonia. The calcium sulfite is air-oxidized into gypsum by means of a rotary atomizer "Topca" invented by JECO.

7. OXIDATION PROCESS (DRY)

7.1 MITSUBISHI ACTIVATED MANGANESE OXIDE PROCESS

Activated manganese oxide $MnOx \cdot nH_2O$ in powder form (5 to 150 microns in size) is charged into an absorber, and is dispersed and carried by flue gas (Figure 7). The powder is caught by a multiclone and electrostatic precipitator with an efficiency of 99.99%.

About 90% of SO_2 in the gas is removed, forming magnesium sulfate. The powder caught is a mixture of the oxide and sulfate. Most of the powder is returned to the absorber. The rest is treated with water to dissolve the sulfate; the unreacted oxide which is insoluble in water is centrifuged and returned to the absorber. The manganese sulfate solution is treated with ammonia and air to precipitate activated manganese oxide. The mother liquor is concentrated to obtain solid ammonium sulfate which has more than 99% purity.

A commercial plant to treat 326,000Nm^3/hr gas (110MW equivalent) has been completed recently at Yokkaichi power station, Chubu Electric Power.

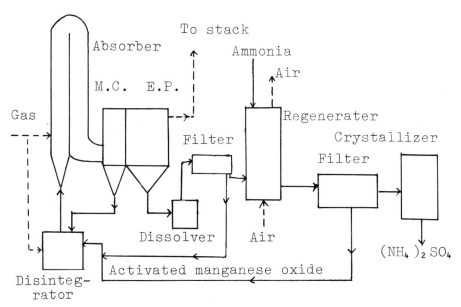

Figure 7   Mitsubishi activated manganese process (326,000Nm^3/hr)

7.2 HITACHI ACTIVATED CARBON PROCESS[8,9]

SO_2 in flue gas is absorbed by activated carbon (charcoal). The carbon is then washed with water to release dilute sulfuric acid (about 20% H_2SO_4). A prototype plant to treat 150,000Nm^3/hr flue gas from an oil-fired boiler was built in 1968 at Goi Works, Tokyo Electric Power (Figure 8). Five towers packed with activated carbon were used alternatively for absorption, water washing, and drying. Tests on concentration of the product acid to 65% H_2SO_4 were carried out by submerged combustion.

A commercial plant to treat 420,000Nm³/hr flue gas from an oil-fired boiler (150MW equivalent) has nearly been completed at Kashima Works, Tokyo Electric Power. Six towers with three compartments each are used in two trains. (Three towers with nine compartments for each train). About 80% of SO_2 will be removed while maintaining the outlet gas temperature above 100°C. Dilute sulfuric acid will be produced which will be treated with powdered limestone to produce salable gypsum.

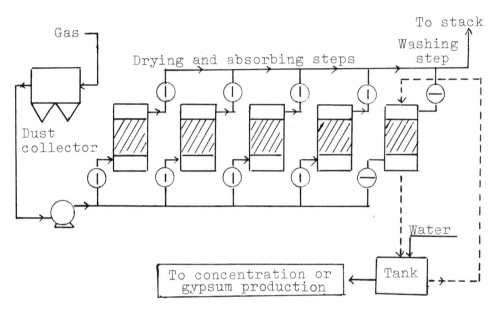

Figure 8 Hitachi activated carbon process(150,000 Nm³/hr) (420,000Nm³/hr plant under construction)

8. REGENERATION PROCESS (WET)

8.1 WELLMAN-LORD PROCESS

SO_2 is absorbed by a sodium sulfite solution to form a sodium bisulfite solution, which is then heated to produce concentrated SO_2 and a sodium sulfite solution. Two commercial plants have been in operation since 1971. One of them was designed by Mitsubishi Chemical Machinery (MKK) to treat flue gas from an oil-fired boiler 200,000Nm³/hr and to use the recovered SO_2 for sulfuric acid production. This plant was built in Chiba Works, Japan Synthetic Rubber (Figure 9). The other was designed by Sumitomo Chemical

Engineering (SCE) to treat tail gas from a Claus furnace 60,000Nm^3/hr and to return the recovered SO_2 to the Claus furnace. This plant was built at Negishi Works, Toa Nenryo. Oxidation of a portion of sodium sulfite into sodium sulfate presents some problem. An oxidation inhibitor is used in the plant designed by Sumitomo. About 90% of SO_2 is recovered.

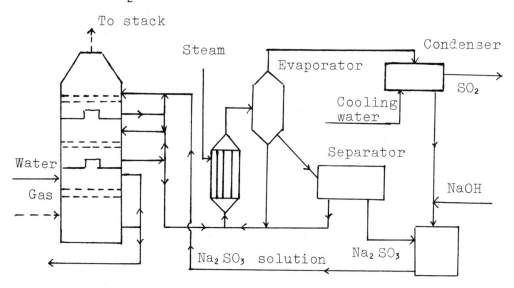

Figure 9 Wellman-Lord(MKK)process(200,000Nm³/hr)

8.2 GRILLO PROCESS

A pilot plant by the Grillo process designed to treat flue gas 1,200Nm³/hr with a slurry containing magnesium and manganese oxides has been operated by Mitsui Shipbuilding Co.

9. REGENERATION PROCESS (DRY)

9.1 SUMITOMO ACTIVATED CARBON PROCESS[8,9]

Moving beds of activated carbon are used for SO_2 recovery. Granular activated carbon descends slowly through an absorber into which flue gas is introduced perpendicularly to the carbon stream. The carbon which absorbed SO_2 is discharged from the bottom of the absorber and is led into a desorber through which the carbon again descends slowly. A recycling gas heated in a heat exchanger is introduced into the desorber perpendicularly to the carbon stream for desorption. An inert gas or reducing gas is introduced to purge the SO_2 gas. SO_2 concentration in the purged gas is controlled between 10 and 20%. A portion of the purged gas is led to a sulfuric acid plant; the rest is recycled to the desorber after being heated in the heat exchanger. Tail gas from the sulfuric acid plant which contains SO_2 is introduced into the absorber. A prototype plant to treat 175,000Nm^3/hr gas was built recently at Amagasaki Works, Kansai Electric Power Co.

9.2 SHELL PROCESS[7]

SO_2 in flue gas at 400°C is absorbed by cupric oxide held by an alumina carrier. The cupric sulfate thus formed is then reduced with hydrogen to release SO_2 and regenerate cupric oxide. A plant with a capacity to treat 120,000Nm^3/hr flue gas from an oil-fired boiler will be built by March 1973 at the Yokkaichi refinery, Showa Yokkaichi Sekiyu Co.

10. OTHER PROCESSES

10.1 REDUCTION PROCESS[9]

The active sodium carbonate process has been studied by the National Research Institute of Pollution and Resources. SO_2 is absorbed by powdered sodium carbonate at 300 to 350°C to form sodium sulfate, which is then subjected to reduction with carbon monoxide and hydrogen to produce sodium carbonate and hydrogen sulfide. The sulfide will be converted into elemental sulfur by a conventional process.

10.2 REMOVAL PROCESS[7]

A desulfurization plant with Chemico scrubbers (two stage venturi) to treat 384,000Nm^3/hr flue gas from a coal-fired boiler (165MW) has been constructed recently by Mitsui Miike Machinery Co. at Omuta Works of Mitsui Aluminum Co. Calcium hydroxide sludge obtained from calcium carbide by the reaction with water is used as an absorbent. The by-produced calcium sulfite will tentatively be used for land filling on a seashore. This plant is the first large-scale desulfurization plant in Japan which does not produce salable by-products.

11. ECONOMIC ASPECTS

11.1 INVESTMENT COST

The investment cost for several desulfurization plants are shown in Table 3. The cost is generally higher with dry processes than with wet processes.

Table 3 INVESTMENT COST FOR DESULFURIZATION PLANT

Process	Amount of gas (1,000Nm^3/hr)	Investment cost (Millions of yen)
Hitachi(Charcoal,dry)	420	1,684
Mitsubishi(MnOx,dry)	326	1,440
Sumitomo(Charcoal,dry)	175	850
Wellman-MKK(Sodium, wet)	200	800
Showa Denko(Sodium, wet)	270	550
Mitsubishi-JECO(Lime,wet)	100	450*
Mitsubishi-JECO(Lime,wet)	52	100

* Fully automatic

11.2 DESULFURIZATION COST

The desulfurization cost, as is shown in Table 4, is about ¥2,000/kl oil (about ¢0.14/kWh) to treat flue gas from a 100MW power plant (300,000Nm^3/hr) and ¥1,200/kl oil for a 350MW plant (1,050Nm^3/hr). The cost includes the expenses for reheating and depreciation, and also the revenue from by-products, namely, ¥2,000/ton for gypsum, ¥6,000/ton for 98% sulfuric acid, ¥10,000/ton for ammonium sulfate, and ¥16,000/ton for solid sodium sulfite. Thus the waste-gas desulfurization may be advantageous over topped-crude hydrodesulfurization of heavy oil by which about 70% of sulfur in the oil is removed at a cost of about ¥2,000/kl oil provided that the recovered sulfur is sold for ¥11,000/ton.

Table 4 COMPARISON OF DESULFURIZATION COST

(INCLUDING DEPRECIATION AND REVENUE FROM BY-PRODUCT)

Process	Sulfur removal(%)	Desulfurization cost (Yen/kl oil)	
		100MW*	350MW**
Wet process	90	2,000	1,200
Dry process	80	2,000	1,200
Topped-crude hydrodesulfurization	70	2,000	

* 300,000Nm^3/hr   ** 1,050,000Nm^3/hr

11.3 SUPPLY OF AND DEMAND FOR SULFUR AND ITS COMPOUNDS

The amount of sulfur and its compounds by-produced from desulfurization of both oil and waste gas has increased remarkably and would reach about one million tons as sulfur in 1974 (Table 5).

Table 5 ESTIMATED SUPPLY OF AND DEMAND FOR SULFUR AND ITS COMPOUNDS (1,000 TONS AS SULFUR)[12]

		1970	1972	1974
Supply	Pyrite	1,475	1,427	1,315
	Smelter gas	1,042	1,338	1,693
	Mined sulfur	109	36	0
	Recovered sulfur	270	570	763 - 1,282
	Total	2,896	3,277	3,766 - 4,285
Demand	Sulfuric acid	2,289	2,525	2,731
	Elemental sulfur	326	342	357
	Others	258	329	415
	Total	2,873	3,196	3,057
Surplus		23	79	263 - 782

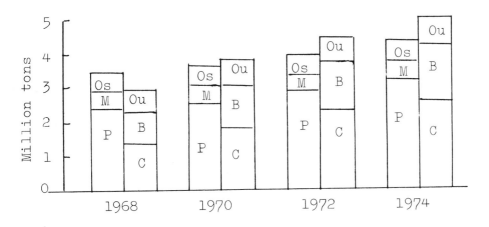

Supply P:Phospho-gypsum M:Mined gypsum Os:Other sources excluding by-product of desulfurization
Demand C:Cement B:Board Ou:Other uses
Figure 10 Supply of and demand for gypsum in Japan

A considerable oversupply of sulfur and its compounds may occur even though the supply of pyrite is reduced and that of mined sulfur stopped.

Among the by-products, gypsum seems to be most promising because of the rapid increase in the demand which would exceed the supply considerably without the by-product of desulfurization (Figure 10). Moreover, gypsum is stable and easy to store or abandon in case of overproduction.

REFERENCES

1) Enerugii Tokei (Energy Statistics), Ministry of International Trade and Industry (1971)

2) Nippon No Enerugii Mondai (Energy problems in Japan) Tsushosangyo Kenkyu, No. 160, MITI (1971)

3) Shoichiro Hori, Publication of Japan Ammonium Sulfate Association (1952). Report of the Government Chemical Industrial Research Institute, Tokyo, Vol. 46 No. 8 (Jan. 1952)

4) Japan Pat. 529,063 (1968), Canada Pat. 849,439 (1970), Brit. Pat. 1,206,326 (1971)

5) Y. Ishihara, C. Asakawa, and H. Furukawa, CRIEPI Technical Report CH-67002 (1968), CH-68001 (1969)

6) T. Uno, M. Atsukawa, and K. Muramatsu, Second International Lime-Limestone Wet Scrubbing Symposium, New Orleans, Louisiana, USA (Nov. 1971)

7) Private communication

8) Haiendatsuryu no Jissai (Practice of Stack Gas Desulfurization) Chuo Netsukanri Kyogikai, Oct. 1971

9) Haiendatsuryu Gijutsu (Technology for Waste-gas Desulfurization) Kagakukogyosha, 1970

10) Y. Ogata et. al., Juyu Haien Datsuryu Gijutsu (Technology for Desulfurization of Heavy Oil and Waste Gas), Nikkan Kogyo Shinbun 1971.

11) K. Kurosawa, Lecture at 26th Annual Meeting of Chemical Society of Japan, April, 1972.

12) A. Matsumoto, Kagaku Keizai, Oct. 1971

SULFUR DIOXIDE REMOVAL FROM WASTE GASES: A STATUS REPORT FOR EUROPE

Werner Brocke

*Landesanstalt für Immissions - und
Bodennutzungsschutz des Landes Nordrhein-
Westfalen, Essen*

From several years ago up to now the control of SO_2-emission in Europe is an environmental problem of high priority. Just recently calculated figures /⎽1⎽/ show that in some Western European countries a considerable increase of the emission has to be envisaged for the years up to 1980.

The dimensions of the problem can be imagined from the following predicted situation in 1980: Rate of increas from 1968 to 1980 for the U.K. 9 %, FRG 41 %, France 51 %, Italy 62 % and Spain 188 % approximately, and an absolut emission in 1980 of 2.6 million tons SO_2 in Spain, 3.2 in France, 5 in Italy, 5.4 in the FRG and 6 in the U.K. respectively. All these figures to the emissions related to fuel combustion in stationary sources only. In the FRG the share on the total is approximately 90 %. Other relevant sources for SO_2 are iron ore sintering with 5 - 6 %, traffic and sulfuric acid production with 2 - 3 % each.

Generally speaking, the consumption of solid fuel will decrease and of fuel oil increase in Western Europe. In a few countries like U.K. and FRG solid fuel remains a relevant factor so far as the SO_2 emissions are concerned, whereas the consumption of solid fuel in some mediterranean countries will increase. Lignites are the main solid fuel in the future. The remaining use of solid fuels has an impact on the type of process needed for the reduction of SO_2-emission because of the fact that a desulphurisation of these fuels to a sufficient extent is rather impossible.

5. SULFUR DIOXIDE

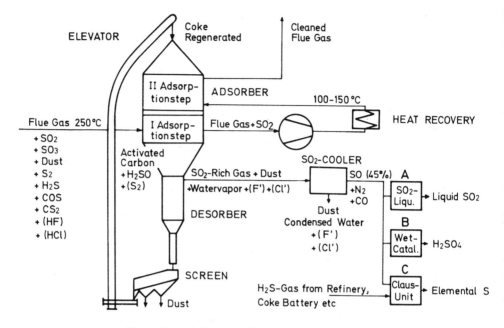

Fig. 1. Scheme of Reinluft-Process

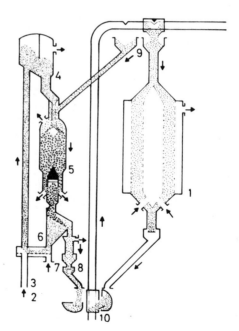

Fig. 2. BF-Process for Flue Gas Desulphurization

Therefore, great efforts have been and are made in developing processes for either the desulphurisation of fuels and waste gases or for the desulphurisation of gases before or during combustion. In addition, activities are under way to increase the efficiency of energy use and thereby to decrease pollution by fuels which cause emissions. Meanwhile there exist nine processes for waste gas desulphurisation in Western Europe which look promising for application from a technical point of view.

1. ADSORPTION

Three physical adsorption processes (Reinluft, Sulfacid, BF) are using carbonacious adsorbents (peat coke, lignite, low temperature coke, activated carbon on hard coal basis). Two of them produce either sulphuric acid or sulphur (Reinluft, BF), one sulphuric acid (Sulfacid). The Reinluft process was tested with a capacity of 33 000 m_n^3/h (1 m_n^3 = Cubicmeter at $0°$ C, 1 atm) at a coalfired power station, the BF-process with 3000 m_n^3/h respectively. Both work with a moving bed of adsorbents at temperatures up to $120°$ C (Reinluft) and $210°$ C (BF) and thermal regeneration and recycling. In contrast, the Sulfacid process used fixed layers at 60 - $80°$ C and wet extraction of the adsorbed H_2SO_4. The latter is already used in 7 commercial plants with different waste gases of very little or no dust content up to 32 000 m_n^3/h capacity.

The essentials of these 3 processes are shown by the next figures:

Fig. 1 shows the principal design of a Reinluft plant $/2/$. The regenerated adsorbent - peat coke or lignite - low temperature coke - is brought mechanically to the adsorption part II, moves through part I and is than desorbed by hot gases at temperatures between 350 and $400°$ C in the smaller part at the bottom of the installation. In the tested unit, an efficiency of 70 % with 4 g SO_2/m_n^3 and up to 2 g dust/m_n^3 in the flue gas from a coal fired boiler was obtained. The gas from coke-desorption contained 25 % SO_2 by volume. Lignite-low temperature coke was found to be superior to peat coke.

From a technical point of view we consider this process to be ready for application, but some modifications are found to be useful, mainly to avoid the danger of ignition. In actual practice one would separate adsorber and desorber.

In the BF-process a specially developed type of coke

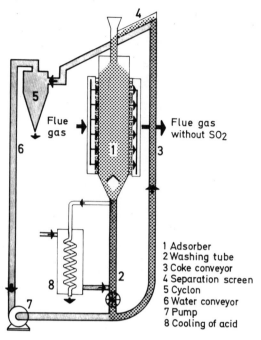

Fig. 3

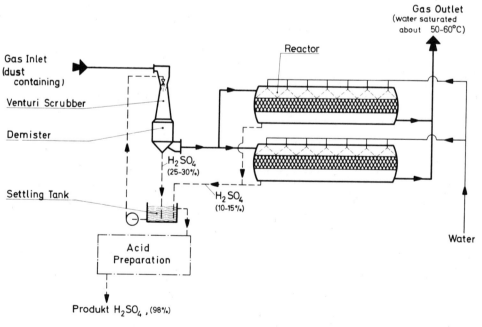

Fig. 4. Sulfacid-Process for Dust Containing Flue Gases

with high resistivity against ignition and attrition is employed in a circuit similar to that of the Reinluft process $\underline{/}3, 4\underline{/}$. The adsorption is practically the same but can be done at higher temperatures. The coke can be regenerated by 2 different methods.

Fig. 2 shows a plant with thermal regeneration. The coke coming from the adsorber (1) is transported to a separate desorber (5) where it is regenerated in a hot moving sand bed. After desorption coke and sand are separated and prepared for the next circuit. The desorption gas contains 30 % SO_2 by volume.

Alternatively, the coke can be regenerated by extraction with water. Fig. 3 shows this modified plant. The coke from the adsorber (1) is put into a washing pipe (2). In a counter current extraction a dilute acid of about 20 % is obtained.

Both alternatives have been tested in pilot plants with flue gas from a coal-fired boiler, which contained up to 2500 ppm SO_2 and 2.5 g $dust/m_n^3$. Up to 95 % of SO_2 and 94 % of dust have been removed.

Fig. 4 shows the plan of a <u>Sulfacid plant</u> for the desulphurisation of dust-containing gases. Fixed layers of a cheap type of activated carbon are in use. The adsorption takes place at 60 to 80° C under nearly water-saturated gas conditions. The adsorbed acid is discontinuously extracted with water, resulting in a weak acid of 10 to 15 %.

There are already some commercially used plants for effluents of different sources with SO_2 contents between 0.13 and 0.18 % by volume, gas temperatures between 50 and 350° C and gas volumes between 750 and 32 000 m_n^3/h per adsorber, working with efficiencies up to 99 % removal of SO_2. The sources are sulfon processes, oil-fired boilers of small sizes, sulphuric acid plants and titanium calcining plants $/5/$. Besides this, pilot plant work is under way for coal fired boilers in Germany and Czechoslovakia.

2. ALKALI ABSORPTION

The injection of lime, hydrated lime and a suspension of hydrated lime has been tested in full scale on hard coal as well as on oil fired boilers at temperatures up to 1500° C. In the range of 30 % absorption efficiency where obtained with 1.2 - 2 times the stoichiometrically necessary amounts

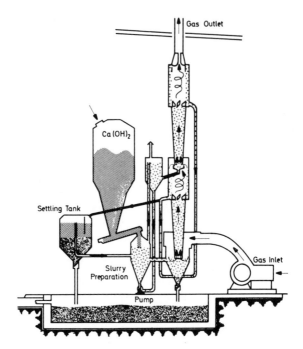

Fig. 5. Absorption Scrubber CTB

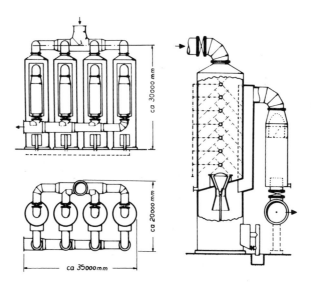

Fig. 6. SO_2 and Dust Removal from Flue Gases. Scrubber Plant for 350 MW.

of absorbent. The throw-away byproduct consists mainly of sulphate $\underline{/\,6\,\underline{/}}$.

Although **this** method is easy to practise and very cheap in investment it has a very limited efficiency and has a tremendous consuption of material and results in enormous quantities of residue when it is applied continously. Therefore, we assume this process only to be a type of intervention method for existing boilers under critical meteorological conditions.

For continous operation lime, linestone or hydrated lime are used in scrubbing processes. The main principles of these processes are well known since decades of years from industrial applications. In Europe there are 2 newer processes, which fall into this category, namely the Bahco and the Bischoff process.

Fig. 5 shows the principal plan of a <u>Bahco-scrubbing-unit.</u> In this two-stage venturi-type scrubber a suspension of hydrated lime is used for SO_2 absorption $\underline{/\,7,\,8\,\underline{/}}$. Since November 1969 the Bahco-process is used on an industrialsize unit behind an oil fired boiler equivalent to 25 Megawatt. Two mor units of the same size are running since 1970 and 1971 respectively. A removal efficiency of 97 to 98 % at a final pH of suspension from 6 to 6.5 is obtained. The throw-away byproduct consists of sulphate and sulphite.

Approximately 1.1 times the stoichiometric amount of lime is consumed. The dust content is reduced simultaneously with an efficiency between 90 and 95 %. The scrubbers have to be cleaned 2 to 3 times a year. No serious problems occured in the commercial units in a plant near Stockholm. This process seems to be ready for application in large oil fired plants.

The <u>Bischoff process</u> $\underline{/\,9\,\underline{/}}$ differs somewhat form the Bahco process, so far as apparatus and addition of absorbent material are concerned and the source of waste gas used in the test. Fig. 6 shows the testing unit at a coal fired boiler near Dortmund, with a capacity of 35 Megawatt or 140 000 m_n^3/h. The gas passes from the top through the spray scrubber part down through the Ringspalt- (ring-slot-) scrubber part and further over a liquid trap to the blower. To obtain a steady dedusting efficiency or to increase or to decrease it, the ring-slot is adjustable during operation. Pulverised lime is added at the gas entrance.

Since early 1971 this full scale testing unit is under

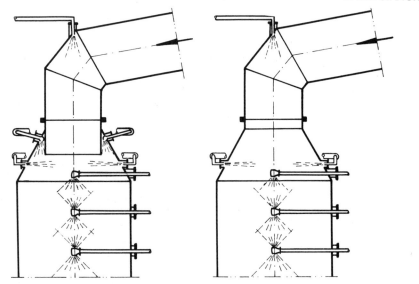

Fig. 7. SO_2 and Dust Removal from Flue Gases. Gas Inlet of Pilot Plant.

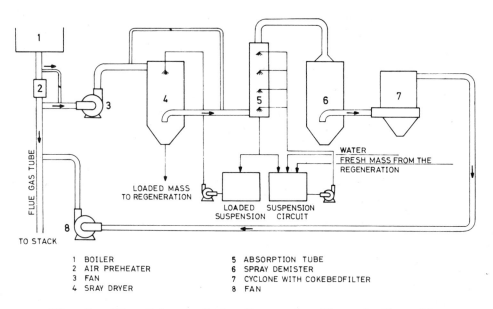

1 BOILER
2 AIR PREHEATER
3 FAN
4 SRAY DRYER
5 ABSORPTION TUBE
6 SPRAY DEMISTER
7 CYCLONE WITH COKEBEDFILTER
8 FAN

Fig. 8. Flow Scheme of the Absorption Plant in Wesseling

operation. With SO_2-concentrations between 3 to 4 g/m_n^3 and dust concentrations between 8 to 17 g/m_n^3 SO_2 were removed by 80 % with approximately 1.15 times the stoichiometric amount of lime. The remaining dust concentration was lower than 50 mg/m_n^3, equivalent to 99.7 % efficiency, for which a pressure drop between 250 to 300 mm water gage in the whole unit is needed.

During the first time a lot of troubles occured, mainly with invrustations in the spray scrubber part and at the blower vanes. Fig. 7 shows the critical part, the gas entry. In the meantime the main troubles are overcome by changing the sprayer characteristics and by a better demisting efficiency. But additional tests are necessary before application in full scale.

Additional tests are conducted concerning the behaviour of deposited residues with regard to possible ground water contamination and the feasibility of the use of the residues in brick production. Also another method of lime addition as lime water will be tested.

German lignite coals give a considerable amount of calcium in the fly ash of boilers using them. First tests with a similar but much smaller installation of Bischoff (2800 m_n^3/h flue gas) in a power plant using lignite resulted in a dedusting efficiency up to 98 % and a remaining dust concentration lower than 70 mg/m_n^3 and simultaneously a SO_2 removal of about 67 % without any additional absorbing material. These tests give a promising outlook on an economic solution for lignite.

The adsorption of SO_2 and SO_3 by the Grillo-AGS-process occurs in two different phases in the gas stream with mixtures and compounds of magnesium- and manganese oxides [10]. The SO_3 absorption takes place upstream before the air preheater at about 350° C with a dry mass in a fluid bed reactor. The aim of the SO_3 absorption is to decrease the acid dew point and therefore to increase the heat recovery. The SO_2 absorption takes place after the boiler and is done by a scrubbing process with a suspension of the same absorption mass. The absorption mass can be loaded up to a sulphur content of about 17 to 21 % wt with respect to the dry substance. The loaded mass is thermally regenerated with coal or fuel oil under reducing conditions in a fluid bed calciner. The calciner-off-gas contains 5 to 7 % vl SO_2 and can be processed in the usual manner to sulphuric acid.

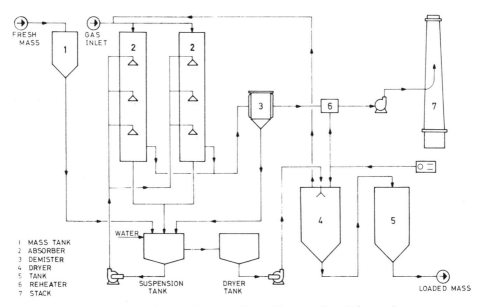

1 MASS TANK
2 ABSORBER
3 DEMISTER
4 DRYER
5 TANK
6 REHEATER
7 STACK

Fig. 9. Flow Scheme of an Absorption Plant in Technical Size

Fig. 10

Meanwhile this scrubbing process was tested in prototype plants of technical size downstream of a sulphuric acid contact plant with gas streams up to 20 000 m^3/h as well as downstreams an oil fired boiler with gas streams up to 27 000 m$_n^3$/h. The SO_3 absorption indeed was tested in a pilot plant only which was installed at the same power plant. The processed gas stream was about 8000 m$_n^3$/h. Any unsolved problems remained during tests.

Fig. 8 shows the pilot plant in the power station. The hot partial flue gas stream was drawn by the blower 3 into the spray dryer in which the suspension is dryed and the flue gas is enriched with water vapour while it is cooled simutaneously. In the dryer there is a partial absorption while in the absorber (5) the actual scrubbing process occurs and in the separators (6) and (7) spray and dust are removed. The regeneration of the absorption mass was done in a special plant which for this reason especially was built in a sulphuric acid plant. Together with a few additional equipments it was identic with a pyrite roasting facility.

Fig. 9 shows the concept for commercial installations, which resulted from experience during development. The main change is that drying of the mass should be done with a separate dryer (4) outside of the flue gas stream, separately heated, combinde with heating the flue gas (6) before the chimney if necessary.

It derived from the test experience that large installations can be built with a garanteed SO_2 removal up to 95 % for a flue gas from oil or coal fired boilers. In the latter case, the flue gas has to be dedusted down to less than 150 mg/m$_n^3$. Fig. 10 shows a picture of a pilot plant in Japan, built by Mitsui, running since June 1971, designed for 1200 m$_n^3$/h. The Japanese built a fluidized bed calciner for 100 kg mass/h. In our opinion the process is ready for application at boilers. A pilot plant for iron ore sintering machine waste gases is under construction in Germany.

The EDF-Process makes use of NH_3 for absorbing SO_2 in a scrubber /¯11_7. With regard to processing the solutions of ammoniumsulfite, bisulfite and sulfate flowing out of the scrubbers one differentiates between 2 methods.

Fig. 11 shows the first possibility. Gaseous NH_3 is added to the flue gas before entering the scrubber and reacts in the gasphase with SO_3 and SO_2. In the scrubbers Nr. 1 and

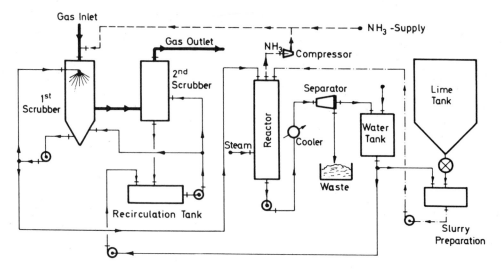

Fig. 11. Scheme for Desulphurization of Flue Gases; E.D.F. Process

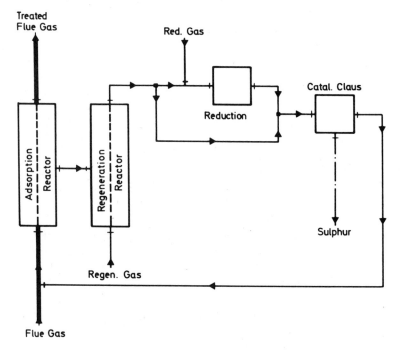

Fig. 12. Flow Scheme of an SFGD-Processing Route

2 the reaction products and further SO_2 naturally will be
scrubbed out while the flue gas is cooled from 120° C to
55° C at the same time. Simultaneously the solutions are evaporated to a desired concentration of soluble ammonium-salts.
Then the solution is led to a reactor, which is heated by
steam up to 98 - 100° C, where a solution of hydrated lime
is added. Calciumsulfite, - sulfate and gaseous NH_3 are
formed. The NH_3 is added in a circuit to the flue gas. After
cooling and separation of the salts the liquid is reused in
the scrubber circuit and the sludge is deposited.

Meanwhile a process route was found of which details
are not yet published, but generally it is known, that about
60 to 70 % of the scrubbing solutions can be processed, so
that about 70 % of the SO_2 scrubbed out of the flue gas can
be recovered for further processing to useful products. The
precondition for this possibility was a method of dividing
the scrubber solutions in a way that the one part contains
the sulfites and the other part the sulfates. The solution
containing sulfates is processed with lime water. This process was tested since 1967 at a boiler which was fired with
flue oil containing 3 % sulphur. The size of the plant
corresponds to 25 MW.

The desulphurization efficiency was about 93 to 97 %.
The French point of view is that plants of any size could
now be built.

3. CATALYTIC ABSORPTION

The last but not least western european process to be
mentioned is the SFGD-process. This process was developed
by Shell in a pilot plant with a flue gas capacity of about
1000 m_n^3/h $/$ 12 $/$. In the SFGD-process (see Fig.12) waste
gases pass over CuO on activated alumina as a carrier at
400° C, forming $CuSO_4$ in the presence of oxygen. At a breakthrough of 10 % SO_2 the reactor is changed to regeneration
by a reducing gas. The SO_2-rich gas can be processed
further to sulphur by first reducing 2/3 of the SO_2 to H_2S
and then handling the total in a Claus-unit. Depending on the
special method for regeneration a double reactor capacity
has to be installed. For further details reference is made
on the paper of Prof. Ando.

4. EASTERN EUROPE

An informative overlook over the eastern european activities gives the documentation of the results of a seminar, held by the United Nations european office in Switzerland in 1970 $\underline{/}$13$\underline{/}$. It indicates, that similar processes, mainly using NH_3, lime or limestone and activated carbon have been and are being developed in different countries. There exist pilot and full scale plants for boilers and sintering machines, up to a waste gas capacity of 2 millions m_n^3/h.

5. RELATIVE ECONOMICS

To say it very briefly, the relative economics depend on so many factors that it is impossible to give an exact answer for each of the above mentioned processes. Generally speaking under german conditions the additional costs on power generation appears to be in a range from 0.06 to 0.15 US-cents/kWh for 80 % sulphurdioxide removal $\underline{/}$14$\underline{/}$. This would mean a 7 to 15 % increase in power generation costs or a 2 to 5 % increase in costs for electricity consumers. If we try to keep the SO_2 emission from german power plants in 1980 on the level attained in 1970 by all available means like fuel and waste gas desulphurisation the total power production by pollutant and non-pollutant processes will cost 1 to 2 % more and the electricity consumer has to pay less than 1 % more on an average basis.

6. REFERENCES

$\underline{/}$1$\underline{/}$ Report of Joint Group an Air Pollution from Fuel Combustion in Stationary Sources, OECD, Paris, 1971.

$\underline{/}$2$\underline{/}$ Abschlußbericht über die Versuchsanlage Lünen zur Entschwefelung von Rauchgasen nach dem Reinluftverfahren. Steinkohlen-Elektrizität Aktiengesellschaft, Essen, Dezember 1970.

$\underline{/}$3$\underline{/}$ Jüntgen,H. und K.Knoblauch: Neuere Entwicklungstendenzen der Rauchgasentschwefelung in den USA und Deutschland.
Mitteilungen der Vereinigung der Großkesselbetreiber, Heft 3, 51. Jahrgang, Juni 1971, Seiten 211/218.

$\underline{/}$4$\underline{/}$ Knoblauch,K. und H.Jüntgen: Das Bergbau-Forschung-Verfahren zur trockenen Entschwefelung staubhaltiger Abgase.
VDI-Berichte Nr. 149, 1970, Seiten 116/121.

/ 5 / Güpner, O.: Sulfacid-Anlagen zur SO_2-Abscheidung.
VDI-Berichte 149, 1970, Seiten 127/133.

/ 6 / Goldschmidt, K.: Praktische Erfahrungen im Bereich der Industrie - Schlußfolgerungen.
VDI-Berichte Nr. 149, 1970, Seiten 133/139.

/ 7 / Gustavsson, K.A.: Bahco SO_2-scrubber CTB-CTD-1,5, Pilot plant connected to Bahco's heating central plant.
Proceedings from lime/limestone wet scrubbing symposium. March 16-20, 1970, Florida, USA.

/ 8 / Bernhoff, R.: Erfahrungen mit dem Einsatz von Kalk bei der Rauchgasentschwefelung.
Aufbereitungstechnik (1970), Nr. 11, Seiten 651/661.

/ 9 / Hausberg, G.: Das Bischoff-Verfahren.
VDI-Berichte Nr. 149, 1970, Seiten 121/127.

/ 10 / Husmann, K. und G. Hänig: Das Grillo-AGS-Verfahren zur Entschwefelung von Abgasen.
Brennstoff-Wärme-Kraft Bd. 23 (1971), Nr. 3, Seiten 85/91.

/ 11 / Mascarello, J.M. und J. Auclair: Ergebnisse der Versuchsanlage zur Abgasentschwefelung im E.d.F.-Kraftwerk St. Quen.
Mitteilungen der VGB 51, Heft 4, August 1971, Seiten 324/328.

/ 12 / Dautzenberg, F.M., Naber, J.E. and van Ginneken, A.J.J.: The Shell Flue Gas Desulphurization Process.
Sixty-Eight National Meeting, Febr. 28 - March 4, 1971, Houston, Texas.

/ 13 / "Desulphurization of Fuels and Combustion Gases".
Proceeding of the first Seminar, St/ECE/Air Poll/1, and Add 1, 2 and 3;
United Nations Publication, Sales number: E.71.II.E/Mim.21, United Nations, Sales Section, 1211 Geneva 10, Switzerland.

/ 14 / Brocke, W., Peters, W. et al.: Air Pollution by Flue Gases from Stationary Boilers and Furnaces in the Federal Republic of Germany.
National Report of the Federal Republic of Germany for OECD, Paris, 1971, available from Bundesminister des Innern, Bonn, Germany, Rheindorfer Straße.

POLLUTION ABATEMENT IN THE MANUFACTURE AND LOADING OF PROPELLANTS & EXPLOSIVES

Irving Forsten

Picatinny Arsenal, Dover, New Jersey

The United States Army Munitions Command, Dover, New Jersey is responsible for the manufacture and load, assemble and pack of explosives for the Army. Because of the nature of the industry and its uniqueness to military needs, there are many pollution problems that arise which are not being addressed by the private industrial complex. (See figures 1 - 3) In the area of pollution abatement, Picatinny Arsenal, Dover, New Jersey, a commodity center under Munitions Command, has the lead responsibility for abatement at the 17 Government Owned Contractor Operated Plants who manufacture and load, assemble, and pack munitions items.

One mission of Picatinny Arsenal is to perform engineering in order to incorporate necessary manufacturing controls in current and future ammunition manufacture, and loading operations and to emphasize pollution abatement in accordance with advance technology and applicable governmental standards. To accomplish this mission, the pollution abatement staff was divided into functional groups. The areas were survey, technology, automated instrumentation and control and pilot plant demonstrations. Each functional group is responsible for air, water, and solid waste pollution in their technical area.

The survey team's chief function is to establish definitively the pollution problem to be solved. Typical manufacturing, as well as loading operations will be discussed together with the pollutants which are to be abated. (See figures 4 & 5) The survey team is taking each plant and defining the problem by chemical analysis and relationship to production levels. Included in this effort is water management. This data is compiled in a report form for each plant and the data is given to the technology team.

TYPICAL POLLUTANTS
EXPLOSIVES MANUFACTURING

WATER	AIR	SOLID
• RED WATER	• NO_x	• EXPLOSIVES
• NITROBODY WASTES	• SO_x	• INORGANIC SALTS
• NITRATES	• ACID MIST	• CONTAMINATED WOOD & PAPER
• SULFATES	• PARTICULATES	
• TOLUENES	• METHYL NITRATE	
• ACIDS	• ORGANIC VOLATILES	
• ACETATES		
• ALCOHOLS		
• CYCLOHEXANONE		

FIGURE 1

TYPICAL POLLUTANTS
PROPELLANT MANUFACTURE

WATER	AIR	SOLIDS
• SULFURIC ACID	• NO_x	• WASTE PROPELLANTS
• NITRIC ACID	• SO_x	• INORGANIC SALTS
• SUSPENDED SOLIDS	• VOLATILE SOLVENTS	• HEAVY METAL SALTS
• TOXIC METALS	• PARTICULATES	
• GRAPHITES		
• SOLVENTS		

FIGURE 2

TYPICAL POLLUTANTS
LAP PLANT

WATER (WASTE WATER & WASH OPERATIONS)	AIR (OPEN BURNING OF SCRAP)	SOLID (CONTAMINATED PACKING MATERIALS)
• EXPLOSIVES	• SO_x	• WOOD
• OILS	• NO_x	• PAPER
• HEAVY METAL WASTES	• HCL	• METAL PARTS
• CYANIDES	• PARTICULATES	
• NITRATES		
• CHROMATES		
• GRAPHITES		

FIGURE 3

TYPICAL GOCO POLLUTANTS AND CONTROLS

POLLUTANTS	CURRENT PLANT CONCENTRATIONS	PROPOSED SYSTEM
AIR		
NO_x	18000 - 104000 PPD	MOLECULAR SIEVE
SO_x	1000 - 2500 PPD	WET SCRUBBER
ACID MIST	1000 - 6000 PPD	MIST FILTER

PPD = POUNDS PER DAY

FIGURE 4

6. PROPELLANTS AND EXPLOSIVES

TYPICAL GOCO POLLUTANTS AND CONTROLS

POLLUTANTS	CURRENT PLANT CONCENTRATIONS	PROPOSED SYSTEM
WATER		
RED WATER	40000 - 60000 PPM	REGENERATION
PINK WATER	70 - 150 PPM	CARBON ADSORPTION
SOLIDS		
INORGANIC (NITRATES & SULFATES)	150 - 1000 PPM	CHEMICAL
EXPLOSIVES AND PROPELLANTS	1000 - 20000 PPD	INCINERATION

PPD = POUNDS PER DAY
PPM = PARTS PER MILLION

FIGURE 5

PINK WATER

SOURCE:
- SCRAP TNT DISSOLVED IN CLEAN UP WATER

AMOUNT:
- MAXIMUM OF 150PPM TNT DISSOLVED IN CLEAN UP WATER

METHODS OF TREATMENT

TECHNICAL APPROACH
- CARBON ADSORPTION WITH REGENERATION
- ION EXCHANGE WITH REGENERATION
- BIODEGRADATION

FIGURE 6

RED WATER WASTE

SOURCE:
- SELLITE WASHING FROM THE MANUFACTURING OF TNT PROCESS

AMOUNT: 250000 POUNDS PER DAY AVERAGE RED WATER PER PLANT

METHODS OF TREATMENT

TECHNICAL APPROACH
- FLUID BED REDUCTION OF SULFATE TO CARBONATE
- ACIDIFICATION OF RED WATER TO RECOVER DINITROTOLUENES
- ADAPTION OF THE TAMPELLA PROCESS TO RED WATER RECOVERY

FIGURE 7

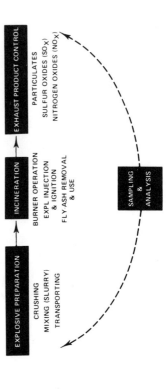

FIGURE 8

Once the pollutants have been defined by type and amount, the technology team has the responsibility for developing solutions. A wide range of solutions is required to satisfy many of the requirements. Such techniques as reverse osmosis, fluid-bed reduction, ion exchange, carbon adsorbtion, denitrification and chemical reaction are all used as the situation warrants. Resolution of proposed solutions into pilot plant demonstrations is accomplished in conjunction with the plants to establish design criteria leading to a representative prototype. Each solution is critically examined to insure that pollution is not simply transferred from one media to another. For instance, removal of particulates from the air by scrubbing only to dump the scrubber water into the river or burning of solid wastes in the open which turns solid waste into air pollution. In all technological solutions maximum use is made of recycle or reuse techniques to reduce the effluent to a minimum. Typical TNT waste considerations are shown in figures 6 & 7.

Once a process is established, it must be maintained at a high rate of efficiency. To insure this will be done, the automated instrumentation team establishes monitoring and controls which will maintain proper operation. Extremely sensitive instruments must be developed which not only can measure the low pollution levels, but which can do so with a minimum of calibration and maintenance. Particular emphasis is being placed on instruments which measure NO, NO_2, SO_2, CO, hydrocarbons and particulates for air pollution. Suspended matter, dissolved ionic and non-ionic materials and temperature are being investigated for water monitoring. Here again, the stream is placed on field applications rather than laboratory operations. The instruments are evaluated under actual operating conditions to determine which system best meets the needs of the industrial operation. Selection is made by a cooperative effort between the plants, Picatinny Arsenal and other concerned federal agencies.

The pilot plant group, establishes the demonstration unit which combines the technology with the instrumentation to meet the stated survey needs. From this demonstration pilot plant, design criteria will be established to allow for full scale development of a facility. Current pilot plant work is being conducted on explosive contaminated inert wastes, explosive and propellant incineration, fluid-bed technology, molecular sieves for NO_2 abatement and SO_x abatement systems. As in all of the previous team's efforts, great emphasis is placed on involvement of the industrial plant in these operations. A typical systems analysis of the work to be performed for explosive waste is shown in figure 8.

Manufacturing and load, assemble, pack operations for which pollution abatement is well underway are TNT, RDX, HMX, lead azide, nitrocellulose and nitroguanidine manufacturing.

INVESTIGATIONS RELATED TO PREVENTION AND CONTROL OF WATER POLLUTION IN THE U. S. TNT INDUSTRY

David H. Rosenblatt

Edgewood Arsenal Chemical Laboratory Edgewood Arsenal, Maryland 21010

INTRODUCTION

TNT (2,4,6-trinitrotoluene), a high explosive, is first among the relatively small number of chemicals produced in the U.S.A. exclusively in government-owned facilities. The production rate has varied, reaching impressive proportions at the height of the conflict in Southeast Asia.

Water pollution from TNT arises in two ways: <u>First</u>, there is the washing of solid or molten crude TNT with 16% aqueous sodium sulfite to remove selectively the 3-5% of TNT isomers containing a nitro group in the 3 or 5 position. The intensely colored high-solids spent wash solutions, known as "red waters," are difficult to dispose of. In the most modern U.S. plant, which manufactures TNT by a continuous process, these solutions contain about 25% dissolved solids; for example, there might be 9% organics, 10.6% sodium sulfite, 0.6% sodium sulfate, 3.5% sodium nitrite and 1.7% sodium nitrate. At present, the "red waters" are concentrated and sold to paper mills when possible or incinerated; the latter is an expensive though relatively successful answer to the problem. <u>Second</u>, wastes from a variety of sources, including bomb and shell washing operations, find their way into streams and rivers; when such TNT-containing waters, at neutral or alkaline pH, are exposed to sunlight, they develop a highly visible color and are called "pink waters." The color chemistry of both the "red" and the "pink" waters is far from simple. The color is due, at least in part, to the presence of Meisenheimer complexes that are very difficult to adsorb from water.

The first step in seeking appropriate responses to the "red waters" challenge was to review the entire technological picture and list available options. The specifications for TNT, in particular the requirement for a solidification point of 80.2 C or higher, were called into question, since deletion of the requirement for removing sulfite-reactive isomers to achieve this specification would obviate formation of "red waters;" provided certain changes in munitions manufacture could be made, it appeared that the largest part of the TNT produced might not need sulfite treatment. Recrystallization of TNT, as practiced in Europe, needed to be reconsidered; the mixture of isomers from the mother liquors might find military application or a place in the growing civilian explosives market. New synthetic approaches to TNT might decrease, if not eliminate, the formation of undesirable products. The removal of isomeric impurities might be carried out at intermediate manufacturing stages, e.g., by distilling mononitrotoluene mixtures. Finally, nucleophiles other than sulfite ion, applied to TNT or the intermediate dinitrotoluenes, might prove more economical.

In a similar manner, the "pink waters" problem was assessed. Improved housekeeping and water reuse could undoubtedly diminish TNT pollution. Chemical or microbiological treatment might provide an answer. Key emphasis was placed on the possibility of treatment with activated carbon.

In addition to the above listing of options available for attacking pollution, we gathered information on current TNT manufacturing processes, analytical methodology, toxicology, and the chemistry and photochemistry of "red" and "pink" waters. The sulfite-hydroxide method (C. C. Ruchhoft and W. G. Meckler, Ind. Eng. Chem., Anal. Ed. $\underline{17}$, 430 (1945)), slightly modified, was chosen for analyzing the relatively pure TNT solutions to be encountered in carbon treatment studies. We recommended that all or several of the options be pursued, both in the realization that some might fail and in the belief that the results of several lines of attack might be employed at the same time. Our own small research group proposed to carry out only those two investigations for which we felt most qualified, namely a study of the reactions of nucleophiles with TNT isomers, and an evaluation of activated carbons for removing TNT from water. Our initial review provided a fairly solid launching platform for the work we subsequently carried out.

NUCLEOPHILE STUDIES

Purification of TNT with aqueous sodium sulfite is based on the powerful nucleophilicity of sulfite towards activated aromatic nitro groups. Since all nitro groups in TNT itself are meta to one another, none of these groups is activated to any extent. In all the isomers of TNT, however, there is one nitro group that is ortho to a second nitro group and ortho or para to a third; the second and third doubly activate the first. Hence, the 5-nitro of 2,4,5-trinitrotoluene, and the 3-nitros of 2,3,4- and 2,3,6-trinitrotoluenes (the three principal isomeric impurities) are readily displaced by sulfite. Such reactions produce the corresponding (and colorless) dinitrotoluenesulfonate and nitrite ions. The red color arises from other reactions of the sulfite ion, with both TNT and some of its impurities. Dinitrotoluenesulfonate salts are extremely water-soluble and cannot be extracted with solvents. If an isomer-selective process is to be used for upgrading crude TNT, it should avoid both the red color and the difficulties of isolating the organic by-products. This might be accomplished by finding a nucleophile other than sulfite ion.

Most of the work on nucleophiles was carried out with 2,4,5-trinitrotoluene in aqueous acetonitrile. The five best nucleophile species at pH 7 were all sulfur-containing, namely dithiophosphate, thiophenolate, alkylxanthate, sulfite and dithiocarbpiperidide. At pH 7, amines were relatively inert, the most reactive being hydrazine and morpholine; at sufficiently high pH, however, some of the more basic amines were quite active, notably pyrrolidine. Hypochlorite ion at pH 7 was quite effective, and appeared to produce the instantly hydrolyzed hypochlorite ester; but the phenolic intermediate that was thus formed then underwent chlorination and/or oxidation. When acted on by nucleophiles of decreasing effectiveness, the isomers of TNT exhibited consistent decreases of reactivity in the order 2,4,5-, 2,3,4- and 2,3,6-trinitrotoluene, regardless of the nucleophile. These decreases ranged from about 10 to 100 fold.

Most nucleophiles that displace a nitro group become o,p-directing, activating substituents for electrophilic aromatic substitutions. (Sulfite is an exception.) Thus, it should be possible to nitrate the isolated product of a nucleophilic displacement on the 2,4,5-, 2,3,4- or 2,3,6-isomer of TNT to give, in each case, 2,4,6-trinitro-3-substituted toluene. We may term the formation of this one compound from each of the three isomers a convergent synthesis. If such a compound were to be found suitable as an explosive, its production might consume whatever isomers of TNT were produced. The convergent synthesis could either be part of a washing operation similar to the present sulfite process or could utilize isomers recovered from recrystallization mother liquors.

Discovery of the ambident nature of nitrite ion in its attack on 2,4-dinitrohalobenzenes was another result of this work, though of little direct application to the problem at hand. Nitrite was found to displace the iodine of 2,4-dinitroiodobenzene, exclusively via the nitrogen of the ion, to give 1,2,4-trinitrobenzene. Another nitrite ion then displaced the 1-nitro group through oxygen attack, presumably forming 2,4-dinitrophenyl nitrite, which must have hydrolyzed at once to 2,4-dinitrophenol (the observed product). By contrast, 2,4-dinitrofluorobenzene underwent direct oxygen attack and eventual formation of 2,4-dinitrophenol, without the intermediacy of 1,2,4-trinitrobenzene. The reaction between nitrite ion and the chloro and bromo analogs followed both reaction pathways. An interesting offshoot of this work was the discovery of a color reaction of potential use in analyzing for TNT isomers contained in crude TNT. Thus, nitrite ion reacted rapidly with the various isomers in N,N-dimethylformamide; on dilution of such mixtures with water, the color initially produced by TNT faded markedly, while the yellow dinitrocresol color that was formed with the isomers persisted. This difference could be of value for process control technology.

ADSORPTION OF TNT FROM WATER

Several commercially produced activated carbons were tested for their ability to remove TNT from water. The Freundlich adsorption isotherms obtained for the powder-form material were not meant to be the only selection criteria; however, they did help to eliminate the least promising carbons. It was concluded that only activated carbons derived from bituminous coal merited further work, i.e., Calgon Filtrasorbs 300 and 400 and Westvaco WV-L, WV-W and WV-G. By these tests, the following were excluded from further consideration: Westvaco Nuchar C-115; Kerr-McGee natural and synthetic bone char; Barnebey-Cheney JV wood charcoal; Barnebey-Cheney YF nutshell charcoal; Atlas Darco S51 lignite charcoal; and Atlas Darco KB sweetgum sawdust charcoal. Most of the charcoals studied had been manufactured in granular form and ground to powder for these experiments.

Lack of time to investigate all bituminous coal granular activated carbons in adsorption columns led to the arbitrary selection of 12-40 mesh Filtrasorb 400, since this was already being used at one U.S. installation. Initial column studies gave good linear plots of volume of effluent versus log $(C_o/C_B -1)$, where C_o was the TNT feed concentration and C_B the concentration of TNT in the effluent at a particular time. For these purposes, 4.09 g (dry weight) of carbon activated at 150° was used, filling a 9 mm diameter glass tube to a height, λ, of 16.5 cm. Feed solution containing about 144 ppm of TNT in distilled water was passed through at a flow rate, R, of 4 ml/min.

An equation was used of the form

$$V_t = \text{total volume of effluent} = A\lambda - \frac{2.303R}{C_o k_v} \log(C_o/C_B - 1)$$

$A\lambda C_o$ = saturation capacity of the entire carbon bed
k_v = kinetic constant

The critical bed depth, λ_c, is defined as that bed depth, λ, for which $V_t = 0$, i.e., the bed depth at which the arbitrarily chosen breakthrough concentration, C_B, is seen immediately in the effluent. It is obvious that

$$\lambda_c = \frac{2.303R}{AC_o k_v} \log(C_o/C_B - 1).$$

By treatment of the data with these equations, k_v was determined to be 0.0133 L min^{-1}g^{-1} and the saturation capacity of the column ($A\lambda C_o$) was calculated as 3.20 g (i.e., 0.78 g of TNT per g of carbon). The latter value compares favorably with a saturation capacity of 3.34 g by influent-effluent analysis. Critical bed depth for whatever breakthrough concentration may be chosen is important, since it represents the portion of the column that cannot be used. It may be seen that critical bed depth is directly proportional to flow rate, according to the model used. Further testing will be needed to validate the model.

It was possible to regenerate part of the original sorptive capacity of the activated carbon with warm toluene, and to reuse the carbon. The toluene solution contained 1.82 g of TNT. In the second use cycle, only 1.51 g (i.e., 0.37 g of TNT per g of carbon) was adsorbed, and a second regeneration of the column with toluene removed most of this, namely, 1.40 g. Thus, it appears likely that the sorptive capacity approaches some constant level on repeated use of the column.

STUDIES IN PROGRESS

The nucleophile investigations are being continued. Additional nucleophiles are under examination for reactivity with TNT. The reactivity of the most potent nucleophiles with o-dinitro and p-dinitro isomers of dinitrotoluene is being studied, as are physical methods for removing such isomers from the desired mixture of 2,4- and 2,6-dinitrotoluene. Experiments on convergent syntheses from TNT isomers are also planned.

The study of adsorption of TNT from aqueous solution by activated carbon columns is likewise being continued; the columns will be cycled several times, with solvent regeneration. The kinetic aspects will be improved by the use of a very constant pumping system and temperature regulation. If time permits, the effect of using different flow rates (with a fresh sample of carbon for each rate) will be observed.

These experiments are coordinated with work by others to improve the quality of aqueous effluents issuing from TNT facilities.

CONTROL OF NITROGEN OXIDE EMISSIONS FOR NITRIC ACID PLANTS

H. G. Rigo, W. J. Mikucki, M. L. Davis

U. S. Army Construction Engineering Research Laboratory, Interstate Research Park, Newmark Drive, P.O. Box 4005, Champaign, Illinois 61820

INTRODUCTION

In December of 1971, the United States Environmental Protection Agency (EPA) published its New Source Emission Standards.[1] One of the industries controlled by this legislation is the nitric acid industry. The regulation specifies that no more than three pounds of nitrogen oxide can be emitted from a nitric acid plant for each ton of acid produced. This is equivalent, under normal operating conditions, to 208 ppm NOx (106 mg/m^3).

Like most munitions manufacturing plants, U. S. Army ammunition plants have four major sources of nitrogen oxides emissions: nitric acid production, concentration, and storage and nitrated explosives manufacture. CERL's responsibility was the abatement of NOx associated with nitric acid manufacture; this operation constitutes the predominant source of NOx at ammunition plants. As an aid to understand the problem, the following is a brief description of the operation of a typical AOP.

Following the schematic process flow diagram (Figure 1), high pressure air is first mixed with pressurized, gaseous ammonia. The mixture is then passed over a platinum catalyst. An exothermic reaction occurs in the air-ammonia mixture which releases nitric oxide and water vapor. The hot gases are cooled in a heat exchange train. The NO oxidizes to NO$_2$ during the cooling step. Further temperature reduction occurs in a condenser where water is condensed out of the gas stream. Weak nitric acid results from the condensation. The cooled gas is introduced into the bottom of a bubble tray column where the nitrogen dioxide

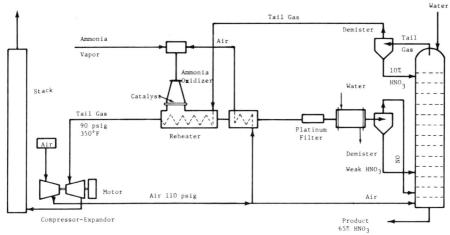

Fig. 1. AOP Process Flow Diagram

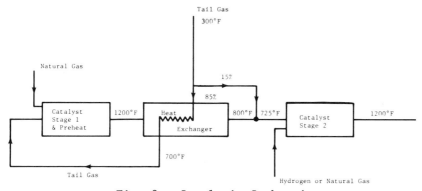

Fig. 2. Catalytic Combustion

is absorbed in water to produce nitric acid. The weak acid removed in the condenser is injected into the tower where it absorbs more NO_2 and becomes a stronger product. Overall, 50-65% nitric acid results from this system. Leaving the top of the absorption tower is a gas stream (tailgas) whose quality is 1500-3500 ppm oxides of nitrogen (NOx) and 100% relative humidity.

If leakages are minimal, then the tail gas is the primary source of nitrogen oxide emitted from an AOP into the atmosphere. As this gas is exhausted to the atmosphere, its nominal quality is 3000 ppm NO_2 and 3% oxygen. However, at the request of CERL, sampling and analysis of the subject ammunition plant's tail gas was performed by the Army Environmental Hygiene Agency (AEHA). They reported a tail gas quality of 4500 ppm NO_2 and 6% O_2. This indicates that the AOP was probably being overdriven. This disparity between nominal and actual tail gas quality points up the need for exact quantification of the gas stream to be treated. Perhaps the most important discovery of the AEHA analysis was the fact that the tail gas at that plant contained 2 ppm of sulfur oxides (SOx).

FINDINGS

Four NOx control techniques (improved AOP design, incineration, scrubbing, and selective catalytic reduction) will be briefly reviewed to establish their limitations. Two other approaches, general catalytic reduction and dry absorption, will be discussed in greater detail. The latter group contains the two control methods determined to be most applicable to AOP's.

Improved AOP design includes enhanced ammonia-to-NO conversion through atmospheric burning, then compression of the NO stream for more complete NO-to-NO_2 conversion. Extending the number of stages in the absorption tower is included in this group. While two-pressure-stage nitric acid processes are common in Europe, due to their large size, they typically are more expensive to build than the completely high pressure DuPont process typical of American plants. Shorter available pay out periods in the United States caused Newman and Klein[2] to conclude that these two stage processes probably are not economical in the United States. Further, when the question is controlling the pollution from an existing facility, constructing a different process is probably not a reasonable alternative. Replacing the absorption tower is feasible, however. The drawback is noted by Newman[3] when he demonstrates that the absorber costs needed to achieve 600 ppm NOx concentration in the tail gas are approximately $80,000 more than that required to achieve 2,000 ppm. Further, to meet the legal limit imposed on new nitric acid plants, Newman's cost curves extrapolate to well over $500,000. The reasonableness of this number can be determined by utilizing the AOP absorber sizing techniques presented by Chilton.[4] More than twenty-one additional trays would be required to reduce the NOx in the tail gas to an acceptable level. For practical and economic reasons, process improvements were eliminated as a viable alternative for the present problem.

The second approach examined was direct flame incineration; a process in which the NOx is reduced to elemental nitrogen by burning in a fuel-rich or reducing atmosphere. For high oxygen levels, this involves a large fuel cost. The incinerator can be located before or after the power recovery train. If the system is to be located after the existing XRD power recovery expanders (piston expanders), the incineration chamber would have to be preceded by a large damping chamber to absorb large pressure pulses between XRD piston strokes. Location of the incinerator before the XRD requires the design of a combustion chamber to operate at 90 psig. Such a chamber could be built, but it would be expensive. Regardless of location, the reduction process would yield carbon monoxide which would require secondary combustion where, unless extreme care is exercised, more NOx will be generated. Economics indicate that this type of system is not applicable to AOP's.

The next approach is scrubbing. This approach suffers the same limitations that taller absorption towers face; i.e., NOx is hard to absorb. Even if any absorbed NOx is immediately complexed with calcium or sodium, NOx scrubbing efficiencies much above 20% have not been obtained to date even in the laboratory.[5] In addition, workable calcium slurry generation and sodium feed techniques are not well developed. Scrubber erosion and plugging are common. Scrubbing is not really a viable alternative.

The most advanced method of selective catalytic reduction involves burning ammonia and the tail gas between 750°F and 950°F. The idea is to use the hydrogen in the ammonia to strip oxygen from nitrogen oxides and leave elemental nitrogen and water vapor. The difficulty is a practical one; i.e., insuring that the temperature of the whole catalyst bed is within the prescribed limits. If the temperature is too low, besides the original NOx problem, the acid plant now faces an ammonia problem. If the temperature is too high, the ammonia will combine with oxygen to produce more nitric oxide. While this system may be perfected in the future, the information presently available to the authors indicates that this control technique is not yet commercial.

The first control technique considered to be commercially available is general catalytic reduction (see Figure 2). Here the tail gas is mixed with a reducing agent such as methane and fed into one or more catalyst beds. The mixed gas temperature is raised to the fuel ignition temperature and nitric oxides are reduced to elemental nitrogen. The tail gases are then directed through a waste heat boiler. If steam can be utilized, a salable by-product is produced.

Since there is insufficient oxygen to burn all the carbon if an organic gas is used as a fuel, carbon monoxide would be expected as a product instead of the nitric oxides. The reports of one emission test where the same catalyst bed was operated to achieve NOx concentrations of 200 and 46 ppm in the tail gas also reported that CO concentrations of 600 and 3,000 ppm resulted.[6]

Study of the kinetics of the reduction reaction also indicates a significant possibility of cyanide gas being generated. Using equilibrium data[7] which predicts the measured CO concentrations from the cited tests, cyanide concentrations of 200 and 600 ppm are indicated.

Hydrogen fuel for catalytic combustion is proposed as a method for avoiding the HCN and CO problem. The quantity of hydrogen consumed for abatement would require onsite hydrogen generation which in turn requires additional operating personnel and presents an explosion hazard.

Further, Reed and Harvin[8] indicate that catalysts sometimes don't perform as designed. In fact, those authors cite an example where two competitors' honeycomb catalysts were used in the same catalytic reduction unit. Neither catalyst performed as specified, yet the catalyst manufacturers could detect no loss of catalyst activity.

Platinum and paladium catalysts can be poisoned. The sulfur analyzed in the AOP tail gas could rapidly inactivate an abater catalyst change. If this measured sulfur is indeed a common constituent of AOP tail gases, application of catalytic abaters should be viewed skeptically.

Although there are problems associated with catalytic reduction, this control technique is the industry standard and was the basis for the 208 ppm EPA standard. Therefore, catalytic reduction was considered as a viable abatement technique.

The final control method is dry adsorption, or the use of a molecular sieve. Before examining the sieve process, it should be noted that there are two basic types of sieve materials, natural and synthetic zeolites. The synthetic zeolites are primarily dehydration structures and as such have a strong preference for water. While the available systems involve two or three beds, the principles of operation are the same. The system described will be based on a two-bed design (see Figure 3). The tail gas is picked up at the top of the adsorption tower before any heat exchangers are encountered. This gas is cooled to both remove moisture (this consumes bed space) and drop the gas temperature to compensate for heat released as NO is oxidized to NO_2. The gas is then fed into the adsorbing bed where residual moisture is adsorbed, NO is oxidized to NO_2, and NO_2 is adsorbed into the sieve material (see Figure 4.a). The clean gas stream is then divided. Part is returned to the power recovery sequence and part is heated to thermally desorb the second sieve bed. The adsorbed material comes off in two concentration peaks (see Figure 4.b); water first, then NO_2. This purge gas is introduced into the base of the adsorbing tower. The net effect is to increase tower gas flow rate 25%. After the NO_2 peak has passed, the sieve bed is cooled by shutting off the heater. When the first bed is saturated, valves are operated and the first bed is desorbed while the second bed adsorbs the NOx from tail gas.

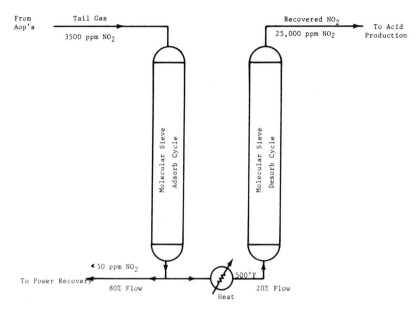

Fig. 3. Molecular Sieve Flow Diagram

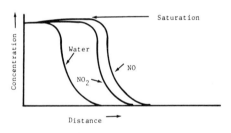

Wave Fronts in Adsorbing Bed

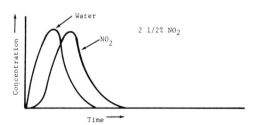

Wave Fronts in Desorbed Gas Stream

Fig. 4. Wave Fronts in a Molecular Sieve

From Figure 4.a, it is clear that zero emissions could be obtained if leakage can be eliminated. This is not a realistic design point since it would require extremely large sieve beds. As a more reasonable criteria, the unit used in the economic cost analysis was selected for a peak NOx emission of 50 ppm.* This theoretically corresponds to an average NOx emission rate of some 7 ppm if there is no leakage.

The degree of air contaminant control possible certainly makes the system technically attractive. One manufacturer[10] has made a presentation to CERL which indicates that the molecular sieve system should be considered proven technology. Extensive laboratory and slip stream tests demonstrate that after the sieve material has been cycled 1,000 times, it still possesses 96% of its original absorptive capacity. Its structural integrity is not affected by use or acids. Unlike the adsorbant in the alkalyzed alumina process for SOx control, this material will not fail physically. The manufacturer is confident enough of his product that he will guarantee a sieve charge for two years. Since molecular sieves have been used to clean gas fractions since 1946, the hardware and cycles can be considered proven. The available sieve material guarantee certainly minimizes any risk involved in this "emerging technology."

As long as the absorption tower on an existing nitric acid plant can tolerate a 25% increase in gas flow rate (the caps are not floated), application of a molecular sieve promises clean air and 100% utilization of the ammonia burned.

COST COMPARISON

Two basic systems were considered for use: catalytic reduction and molecular sieves. The actual systems to be compared are sized for a 200 ton per day, 60% nitric acid plant. The molecular sieve and the catalytic abater are considered to be tied in using bypass valves and fail-safe controllers (see Figure 5). The systems to be compared include a molecular sieve, a natural gas-fired catalytic reduction unit, and a catalytic reduction unit fired with onsite generated hydrogen. Standard estimating procedures were employed. Control system prices were obtained from Union Carbide and D. M. Weatherly.

---

* The EPA has indicated a desire to enhance the current emission standard by lowering it to 50 ppm or less. This is their goal for the next five-year period. In view of this fact, it was deemed advisable to attempt to reach that goal in this design.

The hydrogen plant is a DeMarkus package plant. It was assumed that turn-key contracts would be let.

In compliance with the Federal New Source Standards, all air contaminant control systems are priced as being fully instrumented for emission monitoring. It is assumed that no additional manpower will be required to operate the molecular sieve or catalytic abater. The hydrogen plant will require two operators. Fuel and utility costs are characteristic of an Army ammunition plant and consumption rates are supplied by manufacturers.

Life cycle costs were reduced to an annual cost for comparative purposes using a capital recovery factor based on a five-year payout and 6% interest on the money. All costs are expressed in 1971 dollars.

Table 1 clearly shows that the total annual cost of a molecular sieve installation is approximately one-half that of methane-fueled catalytic reduction. However, the cost of avoiding the secondary pollution with hydrogen combustion is minimal. This is because in the hydrogen generation process selected, a large fraction of the hydrogen generated comes from live steam reacted with carbon monoxide.

The most economical control of nitric oxides from AOP's appears to be afforded by the use of a molecular sieve. This conclusion is reached even though the initial capital cost is almost two to one against the molecular sieve.

CONCLUSIONS

Two conclusions can be reached: life cycle cost effectiveness analysis is mandatory and a NOx specific molecular sieve is the technically and economically preferred control alternative.

It must be noted, however, that the life cycle cost effectiveness of the various control techniques will change with the state-of-the-art, the degree of control desired, the need for various by-products, the cost of energy and water, and the degree to which the control technique can be fit into the process. While the cost comparisons which will be presented at the end of this paper are probably applicable to ammonia oxidation processes today, the findings must be critically reviewed before they are applied to any other basic process.

TABLE 1

Cost Comparisons

Cost	Molecular Sieve	Cat. Red. CH_4	Cat. Red. H_2
First Cost	545,000	265,000	565,000
Annual Cost[a]	129,200	62,900	134,000
Operating Cost			
Catalyst[b]/Sieve[c]	28,000	70,000	70,000
Electricity[d]	10,662	---	1,609
Gas[e]	---	192,960	133,110
Oil[f]	9,504	---	---
Steam[g]	1,451	---	6,825
Cooling Water[h]	1,038	---	430
Annual Operating Cost	50,655	262,960	211,964
Credits			
Acid[i]	-52,648		
Steam[g]		-74,774	-74,774
Total Annual Operating Cost	-1,993	188,664	137,190
Total Annual Cost	27,207	251,564	271,190

a - CRF (5 yr., 6%) = .236
b - 1 yr. guarantee
c - 2 yr. guarantee
d - .007 $/KW
e - .50 $/KCF (1000 Btu/$ft^3$)
f - .088 $/K gal (#6)
g - .44 $/K #
h - .005 $/K gal
i - 17.41 $/ton (100%)

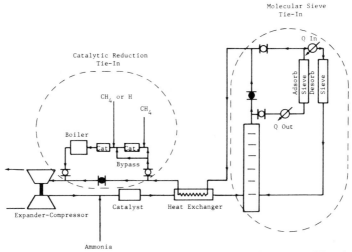

Fig. 5. Molecular Sieve and Catalytic Reduction Tie-ins on 200 TPD Nitric Acid AOP

REFERENCES

1. Federal Register, December 23, 1971, p. 24881.

2. Newman, D. J., and Klein, L. A., CEP, 68, No. 4, 62 (April 1972).

3. Newman, D. J., "The Third Source, Pollution from Nitric Acid Plants," The Third Joint Meeting AIChE and IMIQ, August 31 1970.

4. Chilton, T. H., Chemical Engineering Progress Monograph Series, 56, No. 3, 1 (1960).

5. ESSO Research and Engineering, "Systems Study of Nitrogen Oxide Control Methods for Stationary Sources," Vol. II, Linden, New Jersey, 20, November 1969.

6. Private communication with Povey, Matthey Bishop Company.

7. Private communication with C & I Girdler Corporation.

8. Reed, R. M., and Harvin, R. L., CEP 68, No. 4, 70 (April 1972).

9. Private communication with D. M. Weatherly Company.

10. Private communication with Union Carbide Corporation, Sieve Division.

ABATEMENT OF NITROBODIES IN AQUEOUS EFFLUENTS FROM TNT PRODUCTION AND FINISHING PLANTS

Leo A. Spano, Ronald A. Chalk, John T. Walsh and Carmine DiPietro

U.S. Army Natick Laboratories, Natick, Maine

In the manufacturing of TNT, there are three waste water streams that must be processed in order to comply with federal and local government regulations. These are: (1) "red" water which is produced by the sellite (sodium sulfite) treatment; (2) "pink" water which is produced when the partially purified TNT is washed following the sellite treatment; and (3) wash waters from the TNT finishing process. While all these polluted streams are being investigated at the U. S. Army Natick Laboratories, this paper will cover results obtained with wash waters from TNT finishing processes.

The finishing plant is the final step in the TNT manufacturing process; consequently, the principal pollutant of these wash waters is α TNT. These waters when leaving the finishing plant are acidic (pH 3-4) and colorless. As this effluent flows to the equalization basin, it may be exposed to sunlight, and it acquires a "pink-amber" color. From the equalization basin the effluent flows into a neutralization tank where Na_2CO_3 is added until a pH 6 is achieved. This treated stream is directed into a settling tank, and from there it flows into a second neutralization tank where additional Na_2CO_3 is added to adjust the pH to 7.0. The neutralized stream containing nitrobodies is then directed into a second settling tank and the overflow of this tank is diluted approximately 8-fold with clean process water to reduce the concentration of nitrobodies and color producing compounds, and then it is discharged into waterways or rivers. This treatment scheme is shown in Figure 1.

The concentration of α TNT and/or color producing compounds dis-

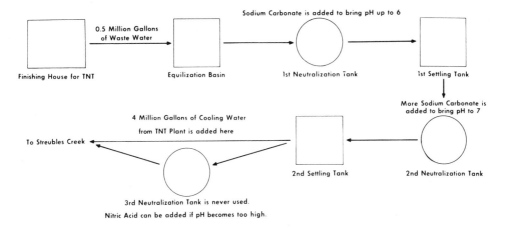

Fig. 1 TNT FINISHING PLANT WASH WATER TREATMENT PROCESS

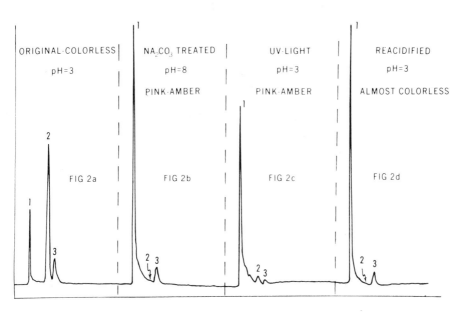

FIG 2 TNT-FINISHING PLANT WASH WATER CHROMATOGRAM

charged into waterways must be limited to less than 2.5 ppm. It has been found that α TNT at concentration of 2.5 ppm is toxic to fish and other aquatic life.

To overcome both the red discoloration imparted by this "pink" water to the receiving river and reduce the concentration of the nitrobodies discharged, to levels below those found toxic to living organisms, two approaches are being investigated at the U. S. Army Natick Labs. These are: (1) direct absorption on polymeric resins and (2) biodegradation processes. This paper is limited to results obtained with a resin adsorbent that is a copolymer of styrene and divinylbenzene. This material is a commercial product produced by the Rohm and Haas Co., and is sold under the trade name of Amberlite XAD-2. It is supplied in the form of white insoluble beads, and is designed for use in columns or in batch operations for the adsorption of water soluble organic substances. The manufacturer of the resin claims that the adsorption on this resin is dependent largely on Van der Waals' forces, consequently, both the adsorption and regeneration of the resin can be controlled by changing the hydrophobic/hydrophilic balance of the molecules adsorbed.

Initially this resin was considered as a potential media to concentrate the nitrobodies in the "pink" water for analytical purposes. However, results of preliminary experiments proved so promising that it was decided to investigate XAD-2 as a potential adsorbent bed to clean the total effluent stream from the α TNT finishing process.

The colorless wash water after leaving the α TNT finishing plant is neutralized, usually with Na_2CO_3, and turns a "pink-amber" color. This color can result also from exposure to sunlight (UV) prior to pH adjustment with Na_2CO_3. The color producing compounds that are formed when the acidity of the waste stream is shifted from a low pH (3-4) to a pH (6-7), have been postulated to be "Meisenheimer complexes" that are formed through reactions of the α TNT.

Figure 2 is a series of high speed liquid chromatograms made for the separation of the various constituents present in the polluted wash water effluent from the α TNT finishing house of an actual ammunition plant.

Chromatogram 2a was obtained from a sample of wash water leaving the α TNT finishing house prior to any treatment. This water sample was acidic (pH-3) and colorless. Chromatogram 2b was obtained from a sample of the same wash water after adjusting its pH to 8. This water sample turned to a "pink-amber" color when treated with Na_2CO_3. Chromatogram 2c was obtained from a sample of the same wash water, however, the "pink-amber" color for this sample was produced by exposure to UV light while its acidity was maintained constant to that of the original sample (pH-3). Chromatogram 2d was obtained with the same sample used to obtain chromatogram 2b except that the sample had been reacidified to the original acidity (pH-3) and the "pink-amber" color removed. It should be noted that while the sample reverted to its original color, the chromatogram shows that the color-causing compound did not change back to α TNT. In other words, the chemical change is irreversible. Peak 2 on these chromatograms represents α TNT in the waste stream. Peaks 1 and 3 represent other polluting constituents that, as yet, have not been identified at the Natick Laboratories. These data confirm that the "pink-amber" color is produced by structural and chemical changes of α TNT in the waste stream, and that the change is not reversible by altering the pH. It has been reported by other researchers that these color producing constituents when adsorbed on activated carbon, have a tendency to reduce its adsorbing effectiveness. Consequently, it is desirable to remove all pollutants from the effluent stream prior to any color formation.

Chromatograms 2a and 2c in Figure 2 indicate that nearly all the α TNT in the original sample undergoes reaction to form color producing constituents. This is confirmed by the data shown in Figure 3. The data shown in Figure 3 were obtained with an aqueous solution of pure α TNT containing 100 ppm of α TNT. Figure 3a shows the concentration of α TNT present in solution, and Figure 3b shows that nearly all the α TNT has been converted into color producing compounds when the original colorless solution was made alkaline with Na_2CO_3. Integration of the peaks shown in Figure 3a and 3b shows that 95% of the α TNT present in the original solution has been transformed into other compounds.

To check the feasibility of using XAD-2 for the elimination or abatement of nitrobodies in acid wash waters leaving the TNT finishing house of a typical ammunition production plant, samples of colorless wash waters from the TNT finishing house of an actual munition plant were passed through an XAD-2 packed column and the eluant analyzed for pollutants by high speed liquid chromatography. The acidic eluant was found clear of α TNT, and upon neutralization did not develop the characteristic "pink-amber" color that is normally produced, when α TNT is present in the solution.

Activated carbon has been used extensively for the adsorption of organic compounds including TNT present in waste water streams. Filtrasorb-400 which is one form of activated carbon is now being used to remove α TNT and RDX in wash waters from munitions loading and packing (LP) operations. At one loading and packing site, two columns packed with Filtrasorb-400 are operated in series, and the effluent of the first column is monitored to determine when the packed bed must be replaced. The effluent is normally analyzed for α TNT by a colormetric technique. The standard operating procedure is that when the effluent from the first packed bed shows a concentration of 2 ppm, the column is repacked with new carbon and the used carbon is incinerated. No attempt is being made for its regeneration.

Since XAD-2 was found to be easily regenerated, studies were conducted to compare the adsorption effectiveness and regeneration characteristics of XAD-2 and Filtrasorb-400 after both have been used to adsorb pollutants present in "pink" water from α TNT finishing plants.

Two one centimeter diameter glass columns were set up holding equal volumes of XAD-2 and Filtrasorb-400. Wash waters obtained from an actual ammunition plant TNT finishing operation were passed through both columns, and the eluants from both columns were analyzed for eluted α TNT. The concentration of α TNT in the wash waters used was 34 ppm. The flow rate through both columns was maintained at 250 ml/hr. The properties of XAD-2 and Filtrasorb-400 are listed in Table 1.

Table 1 Properties Of XAD-2 And Filtrasorb-400

Properties	Amberlite XAD-2	Filtrasorb-400
Surface Area, $\frac{m^2}{gr.}$	330	1125
Bulk Density, $\frac{lbs}{ft^3}$	42	25
Mean Particle Diameter, mm.	0.53	1.00

The concentration of α TNT eluted from each column is listed in Table 2.

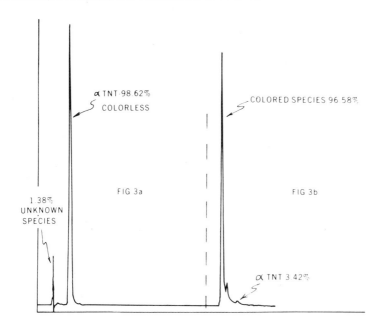

FIG 3 CHROMATOGRAM OF α TNT CONVERSION TO COLORED SPECIES

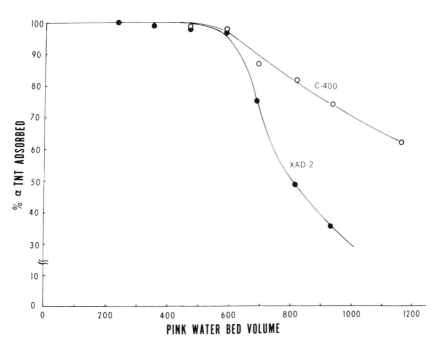

FIG 4 COLUMN ADSORPTION EFFICIENCY

Table 2 Eluted α TNT From XAD-2 And Filtrasorb-400 Columns

Liters	Influent α TNT - 34 ppm Equiv. Adsorb. Bed Volumes	Effluent α TNT - ppm XAD-2	Filtrasorb-400
1	116	0.0	0.0
2	232	0.0	0.0
3	349	<0.5	<0.5
4	465	0.5	0.5
5	581	1.5	1.0
6	697	8.3	4.5
7	814	17.4	6.2
8	930	22.0	8.6
10	1162		12.8

Table 2 indicates that the breakthrough volume is the same for both the XAD-2 resin and the carbon column. The first breakthrough occurred after approximately 349 bed volumes of effluent. The concentration of α TNT leaking through at this stage was less than 0.5 ppm.

The table further shows that XAD-2 is as efficient as the Filtrasorb-400 up to almost 600 bed volumes. At this point the leakage endpoint of both adsorbents has exceeded allowable α TNT limits. Since the allowable limit of α TNT in disposed waters is expected not to exceed 0.5 ppm, both column materials are effective to meet the standards at a throughput equivalent to approximately 465 bed volumes.

To compare the adsorption effectiveness of XAD-2 and Filtrasorb-400 for pollutants present in the wash water, the amount of α TNT adsorbed compared to the total TNT present in the influent processed through each column was plotted as a function of the total flow through the columns. This is shown in Figure 4.

Figure 4 shows that at throughputs exceeding 600 equivalent bed volumes, the percentage of α TNT and other pollutants removed by the XAD-2 column is much less than that removed by Filtrasorb-400.

This is understandable since the fraction of active sites remaining for adsorption on Filtrasorb-400 would be relatively large

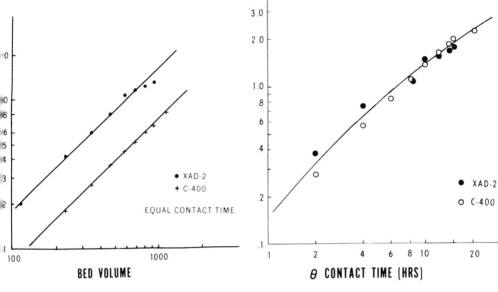

FIG 5 COLUMN ADSORPTION EFFICIENCY
log-log

FIG 6 ADSORPTION RATE

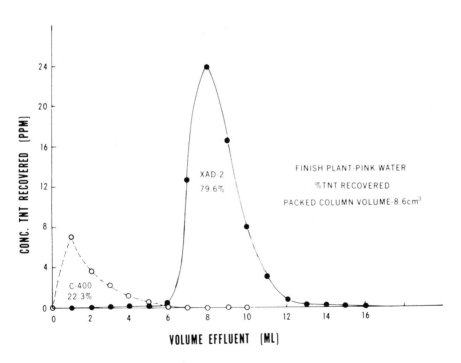

FIG 7 REGENERATION CURVES

while the active surface of the XAD-2 would be nearly saturated and not available for adsorption until regenerated. It should be noted that the initial active surface per gram of Filtrasorb-400 is 2.25 times that of XAD-2.

Because of the significant difference in active surface areas per unit weight of Filtrasorb-400 and XAD-2, it was of interest to compare the adsorption effectiveness of the two adsorbents on a unit surface basis. Figure 5 shows the quantity of α TNT adsorbed per square meter of surface for both adsorbents. These data show that for fixed throughputs of wash water containing 34 ppm of α TNT as contaminant, the XAD-2 appears to have better adsorbing characteristics than Filtrasorb-400.

One other parameter which should be factored into the evaluation of adsorption systems is the contact time between the material to be adsorbed and the active surfaces.

Figure 6 shows our experimental results normalized to reflect the adsorption rate of α TNT in parts per million as a function of contact time with the adsorbent surface in hours. These data indicate that the surface activity of Filtrasorb-400 and XAD-2 for α TNT are comparable.

In the selection of adsorbents to remove nitrobodies in polluted streams discharged from munitions manufacturing facilities, one must give serious consideration to techniques for disposing of the adsorbed pollutants and regeneration of the spent adsorbent bed. As indicated earlier, regeneration of XAD-2 has been found relatively simple. Studies at the Natick Laboratories show that acetone can easily strip the nitrobodies adsorbed from "pink" water quite rapidly.

To compare the regeneration characteristics of XAD-2 and Filtrasorb-400, fixed quantities of "pink" water containing 34 ppm α TNT and other color constituents were passed through equal bed volumes of Filtrasorb-400 and XAD-2. The adsorption cycle was followed by a stripping cycle in which acetone was used to regenerate each column. Figure 7 shows the percent recovery of the adsorbed nitrobodies as measured by the total α TNT in the eluant from each column.

It should be noted that the α TNT recovered from the XAD-2 column by flushing the adsorbent with approximately two bed volumes of acetone is approximately four times greater than the amount recovered from the Filtrasorb-400 column.

While the data thus far appear quite favorable, XAD-2 does allow some of the color components to pass through but Filtrasorb-400

shows that all color can be removed from the influent stream.
Consequently, if the wash water from the α TNT finishing processes
has changed color, it would be advisable to use a combination
adsorption system in which XAD-2 can be used to remove nearly all
pollutants in the effluent stream on a gross scale, and then
finish the cleaning of the stream by removing the remaining color
that may pass through the XAD-2 with activated carbon.

Conclusion

In summary, the results of studies with XAD-2 and Filtrasorb-400
as potential adsorbents for nitrobodies in wash waters from α
TNT finishing processes at ammunition production plants and
loading and packing plants show that:

1. XAD-2 is an effective adsorbent for nitrobodies containing primarily α TNT. This is particularly true for wash waters that have not been neutralized, and have not acquired the "pink-amber" color that results from structural changes of α TNT in the water.

2. The adsorption effectiveness of XAD-2 is comparable to that of Filtrasorb-400 for influent throughputs equivalent to approximately 600 adsorbent bed volumes.

3. Chemical regeneration of XAD-2 saturated with α TNT can be effectively accomplished with acetone. Chemical regeneration of Filtrasorb-400 has not been satisfactorily accomplished.

4. To effectively clean up the nitrobodies in wash waters from α TNT finishing processes and load and pack operations, XAD-2 should be used to adsorb the major fraction of nitrobodies and color producing constituents in the stream. The effluent of the XAD-2 column should then be passed through a polishing activated carbon bed to pick up any residual color that may pass through the XAD-2 column.

EXPLOSIVE INCINERATION

Irving Forsten

Picatinny Arsenal, Dover, New Jersey

The current method of disposing of scrap and waste explosives and propellants, at all the Government installations and manufacturing and loading plants, is by the "open burning" technique; whereby the wastes are manually spread on a concrete pad, ignited by some device, and allowed to burn in the air. This results in air and water pollution, an uncontrolled burning operation, delays and inefficiency of disposition due to weather, and hazardous exposure of personnel. Most states in the United States now have regulations barring open-burning, and these plants are operating under temporary waivers from the Federal Government although the obligation to comply with state standards is an announced Federal Government objective. This problem, coupled with the Presidential Executive Order 11507 that directs the Federal Government of the United States to take the lead "...in the nationwide effort to protect and enhance the quality of our environment", makes it mandatory that the government acquire the necessary technology to be able to incinerate explosives and propellants with minimal pollution and in a safe manner.

A program was instituted to develop, essentially, a pollution free incineration system and provide design criteria for the eventual facility. Three types of incinerators are being evaluated as disposal methods. These consist of a vertical induced draft, fluidized-bed combustor and a rotary kiln.

For initial work, there was available an existing explosive incinerator designed and installed in 1957. The high explosive incinerator is being used to generate pollutants and make inroads on controlled explosive incineration so that designs of anti-pol-

EXPLOSIVE WASTE CONTINUOUS WET GRINDING PROCESS

WATER SUPPLY
MAGNET
WASTE FEED CONVEYOR
OVERSIZE
FINES
WATER SUPPLY
HD CRUSHER
SHAKER TABLE
CONTINUOUS FEED BALL MILL GRINDER
SLURRY MIXING TANK
OVERSIZE AND WATER RETURN
TO INCINERATOR
PUMP
STORAGE BIN
WEIGHING AND BAGGING STATION (ALTERNATE)

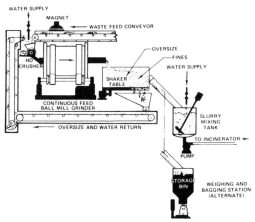

FIGURE 1

EXPLOSIVE WASTE INCINERATOR FEED SYSTEM

FLAKE EXPLOSIVE CONVEYOR
WATER
INCINERATOR
AGITATOR
300 GAL LEVEL CONTROL
SLURRY
STEAM INJECTOR
MIXING TANK (350 GAL)
STEAM
DUMP VALVE
HOLDING TANK (350 GAL)
25 GAL LEVEL CONTROL
VALVE
EXPLOSIVE SLURRY PUMP

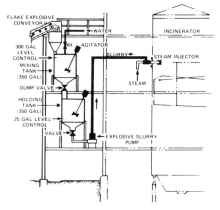

FIGURE 2

POLLUTION ABATEMENT INCINERATOR SCHEMATIC

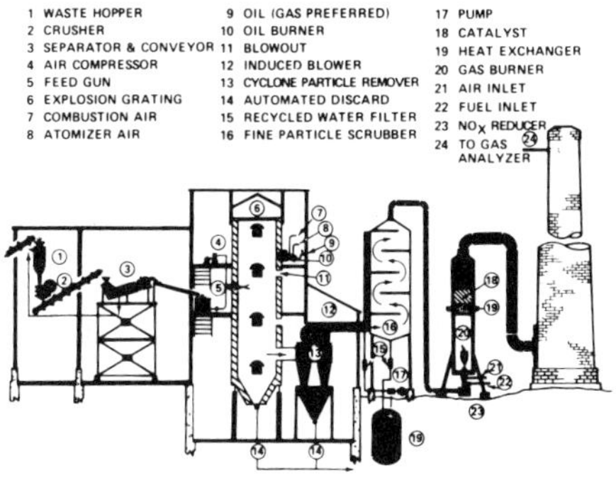

1 WASTE HOPPER
2 CRUSHER
3 SEPARATOR & CONVEYOR
4 AIR COMPRESSOR
5 FEED GUN
6 EXPLOSION GRATING
7 COMBUSTION AIR
8 ATOMIZER AIR
9 OIL (GAS PREFERRED)
10 OIL BURNER
11 BLOWOUT
12 INDUCED BLOWER
13 CYCLONE PARTICLE REMOVER
14 AUTOMATED DISCARD
15 RECYCLED WATER FILTER
16 FINE PARTICLE SCRUBBER
17 PUMP
18 CATALYST
19 HEAT EXCHANGER
20 GAS BURNER
21 AIR INLET
22 FUEL INLET
23 NOx REDUCER
24 TO GAS ANALYZER

FIGURE 3

lution devices and explosive handling equipment can be established and evaluated. Some of the materials to be incinerated, Table I, consist primarily of explosive and propellant wastes from a variety of sources such as (a) shell washout operations, (b) experimental loading operations, (c) surplus explosives from discontinued programs, (d) excessive shelf life ammunition, and (e) losses sustained in the manufacture processes of explosives.

TABLE I

MATERIALS TO BE INCINERATED

1. Comp B
2. TNT
3. HMX
4. RDX
5. Black Powder
6. Single-Base Propellants
7. Double-Base Propellants
8. Tetryl

Waste material delivered by truck to the incinerator storage area will be remotely loaded onto the conveyor supplying the waste hopper. The material will be ground to a uniform small particle size, in the presence of water, and discharged to the separator where the water is removed and the proper size material is fed to the slurry mixing tank. (See Figure 1) In the slurry mixing tank, the waste material is mixed with water to obtain the proper concentrations (19:1 to 3:1; water: solids) and then the slurry is pumped to the incinerator. (See Figure 2) The slurry will be fed into the incinerator through an injector (e.g. steam ejector, swirl nozzle) to atomize and disperse the slurry into the combustion zone which will be controlled to a temperature of 1500° to 1700°F. (See Figure 3) The exhaust will be processed through a cyclone dust separator and an appropriate gas scrubber prior to being released to the atmosphere via the 125 foot stack. Figure 4 shows data on incineration of TNT/Water Slurry run in the incinerator to date.

Stack sampling will be used to determine the quantity and types of pollutants that are contained in the gases that are emitted from the operation. The data obtained from stack sampling surveys will be used to design and evaluate air pollution control equipment, to verify compliance with emission standards and to determine efficiencies of combustion and air pollution control equipment. Sampling principles and characterization of stack flow will be done in accordance with established practices.

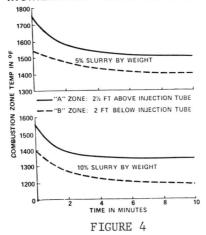

FIGURE 4

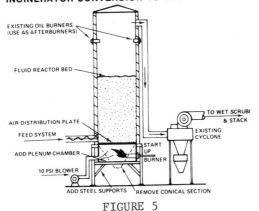

FIGURE 5

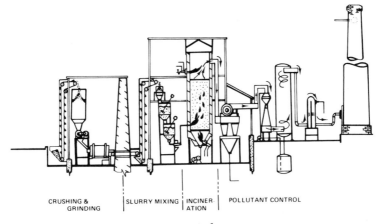

FIGURE 6

After evaluation of the vertical induced draft incinerator portion, this unit will be converted to a fluidized-bed incinerator. (See Figure 5) Fluidized-bed reactors are not new. They have been used for years by the Petroleum and Chemical Industries because of their unique characteristics: (1) each particle is surrounded by a layer of the fluidizing media, (2) the bed is at a comparatively lower temperature, (approximately 1400° F) reducing NO_x formation, (3) the heat transfer rate in the bed is rapid because the movement of solids in the bed is turbulent. The operation of the fluidized-bed can be best described by considering a cylindrical tube containing sand that is held on a porous plate. Gas flow through the plate can be controlled to the desired rate; at low flow rates the bed remains in its original packed state, and pressure drop across the bed increases with flow rate until it is equal to the downward force exerted by the solids resting on the plate. The bed begins to expand at this point (incipient fluidization) allowing more gas to pass through the bed at the same pressure drop. With a further increase in flow rate, the particles become free to move about, air bubbles form in the bed and surround the particles, and mixing action becomes very vigorous. The bed is now fluidized and has many characteristics of a fluid; viz. it has an apparent viscosity, obeys Archimedes' principle, flows from a vessel.

The characteristics of the fluidized-bed made it very desirable for explosive incineration. (See Figure 6) Explosive particles are fed into the bed having a known, controlled temperature and the high heat transfer rates within the bed rapidly raise the temperature of the particle to its ignition point then immediately absorb the heat generated by combustion. Intimate mixing of the fluidizing air with the particles makes the required oxygen for combustion readily available keeping air requirements to a minimum. This minimum air usage plus the uniform bed temperature minimizes the quantity of NO_x (the principal pollutant) formed, while the rapid mixing action prevents any buildup of explosives at any one point.

The third type incinerator to be tested and evaluated is the rotary kiln. The waste propellants will be ground to a uniform size (flying knife grinder), slurried with water, and pumped and injected into the incinerator. The rotary kiln incinerator is a standard type where the combustion chamber is a rotating drum canted at a small angle. The temperature in the chamber will be controlled at 1500° to 1700°F. The slurry will be injected at the high end and will slowly be heated and burned as it travels down toward the low end where the oil burner is located. Any ash generated can be removed at the low end and the exhaust gases will be passed through a secondary combustion chamber and a wet scrubber prior to being emitted to the atmosphere. The stack emissions will be monitored by an eight channel stack monitoring system to be sure that the combustion and pollution control system are operating properly.

THE U. S. NAVY PEPPARD PROGRAM

Bernard E. Drimmer

Naval Ordinance Systems Command Research and Technology Directorate

The U. S. Navy is undertaking an extensive program to minimize, and eliminate, if possible, all of its environment-harming activities. This stems from the world-wide awareness of the accelerating degradation of the environment resulting from mankind's activities. This paper will describe one element of the Navy effort: the PEPPARD Program. "PEPPARD" is an acronym for "Propellant, Explosive, Pyrotechnic Pollution Abatement Research and Development" Program. The PEPPARD program was initiated to develop pollution-free alternatives to the routine (and in some instances, not-so-routine) activities necessary to providing the U. S. Navy with ordnance. In a word, then, this paper will describe the Navy's Research and Development effort supporting the rather extensive on-going "clean-up" program where standard engineering techniques are being employed to eliminate practices which in the past were considered "acceptable", though sometimes annoying.

An example of a need for a new, or more economic alternative is shown in the **figure 1**. During the normal course of supplying the Navy with explosive-filled ordnance, large quantities of unwanted explosive-filled items routinely accumulate. (Parenthetically, please note when I use the term "explosive-filled", I mean items containing any high-energy chemical, such as propellants, explosives, pyrotechnics, tracers, etc.). In the past, the most expeditious means for getting rid of such items was to load them on to a stripped, surplus cargo vessel, tow it to sea, to an international-agreed-upon graveyard for ships, and sink the vessel. This economically attractive technique - it cost approximately one cent per pound of gross cargo weight - was banned by the U. S. Navy some two years ago when

Figure 1. Surplus Ship, Loaded with Obsolete/Dangerous Ordnance, being Sunk.

it was realized that the oceans, previously considered as having an infinite capacity for such materials, may, in fact, have real limits for such activities. And so was born the U. S. Navy's PEPPARD Program.

Now, we _can_ provide the Navy with all required ordnance and get rid of all unwanted ordnance without causing any harmful effects on the environment. The "trick" however, is to do this at a cost and performance penalty that is acceptable. I cannot define quantitatively what are "acceptable" cost penalties, but let me assure you that straightforward application of state-of-the-art engineering will run the costs up to hundreds of millions of dollars for the U. S. Navy alone.

As you can see from figure 2, one of the first things we did in the PEPPARD Program was to assign one of our Laboratories the task of defining our problem, of determining quantitatively how much ordnance scrap is the U. S. Navy generating. Such scrap can arise from numerous sources, ranging from the wash water effluent from our ordnance explosive-loading plants, to the accumulation of obsolescent or dangerously corroded explosive-filled ordnance items. The several thousands of _tons_ of Naval ordnance requiring disposal, being generated annually have now been identified and categorized. Lacking the economic ship-sinking methods of disposal, this accumulation of ordnance is being stored pending development of safe, economically acceptable methods for disposal.

The next several items on the list indicate the efforts under way to develop such methods. For the first three, explosives, propellants, and pyrotechnics reclamation, the orien-

FIGURE 2

THE U.S. NAVY PEPPARD PROGRAM

- REQUIREMENTS AND APPROACHES
- EXPLOSIVES RECLAMATION
- PROPELLANTS RECLAMATION
- PYROTECHNICS RECLAMATION
- CHEMICAL AGENT DISPOSAL
- ORDNANCE PACKAGING DISPOSAL
- MILLILE MOTOR BREAKDOWN
- BIODEGRADATION OF EXPLOSIVES

- PEP DISPOSAL BY FUSED SALTS
- READILY DISPOSABLE EXPLOSIVES
- DESIGN OF WEAPONS
- UNDERWATER EXPLOSIONS
- DEEP WATER DUMP
- CHEMICAL ANALYSIS
- DISPOSAL ALTERNATIVES

tation is to develop recycling techniques. Thus, in some cases the explosive can be removed, say from an obsolete mine by steam-melting, and the explosive, for example TNT, upgraded to military standards by a one, or two-step purification method, and re-used.

Another approach to recycling is to convert the unwanted explosive into a different, usable product: fertilizers, for example. Here, relatively minor constituents tend to play a significant role. For example, lead, a very minor constituent in some propellants, may travel through the conversion process and wind up as one ingredient of the fertilizer. We are investigating the influence of this minor ingredient on plant and animal life: what happens ultimately to the lead? Is it picked up the the plants; if so, what happens next, etc., etc.? Thus, a relatively simple solution to a disposal problem may be frustrated by the presence of an apparently minor component in the original propellant formulation.

Some successes have been achieved in this recycling effort. TNT from mines has been reclaimed; so, too, has magnesium powder from some obsolete flares. Many problems remain, but we believe no so-called technological "breakthrough" is needed: only clever applications of state-of-the-art technology - and we are working on them.

The next two items are, for the Navy, relatively trivial problems. When chemical warfare was still a potential, the Navy set up a small number of training kits, each of which contained tiny amounts of sample agents. Nevertheless, trivial as these amounts are, the Navy will not "dump them down the sink"! We are therefore looking into environmentally acceptable ways of getting rid of these tiny samples.

The pollution attendant to packaging appears to be a problem not specific to ordnance. Thus we find we are confronted with the problem of getting rid of styrofoam packaging, surplus wooden boxes, etc. But normal industry is confronted with the very same problems, and with our limited financial resources we have opted to pursue only those Research and Development efforts relevant to <u>unique</u> Naval ordnance problems. Whatever solutions industry develops for the disposal of styrofoam, etc., will be applied to our problems. However, where such packaging becomes contaminated with explosive by one means or another, then it does become one of our problems. Such problems are normally one-shot affairs, such as a wooden box accidentally contaminated with TNT, and solutions are generated on an ad hoc basis.

The next item, missile motor breakdown methods refers to the problem of how to remove, recycle, etc. the propellants of missile motors. As with other ordnance, dumping these obsolete/surplus motors into the sea was the most economic method for disposal. Surprisingly, one of the most obstinate problems here is how, safely and economically, to reduce the large propellant grains into small, more manageable chunks. No really good solutions have yet been found.

The next two items are part of the high risk portion of our Research and Development program. We would like to put microorganisms to work for our benefit, strictly analogous to the way ordinary sewage disposal plants utilize such "bugs". We have some hints that some of our explosives, such as TNT, are biodegradable. Our problems here are manifold: identify the proper "bugs", optimize their performance, and finally insure by one means or another, that the products of their activity are not harmful (better, beneficial!) to the environment. Here we have our work cut out for us, but the payoff is substantial, justifying the pursuit of such a high-risk effort!

Some people feel that decomposition of explosive can be controlled by "cooking" the explosive in a carefully monitored fused salt bath. Needless to say, one does such "cooking" very carefully, and we are pursuing such research with the utmost caution. To date no uncontrolled reaction has occured, and limited success has been achieved on a laboratory scale. However, we believe there is a long road from such small scale success to full scale plant operation. As in the case of propellant grain disposal mentioned a moment ago, a key factor here is the determination of the optimum size of explosive pellet needed for safe operation.

The next several items look to the future of explosive application. For example, we are looking at explosives which are more readily disposable than our present stocks. One would like to develop an explosive as militarily effective as RDX, as cheap as TNT, and which, when no longer needed, will readily and cheaply turn into some usable, non-explosive material. That's a pretty tall order - but we will never find one if we do not search for it!

Similarly, we are re-studying the design of some of our explosive-filled ordnance to see if we can modify the design to permit easy and safe removal of the explosive. Today's weapon design doctrine is essentially the opposite: ready removal of the explosive is, if anything, discouraged for a variety of reasons. Nevertheless, this doctrine is being re-evaluated in the light of modern day emphasis on pollution abatement. In any event, doctrine or not, the design of a weapon to permit ready and safe removal of the explosive is extremely complex, as our warhead design engineers are learning!

The next item is looking into the question of the environmental effects of underwater explosions. The advancement of the technology of military explosions requires the testing of such explosions under carefully controlled conditions. Ultimately such experiments must be run in "real" bodies of water or air, where various flora and fauna may exist. The resulting disruption of such biota may have significant effects, both ecologically and economically. Since the Navy must maintain the highest expertise in this technology, the question naturally arises as to how to accommodate these two apparently contradictory requirements. To date some limited success has been achieved in developing better guidelines on when and where such experimental test explosions should be permitted. Hopefully, with the newer fish kill models being developed, future, underwater tests can be run with essentially zero permanent effects on the marine environment.

The next item is a small effort to determine the optimum technique for employing the now banned Deep Water Dump. It may turn out that after all is said and done, some very limited Deep Water Dump will be permitted, say one ship every several years. What would be the optimum method for dropping the ship to the sea bottom? Should we simply let it sink? Should we deliberately detonate it on the way down? Should we detonate it on the surface? It is one function of our office to be prepared to answer such questions, and this study should provide us with an assessment of the risks and benefits of all potential Deep Water Dump techniques, which will then be on hand if needed.

The next-to-last item looks to the development of appropriate analytical techniques for explosives - for example, how much explosive is there in certain explosive loading plant effluents, in some wash waters, etc.? The gas chromatograph technique, used in conjunction with a time-of-flight mass spectrograph has given us the capability of quantitative detection of TNT in water down to 2 parts per trillion, RDX to 5 parts per trillion, and tetryl to 20 parts per trillion. This exceedingly refined analytical technique has opened many possibilities for monitoring the movement of exceedingly small amounts of explosive in such areas as lakes, rivers, ground water, etc.

Finally, we have convened a group of highly experienced experts to examine our disposal program, and to suggest areas where Research and Development could provide a good potential for developing new, safer, more economic methods for ordnance disposal. We look forward to their suggestions, just as we are hoping that the broadranging discussions of this very timely Conference may suggest new approaches to some of the problems discussed above.

ELIMINATION OF STYPHNIC ACID FROM PLANT EFFLUENTS

E. KURZ

*Israeli Military Industries, P.O.B. 7055,
Tel Aviv, Israel*

The chief aim of the usual methods for eliminating styphnic acid from plant effluents resulting from the manufacture of lead styphnate is to avoid explosive hazards. They are based on the destruction of styphnic acid, either by oxidation or by reduction. Their main disadvantage is the formation of toxic reaction products and the toxic reagents added in some cases.

Thus in the oxidation process, where chlorine is added, as hypochlorite (1), in gaseous form, or generated by electrolysis, the strong lachrymant chloropicrin is formed as a result of the destruction of the aromatic ring.

In the reduction processes, which lead to the formation of toxic aromatic amines, either hydrogen sulfide generated in situ from ammonium sulfide and acid, or nascent hydrogen developed by action of acid on iron is used. Also electrolytic reduction has been employed.

In order to avoid pollution, which seemed inevitable as a consequence of the destruction of styphnic acid, we radically changed the approach and applied regeneration of the major part of the styphnic acid for its removal from the effluents, instead of its destruction.

We elaborated a simple method for the elimination of styphnic acid from the effluents, based on the decrease of its solubility by addition of mineral acids, e.g. nitric acid. The procedure consists of two steps:

1. Precipitation of styphnic acid from concentrated effluents,

Table 1. Solubility of Styphnic Acid in HNO_3 of Various Concentration

A. Determinations of Knox (2), reported by Aubertein and Emeury (3)

Normality of HNO_3	0	1.8	8.4	12.0	15.6
Styphnic acid dissolved, g/l	5.3	0.34	1.25	3.2	13.7

B. Approx. determinations, from 0 to 0.5 N HNO_3

Normality of HNO_3	0	0.1	0.2	0.3	0.4	0.5
Styphnic acid dissolved, g/l	5.0	0.75	0.45	0.34	0.30	0.26

reducing their styphnic acid content by more than 90%, and

2. practically complete elimination of the rest of styphnic acid in the filtrate and in more dilute effluents by adsorption on activated charcoal.

Solubility of Styphnic Acid in Presence of Nitric Acid.

Determinations of the solubility of styphnic acid in water containing various amounts of nitric acid above 1.8 N were made by Knox (2) and are reported by Aubertein and Emeury (3). As shown in Table 1.A, there is a steep decrease in the solubility of styphnic acid in 1.8 N nitric acid, as compared to its solubility in pure water, and again a slow increase in solubility with further growing of nitric acid concentration.

As we were chiefly interested in the range of lower nitric acid concentrations for which there were no data available, we made some approximative determinations of the solubility of styphnic acid in the range of 0 to 0.5 N nitric acid, shown in Table 1.B.

Composition of Plant Effluents.

In our plant, where lead styphnate is manufactured from lead nitrate and magnesium styphnate, there are three kinds of effluents,

Table 2. Composition of Plant Effluents

Effluent	Concentration, g/l			
	Styphnate, calculated as styphnic acid	Pb^{2+}	Mg^{2+}	Nitrate, calculated as nitric acid
Mother liquor	8.42	48.4	14.8	102.2
Decantation	1.65	8.4	2.6	17.8
Wet sieving	0.45	0.37	0.07	0.61

about equal in volume, and of decreasing concentration:

1. Mother liquor, remaining after the crystallization of the lead styphnate,

2. Solution resulting from washing the lead styphnate with water by decantation,

3. Solution resulting from wet sieving of the lead styphnate.

The composition of these effluents is given in Table 2.

All further determinations and experiments were made on a laboratory scale on one batch of these effluents.

Influence of Nitric Acid Concentration on Solubility of Styphnic Acid in Mother Liquor

Upon addition of nitric acid in the range from 0 to 1.5 N to the mother liquor, containing about 8 g of styphnic acid per liter, the amount of styphnic acid remaining in solution shows the same trend as in the case of solutions of styphnic acid in water containing nitric acid, as can be seen from Table 3: an initial steep decrease in solubility and a flat minimum extending from about 0.5 or 0.75 to 1.5 N nitric acid, where there is only a small variation in the amount of styphnic acid precipitated.

Table 3. Styphnic Acid Dissolved in Mother Liquor after Addition of Various Amounts of Nitric Acid

Normality of HNO_3	0	0.25	0.50	0.75	1.0	1.25	1.5
Styphnic acid dissolved g/l	7.58*	1.06	0.65	0.50	0.41	0.30	0.34
Styphnic acid dissolved as % of original content	100	14.0	8.6	6.6	5.4	4.0	4.5

*900 ml mother liquor containing 8.42 g/l styphnic acid diluted to 1000 ml

The absolute values of the solubility of styphnic acid compared to its solubility in water are greater both in the mother liquor itself without addition of acid (8 g as compared to about 5 g in water), and at low concentrations of nitric acid added. At higher concentrations of nitric acid the effect is reversed and there is a "salting out" of the styphnic acid, as was proven by parallel determinations of the solubility of styphnic acid in less and more diluted solutions of mother liquor, all containing the same final concentration of nitric acid – up to 1.5 N – and also by parallel determinations effected on the effluent resulting from decantation.

REGENERATION OF STYPHNIC ACID FROM MOTHER LIQUOR

In order to keep the amount of nitric acid added at a minimum, we used only the most concentrated effluent – the mother liquor – for regeneration of the styphnic acid, at a nitric acid concentration of 0.75 N.

After adding an amount of 25 ml of concentrated nitric acid to 475 ml of mother liquor, the solution was briefly stirred for mixing. A rise in temperature of about 4 °C and beginning of crystallization of styphnic acid after a few seconds was observed. The styphnic acid settled fast and was easy to filter. In order to minimize its redissolution, it was first washed with dilute (2% v/v) nitric acid, and small portions of water were only used for the last washings. The styphnic acid obtained consisted of pure crystals of rather uniform dimensions and of a light yellow colour. – Samples of solution taken after two hours and after standing overnight, showed a decrease in the amount of styphnic acid left in solution upon longer standing. Precipitation of the styphnic acid could be accelerated by stirring, but this was not desirable, at the initial stage at least, because of the formation of smaller crystals.

Table 4. Styphnic Acid Remaining in Solution after Addition of Various Adsorbents

Adsorbent*	Styphnic acid in solution, ppm
Activated Carbon "DARCO", Grade S 51, "Atlas" Chem Ind.	1
Charcoal, granular, activated for gas absorption, BDH	23
Charcoal, animal, granular for filters, BDH	575
Amberlite resin IRA-401(Cl), Anal. Grade, Rohm & Haas	95

* 20 g of adsorbent added to 1 liter of effluent mixture 0.5 N in HNO_3, containing 790 ppm of styphnic acid

REMOVAL OF THE REST OF STYPHNIC ACID FROM THE EFFLUENTS

Before removing the small amount of styphnic acid remaining in the mother liquor after its crystallization, and that present in the two other effluents resulting from decantation and wet sieving, all solutions were blended in the same proportions as resulting from manufacture, i.e. 1:1:1 . The resulting effluent mixture contains free nitric acid in a concentration amounting to one third of the normality present in the mother liquor, and a mean of about 800 -900 ppm of styphnic acid.

In order to eliminate this remainder of styphnic acid from the mixed effluents without causing serious pollution, adsorption seemed to be the most advantageous procedure.

Choice of Adsorbent

In the preliminary experiments made to establish the most suitable adsorbent to be used, we also took into account the possibility of using a column procedure. Therefore we made trials with activated charcoal, both in powder and in granular form, and with a strong cationic ion exchange resin. The same amount of each adsorbent was added to 1 liter of the mixed effluents. From the results given in Table 4, it can be seen that the activated carbon in powder form is the most effective

Table 5. Styphnic Acid Remaining after Addition of Various Amounts of Activated Carbon Powder

A. Single Treatment

Amount of carbon powder added to 1 liter of mixed effluents*, g	5	10	15	20	25
Styphnic acid remaining in solution, ppm	127	6.4	3.1	2.3	1.7

B. Two Consecutive Treatments

Treatment	First	Second	
Amount of carbon powder added to 1 liter of solution, g	5	2.5	5
Styphnic acid remaining in solution, ppm	127	4.0	2.6

* Mixed effluents: 0.25 N in HNO_3, containing 867 ppm of styphnic acid

adsorbent. It can be used in a batch procedure, which also is more simple than the use of columns.

The presence of free nitric acid is necessary for an effective adsorption on carbon, since free styphnic acid, and not the styphnate ion is adsorbed. This is illustrated by the fact that upon addition of alkali to carbon powder containing adsorbed styphnic acid, the latter is released into the solution.

In the case of the ion exchange resin, the comparatively small amount of styphnic acid adsorbed is chiefly due to the strong competition of nitrates, which are present in a molar ratio of about 200 or 300 to one mol of styphnic acid.

Amount of Activated Charcoal to Be Used

In order to establish the amount of charcoal powder necessary to remove practically all styphnic acid from the solution, various amounts of the powder, from 5 to 25 g, were added to 1 liter of effluent mixture, 0.25 N in nitric acid and containing about 870 ppm of styphnic acid.

As can be seen from Table 5 A, the styphnic acid content of the effluent was decreased to about 3, resp. 2 ppm upon addition of 15 resp. 20 g carbon powder to the effluent.

A reduction of the amount of carbon powder necessary can be achieved by applying two consecutive treatments to the same effluent with two portions of carbon powder, 5 g each, adding the second portion to the filtered solution after the first treatment, as can be seen from Table 5 B.

In this case the carbon powder used for the second treatment and containing only a small amount of styphnic acid - about 130 mg - can be re-used for the first treatment of another batch of effluent, and thereby a further reduction of the quantity of carbon, to about 5 g per liter effluent can be achieved.

Conditions of adsorption

Adsorption takes place very fast upon stirring or shaking the solution for a few minutes. If an amount of carbon powder sufficient for the complete extraction of the styphnic acid is added, the carbon powder settles quickly, leaving a solution devoid of any coloration. In case of incomplete adsorption, as in the first of two consecutive treatments, settling is slower, but filtration can be easily effected.

In order to control whether the treated effluent contains no styphnic acid, a colorimetric field test can be used, instead of spectrophotometric determination. A sample of the effluent is alkalinized with an excess of sodium hydroxide in order to redissolve any precipitate of lead initially formed, and after filtration from the magnesium hydroxide precipitated, the color of the filtrate is compared with standards prepared from styphnic acid by alkalinization. In a Nessler tube, in a height of 20 cm, 0.5 ppm of styphnic acid can be easily detected, when compared to a blank.

The carbon powder remaining after filtration and containing about 5 to 20 weight percent of styphnic acid, depending on the form of treatment chosen, can be stored in the wet state, and from time to time brought to explosion together with waste TNT, or incinerated after addition of some fuel.

NEUTRALIZATION OF EFFLUENT AND PRECIPITATION OF LEAD CARBONATE

After filtration from the active carbon, the effluent free of

Table 6. Salt Concentration in Effluent after Complete Treatment

Ion	Mg^{2+}	Na^+	NO_3^-
Concentration, g/l	5.6	10.8	57 *

* Mean concentration of nitrate in plant effluents before addition of nitric acid: 38 g/l

styphnic acid, contains the amounts of lead (about 18 g), magnesium, and nitrates present in the original plant effluents, besides free nitric acid that was added.

To remove the lead from the solution, the usual method of precipitation as carbonate was used.

A potentiometric titration of the effluent, using a glass measuring electrode, showed that upon addition of a solution of sodium carbonate, crystalline lead carbonate precipitated at a pH of 4.6 . After precipitation of the lead carbonate, there is a steep rise in pH. We continued neutralization to a pH of about 6 to 6.2 . Because of redissolution of lead carbonate in alkaline solutions, it is not advisable to add a large excess of sodium carbonate, which may also cause precipitation of magnesium.

After filtration and washing, the pure lead carbonate can be used for any suitable purpose.

The amount of lead remaining in solution after filtration was not determined by us. In the literature (4) the solubility of lead carbonate in water is given as 1.1 ppm at 20 °C. Because of the constance of the solubility product, the before-mentioned amount should be reduced by adding a small excess of sodium carbonate, but as there is a considerable concentration of salts in the effluent, solubility conditions may be different.

Composition of Treated Effluent

The major constituents of the treated effluent are given in Table 6. The mean concentration of magnesium is the same as in the original solution. Sodium, which was not present, has been added, and the nitrate content raised from about 38 to 57 g/liter.

Table 7. Amount of Styphnic Acid and Lead Carbonate Regenerated and of Reagents Added

	Compound	Amount per liter of effluent *
Regenerated:	Styphnic acid	2.4 g **
	Lead Carbonate	23 g
Added:	Nitric acid, concentrated	17 ml
	Activated Carbon Powder:	
	One Treatment	15 – 20 g
	Two consecutive treatments	7.5 – 10 g
	Re-using carbon of second treatment	5 g
	Sodium carbonate:	
	For neutralization to approx. pH 6	16 g
	For precipitation of lead carbonate	9 g
	Total:	25 g

* after treatment

** 2.4 g of styphnic acid represents 70% of mean of 3.4 g dissolved in plant effluents before treatment

Compounds Regenerated and Added

The balance of the compounds regenerated and of those added is given in Table 7.

Because of the relatively high cost of styphnic acid and the large amount of lead carbonate regenerated, in comparison to the cost of the reagents added, the process seems to be advantageous also from an economical point of view. It is relatively simple, and no substances of high toxicity are generated, as in the procedures for the destruction of styphnic acid.

References:

1. U.S.A. Pat. No. 2,487,627, from 8th Nov. 1949 :
   Process for the Decomposition of Styphnic Acid

2. Knox, Richard , J. Chem Soc. (London) , 115, 522 (1919)

3. P. Aubertein & J.-M. Emeury, Mémorial des Poudres, 39, 7 (1957)
   Préparation et Propriétés de la Trinitrorésorcine

4. Lange's Handbook of Chemistry, 10th Ed., Mc Graw-Hill, p. 277

ENVIRONMENTAL PROBLEMS AND SOLUTIONS ASSOCIATED WITH THE DEVELOPMENT OF THE WORLD'S LARGEST LEAD MINING DISTRICT

J. Charles Jennett, Bobby G. Wixson,
Ernst Bolter and James O. Pierce

University of Missouri-Rolla and Columbia

The world's most modern lead industries have developed in the remote Ozarks region of S. E. Missouri during the past five years. In 1970 the "Viburnum Trend" of "New Lead Belt" ranked first in the world by producing 432,576 tons of lead ore, and in 1971 continued to pace the United States in lead production. This industrial development in a sparsely populated rural forest region has resulted in the release of lead, copper, zinc, cadmium and other metals into a formerly unaffected ecosystem. Due to this abrupt change, the mining district has become an unique area for studying the impact of trace metals on a wide variety of bio-geochemical systems and developing new techniques to control detrimental effects. See Figure 1 for the area location. Pollution abatement studies have been conducted since 1967 and during the past 15 months, an interdisciplinary team supported by the National Science Foundation RANN program with the assistance of local citizens, the mining industry, concerned local, state and Federal agencies, have continued with an extensive investigation on the possible environmental effects of this industrial development. A complete report of this investigation has been published.[1]

SOIL AND GEOCHEMICAL STUDIES

In a previous study, Bolter et al.[2] determined the geochemical background values for heavy metals, cadmium, and manganese in the New Lead Belt. The investigation also outlined the extension of a heavy metals anomaly in the vicinity of the region's lead smelter. In addition, the influence of ore transport by open trucks and uncovered railroad cars was also studied.

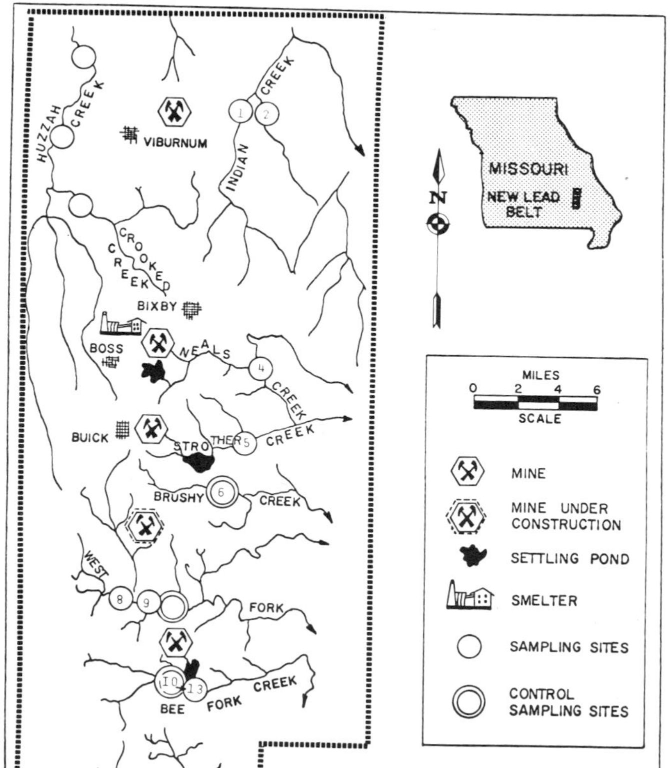

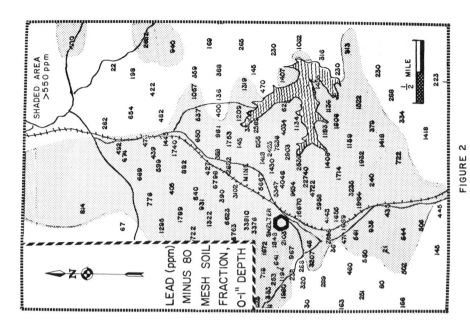

FIGURE 2

FIGURE 1 THE VIBURNUM TREND OR NEW LEAD BELT OF SOUTHEASTERN MISSOURI

During the 1971-72 research period, samples at 125 sampling stations were collected in an area of approximately 12 x 25 miles at a depth of 0-3 and 3-6 inches. The minus 80 mesh fraction (less than 0.177 mm) of the 0-3 inch samples was analyzed for copper, lead, zinc, cadmium, and manganese.

Most lead values were found to be in the 17-32 ppm range, with approximately 22 ppm being the average background. However, some samples contained lead concentrations up to 222 ppm.

The regional survey shows no indication of natural heavy metal anomalies caused by mineralized bedrock; however, the northern part of the district, which includes the oldest mining activity and the smelter, indicates higher lead concentrations with values up to 222 ppm. All data for the southern part of the New Lead Belt are below 50 ppm.

Zinc values range from 10-60 ppm, with most of the data in the 15-35 ppm range. Copper values are from 1.5-27 ppm, with most data in the 2.5-6.5 ppm range. High values for both elements are rather uniformly distributed throughout the district. Most cadmium values are below 0.5 ppm, with a few samples containing up to 2 ppm.

SMELTER AREA

Analysis of the less than 80 mesh (0.177 mm) fraction of 300 samples taken from a depth of 0-1 inch within a radius of about four miles around one smelter show an anomaly for lead within several miles distance from the smelter. Anomalies for zinc, copper, and sulfur are lower but distinct. Averaged data for selected distances from the smelter are given in Table 1.

Table 1
Averaged Concentrations (ppm) in Vicinity of Smelter

Distance from smelter (miles)		Pb	Zn	S
0-0.5	mean	5220	635	1088
	standard deviation	7095	849	1170
	number of samples	32	37	42
0.5-1.0	mean	1285	210	627
	standard deviation	1349	196	577
	number of samples	65	64	70
1.0-1.5	mean	540	113	593
	standard deviation	426	70	388
	number of samples	65	64	70
1.5-2.00	mean	355	79	582
	standard deviation	483	45	773
	number of samples	33	37	33
2.0	mean	188	62	408
	standard deviation	323	33	238
	number of samples	52	62	60

Figure 2 is an illustration of the distribution of lead in part of the smelter area. Soil samples, which were covered by several inches of organic debris, did not show appreciable increases in lead in the top one inch of soil even in areas where samples of exposed soil have lead concentrations varying from hundreds to thousands of ppm. Also, analyses of about 150 samples from a depth of 4-6 inches gave practically background values even in locations with extremely high (thousands of ppm) heavy metal concentrations in the top one inch of soil, indicating a very slow rate of penetration. This tendency of metals to be concentrated in the uppermost layers of soil in areas polluted by smelters has also been observed by other researchers such as Canney[3] and Hawkes and Webb.[4]

HEAVY METALS NEAR HIGHWAYS AND RAILROADS

Soil samples taken next to highways, which are used by open, uncovered trucks to transport ore concentrate, had lead values in the range of several hundred ppm in the 0-3 inch layer.[5] Considering the slow penetration rate, this would indicate possible lead concentrations of more than 500 ppm in the 0-1 inch layer. The extent of anomalies created in such a way has not yet been established; however, data by Bolter et al.[2] from biological samples indicate that the wind transport of ore minerals is possible for a distance of at least 300 yards.

Preliminary data from samples near the railroad tracks used to transport ore concentrates from the area indicate a similar degree of pollution.

MINE-MILL EFFLUENT STUDIES

In order to evaluate the effects of the developing lead mines on water quality and aquatic ecosystems, an extensive sampling program was established. Fortunately each mine, mill or smelter is on a separate stream, usually with similar non-developed streams nearby to serve as controls. Twenty-one principal sampling sites were selected during the initial phases of this study (see Figure 1). Sites 3 and 7 are not shown; 3 is a control stream and 7 is the confluence of all the streams of the region. Samples were taken bi-monthly and analyzed for both heavy metals and organic parameters of pollution.

In addition to these principal stations located on appropriate control streams and all streams known to receive mine and mill effluent, a number of secondary stations were selected to spot check a variety of waste ponds and associated tributary streams in close proximity to the mining, milling and smelting

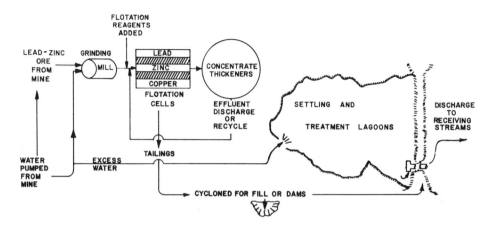

FIGURE 3A FLOW DIAGRAM OF MINING AND MILLING WASTES

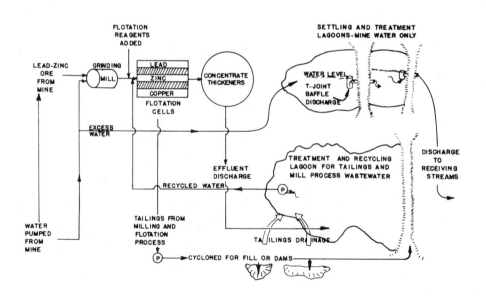

FIGURE 3B MODIFIED FLOW DIAGRAM FOR TREATMENT AND RECYCLING OF MINING AND MILLING WASTES

operations. The prime treatment devices for the milling operations are retention ponds usually constructed by daming one end of a valley with tailing left over from the mining and milling process. These tails are generally less than 200 mesh in size and in this region are generally 99% dolomite and approximately 1% galena. See Figure 3 A.

Quantitative determinations of the metallic content of the water, both dissolved and suspended, of the algae, of the stream sediments, and of the aquatic biological specimens were determined at the Environmental Trace Substances Center of the University of Missouri using Atomic Absorption techniques. Basic water quality measurements other than metals were performed in accordance with procedures as outlined in "Standard Methods for the Examination of Water and Waste Water"[6].

Visual observations were made of the general stream conditions, amount of flow, and dominant plant, and animal life apparent. Preliminary studies were conducted to ascertain heavy metals concentrations in algal mats and other aquatic vegetation at selected areas upstream as well as downstream from selected sites where contamination was indicated.

The concentrations of heavy metals present in a dissolved state in the streams of the New Lead Belt are relatively low. Effluents from mining and milling operations rarely exceeded the allowable limits set by the U. S. Public Health Service (1962) and the proposed "Effluent Guidelines" issued by the Missouri Clean Water Commission with regard to lead, zinc, and copper. The data presented in Table 2 includes figures on filtered and unfiltered samples to allow some conclusions to be made with respect to the amount of heavy metals in the dissolved and suspended states. Copper and cadmium are not shown since they were generally not present at the normal detection limits of 0.010 ppm. The allowable concentrations of the particular heavy metals under study presently stand at 0.1 ppm for lead, 0.1 ppm for zinc. 0.02 ppm for copper, and 0.01 for cadmium, as proposed by the Missouri Clean Water Commission[7]. The suggested maximum values set by the U. S. Public Health Service are:

Metal	Drinking and Culinary	Animal Watering
Pb	0.05 ppm	0.5 ppm
Zn	5.0 ppm	0.3 ppm (fish toxicity)
Cu	1.0 ppm	0.02 ppm (Fish toxicity)
Cd	0.1 ppm	

Table 2

MEAN PHYSICAL AND CHEMICAL PARAMETER VALUES FOR PRIMARY SAMPLING STATIONS

	Station* No. 1	Station No. 2	Station* No. 3	Station No. 4	Station No. 5	Station No. 6	Station No. 7	Station* No. 8	Station No. 9	Station* No. 10	Station No. 11	Station No. 12	Station No. 13	Station No. 14
Lead Pb Mean unfiltered	.022	.014	.013	.012	.083	.015	.019	.015	.014	.014	.030	.024	.016	.014
ppm Mean filtered	0.009	.005	.006	.005	.013	.005	.005	.006	.014	.005	.005	.008	.008	.005
Zinc Mean unfiltered	.023	.022	.011	.085	.037	.010	.019	.010	.010	.010	.132	.129	.035	.011
ppm Mean filtered	.025	.012	.010	.128	.032	.011	.011	<.010	<.010	<.010	.104	.111	.040	.010
pH	8.0	7.4	7.7	7.7	7.7	7.6	7.5	8.2	8.1	8.0	7.6	7.6	8.0	7.4
Temperature	0-25°C	0-25°C	0-25°C	0-25°C	0-25°C	0-25°C	0-25°C	0-25°C	0-25°C	0-25°C	0-25°C	0-25°C	0-25°C	0-25°C
Turbidity	3.7	2.9	1.2	1.0	7.7	1.3	.76	1.0	1.0	1.0	8.0	6.5	1.2	3.4
Dissolved Oxygen (DO)	5.4	4.6	4.7	5.5	5.5	5.6	5.3	5.6	5.2	5.5	5.9	5.9	5.4	5.3
Alkalinity	180	200	198	152	150	150	150	144	156	135	152	144	145	30
Hardness														
Calcium	150	140	210	100	95	100	100	75	80	75	110	110	110	15
Total	300	280	360	250	235	200	210	155	160	135	260	260	260	30
Chloride	0	0	30	20	0	0	0	0	0	0	0	0	30	0
Chemical Oxygen Demand (COD)	30	20	50	40	75	95	35	15.0	20.0	30	31	20.0	50	40
Phosphorus														
Ortho	.1	.1	.1	.1	.1	.1	.1	.1	.1	.0	.1	.1	0	.1
Total	.3	.3	.4	.3	.2	.1	.1	.2	.1	.1	.3	.2	.1	.1
Nitrate	0	0	0	0	0	0	0	0	0	0	0	0	0	0
Nitrate	.1	.1	.2	.1	.2	.1	.1	.2	.1	.1	.2	.1	.1	.1
Nitrogen														
Ammonia	0	0	0	0	0	0	0	0	0	0	0	0	0	0
Total Organic	26.4	8.4	30.2	25.7	16.4	18.0	12.0	15.6	10.0	22.4	16.4	17.6	35.3	25.7

*Indicates control stream; all others receive mine and/or mill wastes.

Table 3
SELECTED HEAVY METALS ANALYSIS IN ALGAE (May 10, 1972)

Sample No.	Aquatic Form	Lead Pb. (µg/g)	Zinc Zn. (µg/g)	Location of Sample
A	Cymbella	3996	94	Just above Tailings Ponds on Indian Creek Station No. 2
B	Nutgrass	118	198	Junction of Tailings Ponds and Indian Creek. Approximately 300 yds. downstream from dam
C	Fescue	1210	626	Station No. 1 2.5 miles downstream from dam
D	Cladophora	457	298	Station No. 1
E	Spearmint	1266	757	Station No. 1
F	Fescue	489	723	Station No. 4
G	Fescue	67	203	3 miles downstream from Station No. 4
H	Watercress	71	468	3 miles downstream from Station No. 4
I	Cladophora	699	336	Junction Strother and Neals Creek 4.2 miles downstream from Station No. 4
J	Cladophora	297	146	5.1 miles downstream from Station No. 4 0.9 miles downstream from Junction of Strother and Neals Creek
K	Cladophora	2342	704	2.3 miles downstream from Sample J and 1.1 mile downstream from Station No. 5
L	Cladophora	4631	1555	3.4 miles downstream from Sample J Station No. 5
M	Myriophyllum	3876	1088	Station No. 5

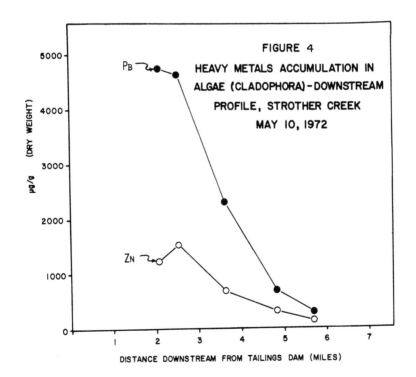

FIGURE 4

HEAVY METALS ACCUMULATION IN ALGAE (CLADOPHORA) - DOWNSTREAM PROFILE, STROTHER CREEK MAY 10, 1972

The results of lead analyses on the 15 principal stream sites over a nine month's period showed sixteen separate samples which were in excess of the 0.05 ppm limit set for drinking water by the U. S. Public Health Service. Six of these high samples, of which the highest contained 0.83 lead, occurred on the same day, December 15, 1971. These high values were concomitant with storm runoff, and the the stream flow during the storm of December 15 was of sufficient quantity and turbulence to achieve complete scouring of the bottom with virtually complete removal of the biological slime layer on all affected streams.

There were some 32 episodes noted among the 15 sampling sites over nine months where the zinc concentrations were above the 0.1 ppm established as the maximum allowable concentration in the "Effluent Guidelines" proposed by the Missouri Clean Water Commission. Of the 32 episodes, seven samples from station No. 4 (Neals Creek) taken after the first of the year showed high values; twelve samples were observed at station No. 11 with the highest value (0.237 ppm) observed on December 15, concomitant with heavy runoff. Ten high samples were observed at the final effluent from station No. 12 with the highest concentration of 0.226 ppm also occurring on December 15. Results therefore indicate that problems with zinc in effluents occur in two mining and milling effluent discharges.

There were seven episodes of copper concentrations in excess of the 0.02 ppm guidelines set by the Missouri Clean Water Commission and the U. S. Public Health Service standards for fish toxicity. The seven episodes were observed at five sites (stations No. 5, 11, 12, 13 and 15) with three stations showing elevated concentrations on December 15.

The results of the physical and chemical analyses are summarized in Table 2. It should be noted that more than 99 percent of the available nitrogen was consistantly in the form of total organic nitrogen at the 15 principal sampling sites. The pH of the streams under study was slightly alkaline, averaging between 7.2 and 8.0. This is undoubtedly due to the high carbonate content of the characteristic streams of the Missouri Ozarks. This alkaline pH and hardness is fortunate since the salts of the heavy metals under study are known to be extremely insoluble under these conditions.

Samples of algal mats, biological slimes, and aquatic vegetation were collected for analysis of heavy metals contamination. Selected results are shown in Table 3.

It was found that heavy metals concentrations were highest (approximately 5000 ppm on a dry weight basis) near the tailings dams of milling operations (after Gale 1972).

Concentrations decreased in a linear fashion with increased distance from the dams, as typified by Figure 4. Additional data and evaluations will be made as to species variation with regard to heavy metals accumulation. When 10 grams (wet weight) of Cladophera, one of the algal forms found in problem streams, was suspended in 250 ml of 20 ppm solutions of lead nitrate or lead acetate in the laboratory, 96 percent of the dissolved lead was taken up by the algae in less than one hour. It was not possible to distinguish between chemical adsorption and absorption, but it was evident that the lead was being removed from solution by the algae. Rates of lead removal by Cladophora were somewhat faster from solutions of distilled water than when lead salts were added to creek water. The content of carbonate salts in the creek water caused immediate precipitation of approximately 40 percent of the added lead (calculated to yield total concentration of 20 ppm), as shown by filtration through Watman No. 50 filter paper and removal of the precipitate. Approximately 70-75 percent of the original concentration of lead was removed from 20 ppm dissolved and suspended lead in creek water by Cladophora under these conditions within four hours, and 95 percent was removed in less than 16 hours.

SOLUTIONS

Using this data and input from the many other research areas involved in this study a number of recommendations have been made to or implemented by the mining companies including:

1. All lead, zinc or copper concentrate should be either covered or sprinkled to prevent blowing into the surrounding woods.

2. All ore and concentrate carrying vehicles should be covered to control soil pollution by heavy metals.

3. Dust containing heavy metals from the tailings dams is suspect as another source of stream pollution and these dams should be covered with soil and planted with grass to prevent future problems.

4. The lagoons used to treat the milling wastes have been redesigned to prevent short circuiting thereby allowing better sedimentation and biodegration of the organic reagents used. Where economically possible, total recirculation of the mill water would be advised. Figure 3B shows the proposed modified mining and milling lagoon waste treatment scheme now employed as opposed to the usual normal mining and milling scheme.

BIBLIOGRAPHY

1. "An Interdisciplinary Investigation of Environmental Pollution by Lead and Other Heavy Metals from Industrial Development in the New Lead Belt of Southeastern Missouri", submitted by the Interdisciplinary Lead Belt Team, University of Missouri-Rolla, to the National Science Foundation RANN program, Vol. I & II, (1 June 1972).

2. Bolter, E., Hemphill, D., Wixson, B., Chen, R., and Butherus, D., "Geochemical and Vegative Studies of Trace Substances from Lead Smelting", Proceedings Sixth Annual Conference of Trace Substances in Environmental Health, University of Missouri (1972-in print).

3. Canney, F. C., "Geochemical Study of Soil Contamination in the Coeur d'Alene District, Shoshone County, Idaho", Mining Engineering (1959).

4. Hawkes, H. E., and Webb, J. S., Geochemistry in Mineral Exploration, Harper and Row Publishers (1962).

5. Purushothaman, K., "Air Quality Studies of a Developing Lead Smelting Industry", First International Meeting of the Society of Engineering Science on Pollution, Tel Aviv, Israel (1972-in print).

6. Standard Methods for the Examination of Water and Wastewater, 13th Ed., American Public Health Association, 1015 18th Street, N. W., Washington, D. C., (1971).

7. Effluent Guideline, Missouri Clean Water Commission, P. O. Box 154, Jefferson City, Missouri 65101 (September 1971).

ACKNOWLEDGEMENTS

The author's would like to acknowledge that this project was supported by the RANN (Research Applied to National Needs) Lead Study Program of the National Science Foundation. The St. Joe Mineral Corporation and the Missouri Lead Operating Companies are also due thanks for their support of research, technical assistance and help. Our appreciation also goes to Mr. Nick Tibbs, David Butherus and Mike Hardie, graduate students at the University of Missouri-Rolla.

A STUDY OF THE EFFECTIVENESS OF BACKFILLING IN CONTROLLING MINE DRAINAGE

Keith G. Kirk

Department of Geosciences, Pennsylvania State University, University Park, Pennsylvania 16802

INTRODUCTION

A major ecological problem in Pennsylvania today is the pollution of our streams and rivers caused by acid mine drainage from strip mining. One of the current solutions to the problem is contour backfilling of the stripped land.

In strip mining, the soil and rock strata above the coal are removed, and the coal is then mined. Afterwards the overburden is replaced in the cut. The problem arises due to the nature of the coal and the minerals in the strata above and below the coal. Certain iron sulfides are present, such as pyrite and marcasite, which when exposed to oxygen form acid salts. When these salts come into contact with water they go into solution and produce acid drainage. The results of this process can be witnessed on many of Pennsylvania's streams, among them the Susquehanna River and Moshannon Creek. The effect of acid drainage of the value of our waters for industrial, public, and private use is severely reduced by the presence of acid mine pollution.

Pennsylvania has recognized this problem for over 20 years, and since 1945 has passed several acts of legislation aimed at easing this pollution problem. In May of 1945 the State Legislature passed the Bituminous Coal Open Pit Mining Conservation Act. The purpose of this act as stated in section 1 is as follows:

"This act shall be deemed to be an exercise of the police powers of the Commonwealth for the general welfare of the people of the Commonwealth, by providing for the conservation of areas of land affected in the mining of bituminous coal by the open pit or stripping method, to aid thereby in the protection of birds and wildlife, to enhance the value of such land for taxation, to decrease soil erosion, to aid in the prevention of the pollution of rivers and streams, to prevent the combustion of unmined coal, and generally to improve the use and enjoyment of said lands."

Hence, according to this 1945 law, the mine must be backfilled within one year after completion to include covering the un-mined coal, pit floor, and lowest mined exposed coal seam in the high-wall with at least five feet of earth, rounding of peaks and ridges to permit the planting of trees and shrubs, and sloping the high-wall to an angle not greater than 70°. Then in September, 1961, an amendment to the law was passed to require contour backfilling. This means that the land mined must be restored to the approximate original surface contours. This leaves no unsightly spoil piles, and in time, with re-vegetation of the land, there will <u>supposedly</u> be no trace of the stripping operation.

The originators of those laws assumed contour backfilling stops the discharge of acid drainage, although this assumption, as far as the writer can ascertain, has never been scientifically tested. It was therefore our purpose to attempt to evaluate the effectiveness of contour backfilling in curbing mine acid drainage.

BRIEF SUMMARY OF THE EXPERIMENT

To test this effectiveness it was decided to sample drainage from as many mines as possible in the various categories of age of mine, geologic age of coal, and method of backfill. These data were then analyzed statistically to determine which of these factors, if any, exerted the greatest influence on the production of acid drainage.

The drainage was analyzed for its pH, titratable acidity, total iron, sulfate, aluminum, and calcium content. State water quality standards were used as guidelines in determining what tests to run on the samples. The age of mine classifications used were 0-5, 6-10, and 11 and over years. Ages were obtained from local residents and mine operators, or from air photographs when there was nobody nearby or when two people differed widely in their estimates of the age. The methods of backfill used were:

<u>Contoured</u>; The mined area has been graded so that the land surface has been restored to the approximate original contours as existed before stripping.
<u>Struck-off</u>; The spoil peaks have been rounded off and the coal covered with at least five feet of earth.
<u>Untouched</u>; The spoils and highwall have been left untouched when the stripping ceased.

All these factors were compared and analyzed statistically as explained in later sections.

SAMPLE LOCATION

The study was carried out in portions of Clearfield and Centre Counties included in the areas covered by the U.S.G.S. Topographic maps of Wallaceton, Curwensville, Philipsburg, Houtzdale, Glen Richey, Clearfield, Ramey, Frenchville, Pottersdale, and Lecontes Mills. This area was chosen on the basis of availability of mines, abundance of geological data, and proximity to laboratory facilities at the Pennsylvania State University. Clearfield and Centre counties have been and are still being extensively stripped for coal and clay, and much of Clearfield county has been surveyed in detail by the personnel of the Pennsylvania State Geological Survey. Their reports included geologic maps and coal reserve maps which made it possible to determine the coals that had been extrected from most of the sample sites.

Locating of sites proved to be the first major undertaking. First random numbers and a grid system were used to locate the sample sites. This technique proved to be quite unfruitful and tedious, as many of these randomly located mines turned out to be either dry or virtually inaccessable. Thereafter location of sample mines was done by consulting topographic maps for possible watersheds. The areas picked were then searched for any measurable water flowing from the mined area.

Once water was located, it was necessary to be certain that this water source was usable in the study. Due to the nature of this study, it was necessary that each stream to be sampled drain only a single mine, or two mines of the same age, backfill, and coal type. To assure this, the streams were traced to their origins to be certain that they were usable in the study. Reasons for rejecting streams were the presence of deep mines in the area, a stream draining dissimilar mines, streams draining predominantly unmined areas, and springs emerging so far below the mine that they were likely to be unaffected by the mine. Many potential sample sties were rejected for the above reasons.

Sites were then marked and located by grid number on the topographic map. The grid was laid out in one minute squares and subdivided into thirty second squares. The sites were then marked with stakes and red flagging so they could be located for later sampling.

FIELD SAMPLING

At the sample sites tests were conducted to determine titratable acidity, pH, and stream discharge. Samples were also taken to be used for the tests in the laboratory, and information was recorded concerning the backfill, geology, and plant cover at each mine.

Titratable acidity is a means to measure the ability of the

water to release protons. The protons come from the un-ionized portions of any weakly ionizing acids and from certain hydrolysing salts, especially ferrous sulfate and aluminum sulfate. The titration measures directly the amount of alkali needed to raise the pH of the sample to 8.3. The titrant used was a sodium hydroxide solution of known normality. The indicator used was phenolphthalein.

The pH, which is a measure of the hydrogen ion concentration in the water was taken using a Beckman model IV portable electrometric pH meter. In taking the pH the meter was adjusted to the temperature of the sample and calibrated using a standard buffer solution.

Discharge in cubic feet per second was determined using a flow meter. The flow was measured to determine if a correlation exists between the volume of water discharged and the other factors. The discharge was also needed to convert the parts per million readings into actual weights of pollutants leaving the mines. The meter was constructed by cutting a 90° notch in one side of a rectangle of 3/4 inch plywood. The flow was measured by placing the board across the stream and damming the stream so that all the water was forced to pass through the notch. The height that the water backed up behind the notch was read from a scale painted on the board to one side of the notch. This height (H) was converted to discharge by the equation:

$$Q = 2.5H^{2.5}$$

Two water samples were collected at each site for use in the laboratory. A 25 ml sample was collected and added to 2 ml of concentrated hydrochloric acid in a 30 ml plastic bottle. This sample was to be used in testing for total iron and aluminum. The acid was to prevent the iron and aluminum from precipitating out of solution between the time of collection and the time of testing in the laboratory. A 500 ml water sample was also collected to be used for the test for sulfate and specific conductivity.

LABORATORY ANALYSIS

The following tests were performed in the Mineral Constitution Laboratory of The Pennsylvania State University.

> Aluminum and Iron These tests were performed on a Perkins Elmer #303 Atomic Absorption Unit. The samples were preserved as previously mentioned. The additional 2 ml of preservative was accounted for when the final determination of the parts per million was calculated.
> Calcium This test was performed on a Perkins Elmer #303 Atomic Adsorption Unit. No sample preservation was necessary for these tests.
> Specific Conductivity This test was done with a Yellow Springs Instrument Company specific conductivity bridge.

Sulfate This test was determined by measuring the specific conductivity of the sample and assuming that the sulfate anion was the major anion present in the sample.

GENERAL OBSERVED TRENDS

Our statistics have shown some important general trends. A first trend is in the relationship of the coal seam to acid production. An examination of the formation row of table 3 indicates that the Kittanning and Clarion formations are considerably greater pollutant producers than the Freeport Formation. In terms of severity, only one of the mean values (Ca) of the Kittanning and Clarion-Mercer formations are within the recommended upper limits of the USPH drinking water standards, while only one value (sulfate) of the Freeport Formation is above the upper limits of these same standards.

In breaking the data down further, (Table 1b) some very interesting observations can be made. The general trends of Freeport coals being cleaner than the others can again be seen. Of even more interest is the comparison of backfill method within the different formations. It seems that backfilling has little effect on acid production in the Freeport simply because this coal doesn't produce acid whether the mine is backfilled or not.

Some of these general trends might be better visualized in terms of precentages of mines that fail to meet the minimum state requirements provided for by article 6 section 4 part B of the Pennsylvania State Water Laws (Table 1a). In some cases these percentages should be looked at with caution due to the small sample population in the different classifications.

STATISTICAL ANALYSIS

The statistical analysis has two main purposes. First it was desired to establish the reproducibility of the measurements for individual mines. Lack of time necessitated that the entire study be based on only a single sampling of the mines. To determine the validity of this single sample technique it had to be determined if the measurements could be expected to remain constant over the period of the study. Four mines were selected at random and sampled, each five times at one week intervals. The data from these samples were then treated with an analysis of variance which tested the hypothesis the mean did not change from one sampling to the next. Since this hypothesis proved acceptable, it then could be concluded that the sampling technique would not cause significant errors.

The second purpose for statistical analysis was to establish what, if any, influence the factors of age of mine, geologic age of coal seam, and method of backfilling had on the quality of water leaving the mine. This was also done using an analysis of variance.

TABLE #1

Percentages of Mines Failing to Meet State Water Quality Standards

Coal Formation and Method	Percentages
Freeport and Upper Kittanning...............................	*Total 21%
Contoured	22%
Struck-off	22%
Untouched	00%
Middle and Lower Kittanning...............................	*Total 87%
Contoured	67%
Struck-off	94%
Untouched	100%
**Clarion-Mercer...........................	*Total 86%
Average of All Formations:	
Contoured	50%
Struck-off	71%
Untouched	64%

* Total is the percentage of mines in that formation summed over all of the methods of backfill.
** Due to the small sample size in the Clarion-Mercer group, percentages for individual backfill methods have not been calculated.

In this case, the analysis was used to show if significant differences occurred in any of the measured variables as the other factors were changed. This would indicate if any of the three factors had a strong influence on the given variable. When significance occurred it was desirable to know between which levels of the mine factors these significant differences occurred. The Sheffe' test was used to establish 95% confidence intervals for the true means and then checked for overlap of intervals between various levels of the mine factors. When no overlap occurred the source of significant difference was determined.

For both purposes the computer program ANOVES, a library program supplied and maintained by the computation center of the Pennsylvania State University was used. All the analyses were set up as two factor analyses using a cross classification design. The combinations of factors used were mine location vs. sample number (to determine if the site location i.e. particular quadrangle, had any effect on the results), geologic age of coal seam vs. method of backfill, and geologic age of coal seam vs. age of mine. One analysis was run for each of the six variables. The generalized

TABLE # 1b
Table of Means by Cell

	1 Contoured	2 Struck-off	3 Untouched
Freeport 1	pH = 6.57 Ta = 10.69 Fe = 1.27 (11) Al = 0.36 Ca = 37.64 Su = 599.8	pH = 6.24 Ta = 7.54 Fe = 0.81 (10) Al = 0.40 Ca = 29.10 Su = 275.5	pH = 6.97 Ta = 5.80 Fe = 1.12 (3) Al = 0.0 Ca = 17.67 Su = 119.7
Kittanning 2	pH = 4.65 Ta = 153.6 Fe = 2.28 (11) Al = 21.47 Ca = 39.64 Su = 1075.7	pH = 3.94 Ta = 141.3 Fe = 10.53 (18) Al = 19.86 Ca = 42.67 Su = 878.4	pH = 3.38 Ta = 970.9 Fe = 297.4 (5) Al = 83.20 Ca = 40.60 Su = 1766.6
Clarion 3	pH= 3.20 Ta = 156.7 Fe = 2.16 (1) Al = 28.00 Ca = 50.00 Su = 1385.0	pH = 3.52 Ta = 340.5 Fe = 20.95 (5) Al = 56.20 Ca = 24.82 Su = 1111.0	pH = 4.35 Ta = 79.46 Fe = 7.05 (4) Al = 7.40 Ca = 13.88 Su = 365.5

All units except pH in PPM
Ta = Titratable Acidity
Su = Sulfate
(x) - No. of observations in cell

set up of the analysis involves a two dimensional grid system with one factor represented by the columns and one by the rows, as follows:

                    Factor I

	Level 1	Level 2	Level 3	Level 4
Level 1	11	12	13	14
Level 2	21	22	23	24
Level 3	31	32	33	34
Level 4	41	42	43	44

Factor II

This gives the number of cells or categories equal to the product of the number of levels of each factor and numbered consecutively, as shown, giving the row number and then the column number. Each cell receives the value of the mean of the variable being tested for that combination of factors. The analysis of variance is carried out by computing the variance for rows, columns, and for all data combined. The difference between the sum of the variance from the rows and columns and the total variance is called the residual of error variance. The residual variance is a measure of the variance that can be expected due to randomness and experimental error. A ratio, the F ratio, is then computed by dividing the row and the column variance each by the residual variance. This gives a ratio of observed variance to any variance that would be expected due to randomness in a homogeneous population. Then, using the F distribution, the program calculates the probability that, given a homogeneous population, the F ratio will be as high as that calculated. That is, the program calculates the probability that the observed variance is due to randomness given that the population mean does not vary over the range of the factor being tested for significance. Usually a 0.05 level of significance is used. Therefore, if the probability is less than 0. or we can conclude that there is a significant variation in the variable due to changes in the factor for which the significance is shown.

The results indicated that, for the replication data, see Table 3, the variation due to the time of sampling was not statistically significant. That is, the variation in samples due to characteristics, such as age of mine, geologic age of coal seam, method of backfill, etc., was not affected by the time that the sample was taken.

The analysis of variance test for significance was used to show which of the factors were significant in causing changes within the variables, and Sheffe's test was used to determine between which levels of these factors the significant difference was to be found. Two analyses were run; geologic age and method, and geologic age and age of mine.

It was found for pH in both analyses, (see Table 2) only geologic age proved to be significant in affecting the pH of drainage waters. For pH, the level of geologic age where the significant change occurred was between the Freeport and the Kittanning Formations. The tests did not show any significant difference between the Kittanning and the Clarion Formations.

The geologic age method analysis showed both geologic age and method have a significant effect on titratable acidity (see Table 2). Sheffe's test for this analysis again showed the difference associated with a change from Freeport to Kittanning to be significant, and the difference associated with a change from Kittanning to Clarion was not shown to be significant. For the method of backfill, the significant change occurred when going from struck-off mines to untouched mines. The difference between contoured and struck-off mines was not shown to be significant. In the geologic age of mine analysis for titratable acidity, neither geologic age nor age of mine showed significance at the 0.05 level, although geologic age showed significance at the 0.1 level. The Sheffe' test was conducted at a level of 0.05 significance, so there was no indication as to between which levels of geologic age the significant difference occurred.

The geologic age method analysis for dissolved iron showed significance only for method of backfill. Iron content of the water can vary considerably with the Eh or oxidation potential. For iron, as for titratable acidity, the significant difference occurred between struck-off and untouched mines, with significance not shown between contoured and struck-off mines. In the geologic age of mine analysis, no significance was found at the 0.05 or 0.1 levels.

The analysis for aluminum content followed the pattern of the pH analyses. Again, only geologic age showed significance at the 0.05 level, and the source of the significant change was between the Freeport and the Kittanning formations. For aluminum it was found the method of backfill was significant to a 0.1 level of significance in the geologic age method analysis.

For calcium, no significance was shown for either geologic age or method of backfill in the first analysis, but showed significance for geologic age at the 0.05 level and age of mine at the 0.1 level in the second analysis. Here, the source of the difference for the Clarion Formation occurred between the Kittanning Formation and the Clarion Formation. There was no significance shown to the 0.1 level between the Freeport and Kittanning formations.

Sulfate, again, followed the pattern of pH. Significance was shown for only geologic age in both analyses, and the source of significant variation was again found to be the change from Freeport to Kittanning formations.

It must be noted that due to imbalance between sub-groups in the samples and due to missing cells, we were able to perform

Table # 2
Statistical Results

Geologic Age and Method

		pH	Ta	Fe	Al	Ca	Su
F-ratio test	Geologic Age	*	*	-	*	-	*
	Method	-	*	*	**	-	-
Scheffe Test	Contour, Plowed	-	-	-	-	-	-
	Plowed, Untouched	-	+	+	-	-	-
	Contour, Untouched	-	+	+	-	-	-
	Freeport, Kittanning	+	+	-	+	-	-
	Kittanning, Clarion-Mercer	-	-	-	-	-	-
	Freeport, Clarion-Mercer	+	+	-	+	-	+

Geologic Age and Age of Mine

		pH	Ta	Fe	Al	Ca	Su
F-ratio test	Geologic Age	*	**	-	*	*	*
	Age of Mine	-	-	-	-	**	-
Scheffe Test	Freeport, Kittanning	+	-	-	+	-	+
	Kittanning, Clarion-Mercer	-	-	-	-	+	-
	Freeport, Clarion-Mercer	+	-	-	+	+	+

*   reject null hypothesis at 0.05 level
**   reject null hypothesis at 0.10 level
\+   significant difference of means at 95% confidence level
\-   accept null hypothesis on basis of our data

Note discrepency between F-ratio test of method for aluminum and Scheffe tests for aluminum. Explained by the fact that F-ratio test for aluminum (method) was close to failure at the 0.05 level, and the analysis was only approximate due to disproportionate cell frequencies.

Ta = Titratable Acidity

Su = Sulfate

TABLE #3
REPLICATION STATISTICS

Variable Number 1, pH
Analysis of Variance Summary Table

Source	Sums of Squares	Df	Mean Squares	F Ratio	Probability
sample	0.2630	4	0.06575	1.550	0.250
mine	40.8935	3	13.63117	321.363	0.000
error	0.5090	12	0.04242		

Variable Number 2, Titac
Analysis of Variance Summary Table

Source	Sums of Squares	Df	Mean Squares	F Ratio	Probability
sample	3195.0	4	798.6	0.384	0.816
mine	2087097.0	3	695699.0	334.249	0.000
error	24977.0	12	2081.4		

Variable Number 3, Fe
Analysis of Variance Summary Table

Source	Sums of Squares	Df	Mean Squares	F Ratio	Probability
sample	441.3	4	110.32	1.931	0.170
mine	16865.2	3	5621.72	98.407	0.000
error	685.5	12	57.13		

Variable Number 4, Al
Analysis of Variance Summary Table

Source	Sums of Squares	Df	Mean Squares	F Ratio	Probability
sample	640.7	4	160.9	1.605	0.236
mine	54348.0	3	18116.01	181.498	0.000
error	1197.8	12	99.81		

Variable Number 5, Ca
Analysis of Variance Summary Table

Source	Sums of Squares	Df	Mean Squares	F Ratio	Probability
sample	167.5	4	41.88	2.119	0.141
mine	6801.3	3	2267.11	114.710	0.000
error	237.2	12	19.76		

Variable Number 6, Su
Analysis of Variance Summary Table

Source	Sums of Squares	Df	Mean Squares	F Ratio	Probability
sample	46004.0	4	11501.0	1.690	0.217
mine	4191627.0	3	1397209.0	205.314	0.000
error	81663.0	12	6805.0		

*Note: The interpretation of this data is explained in the section of the report on statistics.
** Df - is degree of freedom

TABLE #4

Comparison of Titratable Acidity, pH, Fe, and Al by the Type of Coal and Method of Backfill

Titratable Acidity	Freeport			Kittanning			Clarion		
	no. mines	mean	coef. var.	no. mines	mean	coef. var.	no. mines	mean	coef. var.
Contoured	11	10.7	67.7	11	154.	119.	1	157.	---
Struck-off	10	7.55	60.0	18	141.	140.	5	340.	85.4
Untouched	3	5.80	0.00	5	970.	121.	4	79.4	108.
pH									
Contoured	11	6.57	7.63	11	4.64	38.3	1	3.24	---
Struck-off	10	6.20	10.8	18	3.94	26.5	5	3.52	30.6
Untouched	3	6.97	7.07	5	3.38	25.1	4	4.35	36.3
Total Iron									
Contoured	11	1.27	129.	11	2.28	143	1	21.6	---
Struck-off	10	0.81	172.	18	10.5	172	5	20.9	100.
Untouched	3	1.12	24.3	5	297.	139	4	7.05	145.
Aluminum									
Contoured	11	0.36	223.	11	21.5	145.	1	28.0	---
Struck-off	10	0.40	241.	18	19.8	179.	5	56.2	97.5
Untouched	3	0.00	0.00	5	83.2	90.1	4	7.40	115.

only two factor analyses instead of a three factor analysis. This caused the difference due to the third factor, the one not being analyzed, in each case to be lumped in with the error variance. This increase in error varinace could result in significance not showing up where it indeed exists. An examination of the statistical data shows that there were several cases where the probability was very close to meeting the test for significance, but still had to be rejected. For example, in the geologic age method analysis for iron, the probability level associated with formation is 0.111. A three factor analysis, which reduced the error variance may have resulted in the probability meeting the test for significance. Another possible source of error here is the biased distribution of the samples. This made it impossible to run an exact analysis, so the program executed an approximate analysis as explained earlier. A third source of confusion rises due to the confounding of age of mine and method of backfill. Since these two factors are closely related, it is difficult to analyze one without considering the other. Another factor is the small sample size in many of the catagories. Because of this the effect of any one observation that varies widely from the observed mean may obscure any significant differences between the actual and population means.

PROBLEMS ENCOUNTERED

One of the most important parts of an analysis is to relate the problems encountered and to try to suggest the most fruitful areas and methods of future research. The main problem in this study was to try to determine what factors affected mine acid drainage and how to control these factors.

Deep mines are any mining operation which extend beyond the highwall into the coal seam without the removal of overburden. Often, but not always, water flowing from deep mines is quite polluted. Therefore it is necessary to pick strip mines which are not contaminated by deep mine waters, and this required knowing the location of the deep mines throughout the area of study. Deep mine locations can be determined through the help of geologic maps, topographic maps, local residents, mine inspectors, and miners. The maps do not show all of the deep mines, and the people may not remember, and there is always the possibility that ground water from a deep mine may be affecting the strip mine even though they never intersect. Because the entire study area has been deep mined in the past, the danger of contamination is very real, and must be considered for each and every sample site. Auger holes can also be considered as a form of deep mine. Auger extends a couple of hundred feet beyond the highwall and is said to be a common practice in strip mining. This could expose more coal to the weathering process and produce much the same result as regular deep mines.

Weather affects acid drainage through precipitation. Numerous dry streams were encountered and some sites dried up through the summer. Other streams greatly increased in volume after rainfalls. It was felt that rainfall might either dilute the acid or wash more acid from the mines and into the streams. Samples taken from rain run-off generally showed higher acidity than the stream samples. However, the replication sites demonstrated that increased water due to a recent rain had no statistically significant effect on acid production.

Contamination from other strip mines in the area was also a problem. Though the topographic drainage areas of two mines might be different it was still impossible to conclude that the water from one mine was not contaminated from the ground water interactions with another mine. Only a ground water analyses would reveal a source for the ground water.

Two important variables in the study were geologic age and method of backfill and their effects on acid drainage. One complication which arose in determining ages and methods of backfill was cuased by the common practice of remining an old mine. Even though the area had been mined twice, if all the spoil piles had been reworked it was assumed that all the effects of the older mine had been erased. Thus a mine that had been worked once, plowed over, and then reworked and contoured was classified as contoured. In some cases the new mines were rimmed by the old, and the mine had to be rejected. Exceptions to the above procedure are the active mines in which the miners backfill as they mine. In this case the mine was listed as active even though the drainage included some contoured area.

A very important factor controlling acid production in the study area was the type of coal mined. Different coals have different acid producing characteristics and so it is important to identify the type of coal mined in each mine. In the study, coal identification was used to indicate which geologic formation had been disturbed. It has been suggested that the overburden is just as important an acid producer as the coal (Guber, 1972), and this is why the formation was also considered.

CONCLUSIONS

Of the factors investigated in the mines studied it was found that:
1. Contour and Struck-off methods of reclamation reduced acid drainage from strip mines in the Kittanning and Clarion Mercer coal seams.
2. The Freeport and Upper Kittanning coals do not produce acid in excess of state standards regardless of the degree of reclamation.
3. The mines in the Middle and Lower Kittanning, Clarion, and Mercer coals produce acid in excess of the state standards regardless of the degree of reclamation.

4. The factor of geologic age of the coal most consistently related to high acid production.
5. In terms of reclamation, some backfill significantly reduced acid mine drainage.
6. In terms of method of reclamation there is no significant difference between contoured and struck-off methods of reclamation.
7. The age of the mine showed no significance in relation to acid production.
8. Mines in certain coals (Freeport and Upper Kittanning) produce significantly less acid than mines in other geologic aged coals (Middle and Lower Kittanning, Clarion, and Mercer).

SUGGESTIONS FOR FURTHER STUDY

The problems encountered in this study sparked these ideas for further study.

1. A study could be done to determine the effects of auger holes on acid mine drainage.
2. A study might be done to determine if ground cover affects acid production.
3. Certain shales have a high percentage of pyrite in them (Guber, 1972) and thus it seems reasonable to evaluate the composition of the backfill and how it relates to acid production. This would require a very extensive study since different shales contain different types of pyrite and these different types of pyrite have different acid producing potential.
4. A study might be done to determine if there is any relation between acid production and the permeability of the backfill.
5. A study very similar to this study might be done but continued over a much greater time period and with many more sample sites so that more detailed analysis of the data could be done.
6. One thing that was quite evident from the work with the local residents in the study area is that strip mining has greatly affected the lives of those people. The majority of the time this effect seems to be adverse. A study could be done to determine what the sociological effects of strip mining are on these people.

LAND DISPOSAL OF SEPTAGE (Septic Tank Pumpings)

J. J. Kolega, A. W. Dewey, R. L. Leonard and B. J. Cosenza

University of Connecticut, Storrs Agricultural Experiment Station, Storrs, Connecticut 06268

INTRODUCTION

Household sewage systems involving the disposal of septic tank pumpings, namely septage, are used on a world-wide basis. The system commonly encountered in the United States is a septic tank combined with a subsurface disposal field. In some areas of the country, a cesspool or dry well may be used in place of the septic tank, but this is not the general practice. Over a period of time, a septic tank or cesspool develops an accumulation of floating and settleable solids, along with scum. These solids will build up to a point where they must be removed. If this is not done, sewage flow into the septic tank will be blocked or the subsurface disposal field will become plugged from the excess solids discharge. In the first instance, the homeowner experiences an inconvenience requiring immediate attention. For the second situation, corrective action is required that may involve a renewal of the subsurface disposal field. The corrective measure, when required, can be a major one that is expensive to the homeowner.

Either of the household sewage system problems cited can be avoided by a periodic inspection of the solids buildup in the septic tank. The determining factors are: (1) the amount of settleable solids found on the bottom of the tank; and (2) the amount of floating solids found between the liquid surface and the bottom of the discharge baffle or pipe tee. Simplified techniques (Rockey, J. W., 1963) are available for measuring the amount of solids accumulation.

Other reasons for pumping individual household sewage systems are poor site selection for the subsurface drainage field and faulty installation practices.

Without an on-site inspection, the frequency of pumping individual septic tanks will vary and a suggested rule of thumb is to pump the septic tank every three to five years. Septic tank pumpers suggest pumping at more frequent intervals.

Removal and disposal of septic tank pumpings pose a health-oriented problem. How does a pumper dispose of the pumped material which may contain pathogens? Acceptable sites or facilities for the disposal of the septic tank pumpings, hereafter referred to as septage, may not be readily available. If a sewage treatment plant is nearby, and will accept septage, this would be the preferred disposal method. In the State of Connecticut, earth excavated pits have been used as an alternate procedure. The septage depth in these pits varies and is influenced by the ground water table, the volume of septage handled, and the amount of land available. Many of these earth excavated pits are proving to be unsatisfactory because they neither drain nor dry out and merely serve as septage holding pits that may persist even after being abandoned. Recently, shallow septage basins have been introduced in the State. The septage depth in these basins should be no more than 18 inches (46 cm). A number of basins are required and are used on a rotating basis. Septage dumpings in a particular basin are halted after a period of time and the material is allowed to dry. With these installations, it is possible to accumulate the solids material over a period of time. The partially dried solids can be removed with a payloader, or similar equipment, and then disposed of on the land. With either type of open dumping, objectionable odors can present a problem to people who live in the vicinity of the septage disposal site. Additional study of the shallow basin approach is needed to establish the overall design and construction criteria, the optimum surface area to septage volume requirements, and the number of basins required in a septage volume handling system.

A recommendation resulting from the Connecticut study is that septage be disposed of at sewage treatment plants properly designed for this purpose. Conventional sewage treatment plants can be used, and are preferred, provided they are within an economical septage transport distance and the plant is not operating beyond its design capacity. Generally speaking, the success of this operation can be influenced by the caliber of plant operating personnel. An objective of the Connecticut multidisciplinary study was to provide information on the biological, chemical, and physical properties of septage. Thus, personnel at sewage treatment plants should have a better understanding on the acceptability of septage for treatment in their overall plant treatment processes.

Lacking the availability of sewage treatment plants, two other disposal choices are: (1) specialized treatment process, or (2) land disposal. Unless large volumes of septage are to be processed, a specialized treatment facility for septage disposal is likely to be too costly when compared to land disposal alternatives.

When certain disadvantages can be overcome, the use of the land for the disposal of septage becomes an economical solution for small size communities. To overcome objections to septage pits, subsurface disposal of septage was developed and field tested. This paper reports on an operating system for receiving, storing, and applying septage to the land and an evaluation of the effects of the septage application on the soil water.

EXPERIMENTAL

The experimental site was a five-acre (2 ha) parcel of land owned by the White Memorial Foundation, Inc. (WMF) in Litchfield, Connecticut. The plot of land was provided by the WMF as a part of its policy to support research activities. This site was an abandoned field that had been used by nearby farmers for cutting hay. The cleared area was surrounded by woods located approximately one-third of a mile (0.54 km) from a normally travelled road. There was an existing access road which required some improvement. This consisted of an overall road grading, addition of gravel to potential wet spots, and a clearing of tree limbs to permit passage of large size transport vehicles.

The soil at the site where septage was injected was classified by the USDA Soil Conservation Service (SCS) as a Woodbridge fine sandy loam. This soil is described in the SOIL SURVEY MANUAL (Gonick, *et al*., 1970) as being a moderately well-drained soil underlain by a compact layer, or pan, at a depth of about 24 inches (61 cm). Soil samples were also taken at various depths during the drilling of the observation wells and the soil profile was determined down to the bedrock region.

The experimental site had two separate areas and will be referred to as the receiving or unloading area and the disposal site (the field where septage was injected) and are described as follows:

Receiving Area

This was approximately two acres (0.81 ha) in size. It included a storage (receiving) tank for septage; a turn-around area for pumper trucks; a loading area for the tractor-septage trailer unit; and a miscellaneous area for general vehicle parking, tank storage and equipment, and septage subsoil injection machinery.

A 10,000 gallon (37,850 liters), steel reinforced, precast concrete tank, 8 feet (2.4 m) by 8 feet (2.4 m) by 34 feet (10.4 m), was used to receive and hold septage. The liquid depth in the tank was 7 feet (2.1 m). The tank was set on a 6-inch (15.2 cm) compacted gravel bed and projected approximately two feet (0.6 m) above the ground. Normally, the tank would be earth covered, but since this was a temporary installation, the intent was to allow for the removal of the tank from this location upon completion of the study. The tank was made watertight after it was installed to insure against ground water contamination.

Septage discharge into the storage or receiving tank was by gravity. The pumper truck was backed into the receiving area and a discharge line dropped into the tank. Water for clean-up was available from a 1,000 gallon (3,785 liters) storage tank mounted on a 4-wheel trailer. A record of septage loads brought to the site was maintained.

A tractor power-take-off driven liquid manure pump was used to mix the septage prior to loading the septage hauling tank-trailer. A stationary tractor was used to drive the pump. The liquid manure pump manufacturer's discharge rating was 550-700 gpm (35-44 liters/sec).

Septage Field Application Area

A maximum application rate of 300 pounds of nitrogen per acre (337 kg/ha) was selected as the limiting factor in the quantity of septage which could be applied to the land by subsurface disposal techniques. The value was based upon agricultural field practices presently in use; upon research studies (Wengel and Kolega, 1970) on the effects of high poultry manure application rates on soil water; and from a preliminary investigation of subsurface application to a 50 feet (15.2 m) by 208 feet (63.4 m) experimental plot having a subsurface drainage system. Analyses of a limited number of drain water samples collected from the latter installation when approximately 10,000 gallons (37,850 liters) of septage was injected into the soil indicated no noticeable effect on the soil water.

The theoretical calculations for the size of land plot required were based upon free ammonia and organic nitrogen values obtained from analyses of septage samples. A 75 percent design value (Kolega, 1971) was selected, i.e., of the septage samples analyzed, only 25 percent of the data exceeded the design figures selected. The design values used for free ammonia and organic nitrogen in these calculations were 92 mg/l and 37 mg/l respectively.

This was a conservative figure since the respective sample median values for free ammonia and organic nitrogen, respectively, were 62 mg/l and 12 mg/l. Using the initial figures, a theoretical calculation showed that up to 837,000 gallons (3,170,000 liters) of septage could be applied to three acres (1.2 ha) of land under conventional cropping practices for the period June 1 to September 30 without a detrimental effect on ground water quality.

Three pairs of 6-inch (15.2 cm) diameter observation wells were installed on the diagonal of the rectangular field. The wells were used to monitor ground water quality. One of each pair of wells extended into bedrock which started at a depth of about 10 feet (3.1 m), and the other well was less than 10 feet (3.1 m) deep located in the soil water or aeration zone. These wells were capped and water samples were collected by means of a hand-operated pump. The water samples were analyzed for chloride, chemical oxygen demand (COD), nitrate, pH, phosphate, and for the presence of fecal coliforms.

Three basic methods of septage application were employed. These were (1) Terreator; (2) Plow-Furrow-Cover (P-F-C); and (3) Sub-Sod Injector (S-S-I). The field was divided into three areas. The upper section was used for the P-F-C method and the lower section used for subsoil techniques. The middle strip was reserved for special applications such as demonstrations for interested individuals or groups.

Land Disposal Equipment

Septage was transferred to an 800 gallon (3,028 liters) capacity tank-trailer pulled by a tractor and the material was then transported to the field and injected. The tank-trailer design was based upon the earlier studies of Reed (1969). The tank-trailer was a multipurpose unit designed so that septage, dairy or poultry manure, or sewage sludge could be handled in field application. The design of the equipment involved interstate cooperation with Rutgers University (The State University of New Jersey), and assistance from the manufacturer of the prototype tank-trailer. The multipurpose trailer approach offered the manufacturer the advantage of a greater market potential. This concept centered around the fabrication of a basic tank unit which could then be adapted to the final intended use, i.e., a septage trailer unit would not necessarily require an auger type agitator. The description of the tank-trailer unit follows:

<u>Tank-Trailer</u>. The tank configuration was rectangular having

straight sides which then tapered into a U-shaped bottom. Overall tank dimensions were 10 feet (3.0 m) long by 5 feet (1.5 m) wide by 3 1/2 feet (1.1 m) deep. Corten steel was used in tank fabrication with No. 12 gage steel being used for the sides and No. 10 gage steel for the bottom. There was a 30 inch (76.2 cm) by 30 inch (76.2 cm) hinged manhole cap in the tank cover for filling purposes.

The tank was mounted on a 2-wheel trailer chassis having 17 inch (43.2 cm) by 20 inch (50.8 cm) flotation tires. The trailer chassis was adjustable to permit changing trailer wheels from tracking tractor to an offset position. The gooseneck tongue was an integral part of the tank and provided ease of tractor maneuverability when it was used with subsoil injection equipment. A trailer braking system was included.

Slurry agitation and heavy slurry unloading were aided by a 9-inch (22.9 cm) diameter auger in the U-shaped trough. The auger was driven by a hydraulic motor at a speed not to exceed 250 rpm. A hydraulic motor, 10 brake horsepower in size, was mounted on the front end of the tractor and it was also used to operate a 12-inch (30.5 cm) stainless steel knife gate valve for septage discharge control. The position of the gate valve could be regulated by the hydraulic piston to control the flow of the septage. This valve was much larger than what would be normally required in this type of application. The hydraulic motor required the mounting of an additional hydraulic oil reservoir tank which was placed near the seat of the tractor operator. Hydraulic pump hoses, necessary fittings, and a flow control valve, mounted on the gooseneck frame of the trailer, completed the hydraulic system supplemental needs.

Septage discharge was either from the side of the tank for the P-F-C method of soil injection, or from the bottom front center of the tank. The latter discharge point was used with the S-S-I or Terreator techniques. The size of the tank discharge opening was six inches (15.2 cm). An adapter was provided for reducing the discharge opening to four inches (10.2 cm). The septage flowed by gravity into a 4- (10.2 cm) or 6-inch (15.24 cm) flexible hose and then into the furrow or subsoil injection apparatus. Attachment of the flexible hose at the trailer discharge point was by means of a quick coupling attachment. The hose material used was corrugated plastic pipe for the 4-inch (10.2 cm) application, such as that used in tile drainage work, and a corrugated neoprene hose with reinforcing coil for the 6-inch (15.2 cm) application.

Septage Application Equipment

All soil injection units had a standard three-point hitch for attachment to a tractor with a three-point hitch. The P-F-C and S-S-I techniques used were those developed for liquid manure disposal (Reed, 1970).

The Terreator (U.S. Patent No. 2,694,354) was a subsurface injection method developed by a Mr. T. Roberg of Litchfield, Connecticut. This unit could be either hydraulically or power-take-off driven from the tractor. A 3 3/4-inch (9.5 cm) diameter mole-type hole was made by an oscillating chisel pointed device. Attached to the Terreator was a 4 1/2-inch (11.4 cm) diameter curved tube attachment for receipt of the septage and its injection into the soil. The Terreator could inject septage to depths of twenty inches (50.8 cm) at a discharge rate of about two gallons per linear foot (24.8 liters/linear meter) of travel. This unit was in the development stage for slurry application at the time of its use in the study. The weight of the Terreator was 570 pounds (259 kg).

The P-F-C technique involved use of a 16-inch (40.6 cm) single bottom mold-board plow, a furrow wheel, and a 16-inch (40.6 cm) coulter. Septage was applied in a 6- (15.2 cm) to 8-inch (20.3 cm) deep plowed furrow and immediately covered with soil at the same time opening another furrow for the next septage application. The septage application rate was approximately one gallon per linear foot (12.4 liters/linear meter) of travel.

The S-S-I method consisted of a device which could inject a band of septage up to 1 1/2 (3.8 cm) inches thick and 24 inches (61 cm) wide from 6 (15.2 cm) to 8 (20.3 cm) inches beneath sod or a growing crop. The S-S-I had a spring trip release for passage over subsurface obstructions. A Category II three-point hitch (American Society of Agricultural Engineers Yearbook, 1971) was used. The septage application rate was approximately two gallons per linear foot (24.8 liters/linear meter) of travel.

RESULTS

Observations, Table 1, made during the summer and fall on the concentrations of chloride, COD, nitrate-N, phosphate and hydrogen ion (pH) in the soil water show no significant effects of the septage applications. Concentration levels fluctuated but no trends that could be related to septage applications were observed.

Checks were made for fecal coliforms on the ground water samples during this same period. Except in one instance, there were no fecal coliforms found in the water samples analyzed throughout the reporting period. This single incidence of contamination was from one of the two upper observation wells. The

TABLE 1. Observation Well Data for White Memorial Foundation, Litchfield Septage Soil Injection Experimental Site

Observation Period	Wells	Chloride ppm	COD mg/l	pH	Nitrate-N ppm	Phosphate ppm
Weeks preceding application						
6	U	--	--	--	--	--
	C	5.6	10.0	7.1	1.5	0.4
	L	24.6	6.0	7.3	0.4	1.0
4	U	--	--	--	--	--
	C	7.0	19.2	8.7	1.3	1.2
	L	24.8	14.2	8.2	1.3	2.9
2	U	--	--	--	--	--
	C	6.0	19.2	8.0	0.3	2.1
	L	6.0	15.8	8.2	0.3	2.3

--- START OF SEPTAGE APPLICATION ---

Weeks after application						
0	U	4.7	8.4	7.1	0.3	1.1
	C	6.0	12.6	7.7	0.1	1.0
	L	20.8	16.8	7.6	0.1	0.2
2	U	4.5	19.2	6.7	0.5	1.1
	C	6.4	18.6	7.0	0.3	0.7
	L	3.2	15.2	6.9	1.0	2.4
3	U	--	--	--	--	--
	C	5.1	22.7	7.0	1.5	0.2
	L	22.4	19.2	6.9	0.1	0.3
6	U	7.5	12.0	7.1	0.7	1.0
	C	8.0	16.0	7.6	0.3	0.9
	L	4.0	8.0	7.7	0.0	0.4
7	U	5.5	21.0	8.3	0.1	2.5
	C	8.5	25.0	7.7	0.1	0.2
	L	27.0	12.0	7.8	0.7	0.9

Observation Period	Wells	Chloride ppm	COD mg/l	pH	Nitrate-N ppm	**Phosphate** ppm
8	U	6.5	14.0	7.4	0.0	0.6
	C	6.5	8.0	7.8	0.0	0.6
	L	23.0	10.0	7.8	0.9	0.4
9	U	6.0	15.0	7.2	0.3	2.6
	C	6.0	8.0	7.7	0.0	0.6
	L	--	--	--	--	--
10	U	3.0	14.7	7.8	0.4	0.5
	C	5.0	7.3	8.5	0.2	0.4
	L	11.0	9.8	8.6	0.1	0.1
11	U	3.9	21.0	7.6	0.0	0.8
	C	7.3	16.0	8.4	0.1	0.7
	L	12.0	21.0	7.9	0.1	0.3
12	U	3.3	12.2	7.7	0.1	0.5
	C	6.4	19.6	8.5	0.1	0.7
	L	10.3	25.4	8.2	0.1	0.3

--- END OF SEPTAGE APPLICATION ---

13	U	3.4	18.5	7.8	0.1	0.5
	C	6.4	18.5	8.2	0.2	0.2
	L	12.3	14.2	7.9	0.1	0.3
14	U	4.7	17.3	7.6	0.1	0.4
	C	7.5	18.5	8.2	0.1	0.6
	L	14.1	24.0	7.7	0.1	0.3
15	U	5.1	12.0	7.3	0.1	0.2
	C	9.2	12.7	8.0	0.2	0.1
	L	13.6	14.0	7.6	0.2	0.1
16	U	5.3	7.6	7.4	0.1	0.5
	C	7.2	8.1	8.1	0.1	0.3
	L	12.7	7.9	7.6	0.1	0.2
20	U	8.6	7.6	7.5	0.0	0.1
	C	8.2	7.7	8.1	0.0	0.1
	L	11.9	7.8	7.8	0.0	0.1

KEY: U - Upper wells; C - Center wells; L - Lower wells

positive observation occurred during the second week of septage application and may have been due to contamination picked up during sampling.

Averages of observations for the period preceding septage application, for the twelve weeks of application, and for the eight weeks after applications that ended in the late fall were as follows:

Parameter	Unit of Measure	6 Weeks Preceding Application	12 Weeks During Application	8 Weeks Subsequent to Application
Chloride	ppm	11.7	8.5	8.7
COD	mg/l	13.6	15.6	13.1
Nitrate-N	ppm	0.6	0.3	0.1
Phosphate	ppm	1.4	0.8	0.3

Averages for observations made during the following spring--chloride, 5 ppm; COD, 21.4 mg/l; nitrate-N, 0; and phosphate, .7 ppm--showed no significant differences that could be attributed to the earlier septage applications.

During the twelve-week application period, approximately 120,000 gallons (454,000 liters) of septage were received and injected. The nitrogen equivalent of the total septage volume that was applied was about 43 lbs/acre (48 kg/ha). The results shown above are in agreement with the predicted or forecasted result that there would not be any detrimental effect on the ground water. The field equipment proved to be serviceable and no serious equipment problems were encountered.

Thus, the pilot study demonstrated that soil injection of septage can be a feasible disposal method. Further investigations are required both in terms of an increase in the rate of septage application and to discover any long-term effects on ground water quality from continued application of septage to a given plot of land. Additional information should also be obtained on the effects on crop responses on land on which septage has been applied. The findings in the present study provide evidence that if the septage application rate is controlled, this recycling scheme can be used as a means for disposing of septage or other comparable biodegradable materials.

Experiment Station.
Scientific Contribution No. 515 of the Storrs Agricultural

Acknowledgment:

This study was funded, in part, by the U.S. Environmental Protection Agency under Grant No. 17070DKA, "Treatment Processes--Wastes Pumped from Septic Tanks."

The study involved interstate cooperation between Rutgers University, New Brunswick, New Jersey, and the University of Connecticut, Storrs Agricultural Experiment Station. Prof. Charles H. Reed of the Agricultural Engineering Department at Rutgers University was the institutional cooperator and participant. Collaborators from the Connecticut State Department of Health, Environmental Health Services Division were Mr. Joseph Kosman, Chief, Municipal Wastewater Control Section, and Mr. Ronald E. Topazio, Senior Sanitary Engineer.

The authors are indebted to Mr. Ralph C. Gold and Mr. Randall C. May, Research Technicians, for their contributions to the conduct of the field operation of this study; Dr. Frank Chuang, Post-doctoral fellow in Agricultural Engineering for his overall assistance; Mr. Jaihind S. Dhodhi, Graduate Research Assistant, for assistance with laboratory analyses; Dr. R. W. Wengel of the Plant Science Department for nitrate analyses; Miss Linda Floeting, Project Secretary; Mr. Robert H. Shropshire, Superintendent, White Memorial Foundation, Inc., Litchfield, CT; Mr. George F. Sweeney, District Conservationist, USDA Soil Conservation Service, Litchfield, CT; Mr. Richard Gillett of the Connecticut Sewage Disposal Association; and Mr. Theodore W. Roberg, Litchfield, CT for their assistance and cooperation during the course of the study.

Waymark, Inc. (Cortland, N.Y.) was the manufacturer of the prototype tank-trailer. Reference to brand name pieces of equipment does not necessarily mean an endorsement of a product.

REFERENCES

ASAE Standard: ASAE S217.7, 1971. "Three-Point Free-Link Attachment for Hitching Implements to Agricultural Wheel Tractors and ASAE Standard: ASAE S278.2, 1971. "Attachment of Implements to Agricultural Wheel Tractors Equipped with Quick-Attaching Coupler for Three-Point Free-Link Hitch." Agricultural Engineers Yearbook, American Society of Agricultural Engineers, St. Joseph, Michigan, pp 260-265.

Gonick, W. N., A. E. Shearin, and D. E. Hill, 1970. "Soil Survey of Litchfield County, Conn., Supt. of Documents, U.S. Government Printing Office, Washington, D. C.

Kolega, J. J., 1971. "Design Curves for Septage." <u>Water and Sewage Works</u>, Vol. 118, No. 5, pp 132-135.

Reed, C. H., 1969. "Specifications for Equipment for Liquid Manure Disposal by the Plow-Furrow-Cover Method." <u>Animal Waste Management</u>. Cornell University Conference on Agricultural Waste Management, Cornell University, Ithaca, N.Y.

Reed, C. H., 1970. "Recycling and Utilization of Biodegradable Wastes by Plow-Furrow-Cover and Sub-Sod Injection." Paper No. NA 70-206. American Society of Agricultural Engineers, St. Joseph, Michigan.

Rockey, J. W., 1963. "Farmstead Sewage and Refuse Disposal." U.S. Department of Agriculture Agricultural Research Service Information Bulletin No. 274, Supt. of Documents, U.S. Government Printing Office, Washington, D. C.

Wengel, R. W. and J. J. Kolega, 1970. "Effect of High-Rate Manure Applications on the Soil." Paper presented at the Conn.-Mass. Poultry Engineering Seminar and New England Farm Electrification Institute.

RESOURCE RECOVERY FROM MUNICIPAL WASTES--A REVIEW AND ANALYSIS OF EXISTING AND EMERGING TECHNOLOGY

David Bendersky, William R. Park, Larry J. Shannon, William E. Franklin

Midwest Research Institute, Kansas City, Missouri

I. INTRODUCTION

The vast and growing amounts of solid wastes generated daily in our cities, and the present problems and costs of proper disposal of these wastes, has created an interest in systems designed to recover resources from municipal solid wastes. Some of the resources available from mixed municipal wastes are shown in Figure 1. Resource recovery systems offer the possibility of not only solving the waste disposal problem, but also can reduce the depletion of natural resources. It is for these reasons that federal, state and local governments are interested in resource recovery from municipal wastes.

It has been estimated that approximately 200 million tons of mixed municipal wastes are collected per year in the U.S.A.[1/] This is an average of approximately 5 pounds of waste per person per day, and the generation rate per capita is expected to increase. At the present, most of the municipal waste in the U.S.A. is either dumped or incinerated. Recent bans on open dumping, the growing lack of sanitary landfill sites, and the high costs of incineration (without resource recovery) have created serious waste disposal problems in many cities in the U.S.A.

Interest in resource recovery from municipal wastes has been stimulated in the U.S.A. by recent federal legislation. The U.S. Resource Recovery Act of 1970 promotes the demonstration, construction,

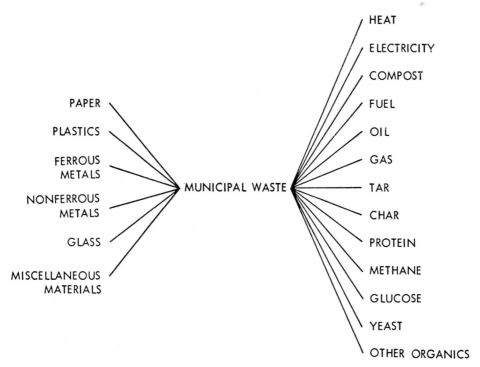

Fig. 1. Resources Available from Municipal Waste

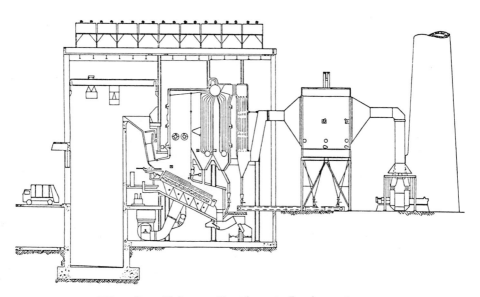

Fig. 2. Chicago Northwest Incinerator

and application of resource recovery systems; provides technical
and financial assistance to the states and local governments in
planning and development of resource recovery programs; and promotes a national research and development program for new and
improved methods of recovery and recycling of solid wastes.

In connection with the United States government's interest in
resource recovery from wastes, the President's Council on Environmental Quality sponsored a technical and economic study of existing
and emerging technology for resource recovery from mixed municipal
solid wastes. The results of this study are summarized in this
paper.

II. EXISTING TECHNOLOGY

Only two methods are presently fully developed and practiced
for the recovery of resources from mixed municipal wastes--(1) heat
recovery incinerators, and (2) composting.

Heat Recovery Incinerators

Heat recovery from the incineration of municipal wastes has
been developed and used in several European countries (notably
France,[2/] Germany,[3/] and Switzerland)[4/] and in Japan.[5/] These
systems are designed to burn mixed municipal wastes and recover
the heat energy, usually in the form of steam. The waste is burned
in large furnaces equipped with moving grates. Conventional tube
boilers and water walls convert the heat into steam. In some cases
part or all of the steam is converted into electricity, using conventional steam power plant equipment.

The first large heat recovery incinerator in the U.S.A. has
recently been installed in Chicago (Figure 2).[6/] This plant is
designed to burn 1,600 tons of municipal waste per day and generate
440,000 pounds of steam per hour. There are actually four incinerator units in this plant, fed from a common refuse storage pit;
the storage pit has a capacity of 1,800 tons of waste. A crane grab
bucket picks up the refuse and dumps it directly into the furnace
feed hopper from which it is automatically fed onto the stoker grates.
The inclined grates have a reverse-reciprocating action producing a
downward movement of the refuse, and at the same time the grates are

7. SOLID WASTE, LAND USE

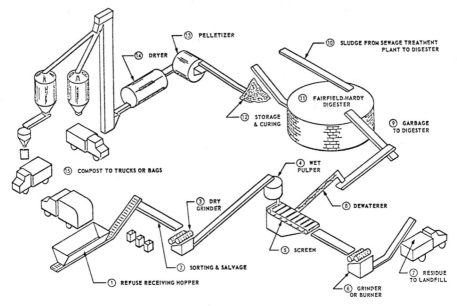

Fig. 3. Fairfield-Hardy Composting Process

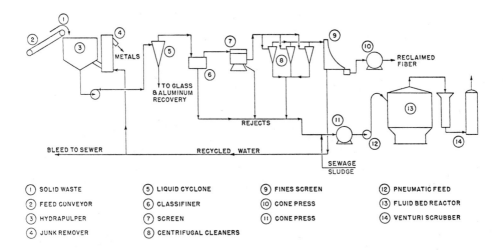

Fig. 4. Black Clawson Materials Recovery System

moving upward, producing a tumbling action. Air jets below and above the grate combined with the tumbling action promote effective burning of the refuse. The hot gases travel through a series of boiler passes in which the heat is converted into steam. The tubular walls of the furnace are also filled with water and contributes to the steam generation. Before entering the stack the cooled gases pass through an electrostatic precipitator where 97% of the particulate matter in the gases is removed. About half of the steam generated is used for plant needs and the rest is available for sale to nearby industries. The ashes are screened and the metals are sold for reuse. The rest of the ashes are to be used in construction projects or placed in a sanitary landfill.

Compost

The production of compost from mixed municipal wastes has been practiced both in Europe and the United States. Several processes have been developed. The most successful compost method used in the U.S. is known as the Fairfield-Hardy process (Figure 3).[7] The refuse is ground in both a dry grinder and a wet pulper, dewatered and placed in a "digester." Sewage sludge is added to the ground refuse and the mixture is stirred and aerated in the digester for 5 days. After the digestion process, the material is pelletized and dried.

Compost is useful as a soil conditioner. It will improve the structure of clay or other hard soils, and increase moisture holding capacity of the soil. However, compost is not a substitute for fertilizer, since it contains less than 1% of nitrogen, phosphorus and potassium, which are the essential elements in fertilizer. Difficulties have been experienced in marketing compost in the United States.

III. EMERGING TECHNOLOGY

There has been a significant increase in the development of new technology for resource recovery from municipal waste during the last few years. Included in this emerging technology are (1) materials recovery systems, (2) energy recovery systems, (3) pyrolysis systems, and (4) chemical conversion systems.

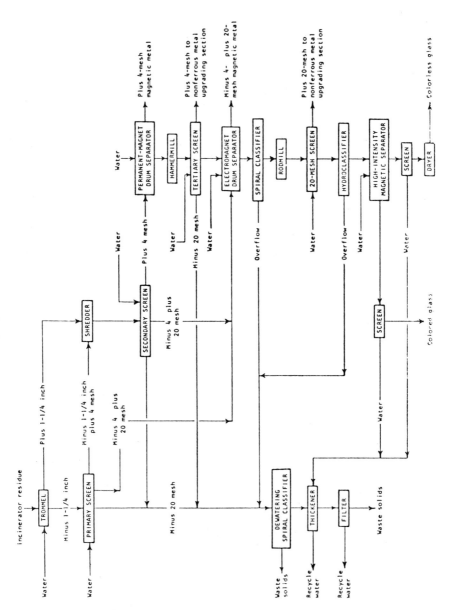

Fig. 5. Bureau of Mines Incinerator Residue Recovery System

Materials Recovery Systems

Materials recovery systems have been designed to recover the three principal raw materials (paper, metals, and glass) contained in municipal solid wastes. Two of the most developed materials recovery systems are the "Hydraposal/Fibreclaim" system developed by the Black Clawson Company,[8] and an incinerator residue recovery system developed by the U.S. Bureau of Mines.[9]

The Black Clawson materials recovery system (Figure 4) is a wet process designed to receive raw municipal wastes and remove the paper and other fibrous materials in the form of reusable pulp, recover the metals and glass, and burn the non-recoverable organic materials. The raw waste is conveyed into a hydrapulper where a high speed cutting rotor converts the pulpable and friable (paper, food waste, plastic, glass, aluminum) materials into a water slurry. Large, heavy objects such as stones, metal cans, etc. settle to the bottom and are removed with a "junk remover." The slurry is pumped through a series of cyclones and screens to remove the glass, aluminum and other non-fibrous material. The recovered fiber is then dewatered in a cone press. The rejects from the various screens and cleaners are essentially non-recoverable organic materials, which are first dewatered and then burned in a fluid bed combuster. A 150 ton per day demonstration plant of this system has been in operation at Franklin, Ohio since June 1971.

The U.S. Bureau of Mines has developed a process for reclaiming the metal and mineral values contained in municipal incinerator residues.[9] The separation process is designed to recover iron, non-ferrous metals and glass from the residues. The process (Figure 5) consists of placing wet incinerator residue in a trommel which separates the plus 1 1/4 inch material from the minus 1 1/4 inch material. The minus 1 1/4 inch material is fed to a primary screen which separates the material into three size fractions. The larger materials from the trommel and the primary screen are fed to a shredder and secondary screen. The larger size fraction from the secondary screen is passed through a permanent magnet drum separator to remove the ferrous metals. The non-magnetic material is fed to a hammermill and a tertiary screen, which removes the larger non-ferrous metals for upgrading. The smaller particles from the primary, secondary and tertiary screens are conveyed to an electromagnetic drum separator to remove the ferrous metals. The non-magnetic particles from the electromagnetic separator are dewatered in a spiral

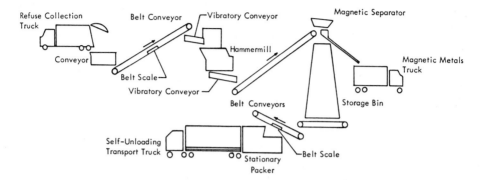

Fig. 6. Horner & Shifrin Fuel Recovery System

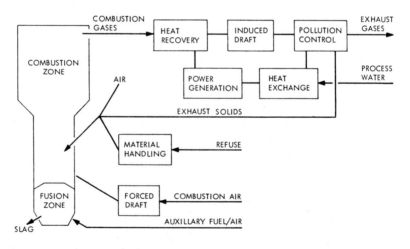

Fig. 7. Melt-Zit Incinerator System

classifier, fed to a rodmill, and screened to remove the larger non-ferrous metals. The fine material is fed to a hydroclassifier to remove slimes and excess water before entering a high intensity magnetic separator which recovers colored and colorless glass. The non-ferrous metals are separated into aluminum and copper-zinc scrap, using a sink-float separator. The Bureau of Mines residue recovery system is presently a pilot plant operation with a capacity of 4 tons per 8 hour day.

Energy Recovery Systems

The emerging energy recovery systems fall into three categories: (1) fuel recovery systems (2) steam generation systems, and (3) electrical power generation systems.

One of the most promising systems for recovering fuel from municipal solid wastes is the Horner-Shifrin system (Figure 6).[10] The system is designed to recover fuel and magnetic metals from mixed municipal wastes. The fuel is to be used as a supplementary fuel for large pulverized coal-fired power plant boilers. The process is simple: The mixed municipal waste is shredded in a hammermill and passed over a magnetic separator to remove the magnetic metals. The shredded waste is then ready for transport to the power plant. The shredded waste is to be used as a supplementary fuel in conventional pulverized coal-fired boilers. A full scale demonstration plant is now being constructed in St. Louis, Missouri. The plant capacity is 650 tons of waste per day (2 shifts). The output is expected to be about 600 tons/day of supplementary fuel (with a heating value of about 5,000 Btu per pound) and 50 tons/day of magnetic metal. The fuel will be burned in an existing power plant boiler and will furnish about 10% of the total fuel supply.

A new incinerator system for generating steam from municipal waste is called the Melt-Zit Incinerator (Figure 7),[11] developed by American Thermogen, Inc. A unique feature of this system is the ultra-high temperature incinerator, which operates at about 3,000°F by burning auxiliary fuel (either coke, oil or gas). Because of the high temperature, the combustibles in the waste are completely burned and all of the metals, glass and other nonburnables are melted, reducing the volume of the original refuse feed to 3 to 5% of its original volume. Secondary air is fed into the incinerator

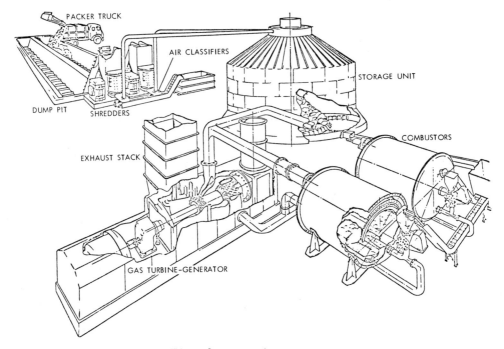

Fig. 8. CPU-400 System

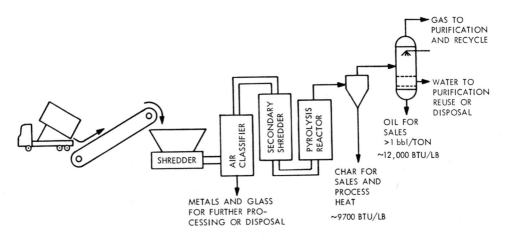

Fig. 9. Garrett Pyrolysis System

just above the melt zone. Air is also admitted through the charging door along with the waste, and the combination of draft control inside the charging door (by induced draft fan) and the secondary air insures complete combustion of all organic materials. The molten materials are continuously drawn out of the bottom of the unit through a melt trap and is water quenched, causing it to shatter into granular lava-like material. The hot flue gases leaving the incinerator are passed through a steam boiler. Some of the steam is used to run the plant equipment and the excess steam is available for sale. A pilot plant has been operated for several years and a 600 ton per day system has been proposed for construction at Malden, Massachusetts, as a central regional waste disposal facility to serve several surrounding communities.

A new system for generating electricity from municipal wastes, called the CPU-400 (Figure 8), is under development by the Combustion Power Company.[12/] This system is designed to burn 400 tons of waste per day and produce 15,000 kilowatts of electricity. The packer trucks discharge the refuse into a receiving pit, from which it is conveyed into shredders. The shredded waste is sent through air clarifiers where the heavy, non-combustibles are separated from the lighter, combustible material, primarily paper and plastics. The shredded combustible material is pneumatically conveyed to a storage unit, which has a capacity of about one day's supply of refuse. Refuse from the storage unit enters a rotary dryer (not shown) where it is dried by hot gases by-passed from the turbine exhaust. The shredded, classified and dried refuse is fed by high pressure feeders into fluid-bed combusters. The bed in the combuster is hot sand, agitated by an upward flow of pressurized air coming from the turbine-driven compressor. The shredded refuse is introduced into the hot sand bed where it is burned, and maintains the fluid bed particles and combustion products at about 1800°F. Particulates in the hot combustion gases are removed by particle collectors prior to flow of the gases through the turbine-generator and out the exhaust stack. A pilot plant with a capacity of 40 tons of waste per day is now under development at Menlo Park, California.

Pyrolysis Systems

Considerable potential exists for reforming, by pyrolysis, the organic portion of municipal solid wastes into lower molecular weight compounds having significant economic value. Because pyrolysis reactions, often termed destructive distillation, can be conducted in the absence of oxygen or in controlled oxygen environments, product composition can be regulated. Unlike combustion in an excess of air, which is highly exothermic and produces primarily heat and carbon dioxide, pyrolysis of organic material is analogous to a distillation process and is endothermic. The high temperatures (1,000-2,000°F) and lack of oxygen result in a chemical breakdown of the organic materials into three component streams: (1) a gas consisting primarily of hydrogen, methane, carbon monoxide, and carbon dioxide; (2) a "tar" or "oil" that is liquid at room temperature and includes organic chemicals such as acetic acid, acetone, methanol; and (3) a "char" consisting of almost pure carbon plus any inerts (glass, metals, rock) that enter the process.

Laboratory and pilot plant pyrolysis units have been successfully constructed and operated, and these units have demonstrated the technical feasibility of the pyrolysis of municipal refuse. Laboratory investigations of the pyrolysis of various organic wastes have been conducted at the University of California (Berkeley), Bureau of Mines, Rensselaer Polytechnic Institute, New York University, and the Utilities Department of the City of San Diego, California.[13/] Batch retorts, fluidized beds, and rotary kilns have been successfully used as reactors in these laboratory studies. Pilot plant studies of pyrolysis systems for municipal wastes have been conducted by Garrett Research and Development Company,[17/] Monsanto's Enviro-Chem Systems,[18/] and Hercules, Inc.[19/]

The pyrolysis system of Garrett Research and Development Company is one of the most developed. Garrett has adapted a pyrolysis process for converting coal to oil, to converting shredded municipal solid wastes to oil, gas and char. A simplified flow diagram of the pyrolysis process is shown in Figure 9. The pyrolysis process has been studied on a continuous bench-scale unit for over two years. Operating parameters and costs are now being confirmed in a 4 ton/day pyrolysis pilot plant at LaVerne, California. A yield of over one barrel of oil per ton of raw waste is reported.[17/] The pyrolysis unit will form the cornerstone of an integrated resource recovery process being developed by Garrett. The integrated system will include glass and magnetic metal recovery in addition to the pyrolysis process.

Chemical Systems

Chemical methods that have been suggested for converting municipal refuse into usable products include hydrolysis, hydrogenation, wet oxidation, photo degradation and anaerobic digestion.

These methods use only the cellulosic portion of municipal refuse, so that separation and pretreatment of the raw refuse is required.

Hydrolysis of cellulosic waste to produce protein and glucose is the only process that has been tested at the pilot plant level.[20] Hydrogenation and wet oxidation have been studied in laboratory equipment, [13, 14, 21] while photodegradation and anaerobic digestion are in the conceptual stage.[13, 15, 22]

A pilot plant for the production of single-cell protein from waste sugarcane bagasse has been designed, constructed, and operated at Louisiana State University. Figure 10 presents a flow sheet for the pilot plant. The pilot plant's equipment can be grouped into five distinct process sections: cellulose-handling, treatment, sterilization, fermentation, and cell harvesting. The plant was designed so that it could operate in both batch and continuous-flow modes. The initial operation of the pilot unit has been limited to a single waste cellulosic substrate, sugarcane bagasse. Purified ground wood pulp has also been used as a control substrate in several runs. Single cell protein with a crude protein content of 50-55% has been produced.

IV. ECONOMICS

The comparative economics of eight basic systems for recovering the resource value from mixed municipal wastes were investigated in this study, along with the primary conventional solid waste disposal methods, based on current conditions within the United States.

In most of the United States, the sanitary landfill offers the lowest-cost environmentally acceptable means of solid waste disposal. The cost of land, though, is increasing rapidly as the availability of suitable landfill sites near major metropolitan areas declines. The effect of land cost on sanitary landfill cost is shown in Figure 11. When land costs are prohibitive, or when suitable landfill sites

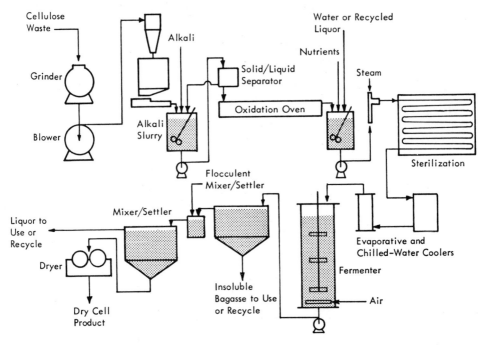

Fig. 10. System for Converting Waste Cellulose into Protein

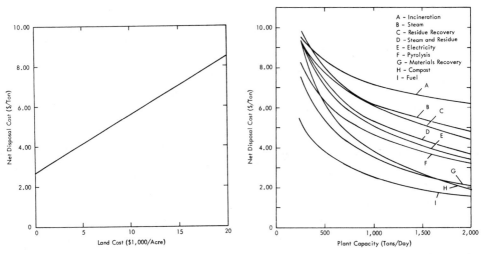

Fig. 11. Sanitary Landfill Costs

Fig. 12. Effect of Plant Capacity on Net Costs

are not available within a reasonable distance, incineration (without resource recovery), which is usually considerably more costly than landfill, has generally been the only alternative for disposing of municipal wastes.

A variety of resource recovery systems described earlier have been proposed as alternatives to conventional incineration and sanitary landfill. For the purpose of this cost analysis, the various resource recovery systems were divided into eight broad categories:*

(1) Conventional incineration, with recovery of materials from the incinerator residue.

(2) Incineration with heat recovery, used to produce steam.

(3) Incineration with heat recovery, used to generate electric power.

(4) Incineration with steam production and residue materials recovery.

(5) Composting with recovery of inorganic materials.

(6) Comprehensive materials recovery, where the mixed refuse is separated into its marketable components.

(7) Fuel production, where magnetic materials are removed from the mixed waste and the balance is used as a supplementary fuel in public utility boiler furnaces.

(8) Pyrolysis, with oil recovery from the combustion gases and recovery of inorganic materials.

Basis of Cost Analysis

Although the characteristics of many different resource recovery systems were examined, the cost analyses presented here are not based on any specific proprietary systems. Standard equipment, processes and manufacturing methods have been assumed throughout. It must be

---

*Chemical conversion systems were not included because of the large variety of systems and their early development state.

TABLE 1

ANNUAL QUANTITY AND VALUE OF RECOVERABLE RESOURCES IN 500 TPD OF MUNICIPAL REFUSE

Resource	Quantity Available	Assumed Unit Price FOB Plant($)	Total Value
Steam	1,226,000 M lb.	0.50	$613,000
Electric energy	126,000,000 kwh	0.01	1,260,000
Compost	45,000 tons	10.00	450,000
Ferrous metals	13,700 tons	10.00	137,000
Ferrous metals (incinerator residue)*	7,600 tons	10.00	76,000
Nonferrous metals	1,080 tons	200.00	216,000
Paper	36,000 tons	12.00	432,000
Glass	10,800 tons	10.00	108,000
Oil	43,200 tons	12.50	540,000
Fuel	1,800,000 mm Btu	0.20	360,000

*Less ferrous metals are available from incinerator residue than from raw refuse because of degradation.

TABLE 2

MARKETABILITY OF RESOURCES RECOVERED IN VARIOUS SYSTEMS

System	Recovered Products	Marketability Factor
Incineration--residue recovery	Glass, aluminum, ferrous metal	0.68
Incineration--heat recovery	Steam	0.35
Incineration--electricity generation	Electricity	0.60
Incineration--heat and residue materials recovery	Steam, glass, aluminum, ferrous metal	0.48
Composting	Humus, glass, Al, Fe	0.49
Materials recovery	Glass, Al, Fe, paper	0.67
Fuel recovery	Fuel; ferrous metal	0.85
Pyrolysis	Oil: inorganics	0.80

recognized that costs will vary widely according to local conditions, and detailed design study would be required to obtain a realistic estimate for a particular plant at a given location. Numerous assumptions were therefore necessary in developing the costs shown here.

It was assumed that a municipality owned and operated the facility. Since most resource recovery systems are capital-intensive, the lower annual fixed or time-related costs associated with municipal plants result in a substantial cost advantage over privately owned plants.

Costs were developed on the basis of a 500 tons per day plant, operating 24 hours per day, 360 days per year, thereby handling 180,000 tons of mixed municipal refuse annually. A 20-year economic life was assumed for the plant.

Municipal refuse composition will vary both regionally and seasonally. In this analysis, a typical waste composition was assumed, consisting of:

	Percent
Paper	50.0
Garbage	10.0
Nonferrous Metals	0.6
Ferrous Metals	7.6
Glass	6.0
All other	25.8
Total	100.0

Since the raw refuse normally contains about 20 to 25 percent moisture by weight, a 500 TPD operation actually has only 375 to 400 TPD of potentially recoverable resources. The quantity and value of the recoverable resources found in 180,000 tons per year of this typical mixed municipal refuse is summarized in Table 1.

The marketable products associated with each of the eight resource recovery concepts are shown in Table 2. To qualify the value of these recovered resources in light of the highly variable and risky markets that would likely be encountered in the United States, a "resource marketability factor" was developed for each system. This factor is a qualitative judgment of both the probability of selling the product output from each system, and the probability of receiving the indicated prices for the products.

Thus, because of widespread demand and relatively stable prices for fuel within the United States, the fuel-producing systems have a high marketability factor. At the other extreme, steam generally requires large, nearby industrial customers, and compost producers in the United States have experienced problems in selling their product in large quantities; these systems, therefore, have low marketability factors. The other systems rank in between these extremes in terms of the marketability of their recovered resources. The inorganic materials (metals and glass) have generally good market potential and fair price stability, though sensitive to local conditions. Paper markets are limited and uncertain; and electrical energy, while universally in demand, is apt to be difficult to sell under existing conditions in the United States.

Table 3 summarizes the capital requirements and annual operating costs for the eight resource recovery systems, and conventional incineration (without resource recovery) for comparison. Using conventional incineration as the base, the percentage returns on incremental investment are shown under two conditions, using gross and expected resource values.

Based on gross resource values, the incinerator residue recovery system, for an additional $870,000 investment, reduced operating costs by $135,000 annually, a percentage return of 15.5 percent. However, on an _expected_ _resource_ _value_ basis, it barely breaks even, returning only 0.6 percent.

The steam recovery and electric generation alternatives both return between 4.0 and 5.0 percent annually if their recovered resources are fully marketable, but neither is attractive when the resource marketability factor is included. The same is true of incineration with steam and residue recovery, and composting systems although they show somewhat higher returns, 8.7 and 6.6, on a gross resource value basis.

Pyrolysis, comprehensive materials recovery and fuel recovery are the most attractive resource recovery systems in an economic sense. All are superior to conventional incineration, even when the marketability factor is taken into account.

TABLE 3

SUMMARY OF RECOVERY SYSTEM ECONOMICS
(500 TPD Plant, Continuous Operation)

	Total Capital Requirement	Based on Gross Resource Value		Based on Expected Resource Value	
		Annual Operating Cost	Return on Incremental Investment *	Annual Operating Cost	Return on Incremental Investment *
Incineration (no resource recovery)	$ 5,415,000	$1,480,000	0	$1,480,000	0
Incineration with Residue Recovery	6,285,000	1,345,000	15.5%	1,475,000	0.6%
Incineration with Steam Recovery	6,840,000	1,415,000	4.6	1,815,000	--
Incineration with Electric Generation	10,360,000	1,265,000	4.3	1,770,000	--
Incineration with Steam and Residue Recovery	7,710,000	1,280,000	8.7	1,810,000	--
Pyrolysis	7,225,000	1,135,000	19.1	1,335,000	8.0
Composting	10,885,000	1,120,000	6.6	1,585,000	--
Comprehensive Materials Recovery	7,045,000	960,000	31.9	1,255,000	13.8
Fuel Recovery	4,680,000	680,000	Infinite**	755,000	Infinite**

* Based on incineration (with no resource recovery).
** Infinite because capital costs for fuel recovery is <u>less</u> than for incineration (no resource recovery).

TABLE 4

EFFECT OF PLANT SIZE ON NET DISPOSAL COST

System	Net Disposal ($/ton)			
	250 TPD	500 TPD	1000 TPD	2000 TPD
Incineration	9.50	8.22	7.12	6.16
Incineration with Residue Recovery	9.11	7.47	6.07	4.88
Incineration with Steam Recovery	9.78	7.86	6.24	4.84
Incineration with Electric Generation	9.34	7.03	5.06	3.38
Incineration with Steam and Residue Recovery	9.28	7.11	5.28	3.72
Pyrolysis	8.22	6.30	4.64	3.22
Composting	9.28	6.22	3.81	1.89
Comprehensive Materials Recovery	7.50	5.33	3.53	2.04
Fuel Recovery	5.33	3.78	2.54	1.54

Effect of Variables

The cost estimates presented in the preceding section represent average or typical conditions. For a particular system at a specific location, the actual costs could vary by as much as 50 percent from those shown here, although their <u>relative</u> standings should not change much with regard to capital and operating costs. Resource values are the key to the success of any recovery system, however, and these can easily range from zero (where there is simply no market for the recovered materials) up to the amounts defined as gross values, which represent current market prices on the various products, and higher if market conditions are extremely favorable.

Table 4 and Figure 12 indicate the economies of scale associated with the various systems, on net costs, based on gross resource values. Incineration (without resource recovery) varies from $9.50/ton at a 250 TPD plant size, down to $6.16/ton for a 2,000 TPD plant. The resource recovery systems all show substantially greater economies of scale, since revenues increase in direct proportion to capacity while costs increase at a lesser rate. Furthermore, all the resource recovery systems are less costly than incineration without resource recovery. Fuel recovery has the lowest net disposal costs at all scales of operation.

V. CONCLUSIONS

The vast amounts of solid wastes generated in the United States, and the growing problem of proper disposal of these wastes have stimulated an interest in resource recovery systems. There are at present only two types of resource recovery systems which are fully developed--heat recovery from incinerators and composting. There are a considerable number of new resource recovery systems now under development, including materials recovery systems, energy recovery systems, pyrolysis systems and chemical conversion systems.

Of the various resource recovery concepts investigated, three appear to be economically attractive--fuel recovery, comprehensive materials recovery and pyrolysis. Each of these systems can recover marketable products of sufficient value to substantially reduce a municipality's net solid waste disposal cost, compared with conventional incineration. None of the systems examined, though, can be expected to generate a profit for the plant's operator; the best that can be expected is to minimize the over-all costs of solid waste disposal.

Minimum disposal costs--maximum returns from the resource recovery facility--will be achieved with a municipally-owned plant, operating on a large scale, continuous basis. Even so, it is unlikely that the resulting net disposal cost will offer a savings over sanitary landfill, except in large metropolitan areas where satisfactory landfill sites at reasonable cost are not available.

The primary obstacle to the economic success of any resource recovery system lies in the marketability of its recovered products. Unless adequate markets are known to exist or can be firmly established for the recovered materials, capital committed to any resource recovery system must be considered a high risk investment. Nevertheless, carefully considered placement of the appropriate resource recovery technology could be economically favorable under today's waste disposal conditions in many parts of the United States, and in other countries with similar waste disposal problems.

REFERENCES

1. 1968 National Survey of Community Solid Waste Practices, U.S. Department of Health, Education, and Welfare, Public Health Service PHS Publication No. 1867.

2. Rauseau, H., "The Large Plants for Incineration of Domestic Refuse in the Paris Metropolitan Area." Proceedings of the 1968 National Incinerator Conference, New York, May 5-8, 1968.

3. Power from Refuse, Mechanical Engineering, March 1970.

4. "Solid Waste Disposal in Switzerland," 4th International Congress of the International Group on Refuse Disposal, Basel, Switzerland, June 2-5, 1969.

5. Matsumato, K., et al, "The Practice of Refuse Incineration in Japan Burning of Refuse with High Moisture Content and Low Calorific Value," Proceedings of 1968 National Incinerator Conference, New York, New York, May 5-8, 1968.

6. Northwest Incinerator Plant, Solid Waste Management, May 1971.

7. "Composting of Municipal Solid Wastes in the United States," U.S. Environmental Protection Agency, 1971.

8. Herbert, William, "Solid Waste Recycling at Franklin, Ohio," Proceedings of the Third Mineral Waste Utilization Symposium, March 14-16, 1972.

9. Heren, J. H., Peters, F. A., "Cost Evaluation of a Metal and Mineral Recovery Process for Treating Municipal Incinerator Residues." U.S. Bureau of Mines Information Circular 8533, 1971.

10. "Use of Refuse as Supplementary Fuel for Power Plants," Summary Report, Horner and Shifrin, Inc., St. Louis, Missouri, October 1970.

11. "Ultra-High Temperature Total Reduction of Municipal Solid Wastes," American Thermogen Incorporated, Whitman, Massachusetts, July 1971.

12. "CPU-400 Development Status Report," Combustion Power Company, Menlo Park, California, 1970.

13. "Comprehensive Studies of Solid Waste Management, 2nd Annual Report," University of California (Berkeley), SERL No. 69-1, January 1969.

14. "Pyrolysis, Hydrogenation, and Incineration of Municipal Refuse--A Progress Report," Proceedings of 2nd Mineral Utilization Symposium, Chicago, March 1970.

15. Hoffman, D.A. and R.A. Fitz, "Batch Retort Research on Pyrolysis of Solid Municipal Refuse," Environmental Science and Technology, 2 (11), November 1968.

16. Partial Oxidation of Solid Organic Wastes, P.H.S. Publication 2133, 1970.

17. Mallan, G.M., "A Total Recycling Process for Municipal Solid Wastes," presented at 161st ACS National Meeting, Los Angeles, California, 31 March 1971.

18. Stephenson, J.W., "Some Recent Developments in Disposal of Solid Wastes by High Temperature Incineration, Pyrolysis, and Fluid Bed Reactor," presented before New York State Action for Clean Air Committee, Schenectady, New York, 7 May 1971.

19. The Hercules Solid Waste Reclamation Concept, Hercules, Inc., 16 August 1971.

20. Construction of a Chemical-Microbial Pilot Plant for Production of Single Cell Protein from Cellulosic Wastes, EPA Report SW-24C, 1971.

21. "Comprehensive Studies of Solid Waste Management 3rd Annual Report," University of California, Berkeley, SERL Report No. 70-2, June 1970.

22. <u>Solid Waste Report</u>, Volume 2, No. 4, February 1971.

REMOVAL OF TRACE METALS FROM WASTEWATER BY LIME AND OZONATION

A. Netzer, A. Bowers and J. D. Norman

Department of Chemical Engineering, McMaster University, Hamilton, Ontario, Canada

INTRODUCTION

Various methods of treatment are generally available for metals removal from wastewater. This paper presents data which shows that two methods of treatment, namely, Lime and Ozonation can be combined to offer a better method of treatment than either method used alone.

Chemical precipitation with lime has recently been surveyed to determine its effectiveness in precipitating metals such as Aluminum, Barium, Beryllium, Cadmium, Cobalt, Copper, Chromium, Iron, Lead, Manganese, Mercury, Nickel, Silver, Tin and Zinc (1-4). The advantages of applying lime for this use are low cost, availability and low dosage requirements below pH 9.0. The disadvantages of lime are high residual levels of metals especially in the pH range 7.0 - 9.0, a logarithmic increase in dose requirements (a strong consideration above pH 9.0), sludge dewatering, i.e., filtering and settling characteristics, and sludge disposal.

A survey of the literature (5) indicated that information was not available on the precipitation of metals by ozonation, except for Iron (6-8), and Manganese (9-13). Preliminary experiments with eleven metals (Aluminum, Cadmium, Chromium, Copper, Iron, Lead, Manganese, Mercury, Nickel, Silver and Zinc), demonstrated that in the presence of ozone these metals precipitate in aqueous solutions (5) as a result of formation of insoluble hydroxides and oxides (14, 15). However, for optimum removal by ozone, pH adjustments of the metal solutions were required. Consequently, lime and ozone were combined in extending the work to a bench scale plant to explore the possibility that a large percentage of each metal will be removed

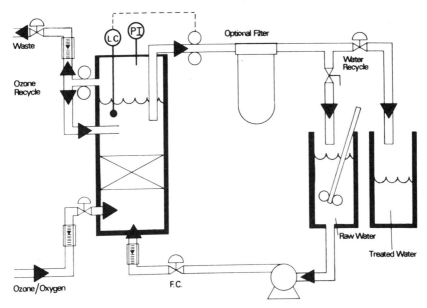

Fig. 1. Schematic of bench scale apparatus.

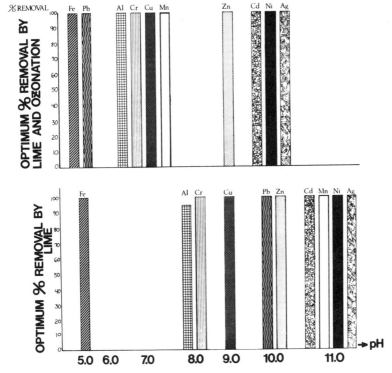

Fig. 2. Comparison of lime treatment with combined lime and ozonation.

by lime adjustment to pH 7.0 - 9.0, and that the remaining dissolved metal will be precipitated by ozone at these pH values.

EXPERIMENTAL

Bench Scale Plant Description

Ozonation was carried out in a 120 cm high by a 6.5 cm diameter perspex column. Ozone was produced from a stream of oxygen by a PSI Model LOA-2 ozone generator. Since this had a fixed power output, i.e., fixed ozone generating capacity, the outlet ozone/oxygen mixture was split into two streams; one by-passed the column and vented to the atmosphere, the other went to the column base, where it was distributed to the column via a porous stone. In this way, both the gas flow to the column and the ozone concentration may be varied. The inlet gas was analyzed by a Welsbach H-82 ozone meter and the ozone concentration was recorded on a standard potentiometric recorder. The gas leaving the column was either vented to the atmosphere or recycled to the column midpoint.

The metal solution to be treated was kept in a continuously stirred 20 litre supply tank and was supplied to the column via a 1/2" centrifugal pump equipped with a flow control valve (F.C.) and rotameter. The liquid inlet was directly below the ozone inlet point.

Liquid removal from the column was intermittent, being controlled by a set of level probes operating a positive displacement pump. This stream was passed through a gas-liquid separator (to prevent ozone emissions in the working area), and then either to the supply tank to be recycled or to a 'treated water' tank. The percentage recycled was controlled by a 'reflux divider' operating two valves for difference time intervals (Figure 1).

PROCEDURE

Stock solutions of each metal were prepared to a concentration of approximately 100 ppm. Samples were taken, filtered through "Whatman No. 1" filter paper, and stored for later determination of exact metal concentration on a Perkin Elmer 303 Atomic Absorption Spectrometer. 12 litres of stock solution were placed in the supply tank. If necessary, lime was added to adjust the pH to the desired value and a sample was taken, filtered and stored for analysis.

The column was partly drained to allow room for the ozone, and oxygen was passed through the column at the desired rate by adjusting the gas by-pass valve. The liquid supply pump was started and its rate adjusted to the desired value. When the column level controller

Table 1. Removal of Trace Metals from Aqueous Solutions by Lime and Ozonation

Metal	pH	Conc. After Lime Addition in ppm	Conc. After Ozonation in ppm	% Removal by Lime	% Removal by Ozonation	% Removal by Lime and Ozonation	Remarks
Aluminum as: $AlCl_3 \cdot 6H_2O$	*3.9	98	98	-	-	-	
	7.0	6	1	93.9	83.3	99.0	
	8.0	5	3	94.9	40.0	96.9	
	9.0	31	28	68.4	9.7	71.4	
	10.0	96	89	2.0	7.3	9.2	
	11.0	98	88	-	10.2	10.2	
Cadmium as: $Cd(NO_3)_2 \cdot 4H_2O$	*6.4	92	88	-	4.3	4.3	
	7.0	68	65	26.1	4.4	29.3	
	8.0	58	55	36.9	5.2	40.2	
	9.0	18	15	80.4	16.7	83.7	
	10.0	0.45	0.40	99.5	11.1	99.6	
	11.0	<0.1	<0.1	>99.9	-	>99.9	
Chromium as: $CrCl_3 \cdot 6H_2O$	*3.5	104	104	-	-	-	**
	7.0	0.6	<0.1	99.4	>83.3	>99.9	
	8.0	<0.1	<0.1	>99.9	-	>99.9	
	9.0	0.7	<0.1	99.3	>85.7	>99.9	
	10.0	0.2	<0.1	99.8	>50.0	>99.9	
	11.0	<0.1	<0.1	>99.9	-	>99.9	
Copper as: $CuCl_2 \cdot 2H_2O$	*4.8	95	91	-	4.2	4.2	
	7.0	1.6	0.3	98.3	81.2	99.7	
	8.0	0.7	0.1	99.3	85.7	99.9	
	9.0	<0.1	<0.1	>99.9	-	>99.9	
	10.0	<0.1	<0.1	>99.9	-	>99.9	
	11.0	<0.1	<0.1	>99.9	-	>99.9	
Iron as: $FeCl_3 \cdot 6H_2O$	*2.8	91	85	-	6.6	6.6	
	7.0	<0.1	<0.1	>99.9	-	>99.9	
	8.0	<0.1	<0.1	>99.9	-	>99.9	
	9.0	<0.1	<0.1	>99.9	-	>99.9	
	10.0	<0.1	<0.1	>99.9	-	>99.9	
	11.0	<0.1	<0.1	>99.9	-	>99.9	
Lead as: $Pb(NO_3)_2$	*5.1	98	<0.5	-	>99.5	>99.5	
	7.0	35	<0.5	64.3	>98.6	>99.5	
	8.0	12	<0.5	87.8	>95.8	>99.5	
	9.0	2	<0.5	97.9	>75.0	>99.5	
	10.0	<0.5	<0.5	>99.5	-	>99.5	
	11.0	4	<0.5	95.9	>87.5	>99.5	
Manganese as: $MnCl_2 \cdot 4H_2O$	*6.4	94	1.5	-	98.4	98.4	***
	7.0	94	0.8	-	99.1	99.1	
	8.0	90	0.3	4.3	99.7	99.7	
	9.0	17	<0.1	81.9	>99.4	>99.9	
	10.0	0.15	<0.1	99.9	>33.3	>99.9	
	11.0	<0.1	<0.1	>99.9	-	>99.9	
Mercury as: $Hg(NO_3)_2 \cdot H_2O$	*3.1	23.5	81	62	-	23.5	
	7.0	13.6	70	70	13.6	-	
	8.0	11.1	72	72	11.1	-	
	9.0	11.1	72	72	11.1	-	
	10.0	9.9	73	73	9.9	-	
	11.0	2.5	79	79	2.5	-	
Nickel as: $Ni(NO_3)_2 \cdot 6H_2O$	*6.3	-	104	104	-	-	
	7.0	4.8	104	99	-	4.8	
	8.0	14.4	104	89	-	14.4	
	9.0	96.8	35	3.3	66.3	90.6	
	10.0	99.1	4.1	0.9	96.1	78.0	
	11.0	>99.9	<0.1	<0.1	>99.9	-	
Silver as: $AgNO_3$	*7.2	5.5	91	86	-	5.5	
	7.0	3.3	91	88	-	3.3	
	8.0	9.9	88	82	3.3	6.8	
	9.0	98.5	37	1.4	59.3	96.2	
	10.0	>99.9	12	<0.1	86.8	>99.2	
	11.0	>99.9	0.4	<0.1	99.6	>97.5	
Zinc as: $ZnCl_2$	*6.5	-	86	86	-	-	
	7.0	-	86	86	-	-	
	8.0	94.5	6.5	4.7	92.4	27.7	
	9.0	99.9	0.30	0.10	99.6	66.7	
	10.0	99.9	0.15	0.10	99.8	33.3	
	11.0	99.8	0.15	0.15	99.8	-	

* Stock solution
** Excess ozone will convert Chromium to chromate
*** Excess ozone will convert Manganese to permanganate

(L.C.) was operating, i.e., the column was full, the reflux controller was set to the desired percentage and the ozone generator was switched on. On completion of ozonation, samples of the column outlet liquid were taken, filtered through "Whatman No. 1" filter paper, and stored for later analysis. The generator was turned off and purged with oxygen for a further five minutes. The column was drained and the solids were separated by vacuum filtration, and the filtrate being brought to pH 7.0 ± 0.5 before discharging it to waste. A weakly acidic solution was circulated through the whole liquid system to remove metal oxide and hydroxide precipitates and discharged. The column was finally rinsed with tap water in preparation for the next experiment. Overall process efficiency was defined as metal concentration after ozonation divided by the concentration of stock solution. Ozonation efficiency was defined as metal concentration after lime addition divided by concentration after ozonation.

RESULTS

Greater than 99.5% removal was achieved for all metals (except Mercury) by combined treatment of lime and ozonation.

In the case of Iron, lime addition alone to pH 7.0 resulted in complete removal. In all other cases ozone or increased pH were required for complete removal. For example, at pH 7.0 lime addition followed by ozonation resulted in complete removal of Aluminum, Chromium, Copper, Lead, and Manganese, whereas, with lime alone, the pH necessary for complete removal of these metals was 8.0, 8.0, 9.0, 10.0 and 11.0, respectively. Higher pH values and ozonation were required for removal of Zinc (pH 9.0), Cadmium (pH 10.0), Nickel (pH 10.0), and Silver (pH 10.0). Comparing with lime alone, complete removal was obtained at pH 10.0, 11.0, 11.0 and 11.0, respectively. The results are presented in Table I.

CONCLUSIONS

By using a combined treatment of lime and ozonation, precipitation of most of the metals occured at pH 7.0. For the metals which precipitated at higher pH values, the values were at least one pH unit lower with lime and ozone treatment than with lime treatment alone (Figure 2), thus reducing lime dosages considerably (4). In addition, the settling and filtering characteristics of the precipitates after lime and ozone treatment were much improved.

In these studies, ozonation was always carried to saturation, with samples taken before and after the experiment. The effect of pH was examined, but further studies considering factors such as metal concentration, ozone concentration, ozone recycle percentage, column residence time, gas/liquid flow ratio, column operating pressure and effect of organics are required to optimize the process for application to waste streams. The final goal will be to produce a set of operating conditions giving optimum ozonation efficiency for any given metal of those studied, and, if possible, an indication of the general operating conditions to achieve optimum process efficiency for various metal solution mixtures. These parameters are currently being quantified in our laboratory.

ACKNOWLEDGEMENT

The excellent technical assistance of Mr. H. Behmann in the analysis of metals was greatly appreciated.

REFERENCES

1. Nilsson, R., "Removal of Metals by Chemical Treatment of Municipal Wastewater", Water Research, $\underline{5}$, pp. 51-60, (1971).

2. Linstedt, Daniel K., Houck, Carl P., and O'Connor, John T., "Trace Element Removals in Advanced Wastewater Treatment Processes", J.W.P.C.F., $\underline{43}$(7), pp. 1507-1513, (1971).

3. Lanford, Charles E., "Trace Minerals Affect Steam Ecology", The Oil and Gas Journal, pp. 82-84, March 31, 1969.

4. Netzer, A., Norman, J. D., and Vigers, G. A., "Removal of Metals from Wastewater by Lime", unpublished results.

5. Netzer, A., Norman, J. D., and Vigers, G. A., "Removal of Trace Metals from Wastewater by Ozonation", The Seventh Canadian Symposium on Water Pollution Research, Proceedings, 1972, (in press).

6. Sergeev, Y. S., "Use of Alluvial Waters for Drinking Water", Sanit. Tekhu. Vodosnabshi i Kanaliz, $\underline{2}$, 11, (1965).

7. Sergeev, Y. S., "Iron Removal and Quality Improvement of Alluvial Water", Kommun Khoz. sb., $\underline{2}$, 72, (1964)

8. "Treatment of Acid Mine Drainage by Ozone Oxidation", Brookhaven National Laboratory, EPA Report 14010 FMH, December, 1970.

9. Marcy, J., and Matthes, F., "The Reaction of Manganese (II) Salt Solution with Ozone", Chem. Tech., $\underline{19}$ (7), 430, (1967).

10. Senzaki, T., and Ikehata, A., "Oxidation of Manganese (II) in Aqueous Solutions by Ozone", Kogyo Yosui, $\underline{116}$, 46, (1968).

11. Rohner, E., "Removal of Manganese in Water with Ozone", Swiss Patent 481, 020, December 31, 1969.

12. Yakobi, V. A., Gavrilov, M. S., Plakidin, V. L., Slezko, G. F., and Ponomarev, B. A., Vodosnabzh, Sanit. Tekh., 23-26, (1968).

13. Whitson, J. T. B., "The Effect of Ozone on Waters Containing Manganese", J. Inst. Water Engrs., $\underline{1}$, 464, (1947).

14. Handbook of Chemistry and Physics, 51st Edition, published by The Chemical Rubber Company, Cleveland, Ohio, (1970-71).

15. Sillen, L. G., "Stability Constants of Metal-ion Complexes", Special Publication No. 17, The Chemical Society, London, (1964).

A FLOATING SETTLER FOR LOW COST CLARIFICATION

S.C. Reed, T. Buzzell, S. Buda

U. S. Army Cold Regions Research and Engineering Laboratory, Hanover, NH

During recent years a very significant level of research activity has been concerned with improving clarification techniques. The basis for much of this work has been classical shallow-tray sedimentation theory. The advantages of this concept have been discussed in the literature for over 65 years. In 1904, Hazen suggested shallow basins with a depth approaching 1 in. (1). Since the vertical travel distance for a settling particle is so small in such a basin the retention time could be a few minutes rather than a few hours. Many attempts have been made to develop a practical technology for implementation of this theory. Two major problems impeded progress: the construction or fabrication cost for installing a large number of shallow trays in a basin, and development of an efficient technique for removal of settled sludge. In addition it was found that very wide shallow trays had insufficient lateral constraints and led to unstable hydraulic conditions. (2).

Papers by Culp and Hansen (2, 3) introduced tube settlers which retain the advantages of tray sedimentation while overcoming most of the problems. The lightweight plastic tube modules can be easily installed in tanks or basins and the small diameter tubing provides laminar, highly stable flow conditions with very low Reynolds numbers. Based on their development a large number of tube settler modules have been installed in a variety of water and waste treatment facilities. The modules are built-up from plastic sheets and the fabrication process produces a tube with either a square or a hexagon shaped cross-section. Other manufacturers offer tubes with a chevron (a laterally expanded V) shape and a system developed in Sweden combines steeply inclined plates and counter-current flow.

The most successful applications of tube settlers have been in water treatment systems where the settler is installed in series ahead of the filter beds. Backwash water from the filter is channeled back through the settler and provides positive cleaning of the tube elements. The tubes in such a concept are inclined slightly to insure positive drainage at the end of the cleaning cycle. In waste treatment applications the tubes are more steeply inclined (45° - 60°) to promote self-cleaning. It is hoped that the accumulating settled sludge will slide out of the tube because of density gradients and natural gravity forces.

In the smaller waste treatment applications the tube modules are generally installed as a fixed element in a hopper bottomed settling compartment. The unit can be more compact than conventional clarifiers but two problems remain: (a) a mechanical or hydraulic element must be provided to return the settled sludge to the aeration tank, and (b) since retention time in the tubes is so short, the flow velocity becomes a critical design parameter and the number of tubes provided is based on peak flow conditions. The authors have overcome both problems by developing a settler module which floats directly in the aeration compartment.

The floating settler retains all of the advantages of high rate tube sedimentation, but in addition provides some very significant benefits: (a) Its utilization in the aeration unit eliminates the need for separate clarifier basins and tanks; (b) Since the settled sludge leaving the module is already in the aeration compartment, the need for sludge collection and return equipment is eliminated; (c) Since the module floats, it is always in the same position with respect to the general liquid surface in the tank. This permits design for a constant steady-state discharge equivalent to the average daily flow rate. The module can then be one-third the size of a fixed tube bundle which has been designed for peak flow conditions, a small additional freeboard in the containing tank provides for temporary storage of surges and peak flows. The resulting very compact floating settler can be installed in the aeration tank and not impose excessive space requirements; (d) The floating settler permits more effective self-cleaning, thereby eliminating the need for backwashing and reducing the frequency of air sparging or other cleaning techniques. Two factors contribute: since the settler is in the turbulent aeration tank it is exposed to forces which induce some lateral oscillation of the unit, this helps settled sludge to slide from the tubes; in addition the tubes are circular in shape so there are no sharp corners where fillets of sludge can collect and provide a base for more extensive accumulations.

The characteristics described above are inherent features of the floating settler when it is used in a turbulent mixture of

liquid and settleable solids. It should not matter whether the
system is activated sludge, a physical-chemical treatment unit or
some industrial process. However, actual performance efficiency
will strongly depend on sludge characteristics and concentration.
Critical design parameters are overflow rate, tube size, angle of
tube inclination, tube length and possibly tube shape. The
authors have considered these parameters in the context of extended
aeration, a variation of the activated sludge process. The sludge
developed is a hetrogeneous flocculant mixture of organic compounds,
biological growths, inert fractions and inorganic materials. Due
to the biological responses the concentration and settling characteristics are often an operational variable in such systems. The
criteria developed by the authors using this source material are
believed to be generally applicable to other mixtures as well.
Specific studies with other mixtures are recommended in that they
might provide a basis for further optimization of certain parameters.

The authors' study considered both the performance of a full
scale prototype operating in a 1700 gpd extended aeration system
and individual experiments to identify critical parameters. These
were concurrent efforts, a basic intent was to develop a unit composed of relatively cheap easily available materials which could if
necessary be assembled on-site with simple tools by inexperienced
personnel.

Factors such as cost, weight, ease of fabrication, corrosion
resistance and frictional hydraulic properties led to the choice
of polyvinyl chloride (PVC) pipe as the tubing elements. A thin
wall (0.05 inches) type is available in the United States which
has the lightweight characteristics desirable for a floating unit
(18 lbs. per 100 linear feet for a 2-inch size). Such pipe can be
cut and assembled into a rigid module with only hand tools and the
proper solvent cement or glue.

Choice of pipe size is dependent on a number of factors.
Large diameters have a long particle settling distance which
requires more retention time and they are not as effective in
damping out entrance turbulence. Very small diameters have very
short retention times but they increase the risk of clogging and
large numbers of tubes are required to provide the necessary overflow areas. The latter is a very important factor in that it
affects assembly costs and the total amount of space required for
the tube settler in the aeration tank. Wall thickness for such
pipes and tubing is generally a constant factor and not proportional to the internal diameter. Based on this fact it can be
shown that a 2 inch I. D. pipe is the optimum size for tube settler
construction. Smaller sizes require more space for a given overflow
area and the benefits are not significant for pipes with internal
diameters larger than 2 inches. A two-inch diameter tube through

which water is passed at a rate of 1000 gallons per day per square foot of cross-sectional area would have a Reynolds number of less than 4.

Angle of inclination is a critical design parameter. It must be steep enough to promote self-cleaning but cannot be so steep as to lose the benefits of shallow settling distances. A vertical tube is one extreme case in that the settling distance is equal to the tube length. Such a unit would function as an upflow clarifier, it would optimize sludge removal characteristics but retention times would be equivalent to conventional clarifiers. For tubes at any other angle the settling distance is a function of tube diameter. The optimum case for settling distance is a horizontal tube but this sacrifices any self-cleaning capability. In his theoretical study of tube and tray settlers Yao (4) indicates little effect on performance efficiency up to angles of 50°, but beyond 60° performance deteriorated rapidly. Earlier experimental results from Culp et al (3) generally demonstrates the same responses, they noted a slight decrease in efficiency as the angle of inclination approached 60° but that self-cleaning was significantly improved as the angle was increased from 45° to 60°. Current experiments by the authors confirm these findings. Angles of 40° or less did not allow effective self-cleaning. It is possible that some angle between 40° and 60° is the optimum point for self-cleaning and sedimentation considered together. There is, however, a third constraint which makes the steeper angles desirable. Since the unit is to be installed directly in the aeration tank a sacrifice of tank surface area will be necessary. The area required will be a function of inclination angle of the tubes. The minimum requirements will result from the steepest angles. Work by the authors and others seems to indicate that 60° is the maximum angle at which acceptable performance can still be realized. It was therefore selected as the basic design angle for pilot scale work by the authors and is recommended for all future applications.

The length of tube selected is directly proportional to design retention time. In a 2-inch tube, inclined at 60° the linear travel distance from crown to invert is only 2.3 inches. However, there is steady state flow up the tube so an additional velocity component must be considered. At a hydraulic loading rate of 1000 gallons per day per square foot the flow velocity in a 2-inch tube would be approximately 0.001 feet per second. As a result the sludge particle will follow a curvilinear path prior to impinging on the tube invert. For activated sludge mixtures the settling velocities are dependent on concentration and a number of other factors. Tube settler design should be based on the worst case which might be high concentration hindered settling or a bulking sludge. Such material might have a settling velocity of only one foot per hour or less (.00028 ft per second). At the flow velocity cited above it can be shown that a tube length of approximately 8 inches should provide total removal. Any particles with slower velocities than specified would be lost from the system. To compensate it is necessary to provide additional tube length. According to Yao (4) the advantages become insignificant beyond a certain length. His calculations show that for 2-inch tubes this point occurs at about a 40-inch length. An operational factor also imposes constraints on the selection of tube length. Tank depth for smaller scale treatment systems is generally between 8 and 10 ft. It is first necessary to provide freeboard in the tank for temporary storage of surges and peak flows. Using a floating settler will result in a varying liquid surface in the tank as the system responds to the variable input rate. This daily cycle will have an amplitude of approximately 8 to 12 inches in response to a typical flow pattern whose peak is 200 to 300% of the average rate. When the liquid surface is in the lowest position it is absolutely essential that there still be sufficient clearance beneath the tube settler to permit resuspension of the settled solids in the mixed liquor. If the settler is too close to the tank bottom solids may settle to the floor and remain as a benthic deposit. Based on observations of the authors, a clearance of 2 to 3 ft for this purpose is recommended. The total submerged depth of the tube settler unit should therefore not exceed 4 ft in an 8-foot deep tank. Use of 3-foot long tubes then leaves adequate space for the effluent collector box and float attachments. Performance of the 3-foot long tubes confirms their capacity for adequate solids removal efficiency. Comparative tests with 4-foot long tubes confirms the previous theoretical discussion in that little gain in performance efficiency was noted.

The shape of the tube very strongly influences fabrication technique and costs. It also can affect short and long term performance efficiency of the tubes. According to Yao's theoretical treatment (4) square and circular tubes are roughly equivalent and are both more effective than parallel plates or shallow trays. The authors concentrated their efforts on experimental comparison of square and circular tubes. The same basic experimental procedure was used to investigate all of the parameters. Two individual tube units were floated in a 140 gallon tank. Fresh mixed liquor from the full scale extended aeration unit was transferred to the tank at the start of an experimental run. An air pipe in the tank kept the contents mixed and in an aerobic state. Effluent was withdrawn from the collector box at the top of each tube with a perestaltic pump. This carefully controlled steady state withdrawal established the hydraulic loading rate on each tube. Flow rates up to 2620 gpd per square foot were used, depending on the particular experiment. The effluent was returned to the 140 gallon tank to maintain the original sludge concentration during the experimental run.

For the tube shape experiments a square and a circular tube, each 3 ft long and inclined at 60° were floated in the test tank. The tubes were constructed to have the same cross-sectional areas (3.14 square inches). The circular tube was 2 inches in diameter and the square was 1.77 inches on a side. Mixed liquor concentration during the experiments was approximately 3600 mg/ℓ . The hydraulic loading was varied from 100 gpd per square foot to over 2000 and was identical for both tubes. At very low flow rates there was little difference in performance. At rates of 1000 gpd per square foot and higher the square tube exhibited a small but consistently better level of performance than the circular tube. At the higher flow rates retention time in the system becomes critical and the performance efficiency may be related to the different settling distances in the two tubes. Particles entering at the crown of the circular tube have to settle approximately 11% further than they would in the square tube. Performance efficiency at the higher flows differed by approximately the same percentage.

Square tubes provide several benefits for fabrication and assembly. Adjacent tubes in a bundle can share a common wall while each circular tube must have its own wall. As a result a bundle of square tubes will be lighter weight and more compact than a bundle of circular tubes designed for the same overflow rate. The savings in space and weight will be approximately 27%. Assembly of circular tubes into a bundle leaves small cusp-shaped openings between the tangent points of adjacent circles. These extend the full length of the settler and theoretically should function as a miniature tube and be even more effective than the larger tubes in the bundle. However, it has been the authors' experience that these cusp-shaped tubes must be closed. Gas bubbles and floating sludge particles have been observed exiting these spaces in an operational settler. Apparently the self-cleaning characteristics of these spaces are not adequate. The sharp-corners may retain the entering sludge and permit denitrification to occur.

A further set of experiments with single square and circular tubes was performed to evaluate the self-cleaning characteristics of these shapes. Each tube was fitted with a hinged stopper at the bottom which could seal the tube entrance when the control wire was pulled. Both tubes were run at the same overflow rate and effluent characteristics during the experiment were similar. At the end of a week's operation the stoppers were closed to retain tube contents, the tubes removed from the tank and the contents analyzed for solids content. The square tube contained 43% more solids .

than the circular unit. Some of the contained sludge in the square tube was dark and anaerobic. It is suggested that the corners at the invert of the square tube permit the accumulation of fillets of sludge, these may then "bridge" across the tube bottom providing a base for further accumulations which would eventually lead to failure of the tube. A frequent cleaning technique must be included in the design of tube units with sharp corners.

The authors examined various tube lengths at different flow rates to optimize both parameters. Two-inch diameter circular tubes with lengths of 1.5, 2, 2.5 and 3 feet were compared. In every case there was very close to a direct linear relationship between settling efficiency and overflow rate. The 3-foot long tube gave the best results at all flow rates. The 1.5-foot long tube was unacceptable except at very low and uneconomical flow rates. Tubes between 2 and 3 feet long, gave generally acceptable performance at a flow rate of 1000 gpd per square foot. This is the hydraulic loading rate and is based on the tube area perpendicular to the flow direction. Since the tubes are inclined at 60° it is necessary to consider the horizontal projection of this area when comparing "overflow rates" with conventional clarifiers. Such an overflow rate for a tube settler at 60° would be 866 gpd per square foot of horizontal surface area. The smaller scale "packaged" extended aeration systems available in the United States are generally designed for 100 to 250 gpd per square foot with retention times of two to six hours.

As a result of these experiments the authors would recommend 3-foot long tubes operated at a hydraulic loading rate of 1000 gpd per square foot. The fact that this length is close to Yao's theoretical optimum of 40 inches is believed to be coincidental. Yao's analysis cannot be strictly applied to a floating settler where the whole unit is subjected to lateral oscillations. An undefined portion of tube length will be required to damp out this extra turbulence. It was not possible in the authors' experiments to identify sludge interface position or zones of utilization in the tube since they were always submerged in the opaque mixed liquor. Fixing the tube to the wall of a transparent tank or introducing probes or samplers would alter the flow patterns which are the basis for performance.

The upper end of the tube settler module must be enclosed with perimeter walls to form an effluent collector box. These walls must be high enough to exclude splashes of the turbulent mixed liquor. The liquid surface inside this box serves as the collection point for scum and floatables entering the system. It is necessary to provide sufficient time for these materials to rise to the surface to avoid their loss with the final effluent discharge. Design for this factor is analogous to a grease trap and an internal liquid depth of 9 to 12 inches over the tubes should provide sufficient retention time. A simple, manually operated airlift skimmer can be used to periodically remove these materials.

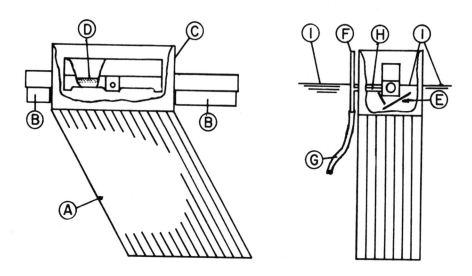

Figure 1. Tube Settler Details: A. Tubes, B. Floats, C. Collector Box, D. Perforated Manifold, E. Baffles, F. Vent, G. Flexible Hose, H. Discharge Tube, I. Water Surface.

As shown on Figure 1 the floats are attached to the outside of the collector box. The authors used sealed rectangular shaped polyethylene carboys for floats. Three 2 gallon carboys were more than sufficient to provide flotation for the full scale prototype settler fabricated by the authors. For the materials and dimensions used in their settler the design would require 1 cubic foot of flotation volume for every 2.3 square feet of cross-sectional tube area. This ratio provides an excess of flotation capacity. Such an excess is desirable since final depth of submergence can then be carefully controlled by adding weights.

Final effluent discharge is through a flexible hose connecting the effluent collector box and a port in the wall of the aeration tank. This port should be located at a depth which is equal to the minimum liquid level desired for the tank. This provides automatic control; when the liquid surface reaches this point discharge through the settler will stop. The flow pattern into this hose from the tube settler is a critical problem. A single entrance point must be avoided in that this would receive high velocity flow from the immediately adjacent tubes. The authors used a perforated manifold to equally distribute the flow contribution from all of the tubes in the bundle. As shown on Figure 1 the manifold extends the full length of the unit and is baffled to reduce intake of scum and floatables. The manifold is connected to the flexible hose by a short length of small diameter tubing. Since the hose is vented to the atmosphere by a riser the depth of submergence of this horizontal tube automatically controls the discharge rate of the settler unit. In the authors' prototype a depth of less than an inch over the 0.75 inch diameter tube was sufficient to maintain the desired 1.2 gpm flow.

Figure 1 illustrates the basic components in a settler module. Prototypes on this model have been tested on a pilot basis at the authors' laboratory and in an extended aeration unit in Fairbanks, Alaska. Performance of both units has confirmed the capability to produce effluent which is at least comparable to that normally expected from extended aeration systems. Table 1 presents operational data from a typical two week period. Raw sewage data are based on a number of 24 hour flow proportional composite samples.

Table 1
Typical Performance Data

Raw Sewage Characteristics
BOD 375 mg/ℓ, Suspended Solids (SS) 263 mg/ℓ

MLSS mg/ℓ	Mixed Liquor SVI	Zone Settling Velocity (ft/hr)	mg/ℓ	Effluent BOD Removal %	mg/ℓ	S.S. Removal %
1436	76	-	8	98	14	95
1810	61	-	26	93	17	94
1830	66	-	17	95	32	88
1812	77	-	29	92	14	95
2262	75	-	48	86	63	76
2324	78	-	-	-	72	73
2410	87	-	30	92	69	74
2182	96	11	29	92	48	82
2386	99	9	17	96	38	86
2244	111	8.5	36	90	30	89
2500	104	7	-	-	33	87
2735	106	6.8	61	86	31	88
2434	133	5.5	29	92	25	90
		Average	27.4	92	37.4	86

The deterioration in effluent quality noted in the middle of the period was due to experimental manipulation of the aeration equipment which resulted in a slight oxygen deficiency in the mixed liquor.

The authors' pilot scale floating settlers have been all plastic construction. The prototype in the 1700 gpd extended aeration unit contains 85, two-inch diameter tubes 3 feet long, inclined at 60°. These are arranged in nested rows with 7 and 6 tubes per row alternating. Based on a 1000 gpd per square foot criteria only 78 tubes would be required. However, fabrication of the collector box is facilitated if the first and last rows of tubes contain the same number. An extra seven tubes were therefore added bringing the total to 85. Cost of materials for this tube settler was approximately $200 and labor requirements were approximately 2 man-days. If $10 per hour is assumed as the labor rate, the total cost of the tube settler would be approximately $200 per 1000 gallons of daily treatment capacity. This is significantly less than the clarifier cost for equivalent scale packaged extended aeration units. This cost is based on hand fabrication of a single module by the authors. Production line factory assembly should considerably reduce unit costs for labor and materials.

Further testing of the concept is intended. The authors in cooperation with the U. S. Navy and others will be conducting full scale performance experiments at 50,000 and 75,000 gpd during the 1972 summer. Basic configuration of the tube settler module is similar to that described in this paper.

Floating tube settlers can be effectively used as the clarification step in new waste treatment systems. They can also be advantageously employed as an additive step to many existing systems. Requirements for further effluent polishing are increasing. This usually means the addition of a filter or a polishing lagoon after the secondary treatment unit. Installation of a floating settler in the aeration tank permits utilization of the existing clarifier as a tertiary unit for final polishing at a considerable savings in cost and space requirements. It would be necessary to lower the discharge port from such a clarifier to a point equivalent to the minimum liquid design level in the aeration tank. Such a sacrifice of 9 inches to a foot of head would be acceptable in many locations and the net savings realized by using the existing tank as a tertiary unit would more than compensate.

In summary, the authors recommend the following criteria for design of floating settler units: hydraulic loading rate of 1000 gpd per square foot of cross-sectional tube area, two inch diameter circular tubes with the thinnest available wall thickness, three-foot tube length and an inclination angle of 60°. Approximately

1 cubic foot of flotation volume will be required for each 2.3 square feet of tube surface (This is based on materials used by the authors and should be used with caution for other designs). Discharge manifolds and other details should be as described in this paper.

References

1. Hazen, A., 1904, On Sedimentation, Transactions American Society of Civil Engineers, Vol. 53, p. 45.

2. Hansen, S. P., Culp, G. L. (1967), Applying Shallow Depth Sedimentation Theory, Journal American Water Works Association, Vol. 59, p. 1134.

3. Culp, G. L., Hansen, S., Richardson, G. (1968), High Rate Sedimentation in Water Treatment Works, Journal AWWA, Vol. 60, p. 681.

4. Yao, K. M. (1970), Theoretical Study of High Rate Sedimentation, Journal, Water Pollution Control Federation, Vol. 42, p. 218.

MEASUREMENT AND COLLABORATIVE TESTING FOR IMPLEMENTATION OF AIR QUALITY

A. P. Altshuller

U. S. Environmental Protection Agency

Instruments have been available for over 20 years in the U. S. to measure the concentrations of air pollutants. Much of the equipment was fabricated in the 1940's to meet the need for measurement of pollutants such as oxidants and nitrogen oxides in Los Angeles. Conductivity analyzers have been available for about as many years for air quality measurements of sulfur oxides. Carbon monoxide was measured in the ambient atmosphere and vehicular exhaust by means of non-dispersive infrared analyzers. Within the last ten years flame ionization hydrocarbon analyzers have become available. Several instruments for sulfur oxides, oxidants and a few other pollutants have been developed based on electrometric principles. However, the number of instruments based on new principles has been small up to recent years. The use of such instruments also was limited to a small number of federal, state or local monitoring networks.

The reasons for this situation were obvious. No governmental laboratory was funded sufficiently to encourage new instrument developments. The instrument industry had little incentive to do R/D because of a small market for air quality instruments. Few university investigators felt justified in involving themselves in instrumentation R/D. The situation began to change with the passage of the 1967 amendments to the Clean Air Act in the U. S. Specific recognition was given to the need for R/D on instrumentation for air pollution applications. In section 104 of the 1967 Act it stated "The Secretary (Administrator) may conduct and accelerate R/D of low cost instrumentation to facilitate determination of quantity and quality of air pollution emissions, including, but not limited to, automotive emissions". The ex-

pectation that funds would be committed to such R/D efforts was
emphasized in section 133 which required a group of annual reports including a report of progress on "the development of quantitative and qualitative instruments to monitor emissions and air quality". In the same amendments the development of a series of criteria documents were required with subsequent establishment of air quality standards by the States. These various requirements stimulated funding by the Federal government to improve air quality instrumentation which would be needed to promulgate air quality standards.

The air quality act of 1970 made the Environmental Protection Agency responsible for promulgating air quality standards. Various sections of this Act also required specific measurement techniques be prescribed for mobile and stationary source emission standards. Implementation plans to meet air quality standards also involved requirements for monitoring compliance with standards for stationary source emissions. Overall these various statutory requirements have stimulated accelerated funding for measurement research by the Environmental Protection Agency.

The present discussion will be concerned with air quality measurements. However, the same principles and often very similar instruments can be utilized with the appropriate sampling interface for source emission measurements. Air quality measurements include those needed for research on health effects and transport processes, trend monitoring, episodes, and enforcement actions. However, the corresponding requirements for instrumentation have considerable commonality, so a single integrated R/D program can meet all of requirements generated by these applications most efficiently.

A discussion of the air quality standards promulgated in 1971 by the Environmental Protection Agency[1] will serve to relate these standards to the present status of measurement R/D. The six pollutants included in these regulations are as follows: sulfur oxides, particulate matter, carbon monoxide, photochemical oxidants, hydrocarbons and nitrogen dioxide. By interval of measurement the following averaging times were specified for the primary (health) or secondary (welfare) standards: sulfur dioxide, 3 hour, 24 hour, annual; particulate matter, 24 hour, annual; carbon monoxide, 1 hour, 8 hour; photochemical oxidant, 1 hour; hydrocarbons, 3 hour; nitrogen dioxide annual. For these pollutants requiring 1 hr, 3 hr or 8 hr measurement intervals instrumental methods are practical necessities. For those pollutants requiring 24 hr or annual averages, time integrated sampling in the field with subsequent laboratory analysis or instrumental analysis can be utilized. Very specific reference methods are listed for each of the six pollutants with procedures detailed in a series of appendices.

PARTICULATE MATTER

The High Volume Method has been used more extensively than any other in the U. S. The relatively inexpensive equipment was developed during the mid-1950's within Federal Air Pollution Technical program. The equipment has been used extensively during the last 15 years by the National Air Sampling Network and by many state and local agencies in the U. S. This method is the reference method for total suspended particulate matter.[1] The primary and secondary standards for maximum 24-hour concentration not to be exceeded once a year are 260 and 150 $\mu g/m^3$ and the corresponding annual means are 75 and 60 $\mu g/m^3$.

In the High Volume Method air is drawn into a covered housing and through a filter by use of a blower operating at 1.13 to 1.70 m^3/min: 40 to 60 ft^3/min. The design and location of the shelter are important. Since a measure of community-wide loadings is the usual objective, the influence of local sources should be avoided. Nominally particles within the size range of 100 to 0.1 micron diameter are collected on the commonly used glass fiber filters. Mass concentration of suspended particulate matter in $\mu g/m^3$ is computed by weighing the filter before and after collection of the particulate matter and determining the volume of air sampled. An EPA sponsored collaborative test[2] in 1971 of this method provided the following results. The relative standard deviation for single analyst variation (repeatability) and for multilaboratory variation (reproducibility) were 3.0 and 3.7 percent, respectively. The minimum detectable amount of particulate matter is 3 mg which is equivalent to the extremely low ambient air loading of 1-2 $\mu g/m^3$ for a 24-hour-sample. The method should be standardized on the 24-hour time limit to insure comparability between series of results. Calibration of flow rate is necessary and this calibration should be conducted by use of a series of perforated plates to vary resistance to flow. The method will give good precision when the procedure is followed regorously.

Many measurement techniques are available for particles in the 0.1 to 10-micron size range but these techniques either do not permit a quantitative measurement of mass concentration or they are research tools, not routine procedures.[3] The tape sampler is the only other method aside from the High Volume method to have received considerable routine use in the measurement of suspended particulates. Both transmittance and reflectance measurements are used, but both suffer from complications related to variability in the characteristics of the deposits. Careful standardization is essential, but even with such standardization the results are more likely to be useful on a relative rather than absolute basis.

Particles equal to or below 1 micron are of particular concern in terms of irritation of the respiratory tract and reduction in visibility. Therefore, a separate measure of such fine particles would be desirable on a routine basis. The integrating nephelometer is a useful device for this application.[4] In this range of particle sizes, particles formed from atmospheric reactions become very important compared to those emitted directly from sources of pollution.

Although particles between 100 and 0.1 micron diameter can be collected on the high volume sampler, it does not follow that particulates throughout this range are collected with equal efficiency. At the upper end of the range large particles emitted primarily from stationary sources can settle out before reaching the vicinity of the sampler. Therefore, control of large particles from such sources need not be reflected in significant reductions in High Volume mass measurements. At the lower end of the range, the collection efficiency curve is not at all well defined for a High-Volume sampler.

No instrument for continuous monitoring of mass concentration in the atmosphere with satisfactorily proven quantitative reliability is available at present. Instruments varying from point sampling instruments based on the microbalance principle to β-ray mass detection and remote type lidar type optical instruments are at varying stages of development and evaluation. Lidar techniques are attractive because of the ability to measure averaged particle masses in volume elements in three dimensions. Since the conversion of signal to mass concentration depends on assumptions about particle size distribution and other particle characteristics, the mass concentrations derived are more reliable on a relative than absolute basis. Such equipment is needed, and continued R/D on such instruments is in progress in the United States.

SULFUR OXIDES

Both primary and secondary air quality standards were established for sulfur oxides measured as sulfur dioxide.[1] The standards are as follows:

Time Interval	Primary	Secondary
(a) Annual arithmetic average	80 µg/m^3 (0.03 ppm)	60 µg/m^3 (0.02 ppm)
(b) Maximum 24-hour concentration - not to be exceeded more than once a year	365 µg/m^3 (0.14 ppm)	260 µg/m^3 (0.10 ppm)
(c) Maximum 3-hour concentration - not to be exceeded more than once a year	not applicable	1,300 µg/m^3 (0.50 ppm)

The reference method for the determining sulfur dioxide in the atmosphere is the colorimetric method involving the formation of pararosaniline methyl sulfonic acid by the reaction of the dichlorosulfitomercurate complex with pararosaniline and formaldehyde. The sulfur dioxide is complexed by absorption from air into an aqueous solution of potassium tetrachloromercurate in all glass midget impingers. Interferences by oxides of nitrogen are eliminated by use of sulfamic acid, ozone by time delay before measurement of absorbance, and heavy metals by EDTA and phosphoric acid. Care in preparation of reagent solutions, daily preparation, and careful time and temperature control are essential. Calibration is provided by use of permeation tubes of known permeation rates originated by O'Keeffe and Ortman[5] of EPA (certified tubes are available from the U. S. National Bureau of Standards).

A collaborative test has been completed on the short time version of the reference method for sulfur dioxide[6]. Fourteen laboratories representing federal, state and local agencies, industry and universities participated in this collaborative test. Calibrations were accomplished by use of NBS certified sulfur dioxide permeation tubes in a complete system specially fabricated and supplied to each participant. The concentrations generated were nominally 150 (0.05 ppm), 275 (0.11 ppm) and 820 (0.31 ppm) $\mu g/m^3$ of sulfur dioxide in air.

The conclusions from this collaborative test are summarized as follows:

1. The replication error varied linearly with concentration from 8 $\mu g/m^3$ at 150 $\mu g/m^3$ to 15 $\mu g/m^3$ at 820 $\mu g/m^3$ and can be extrapolated somewhat beyond this range.

2. The repeatability varies linearly with concentration from 18 $\mu g/m^3$ at 150 $\mu g/m^3$ to 32 $\mu g/m^3$ at 820 $\mu g/m^3$ and can be extrapolated beyond this range.

3. The reproducibility varies linearly with concentration from 35 $\mu g/m^3$ at 150 $\mu g/m^3$ to 63 $\mu g/m^3$ at 820 $\mu g/m^3$.

4. The lower limit of detection is 25 $\mu g/m^3$.

Continuous monitoring instruments for sulfur dioxide are available from at least 29 manufacturers based on 9 different principles (Table I). At least five of these principles involving instruments in increasing common use - colorimetric, conductiometric, coulometric, electrochemical transducers, and flame photometric. The cost of these instruments range from about $1,000 to over $5,000. However, differences in flexibility and accessories explain part of these differences.

None of the commonly used instruments for measurement of sulfur dioxide are specific to sulfur dioxide (Table I). The conductio-

TABLE I

Commercial Sulfur Oxide "Automatic" Instruments Available in the United States for Ambient Air Monitoring [a,b]

Principle	No. of Manufacturers	Cost Range	Remarks
Colorimetric	7	1,895 - 5,455	Most based on West-Gaeke type reaction
Condensation nuclei	1	3,990	
Conductiometric	8	1,500 - 4,900	Interferences from Cl_2, Hcl, NH_3, NO_2 Some types CO_2 also
Correlation Spectrometry	1	19,000	Remote, measures ppm X meters
Coulometric	3	1,775 - 5,290	Interferences from other sulfur compounds oxidants, unsat. compounds
Electrochemical transducers (semi-permeable membranes)	4	950 - 2,250	Interference from H_2S, NO
Flame photometric	2	3,250 - 5,400	Responds to all S compounds
Flame Photometric-gas chromatographic	2	5,275 - 6,000	Good specificity because of gas chromatographic separation-multi-pollutant
Second derivative Spectrometer	1	12,500	Multi-pollutant, but responds variable

[a] availability does not necessarily constitute acceptability, criteria for listing mainly on sensitivity parameter.

[b] condensed from compilation of Lawrence Berkeley Laboratory compilation

metric and coulometric instruments suffer from a wide variety of interferences. Although prefilters can be used with some degree of effectiveness to minimize some types of interferences, the problems of ascertaining time to breakthrough of interferences and providing prompt replacement can be serious in routine monitoring network operations. The use of semi-permeable membranes in the electrochemical transducers confer some additional degree of specificity but certain interferences can penetrate these membranes. The flame photometric analyzer is specific to sulfur compounds, but cannot distinguish hydrogen sulfide from sulfur dioxide. It is claimed on the basis of evaluations in several cities that over 90% of the sulfur in urban areas is likely to be sulfur dioxide. This result was obtained by concurrent use of the flame photometric analyzer and the gas chromatographic analyzer with flame photometric detector. This latter instrument separates sulfur compounds before the sulfur specific detector so it is specific for sulfur dioxide and also for other gaseous sulfur compounds present. This latter instrument as is the case for the second derivative spectrometer listed in Table I is a multi-pollutant analyzer. This factor has to be taken into consideration in terms of cost and operational convenience.

Several years ago a group of commercially available sulfur oxide analyzers were evaluated over a three month period during the winter in New York City. The mean concentration of sulfur dioxide under these field conditions was 0.2 ppm. Analysis of the data from these measurements indicated satisfactory agreement for most analyzer responses when the sulfur dioxide concentration was above 0.16 ppm.

Since the sulfur dioxide concentrations during the period of study in New York City were well above the air quality standards, field evaluation was needed to judge performance in urban areas in which the sulfur dioxide concentration levels would be closer to these required in the air quality standards. Because sulfur dioxide levels are low at most Los Angeles sites, field evaluation was done in downtown L.A. The mean SO_2 concentration during the almost three months period was less than 60 $\mu g/m^3$. The standard deviation in the calibrations was two to four times higher at 120 $\mu g/m^3$ (0.04 ppm) and 240 $\mu g/m^3$ (0.08 ppm) for the conductiometric and coulometric analyzers than for the colorimetric and flame photometric analyzers. Analysis of aerometric results resulted in four conclusions: (1) the conductiometric and coulometric analyzer measurements correlated poorly with the colorimetric and flame photometric results (2) the conductiometric and coulometric analyzers measured as much as twice the SO_2 concentrations of the colorimetric and flame photometric analyzers (3) on the average over 90% of the gaseous sulfur in downtown Los Angeles was sulfur dioxide (on a few occasions the flame photometric-gas chromatographic analyzer detected hydrogen sulfide and methyl mercaptan for a short time interval) (4) valid data was collected 90% or more of the time by four of the five analyzers, but the colorimetric analyzer collected valid data only 74% of the time.

Another recent field evaluation of sulfur dioxide instruments was carried out in the St. Louis urban area.[9] Performance of a coulometric and flame photometric analyzer were compared with the reference method. The two analyzers had correlations of about 0.9 with the 24-hour reference method. The coulometric analyzer had very good zero and span stability.

CARBON MONOXIDE

A primary air quality standard has been established for carbon monoxide based on health considerations.[1] The standard has two aspects:
(1) 10 µg/m^3 (9 ppm) - maximum 8 hour concentration not to be exceeded more than once per year (2) 40 µg/m^3 (35 ppm) - maximum 1-hour concentration not to be exceeded more than once per year. Because of the time intervals specified continuous instrumental monitoring is the only practical approach to measurement.

The reference method specified in the standard is non-dispersive infrared (NDIR) spectrometry.[1] Since a number of commercial analyzers have been available and in use in the U. S. for monitoring for some years, the application of this reference method is straightforward. Certain specifications for sensitivity, specificity and instrument performance - lag time, time to 90% response, drift characteristics were suggested in the Standard. As discussed in the Standard, the NDIR instruments do respond to water vapor so removal or compensation for water vapor is essential.

A collaborative test of this reference method has been conducted with participation from 15 laboratories including federal, state and local air pollution agencies, industry, and a research institute.[10] Test mixtures were prepared containing 8, 30 and 53 µg/m^3 of carbon monoxide in air and the gas stream was saturated with water vapor prior to the inlet to the analyzers. The conclusions from this collaborative study are summarized as follows:

1. The replication error (based on successive determinations with the same operator, instrument and sample within a few minutes of each other) of 0.17 µg/m^3 (0.15 ppm) is independent of concentration and humidity. The checking limit (maximum permissible range) for duplicates is set at 0.5 µg/m^3 (0.45 ppm) and two replicates differing by more than this amount should be considered a matter for concern.

2. The repeatability (the quantity that will be exceeded only about 5% of the time by the difference, taken in absolute value, of

two randomly selected test results obtained in the same laboratory on a given material) of 1.6 µg/m^3 (1.45 ppm) was independent of concentration.

3. The reproducibility (the quantity that will be exceeded only about 5% of the time by the difference, taken in the absolute value, of two single test results made on the same material in two <u>different, randomly selected</u> laboratories) was variable over the concentration range of interest. At 8µg/m^3, the reproducibility is 2.6 µg/m^3, at 30 µg/m^3, the reproducibility is 2.6 µg/m^3 while at 53 µg/m^3, the reproducibility is 3.8 µg/m^3.

4. The minimum detectable sensitivity is 0.3 µg/m^3.

5. Water vapor interference is adequately compensated for by use of drying agents and refrigeration. The results from laboratories using only narrow-band optical filters suggest that this technique may not be adequate for compensation.

6. Accuracy is dependent upon dependable calibration standards. The standards used in this study were carefully prepared in master bylinders with subsequent analysis by gas chromatography with primary calibration with gravimetric standards prepared in glass. Small cylinders with a chromium - molybdenum alloy inside surface of low iron content were filled from the master cylinders and sent to participants. Values between initial and final analyses on 51 cylinders were in good agreement except for one small cylinder. Based on the study results, this procedure produced results, on the average, 2.5% high.

An infrared fluorescent analyzer has been developed for carbon monoxide which has higher sensitivity and does not suffer from water vapor interference.[11] Infrared radiation from a black body source is used to excite carbon monoxide gas in a sealed fluorescent cell.

Another instrumental technique recently available as a commercial analyzer is that involving gas chromatographic analysis of carbon monoxide after its separation from methane and conversion to methane subsequently by hydrogenation. The detector used is the flame ionization analyzer. Considerable work has been done in EPA laboratories and elsewhere on development and field evaluation of this method.[12-15] The commercial analyzers combine the carbon monoxide, methane and total hydrocarbon measurements using the flame ionization detector. These instruments can be used as dual analyzers for both carbon monoxide and nonmethane hydrocarbons. Such instruments analyze periodically rather than continuously but provide many analyses per hour. This gas chromatographic technique also has the advantage of better sensitivity than the NDIR technique. No known interferences can be associated with the gas chromatographic analysis for carbon monoxide.

Another technique which is commercially available is the so-called ultraviolet mercury replacement method.[16] The principle involves the reduction of mercuric oxide by carbon monoxide at $400°F$ to form mercury vapor which is determined in absorption at 2537A.

OZONE AND OTHER OXIDANTS

A single primary and secondary standard for photochemical oxidants has been promulgated of 160 ug/m^3 (0.08 ppm) maximum 1 hour concentration not to be exceeded more than once a year.[1] The measured oxidant must be corrected for interferences due to nitrogen oxides and sulfur dioxide. However, the reference method given is the instrumental technique using the chemiluminescence from the ozone-ethylene reaction as measured by a photomultiplier. This technique is specific for ozone. In the procedure as given this instrumental technique was calibrated in the laboratoy by the neutral KI method using an ozone in air stream from an ultraviolet type of ozonizer. The neutral KI method is interfered with by sulfur dioxide and nitrogen dioxide and responds to other oxidants other than ozone. This colorimetric method also is marginal with respect to reproducibility in a concentration range just below the Standard. A stable ozone source now is available for routine calibration of ozone or oxidant analyzers which can be standardized by gas titration rather than the colorimetric KI method.[17,18]

Three types of chemiluminescent ozone analyzers have been evaluated.[18] All three are specific to ozone and they have sufficient sensitivity for atmospheric analysis. The analyzer based on the Regener procedure measures the chemiluminescent emission produced by reaction on Rhodamine-B absorbed on activated silica gel.

The other chemiluminescent reactions involve reaction of ozone either with nitric oxide or ethylene.[18] The technique involving reaction of nitric oxide with ozone is the inverse of the chemiluminescent reaction to determine nitric oxide. The third technique involves reaction of ethylene with ozone results in chemiluminescence in the 400 um region. The instrumentation can be operated at atmospheric pressure and room temperature. The time constant is 1 second and the linear range extends from 0.003 to above 1 ppm. The chemiluminescent ozone-ethylene instrument is of high sensitivity and specificity, rapid response, good linear range and it is simple to fabricate and operate. Therefore, this particular chemiluminescent technique was selected as the reference method for photochemical oxidant.

Other methods for measurement of oxidant can be considered provided correction is made for sulfur dioxide and nitrogen dioxide interference. Colorimetric and electrometric methods would require such corrections.[19, 20] Sulfur dioxide interferes negatively with ozone measurement on these instruments on about a one to one basis. When sulfur dioxide is the same or in a higher range of concentrations than ozone, oxidant instruments can be driven to zero. Therefore, removal of sulfur dioxide from the air stream must be accomplished before the instrument inlet. The nitrogen dioxide is a positive interference with colorimetric and coulometric methods. Approximate correction is possible using a separate measurement of nitrogen dioxide and the interference equivalent of nitrogen dioxide for the method. Unfortunately, an additional complication can develop. If sulfur dioxide is removed by the chromium trioxide-type absorbent procedure, oxidation of nitric oxide to nitrogen dioxide also occurs but the oxidation is not necessarily complete. Under these circumstances, the correction becomes somewhat uncertain being intermediate between the NO_2 and $(NO + NO_2)$ concentrations times the interference equivalent for NO_2. Because of this problem, removal of sulfur dioxide may not be desirable when the ozone concentration is large compared to the sulfur dioxide concentration (10 to 1 or greater). Under these conditions, sulfur dioxide can be analyzed separately and this concentration substracted.

In a recent field evaluation of instruments in Los Angeles, two different chemiluminescent analyzers (Rhodarine-B coated silica gel disks and ozone-ethylene reaction), an automated colorimetric 10%[8] KI reagent analyzer and an amperometric analyzer were compared. The calibration stability of the colorimetric analyzer was the poorest among the analyzers. Even in Los Angeles, sulfur dioxide interference was observed with the amperiometric analyzer on several occasions. This interference was equivalent on these occasions of 0.05 ppm and more of ozone. Diurnal averages over September 1970 with correction of the oxidant analyzers for NO_2 interferences indicated satisfactory agreement at peak oxidant (1100 to 1400 hours, average about 0.09 ppm) between the chemiluminescent and colorimetric analyzers, but the amperometric analyzer was 24% low. Off peak oxidant, the colorimetric analyzer was only about 0.01 ppm higher than the chemiluminescent analyzer. The statistical correlations were not good probably because of difficulties in making the NO_2 interference correction.

In a recent field evaluation[9] of instruments in St. Louis the same group of analyzers was used. The diurnal averaged concentrations between October and December 1971 for the chemiluminescent and amperometric analyzers were very close between 1100 and 1700 hours (average about 0.05 ppm). The values from the colorimetric analyzers was about 0.01 ppm higher during this time interval. Of concern were the morning and evening peaks showing on the colorimetric and amperometric analyzers while the chemilumescent analyzer was showing minimum response to ozone. It should be noted that nonphotochemical oxidants such as chlorine could cause response on the colorimetric and coulometric analyzers.

HYDROCARBONS

The primary and secondary air quality standard for hydrocarbons is 100 µg/m^3 (0.24 ppm) which is not to be exceeded more than once a year during the 3-hour period between 6 to 9 AM.[1] The measurement is for total hydrocarbons excluding methane. This standard is for use as a guide in the development of implementation plans by the states to control photochemical oxidant. This approach is taken because there is no health or welfare effect directly associated with hydrocarbons in the standards.

The reference method[1] involved periodic measurement (4 to 12 times an hour) of the hydrocarbon content of volumes of air delivered to a hydrogen flame ionization detector. A portion of each of the same volumes of air is passed through a stripper column into a gas chromatographic column which separates the methane from carbon monoxide.[13-15] The methane is determined with the hydrogen flame ionization detector. Hydrocarbon concentrations corrected for the concurrent methane content are determined by substracting the methane concentration from the total hydrocarbon concentration. This method is specific for methane. The stripper or precolumn protects the gas chromatographic column against deterioration by removing water, carbon dioxide and hydrocarbons other than methane. The stripper column is backflushed after each determination.

NITROGEN DIOXIDE

The primary and secondary ambient air quality standard for nitrogen dioxide is 100 µg/m^3 (0.05 ppm) as an annual arithmetic average.[1] The reference method involves 24-hour collection of nitrogen dioxide into 0.1 N aqueous sodium hydroxide solution using a gas dispersion tube with a frit of 70 to 100 µm. The sampling rate is 200 ml/minute. The nitrite ion produced is determined by reaction with N-1 naphthylethylenediamine dihydrochloride, sulfanilamide and phosphoric acid. The colored product after a 10 minute wait for color development is measured at 540 mµ. Calibrations are made using standard nitrite solutions. Sulfur dioxide interference is minimized by oxidation to H_2SO_4 by H_2O_2 before analysis.

Since the standard was promolgated in April, 1971, the reference method has been criticized in several publications.[21,22] The method has been reported to suffer from both a variable collection efficiency and from interference by nitric oxide. These problems have been under active investigations in EPA research laboratories. Both effects have been confirmed although our quantitative results differ somewhat from some of the published work.

The stoichiometric factor in the Griess-Saltzman reaction has been variously claimed to be 0.72 to 1.0.[23] More recently a carefully study using nitrogen dioxide permeation tubes over the range from 0.02 to 1.6 ppm resulted in a factor of 0.75 ± 0.037 (10).[24] Attempts to reproduce the experimental conditions reported to produce a factor of 1.0, resulted in a factor of 0.75.

Automated versions of the Griess-Saltzman reactions or faster reacting reagent modifications have been in use in California and in some Eastern U. S. cities for some years.

Chemiluminescent analyzers for $NO-NO_x$ have been developed.[25-29] The chemiluminescent reaction of nitric oxide with ozone to form excited NO_2 (0.6 to 3.0 microns) is a highly specific and sensitive reaction (0.005 ppm) with a wide linear range (0.005 to 15 ppm of NO). Nitrogen dioxide can be determined by difference between $NO + NO_2$ and NO.

CALIBRATION TECHNIQUES AND REFERENCE MATERIALS

Calibration techniques are an essential part of any research on air measurements.[5] New or existing methods can be quantitated only when accurate calibration techniques are available. Many approaches have been investigated including preparation in gas cylinders, large rigid containers of other forms, plastic bags serial dilution, diffusion cells, and presentation tubes. For the purposes of the individual investigator any one or a combination of these techniques can be exceedingly useful.

The air quality standard for sulfur dioxide already contains a sulfur dioxide permeation tube technique.[1] The collaborative test for SO_2 utilized the permeation tube technique. A nitrogen dioxide permeation tube technique has been in recent development. For carbon monoxide gas cylinder standards are satisfactory and they were used in the collaborative test for carbon monoxide.[10] Cylinder gas standards should be acceptable for calibration of hydrocarbons in air also. Ozone is a special problem. There are ozonizers whose output can be calibrated by titration with the calibration standards for nitric oxide or nitrogen dioxide.[17]

Reference materials are nothing more than calibration techniques which can be used as primary standards and which are convenient for widespread use for standardization and quality control. EPA has funded work at the National Bureau of Standards to follow through on the research in EPA laboratories on permeation tubes for sulfur dioxide and nitrogen dioxide. The Bureau of Standards has issued certified sulfur dioxide permeation tubes and is well along towards certification of nitrogen dioxide permeation tubes. Work also has been in progress on the ultraviolet ozonizers for ozone calibration.

STANDARDIZATION OF METHODS

This work is directed towards determining the acceptability of methods in the hands of knowledgeable analysts. The method

is being tested not the analyst. Collaborative tests conducted by
EPA with its contractors have already been briefly discussed. Col-
laborating laboratories are volunteers from industry, government and
universities. These tests have involved both synthetic mixtures
for sulfur dioxide and carbon monoxide and actual atmospheric
particulate samples for mass of suspended particulates. Three
collaborative testing reports have been released by EPA.^{2,6,10}
A separate program operated by ASTM has been evaluating ASTM methods
which usually are not the same as EPA reference methods in part
because many ASTM methods were developed many years ago. Testing has
been done with atmospheric samples with apiking with the pollutant
of interest. Collaborative testing reports from ASTM have not yet
been released for general distribution to contract laboratories
from the basis of the collaborators in the ASTM system of testing.

EQUIVALENCY METHODS

The air quality standards explicitly state that each of the
pollutants shall be measured by the reference method or by an equivalent method. Equivalent methods include any methods of sampling
and analysis for an air pollutant which can be demonstrated to the
Administrator's satisfaction to have a consistent relationship to
the reference method. However, no complete definition of "equivalent method" was set forth. A document is in preparation which will
do so in terms of set of statistical defined performance specifications
for instruments which must be satisfactorily conducted by the instrument manufacturer and submitted to EPA for evaluation and approval.
Development of such specifications for each type of instrument for
each of the six pollutants is a difficult task. Not only are there
many technical problems in terms of defining the specifications, their
limits and the statistical procedures, but also there are problems
in achieving a balance of fairness to manufacturer and user alike.
The size of the market involved is not such that completely definitive
testing is possible because of time and cost, but the user is entitled to reliable instrumentation to determine compliance with the
standards promulgated by EPA.

REFERENCES

1. Environmental Protection Agency, National Primary and Secondary
Air Quality Standards. Federal Register, Vol. 36, No. 84, P. 8186-
8201, April 30, 1971.

2. McKee, H. C., Childers, R.E., Seenz, O., Jr., "Collaborative
Study of Reference Method for the Determination of Suspended Parti-
culates in the Atmosphere (High Volume Method)". Contract CPA 70-40,
June 1971, Prepared for Division of Chemistry & Physics, National
Environmental Research Center, RTP, N. C. Environmental Protection
Agency

3. Air Quality Criteria of Particulate Matter, Chapter 1, "Atmospheric Particles, Definitions, Physical Properties, Sources, Sources, and Concentrations". U.S. DHEW, PAS, National Air Pollution Control Administration, Washington, D. C. Publication No. AP-49, January 1969.

4. Ibid, Chapter 3, "Effects of Atmospheric Particulate Matter on Visibility".

5. O'Keeffe, A. E., and Ortman, G. C., "Primary Standards for Trace Gas Analysis", Anal. Chem. 38, 760 (1966).

6. McKee, H. C., Childers, R. E., Saenz, O., Jr. Collaborative Study of Reference Method for Determination of Sulfur Dioxide in the Atmosphere (Pararosaniline method) Contract CPA 70-40. Sept 1971. Prepared for Division of Chemistry & Physics, Natl. Environmental Research Center, RTP, N. C., Environmental Protection Agency.

7. Palmer, H. F., Rodes, C. E., Nelson, C. J., "Performance Characteristics of Instrumental Methods for Monitoring Sulfur Dioxide; Part II; Field Evaluation", J. Air Pollution Control Assoc. 19, 778 (1969)

8. Stevens, R. K., Hodgeson, J. A., Ballard, L. F., Decker, C. E., "Ratio of Sulfur Dioxide to Total Gaseous Compounds and Ozone to Total Oxidants in the Los Angeles Atmosphere - an Instrument Evaluation Study in Determination of Air Quality, Mamantov, 6. and Shults, W. D., eds., Pleuum Publishing Co., New York (1972).

9. Stevens, R. K., Clark, T. A., Decker, C. E., Ballard, L. F., "Field Performance Characteristics of Advanced Monitors for Oxides of Nitrogen, Oxides of Nitrogen, Ozone, Sulfur Dioxide, Carbon Monoxide, Methane and Nonmethane Hydrocarbons". Presented at 1972 Air Pollution Control Association Meeting, Miami, Fla., June 1972.

10. McKee, H. C. and Childers, R. E. "Collaborative Study of Reference Method for the Continuous Measurement of Carbon Monoxide in the

11. McClatchie, E. A. "Development of an Infrared Fluorescent Gas Analyzer" Contract CPA 70-152, 1972. Prepared for Division of Chemistry and Physics, Natl. Environmental Research Center, RTP, N. C. Environmental Protection Agency.

12. Altshuller, A. P., Kopczynski, S. L., Lonneman, W. A., Becker, T. L., and Slater, R. "Chemical Aspects of the Photooxidation of the Propylene-Nitrogen Oxide System". Environ Sci. Technol. 1, 899 (1967).

13. Stevens, R. K., O'Keeffe, A. E. "Modern Aspects of Air Pollution Monitoring". Anal. Chem. 42, 143 143A (1970).

14. Stevens, R. K., O'Keeffe, A. E., Ortman, G. C., "A Gas Chromatographic Approach to the Semicontinuous Monitoring of Atmospheric Carbon Monoxide and Methane". Proceedings of the 11th Conference on Industrial Hygiene Studies, Berkeley, California (March 30 - April 1, 1970).

15. Villalobos, R. & Chapman, R.L. "A Gas Chromatographic Method for Automatic Monitoring of Pollutants in Ambient Air". Air Quality Instrumentation, Vol. 1, John Scales, editor pp 114-128, Instrument Society of America, Pittsburgh, Pa., (1972).

16. Robbins, R. C., Borg, K. M., Robinson, E., "Carbon Monoxide in the Atmosphere". J. Air Poll. Control Assoc. 18, 106 (1968)

17. Hodgeson, J. A., Stevens, R. K., Martin, B. E., "A Stable Ozone Source Applicable as a Secondary Standard for Calibration of Atmospheric Monitors" in Air Quality Instrumentation, Vol 1, John Scales, ed., pp 149-150, ISA, Pittsburgh, Pa., 1972.

18. Hodgeson, J. A., Martin, B. E., Baumgardner, R. E., Comparison of Chemiluminescent Methods for Measurement of Atmospheric Ozone", Progress in Analytical Chemistry, Vol. 5, Plenum Press, 1971.

19. Methods of Air Sampling Analysis. Intersociety Committee Tentative Method 406 Manual, Analysis for Oxidizing Substances in the Atmosphere" p. 351-55, American Public Health Assoc., Washington, D. C. 1972.

20. Methods of Air Sampling and Analysis, Intersociety Committee Tentative Method 405 Continuous Monitoring of Atmospheric Oxidant with Amperometric Instruments". p. 341-350, American Public Health Assoc., Washington, D. C. 1972

21. Blacker, J. H., and Brief, R. S. "Evaluation of the Jacobs-Hochheiser Method for Determining Ambient Nitrogen Dioxide Concentrations." Chemosphere 1, 43 (1972).

22. Heuss, J. M., Nebel, G. J., Colucci, J. M., National Air Quality Standards for Automative Pollutants. J. Air Pollution Control Assoc. 21, 535 (1971).

23. Methods of Air Sampling and Analysis. Intersociety Committee Tentative Method 403. Analysis of Nitrogen Dioxide Content of the Atmosphere (Griess - Saltzman reaction). p. 329-335, American Public Health Assoc., Washington, D. C. 1972.

24. Scaringelli, F. P., Rosenberg, E. and Rehme, K. A., "Comparison of Permeation Devices and Nitrite Ion as Standards for the Colorimetric Determination of Nitrogen Dioxide". Anal. Chem.

25. Fontiju, A., Sabadell, A. J. and Rosco, R. J., "Homogeneous Chemiluminescent Measurement of Nitric Oxide with Ozone". Anal. Chem. 42. 575 (1970).

26. Stedman, D. H., Daby, E. E., Stuhl, F. and Niki, H., "Analysis of Ozone and Nitric Oxide by a Chemiluminescent Method in Laboratory and Atmospheric Studies of Photochemical Smog." J. Air Pollution Control Assoc. 22, 260 (1972).

27. Breitenbach, L. P., Shelef, M. "Development of a Method for the Analysis of NO_2 and NH_3 by NO-measurement instruments". Scientific Report No. SR 71-130 (1971).

28. Hodgeson, J. A., Bell, J. P., Rehme, K. A., Krost, K. J. and Stevens, R. K. "Application of a Chemiluminescent Detector for Measurement of Total Oxides of Nitrogen and Ammonia in the Atmosphere". Joint Conference on Sensing of Environmental Pollutants Paper No. 71-1067, American Institute of Aeronautics and Astronautics, New York, N. Y. 1971.

29. Hodgeson, J. A., Rehme, K. A., Martin, B. E., Stevens, R. K. "Measurements for Atmospheric Oxides of Nitrogen and Ammonia by Chemiluminescence", 65th Annual Meeting Air Pollution Control Assoc. Miami, Fla., June 1972.

THE USE OF AIRBORNE SENSOR SYSTEMS FOR ENVIRONMENTAL MONITORING

Howard A. Friedman and Howard J. Mason, Jr.

*National Oceanic and Atmospheric Administration
Environmental Research Laboratories Research
Flight Facility, Miami, Florida*

NOAA's COMMITMENT TO ENVIRONMENTAL AIR QUALITY (POLLUTION) MONITORING, PREDICTION AND CONTROL

The National Oceanic and Atmospheric Administration (NOAA), operating as an important agency within the United States Department of Commerce, is comitted to provide leadership for the environmental air quality (pollution) program of the nation by supporting detection, measurement, forecast, and control functions of NOAA and other government agencies.

Many of these programs were devised in response to the awareness of the importance of air quality as an endangered global resource. In the United States, this awareness has taken form in the Clean Air Act of 1967 and more recently, in the President's plan for Protection and Enhancement of Environmental Quality and the National Environmental Quality Act of 1969.

An important key in both programs is to improve local short-range forecasts issued to the public, including the forecast of parameters in the atmospheric boundary layer that, in fact, determine the ability of the atmosphere to dilute both natural and man-made pollution. To achieve significant improvement for these and longer range pollution forecasts, meteorologists must successfully diagnose and predict the evolution of small-scale meteorological events, such as turbulent exchanges.

The ultimate goal of NOAA's efforts, which includes the establishment of Urban Air Pollution Support Units, is to be more responsive to the needs of the public and to provide a significant contribution to the safety, comfort, and well being of the nation. Bulletins and warnings of high air pollution potential, provided by specialized units in NOAA, activated the first stages of urban emergency alerting procedures, and continuous pollution monitoring and forecasts provide various urban agencies information needed to establish restrictive levels of pollution control actions designed to protect the health and well being of the urban populace.

Special areas of research interest within NOAA, in general, and the Environmental Research Laboratories, in particular, involve studies of boundary layer interactions to determine the natural mixing and pollution transport capability of the atmosphere. Valley ventilation pollution in highly industrialized areas is also being studied. These problems point to the need for new knowledge about pollutants behavior and the prediction of the small-scale behavior and interactions in the atmosphere.

Numerical models are being developed to test pollution prediction guidance procedures for short-range forecasts. New knowledge in this area will also provide more realistic lower-boundary and initial conditions for multi-level, global-scale, long-range prediction models.

Research is also being undertaken to improve measurements of the parameters that affect atmospheric stability and the transport of pollutants. The use of lidar for profile measurements of aerosol concentrations, continuous monitoring of carbon dioxide levels, multi-channel IR radiometers, aerosol spectrometers, electrical conductivity measurement systems and airborne dispersive gas analyzers are just some of the techniques and/or systems being employed.

Since we must be certain that all relevant parameters are considered, the life history and residence times of many particulate pollutants are being studied. To measure and describe the time histories of cloud condensation and freezing nuclei-those particles which affect cloud and precipitation microphysics-which, in turn, affect the earth's radiation and moisture budgets-is planned and NOAA is trying to assess the effects of natural and man-made pollution on the global heat balance and moisture distributions. Through boundary layer models, this information will provide a necessary link to more sophisticated models of global weather prediction and climatic evolution.

There are many other individual programs being supported by NOAA which we have not mentioned above. From the "generalized" philosophy that we have presented, one can see that NOAA is realistically involved, both research and operationally oriented, to detect, monitor, predict and control environmental pollution, and to assess the effects of pollution on our global ecology.

PLATFORM AND BASIC SENSOR CAPABILITIES OF THE RESEARCH FLIGHT FACILITY

It is beyond the scope of this paper to report on the physical principles underlying each measurement subsystem on the RFF aircraft. Detailed accounts of RFF aircraft capabilities, instrumentation systems, and data processing procedures have, however, been reported in the literature (Friedman et. al., 1969a; Friedman et. al., 1969b; Conrad et.al., 1972; and, Reber and Friedman, 1964); therefore, for the purpose of this report, these items will only be summarized in tables 1 and 2.

Table 1. RFF aircraft operational capabilities.*

TYPE AIRCRAFT	DC-6 A/B	WC-130B	B-57A
Registration number(s)	N6539C N6540C	N6541C	N1005
Number of each type a/c	2	1	1
T/O runway length req. at max. gross weight	6000 ft	4000 ft	6500 ft
Long range endurance (at optimum altitude)	12 hr	< 8 hr	5 hr
Practical service ceiling	18,000 ft	28,000 ft	43,000 ft
Max. mission radius	1200 n mi	1200 n mi	950 n mi
Max. range	2400 n mi	2400 n mi	1900 n mi
Cruising speed (normal)	220 kt	290 kt	430 kt

*Figures shown are approximate; to obtain exact figures one would have to specify mission profiles and required fuel reserves.

8. SENSING, INSTRUMENTATION, AND MEASUREMENT

Table 2. RFF Aircraft Instrumentation.

PARAMETER	INSTRUMENT TYPE	RANGE	AIRCRAFT DC-6A/B	WC-130B	B-57A
Position*	OMEGA, Doppler	$90°$ N/S	X	X	X
	Loran	$180°$ E/W	X	X	X
	Inertial		X		
Cloud reflectivity	Radar 3.2 cm RHI	20 n mi	X		
	3.2 cm PPI	150 n mi		X	X
	5.6 cm PPI	150 n mi	X		
	10.2 cm PPI	200 n mi	X		
Aircraft pitch/roll	Doppler gyro	$\pm 30°/\pm 45°$	X		X
Magnetic heading	N-1 flux gate compass	$360°$	X	X	X
Magnetic variation	Published values**		X	X	X
Ambient pressure	Pressure transducer	1050 to 50 mb	X	X	X
Differential pressure	Pressure transducer	0 to 400 mb	X	X	X
Indicated airspeed	Airspeed meter/transducer	50 to 400 kt	X	X	X
True airspeed	TAS transducer	50 to 700 kt	X	X	X
Groundspeed	Doppler	70 to 700 kt	X	X	X
Drift angle	Doppler	$\pm 45°$	X	X	X
Absolute altitude	Radar altimeter	200 to 50,000 ft	X	X	X
Total temperature	Resistance thermometer	-60 to +40 C	X	X	X
Wind velocity	NAV systems	$360°$	X	X	X
		0 to 240 kt	X	X	X
Distance travel count	Doppler	0 to 999.999 n mi	X		X
Pressure altitude	Altimeter/transducer	0 to 50,000 ft	X	X	X
Ice detector	Cyclic sensor		X		
Ambient temperature	Vortex sensor	-60 to +40 C	X		X
	Thermocouple sensor	-50 to +50 C	X		
	Diode sensor	-50 to +50 C	X	X	
Liquid water content	Hot-wire sensor	0 to 6gm/m^3	X	X	
	Hot-wire/ceramic cone	0 to 10gm/m^3	X		
Absolute humidity	IR hygrometer	0 to 20+gm/m^3	X	X	
Dew/frost point	Hygrometer	-50 to +50 C	X	X	
Sea surface temperature	IR radiometer	-40 to +40 C	X	X	
Free air temperature	IR (CO_2) radiometer	-70 to +50 C	X		
Vertical profile data	Dropside (P,T,RH)		X	X	
Cloud seeding	Dry ice		X	X	
	Pyrotechnic flares		X	X	X
	Steady-state generator		X		
Chaff dispenser	Foil chaff dispenser		X	X	
Radiation detector	Air sampler		X	X	X
Time	Crystal oscillator	24 hr	X	X	X
Communications	HF, VHF, UHF, SSB		X	X	X
Aitken nuclei count	Nuclei counter		X		
Vertical velocity	Accelerometer		X	X	X
Refractive index	Microwave refractometer		X		
Water vapor flux	Turbulence system		X		
Solar radiation	Spectral pyranometer		X	X	
Recording and photographic systems	Digital/(magnetic tape)		X	X	X
	Digital/analog FM		X	X	
	Visicorder/strip chart		X	X	
	16 mm camera systems		X		X
	35 mm camera systems		X	X	X
	70 mm camera systems		X	X	
Wave height	Laser system		X	X	
Cloud construction	Formvar/Foil systems		X	X	
			X	X	X
Ice nuclei concentration	Cold-box ice nuclei counter		X	X	

* Includes measurement of parameters required to determine position (groundspeed and true airspeed vectors: $\overline{GS}$, $\overline{TAS}$).

** Manually set in by navigator.

SENSORS FOR THE DETECTION AND MEASUREMENT OF POLLUTANTS

During the past decade there has been a renewed interest in cloud physics and its related science, cloud/weather modification. Environmental pollution is closely tied to cloud and particulate microphysics. The success of numerical models, constructed to simulate the behavior of one or more features of real clouds in the atmosphere or of the global-scale atmosphere, itself, has been somewhat limited by a lack of understanding of the complicated interactions which take place on the microphysical scale; therefore, both cloud physicists and pollution scientists can benefit from new knowledge gained through the measurement of atmospheric pollutants. However, the state-of-the-art of in situ measurement systems, in general, is such that precise and accurate data are not readily available.

Some of the parameters that scientists need to describe are: the distribution of ice nuclei, cloud concentration particles, contact nuclei, other "pollutants", as well as the moisture and thermal structure of the atmosphere. All are necessary for a more complete description and understanding of the entire process of cloud growth and precipitation physics. These processes are important to our understanding and prediction of the effects of atmospheric pollution on the global radiation and moisture budgets, or, in short, the evolution of our global weather and climate.

With this in mind, we would like to briefly discuss some of the airborne instruments that are presently employed for particulate measurements by the RFF.

A. <u>Ice nuclei measurements</u>. The RFF obtains ice nuclei "Counts" with Bigg-Warner cold-box systems, adapted for airborne use. The units, when flown on RFF's DC-6 aircraft, are mounted in such a manner to allow them to swing freely fore and aft around an axis parallel to the latitudinal axis of the aircraft.

The "tank" is refrigerated with a 208 vac/400 Hz, 0.25 hp F-12 refrigeration system, which is regulated by an automatic expansion valve. Operating temperatures in the chamber can be varied and controlled over the range from zero to -25C.

A sample of ambient air is compressed in the chamber by a 0.25 hp - 28 vdc air pump. A tray containing a thin layer of supersaturated sugar solution with a wetting agent is placed in the bottom of the chamber. The walls of the chamber are coated with a glycerin solution to surpress frost. The unit is sealed, and a discrete sample of ambient air is brought in by the pump and then pressurized in the chamber. The air sample is allowed to stabilize to the chamber temperature, and then the chamber is explosively decompressed to cabin altitude. This process causes a supercooled fog to form and an additional amount of pre-determined cooling, (through adiabatic processes), to take place.

The supercooled fog droplets freeze to any available ice nuclei in the air sample. The resulting ice crystals then precipitate on to the surface of the supersaturated sugar solution in the tray, which is at the chamber temperature. The crystals grow rapidly to a size clearly visible to the naked eye, thereby permitting the number of such ice crystals to be determined by a simple counting process. The tray may be photographed and crystals counted after the fact, or, real-time counting may be employed. The resulting ice crystal concentration (number density) is actually count per 10 liters, since the volume of the chamber is 10 liters.

As soon as the number of ice crystals is determined or the sample has been photographed, the tray is removed from the chamber to allow the ice to melt, thereby permitting reuse of the tray for subsequent sampling. This discrete sampling process with multiple trays available and potential reuse allows for a maximum of six to eight samples to be completed in a one-hour time period.

The cold-box appears to perform satisfactorily, however, RFF is hoping someday to replace this discrete sampler with a system which would provide for continuous sampling of ice nuclei in the atmosphere. Unfortunately, such systems are not commercially available at present nor have developmental systems demonstrated the ability to make consistent (repeatable) measurements even under controlled environmental conditions, such as in cloud chambers.

B. <u>Cloud particle measurements</u>.

1. <u>Airborne continuous cloud particle replicator</u>. At present, the RFF utilizes an airborne continuous cloud particle replicator system to obtain spectra of particles from approximately 5 to 200 microns.

The system, which is installed on the RFF's DC-6 and WC-130B aircraft, was manufactured by Meteorological Research, Inc., of Altadena, California. (Another version of the system was manufactured by Mee Industries, also of California).

Basically, the replicator consists of film transport (similar to a complex 16 mm picture projector), a pump and reservior system capable of extruding a thin coating of a plastic solution on the film just prior (in time and space) to the exposure of this film to the cloud air, and the necessary boom or probe to project the sampling film out into the undisturbed air well beyond the boundary layer of the aircraft. The process of cloud particle sampling with this system is not a photographic process; the 16 mm clear mylar film simply acts as a convenient transport mechanism for the encapsulating formvar solution.

Since the system is utilized on RFF's pressurized aircraft, the entire device is enclosed in a pressure sealed housing with the boom (or probe) and sampling slit extending through a pressure sealing gland (made of a resilient silicone rubber) in the skin of the aircraft. The length of the protruding boom is about three feet beyond the skin of the aircraft into the ambient, nearly undisturbed air. The exposure slot, located near the end of the probe, is 20 mm high by 3.5 mm wide; however, the area covered by the formvar is only about 4 mm by 3.5 mm.

At typical aircraft and instrument speeds, the effective volume sampled in a one-second time frame is approximately 1.5 liters. This corresponds to a sample volume of 0.04 liters per frame.

The coating is a 4 percent solution of formvar (a trade name by Shawanigan Chemical Corporation), a polyvynil formal resin, and chloroform (about 96% $CHCL_3$). There are several other solvents for formvar all of which have properties that make them undesirable for use on an aircraft. This coating solution is applied to the mylar film just prior to the time that it passes in front of the exposure slot.

Cloud particles (ice and/or water) which are propitiously located in the stagnation region of the airfoil shaped boom at the sampling slit are impacted into the still fluid formvar solution, which is soon dried leaving a permanent replica of the particles impacted. The film then goes through a drying process, and the contained water substance eventually diffuses through the thin plastic skin, leaving an image or replica of the ice particles or water droplets.

Data retrieval from this instrument is accomplished by projecting the exposed film through a standard 16 mm time-motion study projector, incorporating telemicroscope optics to enlarge the images of the particles up to 10^3 times their actual size, or by photographing samples through a microscope for subsequent analysis. Cloud droplet spectra may be obtained by employing these techniques, but only after a considerable effort on the part of the data analyst.

Ice crystal habit can sometimes be ascertained if the samples are from relatively dry, cold clouds. Seldom are definite crystal shapes other than columns discernable from wet (dynamic) tropical cumulus clouds.

The most common use for the data from this instrument is the determination of the ice-water ratio (with respect to time) in a cloud, or the droplet spectra in "warm" clouds or fog.

2. <u>Airborne continuous hydrometeor sampler</u>. The airborne continuous hydrometeor sampler is used in conjunction with the airborne continuous cloud particle sampler described above. The range of sizes covered by this instrument extends from 200 microns up to the limit of the aperature at the sampling window, or about 3.8 cm by 3.8 cm. This system, manufactured by Meteorology Research, Inc., consists of a transport system that continuously pulls an aluminum foil ribbon from a supply spool over a finely grooved drum behind the sampling aperature, then respools the exposed foil onto another core.

Exposure time for the sampler is controlled by a shutter which is automatically tripped open when the previous exposure has advanced to a position that inhibits any future hydrometeors from making impressions in the foil at the wrong time. The typical volume sampled in a one-second time frame is 145 liters.

Data from this instrument is in the form of impressions of the impacted particles with striations superimposed on the impression every 250 microns. Data retrieval is accomplished by viewing the foil on a flat surface illuminated by strong oblique lighting. The impressions appear as a series of lines, the outlines of which define the size and shape of the particle. For liquid water drops, the shape of the impressions are invariably round or ellipsoidal, and a slight correction factor (a function of size and air speed) is required. In the case of ice, graupel, or hail, no correction factor is necessary.

The combined utilization of the formvar and foil instruments with measurements of the total water content (hydrometeor and vapor), can yield significant cloud physics data. So called "splash" problems can exist with the formvar sampler system, but, not with the foil sampler. In the latter, splashes are generally of insufficient size (mass) to leave imprints on the exposed foil.

Although the formvar and foil systems are available from commercial sources, data retrieval problems and other fundamental drawbacks of these systems clearly indicate a need for more developmental work in this important measurement area.

C. Radiation measurements. All RFF aircraft are equipped with military type radiation detection units, including peripheral instrumentation to investigate electrical and cosmic phenomena. The systems, called the FI-2A (or -2B) Foil Assembly, employ the B/200A or B/400 count rate meter (CRM) and operate with an Ampex 90 NB Geiger Mueller (GM) Tube as the radiation detector.

Both the B/200A and B/400 CRM systems are designed to operate with the GM detector at an average of 900 v. Eight selectable count ranges are available, in the intervals from zero to 10^3 up to zero to 2×10^5.

The collection unit is an internal foil with an externally open intake and exhaust port. The screen for the foil uses a 4.25 inch (10.8 cm) diameter filter paper which is exposed for an appropriate time interval, and then replaced to obtain multiple discrete samples. In practice, the paper filter is exposed to the ambient air and radioactive particles or other particulates impinge on its surface. The system is then closed, paper removed, and radioactive particles counted with the GM system.

A typical use for the system has been to measure peak concentrations of radioactive fission products released to the atmosphere as a result of nuclear rocket tests (see below). Dust (particulates) have also been collected for subsequent analyses with these systems over the open ocean near Barbados to determine the source regions and atmospheric transport time for such particles.

SENSORS FOR THE MEASUREMENT OF THE TRANSPORT OF POLLUTANTS

A. <u>Horizontal Winds</u>. One of the most important meteorological measurements obtained in flight is the ambient vector wind (speed and direction). The so called spot or instantaneous wind ($\vec{W}$) is determined by measuring the ground speed of the aircraft ($\vec{GS}$) and true airspeed ($\vec{TAS}$) vectors and then solving the "wind" triangle equation ($\vec{W} = \vec{GS} - \vec{TAS}$). The $\vec{TAS}$ lies in the direction of the aircrafts true heading (TH) and has the magnitude of the true airspeed (TAS) of the aircraft. The $\vec{GS}$ is determined from Doppler radar measurements of the ground speed and drift angle.

In addition to Doppler systems, the RFF aircraft also employ OMEGA navigational systems (VLF) and will soon utilize inertial platforms for more precise navigation and wind determination.

The horizontal wind measurements obtained with these systems, provides us with an indication of the meso-and synoptic scale motions of particulates in the atmosphere, since these move at a significant fraction of the total horizontal wind speed.

B. <u>Vertical winds</u>. The determination of the vertical wind and therefore the vertical transport of pollutants is a bit more complicated. On a large scale, the mean vertical motion over an area can be computed from continuity considerations. However, in the boundary layer, small-scale turbulent motions are of particular interest (and probably more important) to our understanding of the transport of particulates (smoke, dust and other pollutants).

To measure these micro-scale vertical motions, RFF has developed a turbulence measuring system. To date, this system has been used in conjunction with a microwave refractometer primarily to determine the water vapor flux over the open ocean; however, coupled with additional fast response sensors, it could be utilized to obtain heat and momentum fluxes as well (Friedman et al., 1970b).

Here again, we determine a field of turbulent motion and assume that particulates trapped in the boundary layer are transported by turbulent eddies.[1] Actually, the motion of pollutants in the atmosphere is in part determined by the mean and turbulent motions (eddy diffusivities and conductivities) and the thermal structure (stability) of the atmosphere. Of course, we realize that if the particles grow large so that the vertical motion will not sustain their mass, they will precipitate out of the atmosphere.

The RFF turbulence measuring system is basically comprised of an angle of attack probe (Lockheed), an inertial platform (Litton LTN-51), and supporting electronic and recording systems. The Lockheed probe consists of vertical and lateral fixed-vane sensors arranged about a central pitot-static tube. These sensors consist of a light-weight wedge-shaped vane with a 4-inch span and 2-inch chord attached to a specially constructed strain-gage beam of slotted construction (McFadden et al., 1970; Conrad et al., 1972).

Loads on the vane cause the beam to deflect parallel to itself. The lift is measured in terms of shear as opposed to a bending moment as would be the case if the vane were attached to a simple beam. Semiconductor strain gages, in a Wheatstone bridge configuration are bonded to the fore and aft portions of the beam. Electrical signals caused by the deflection of the beam are amplified and conditioned to levels suitable for input into a high speed recording system.

The probe housing also contains accelerometers to measure normal and lateral accelerations to the boom, which are also recorded.

A pitot-static head and Stratham strain-gage transducer measure the static and dynamic pressures used for airspeed determinations.

---

[1] Note: If w is the instantaneous vertical velocity, $\bar{w}$ the mean vertical velocity, and, w' the eddy velocity, then, $w' = w - \bar{w}$.

The inertial platform is used to furnish aircraft attitude and vertical acceleration data; platform stabilization is accomplished by two, two-degree-of-freedom gyros and three accelerometers. Roll and pitch angles are obtained from the platform's follow-up servos, and its computer electronics furnishes vertical acceleration information in the form of dc voltages.

The measurement of the vertical component of the wind is obtained by determining the motion of the air relative to the aircraft and the motion of the aircraft relative to the ground. A generalized equation for w can be given by

$$w = V_{TAS}(\alpha-\theta) + \int_0^t a_z \, dt + l\dot{\theta}$$

where w is the vertical component of the wind; V_{TAS} is the true airspeed; α is the angle of attack; θ is the pitch angle; a_z is the normal acceleration of the aircraft; l is the distance from the aircraft's center of gravity to the tip of the boom; and, $\dot{\theta}$ is the pitch rate of the aircraft ($\dot{\theta} = \frac{\partial \theta}{\partial t}$).

The first term $[V_{TAS}(\alpha-\theta)]$ represents the motion of the air relative to the aircraft; $[\int_0^t a_z \, dt]$ represents the motion of the aircraft relative to the ground; and, $[l\dot{\theta}]$ represents the vertical velocity increment at the probe caused by the pitching motion of the aircraft.

This system, which has been improved for use in the Internation Field Year for the Great Lakes (IFYGL), is described in more detail by Lappe (1969) and Conrad et al., (1972).

RFF's PARTICIPATION IN ENVIRONMENTAL MONITORING PROGRAMS

The Research Flight Facility has supported many programs over the past decade that could be loosely categorized as "environmental pollution monitoring." These have included: air mass modification research and studies to determine the spatial distribution of carbon dioxide in the equatorial troposphere; studies which utilized (natural) tritium and injected sulfur hexaflouride as atmospheric tracers; and, still others which were designed to measure and describe peak concentrations of man-made radioactive effluents in the atmosphere as functions of both time and space.

Several of these projects will now be discussed briefly.

A. Support of NOAA's Air Resources Laboratory

1. An air mass modification project was conducted in both 1965 and 1966 in support of NOAA's Air Resources Laboratory (ARL). The program, under the direction of Dr. L. Machta, was in part, supported by funds from the United States Atomic Energy Commission (AEC). Sampling missions measured the gradients of artificial and natural radioactivity, temperature, moisture, wind, carbon dioxide, tritium, deuterium, and ozone in the environment as suitable air masses moved from continental to oceanic areas along the east coast of the United States. A correlation between changes in the gradients of atmospheric particulate radioactivity and the oceanic uptake of the radioactivity was sought.

Vertical and horizontal profiles were obtained over land and at several locations over the sea. Flight patterns were generally confined to the lowest 1 km of the atmosphere. Time history data and replicate samples were obtained by utilizing two aircraft flying essentially similar patterns.

2. In the spring of 1966, the RFF and the Institute of Marine Science (IMS) of the University of Miami performed several missions in support of ARL and the University of Stockholm. Here the primary objective was to explore, for the first time, the spatial distribution of carbon dioxide in the lower equatorial troposphere. Later in the study, upper tropospheric carbon dioxide sampling was accomplished in accordance with an agreement between the University of Stockholm and commercial airlines. Also, ozone, and natural radioactivity measurements were obtained. A special gamma sensing device, developed by the AEC, was tested by the RFF during this program.

In general, flights were conducted through the Inter-Tropical-Convergence-Zone (ITCZ) at or below the 700 mb level (about 3 km). High background noise on the aircraft precluded the acquisition of "clean" data from the AEC device, although other systems used did provide adequate data to successfully complete the primary mission objectives.

Some 200 air samples were collected by the RFF with a system devised by Dr. G. Ostlund of IMS. Samples were subsequently analyzed for carbon dioxide content, a parameter then used to study large-scale mass exchanges in the atmosphere.

B. Support of the Institute of Marine Science, University of Miami

For many years the RFF and NOAA's National Hurricane Research Laboratory (NHRL) supported research efforts of IMS. Extensive samples of liquid-water, precipitation and atmospheric water vapor were obtained and analyzed for tritium content with the following objectives in mind: (a) to study the air-sea exchange of water in hurricanes; (b) to investigate the possible occurrence of stratospheric subsidence into the hurricane eye; (c) to examine the natural modification of a continental air mass over the ocean; and, (d) to document the vertical distribution of tritium in the tropical atmosphere. The natural tritium was essentially used as an atmospheric tracer.

RFF flew many different flight patterns. In general, profiles were selected on the basis of the synoptic (meteorological) situation and program objectives.

1. In studying the air-sea exchange of water in hurricanes, long diametrical traverses of hurricane storm centers were accomplished. Data collection runs started well outside the storm envelop on one side of the storm, and were completed well outside the storm on the opposite side. If two or more traverses of the center were made, they were set up to eventually cross through all quadrants of the storm. These patterns were effectively employed in hurricanes Celia, Faith and Inez during 1966 and on two storms in one flight during 1967 (Chloe and Doria). Besides being a first in the hurricane research program (two storms investigated by the same aircraft on the same flight), we also were afforded the opportunity to investigate a coupled storm system (i.e., exchanges and other influencing factors).

2. During flights to investigate the possibility of stratospheric subsidence in the hurricane eye, sampling patterns employed consisted of simple orbits in the eye, well away from the eye wall.

3. Missions accomplished while studying the modification of a continental air mass over the ocean consisted of flying a series of box patterns at altitudes ranging from near the surface to 2 km, oriented along a line perpendicular to, and behind a cold front.

4. To investigate the vertical distribution of natural tritium in the undisturbed tropical atmosphere, two RFF aircraft sampled the atmosphere at levels from near the surface to approximately 200 mb (12 km). Each pattern at the varying altitudes was defined to cover the same geographical area. Information concerning the special liquid-water and water vapor sampling systems can be found in Friedman et al., 1970a.

C. In support of ESSA[1] Radiation Branch[2]

In 1966, RFF aircraft were requested to participate in three programs involving nuclear reactor experiments being conducted by the Westinghouse Astronuclear Laboratory and the Space Nuclear Propulsion Office at the Nuclear Rocket Development Station, Jackass Flats, Nevada.

Nuclear rocket experiments of the type considered, produce a radioactive effluent which, when released into the atmosphere, becomes mixed in a deep layer four to five kilometers thick, which extends upward from the surface and is transported for great distances by the prevailing winds. One of the functions of the ESSA group was to predict the trajectory and transport speed of the effluent and to estimate the downwind concentration of the radioactive fission products released to the environment.

Both the scale of this problem and the paucity of the routine meteorological observations in the western United States made the problem of predicting trajectories and concentrations and of evaluating the observed

---

[1] Now NOAA.
[2] From personal communication with Dr. Mueller

radiological data a difficult task. Meteorological sounding data obtained by the RFF aircraft in real-time was diagnostically useful in establishing post-facto trajectories of the effluent clouds. Radiological data from ground monitoring stations, was, at the time, quite sparse, and therefore, it was very important to determine the representativeness of each piece of information to evaluate the downwind concentration predication capabilities. RFF aircraft data, both radiological and meteorological, assisted in accomplishing this type of evaluation by establishing the aerial extent of the cloud, the vertical and horizontal distribution of radioactivity, and the wind and thermal structure of the environment affecting the cloud.

Measurements of peak concentrations of radiological activity were used to establish long range dillution rates. Data collected during this program was also used to study the effect of mountain waves on the distribution of radioactivity as these radioactive areas crossed mountain barriers.

D. In support of hurricane research

Small-scale tropical cloud systems and experiments using sulfur hexaflouride (SF_6) as an atmospheric tracer in hurricanes have been supported by the RFF working in conjunction with NOAA's National Hurricane Research Laboratory for several years. In 1969, hurricane Laurie was selected for an experiment with SF_6 designed to define the inflow field in a hurricane. This experiment, performed in conjunction with the IMS, was accomplished while the storm was located in the Gulf of Mexico.

One of RFF's aircraft flew a pattern approximately 75 nautical miles from the center of the circulation and dispensed the SF_6 gas along a track parallel to the storm's path. The aircraft flew at approximately 305 m absolute altitude. After dispensing the gas, the aircraft performed a circumnavigation of the storm, maintaining a 70 nautical mile radius.

Before and after the gas was dispensed, two other RFF aircraft accomplished monitoring patterns at several altitudes ranging from about 1500 to 5500 m, and collected air samples with a special air-intake device connected to their cabin pressurization systems. These samples were processed and analyzed by the Taft Engineering Center, Cincinnati, Ohio with the aid of highly sensitive gas chromatograph (better than 1 part in 10^{13}).

The results of this successful operational experiment show that atmospheric tracers (or pollutants that are present either naturally or introduced intentionally) can be used in a constructive manner for meteorological research. (In this case, SF_6 was used to delineate the inflow field in a hurricane). This pilot project also demonstrated the need for further research and development in the detection, measurement, and analysis of atmospheric pollutants (Friedman and Callahan, 1970a).

More recently, a portable gas chromatograph designed for real-time studies in diffusion was successfully tested by the RFF, NHRL and IMS during an airborne mission (Hawkins et al., 1972).

This program was accomplished as a direct result of the successful operation in hurricane Laurie. The opportunity of utilizing an airborne gas chromatograph to analyze samples in almost real-time was taken, and RFF's WC-130B aircraft dispensed a 10 nautical mile-long, low-level plume of SF_6 over an area northwest of Andros Island, Commonwealth of Bahama Islands. The accompanying RFF DC-6 aircraft, which carried the portable chromatograph, completed a total of 81 samples in a period of less than three hours. The system utilized in the test was not as sensitive as the larger ground-based unit at the Taft Engineering Center, which had a sensitivity of better than 1 part in 10^{13}. Actually, the airborne analyzer has a sensitivity of 3 parts in 10^{11}. It does, however, complete the analysis for SF_6 in less than two minutes after the sample is injected into the system. A confirmation of the in situ analyses was provided by a follow up laboratory analysis using grab samples obtained for this purpose.

While researchers (Hawkins et al., 1972) believe that the "portable airborne gas chromatograph is a viable instrument that has many potential meteorological applications", such as experiments in small scale cumulus dynamics, its use for hurricane dynamics may be limited. A system with greater sensitivity (such as 1 part in 10^{13}) in a laboratory "where conditions can be more rigidly controlled and enrichment techniques can be used, if necessary", would be more useful for hurricane and larger scale operations.

In keeping with the United States' commitment to environmental monitoring, RFF expects to participate in future experiments and programs in support of the Environmental Research Laboratories of NOAA and the scientific community at large.

FUTURE PLANS

The Research Flight Facility is presently engaged in feasibility studies of the state-of-the-art methods available for the measurement of pollutants in the atmosphere.

For our purposes we may classify pollutants as either gaseous or solid. The methods involved in the detection or measurement of these pollutants are spectral techniques for gases and collection techniques for solid materials.

An in-house program to improve the efficiency and reliability of present detectors is underway. The initial direction that we are following will lead us to techniques to trap larger volumes and at the same time improve mechanical filtering methods.

Spectral methods may be performed in two generally different ways, namely: on-board continuous sampling (grab sampling) using typically, absorption techniques. Another experimental method involves remote sensing of the scattering (Raman or resonance) properties of pollutants in the same volume illuminated by a high intensity source such as a laser. Absorption techniques such as remote sensing context from aircraft are probably unreliable due to background noise, optical alignment and other more fundamental problems.

The unique mission of the RFF is such that the research aircraft are involved in expeditions remotely located from areas where routine atmospheric monitoring occurs. It is important to make measurements in these otherwise data void areas on a continuing basis in order to document the incursion of man-made and natural pollutants and to study their effects on our ecosystem. These objectives are in accordance with the overall mission of NOAA.

ACKNOWLEDGEMENTS

The authors would like to acknowledge the many helpful comments and suggestions received during the preparation of this paper from the following fellow colleagues: Dr. G. Conrad, H. W. Davis, T. E. Apstein and R. D. Decker of RFF; A. Miller of NOAA's Experimental Meteorological Laboratory; and, R. Sheets and Dr. W. Scott of NHRL. Special thanks to Dr. H. Hawkins of NHRL and Dr. J. McFadden of RFF for reviewing this paper prior to presentation.

REFERENCES

1. Conrad, G., P. G. Connor and H. A. Friedman, 1972: The NOAA Research Flight Facility: Instrumentation Systems. (In press; to appear as part of the NOAA/ERL Technical Report Series).

2. Friedman, H. A., F. S. Cicirelli and W. J. Freedman, 1969a: The ESSA Research Flight Facility: Facilities for Airborne Atmospheric Research. ESSA/ERL Technical Report, ERL 126-RFF 1. August 1969, 89 pages.

3. Friedman, H. A., M. R. Ahrens and H. W. Davis, 1969b: The ESSA Research Flight Facility: Data Processing Procedures. ESSA/ERL Technical Report, ERL 132-RFF 2. November 1969, 64 pages.

4. Friedman, H. A. and W. S. Callahan, 1970a: The ESSA Research Flight Facility's Support of Environmental Research in 1969. Weatherwise, 23-4, August 1970, pages 174-181, 185.

5. Friedman, H. A., G. Conrad and J. D. McFadden, 1970b: ESSA Research Flight Facility Aircraft Participation in the Barbados Oceanographic and Meteorological Experiment. Bulletin of the American Meteorological Society, 51-9, pages 822-834.

6. Hawkins, H. F., K. R. Kurfis, B. M. Lewis and G. Ostlund, 1972: Successful Test of an Airborne Gas Chromatograph. JAM, 11-1, pages 221-226.

7. Lappe, U. O., 1969: Preliminary Analysis of the RFF Turbulence System. Final report to the RFF under ESSA contract nos. E 22-21-69(N) and E 22-75-69(N), 122 pages, April 1969.

8. McFadden, J. D., R. O. Gilmer and R. E. McGavin, 1970: Water Vapor Flux Measurements from ESSA Aircraft. Proceedings, Sumposium on Tropical Meteorology, Honolulu, Hawaii, pages B III 1-6, August 1970.

9. Reber, C. M. and H. A. Friedman, 1964: Manual of Meteorological Instrumentation and Data Processing. U. S. Weather Bureau, Miami, Flordia.

10. United States Code, 1970 Edition, Title 42--The Public Health and Welfare, chapter 55, National Environmental Policy, sec. 4321-4395.

RADIO WAVE MONITORING OF THE DEPTH AND THE SALINITY OF THE WATER TABLE

E. BAHAR

Electrical Engineering Department, University of Nebraska, Lincoln, Nebraska 68508

The extensive exploitation of the earth's natural reserves in recent years and the ever-present danger of their pollution and ultimate destruction, make it imperative to implement new effective methods to monitor repeatedly, the quality and quantity of our indispensable resources.

In this paper, we consider the feasibility of employing a radio wave method to monitor the undulations in the depth of the water table. Since the conductivity of water is critically dependent upon its salt content, we also examine the effects of variations in the salinity of the water table upon radio wave propagation.

On the basis of a full wave solution to the problem of radio wave propagation over nonuniform stratified media [1-5], it is shown that it is necessary to operate in the low to medium frequency range (0.3 M Hz), in order to distinguish effectively between undulations in the overburden depth and variations in the complex permittivity of the overburden. The full wave analysis, which also accounts for the propagation of surface waves, is particularly suited to problems in which the transmitter and receiver are near the earth surface. It is shown that these full wave solutions satisfy the reciprocity relationships in electromagnetic theory.

Recent computation of the electromagnetic fields propagating over a nonuniform overburden (see Fig. 1) indicate that the scattered fields are very sensitive to small changes in the gradient of the substratum (the water table). Furthermore, these

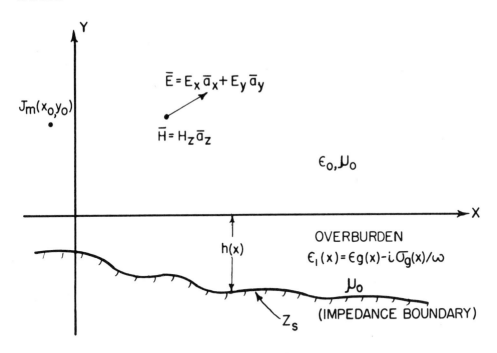

Fig. 1

Nonuniform overburden of depth h(x)

computations indicate that the primary scattered fields are approximately proportional to the Fresnel reflection coefficient at the overburden-substratum interface. Hence, the amplitude of the scattered field is also dependent upon the conductivity of the substratum (particularly for the low frequency range). Thus, variation in the salinity of the water table could also be deduced from measurements of the scattered field.

As a result of the finite conductivity of the overburden, the radiation patterns of the scattered electromagnetic fields are also very sensitive to the average overburden depth. The relative amplitudes of the side lobes decrease more rapidly as the average depth of the overburden increases. Using the full wave analysis, it is also possible to determine the effects of the roughness of the overburden-substratum interface upon the scattered radio waves.

The full wave technique used in the analysis of the problem of propagation over nonuniform layered medium, is also very useful in resolving several difficulties that had arisen in

earlier solutions based on geometrical-optical methods [5,6]. An
interesting feature uncovered in these earlier solutions to the
problem, is that the equivalent surface impedance (that is used
to characterize the air-overburden interface) is not only a
function of the coordinate along the surface of the structure, but
also depends upon the direction of propagation of the incident
wave. The surface waves, which were disregarded in the geometrical-optical solutions, play a very important role in the
full wave solution of this problem. The excitation of these
surface waves by the nonuniform layered structure, is critically
dependent upon the direction of propagation of the incident wave.
In addition, there is a distinct difference in the scattered
field radiation patterns due to incident plane waves of grazing
incidence and incident surface waves.

Other limitations of the geometrical-optical technique
involve the finite conductivity of the overburden, the distance
between the receiver and the transmitter and the frequency of the
radio waves.

The removal of these severe limitations to the analysis of
radio wave propagation over nonuniform geological structures,
should contribute considerably to resolving the complex problem
of remote sensing of nonuniform and inhomogeneous environments.

The advantage of using radio wave methods for monitoring the
changes in the depth and salinity of the water table is that
field measurements can be repeated frequently since they are
conducted above the surface of the earth and not in test wells.

References

1. Bahar, E., (1970),"Propagation of Radio Waves Over a Nonuniform Layered Medium," Radio Science, Vol. 5, No. 7,
pp. 1069-1076.

2. Bahar, E., (1970),"The Relationship Between Geometrical-
Optical and Full Wave Solutions to the Problem of Propagation
Over Non-Parallel Stratified Media," Journal of Mathematical
Physics, Vol. 11, No. 9, pp. 2764-2770.

3. Bahar, E., "Radio Wave Method for Geophysical Prospecting,"
Journal of Geophysical Research, Vol. 76, No. 8, pp. 1921-1928.

4. Bahar, E., (1971), "Radiation From Layered Structures of
Variable Thickness," Radio Science, Vol. 6, No. 11, pp. 1109,
1116.

5. Bahar, E., (1971), "Radio Wave Propagation Over a Nonuniform Overburden," chapter in Electromagnetic Probing in Geophysics, editor J. R. Wait, Golem Press, Boulder, Colorado.

6. Schlak, G. A., and J. R. Wait, (1967), "Electromagnetic Wave Propagation Over Nonparallel Stratified Conducting Medium," Canadian Journal of Physics, Vol. 45, pp. 3697-3720.

7. Schlak, G. A., and J. R. Wait, (1968), "Attenuation Function for Propagation Over a Nonparallel Stratified Ground," Canadian Journal of Physics, Vol. 6, pp. 1135-36.

The research reported in the paper was supported by the National Oceanographic and Atmospheric Administration, National Science Foundation and the Engineering Research Center of the University of Nebraska.

REMOTE MEASUREMENT OF AIR POLLUTANTS UTILIZING THE RAMAN EFFECT

S. LEDERMAN AND M. H. BLOOM

Polytechnic Institute of Brooklyn, Farmingdale, New York U.S.A.

ABSTRACT

The application of Raman scattering to air pollution detection monitoring and measurement is considered. Some theoretical as well as practical aspects of the problem are discussed. Positive as well as negative features of the Raman scattering diagnostic techniques are pointed out. The problems associated with the utilization of lasers, the power requirements, frequency selection, pulse repitition rate and reliability are discussed. It is concluded that the diagnostic technique utilizing Raman scattering has sufficient outstanding features to make it particularly suitable for the task at hand.

I. INTRODUCTION

It is generally recognized that pollution of the environment on a global scale represents one of the most serious hazards confronting modern society. This pollution ranges from poor quality air in most urban communities to increasingly contaminated rivers and lakes everywhere, from nerve-wracking noise levels in many cities and towns to increasingly contaminated oceans and seashores. From recent innumerable pronouncements of ecologists, it became clear that means and methods have to be found to eliminate or at least reduce environmental pollutants generated by our advanced technological civilization if life as we know it is to survive. To accomplish this, methods and techniques have to be devised to diagnose, measure, and monitor contaminants, not only in the environment, but also at their source of generation. In this work, the problem of air pollution and in particular the aspect of measurement monitoring and control is being examined. Based upon surveys of available methods and techniques, it is clearly of extreme importance to have a method which would permit a remote quantitative analysis of air samples which are not easily accessible, a method which would permit three-dimensional mapping of concentrations of a number of constituents of the atmosphere simultaneously, continuously, and instantaneously. Such a method may be possible utilizing the Raman effect. As pointed out later, the diagnostic method based upon the Raman effect could provide several outstanding features. Among these would be the single-ended remote sampling ability, simultaneous and almost instantaneous identification of specific components of a mixture, quantitative analysis of a mixture, and the ability of spacial and time resolution of concentration of given components of a mixture. In this work an attempt is being made at evaluating the feasibility of obtaining quantitative data of air pollutants utilizing the Raman effect.

II. THE RAMAN EFFECT

The Raman effect, known for the last half a century, has been discussed extensively in the literature. This effect has been treated by classical as well as quantum mechanical methods and is well-documented both theoretically and experimentally. The reader might be referred to very extensive references, and in particular, to Refs. [1-5] which contain further references. However, for the sake of completeness, a short qualitative description of this effect is given here, including some of the features which make this effect particularly suitable for this special application.

When a photon of arbitrary energy encounters a molecule, the resulting collision may be of the elastic or inelastic kind. In both cases, the molecule undergoes a double transition, by first absorbing and then emitting a photon. In the case of an elastic collisions, the emitted photon is of the same energy as the absorbed photon, and the resulting scattered radiation is of the same frequency as the incident radiation. In the case of inelastic collision, the energy of the re-emitted photon is changed resulting in a change in frequency of the re-emitted radiation. The former kind of scattering is known as Rayleigh scattering, while the latter as Raman scattering. The frequency shift of the scattered radiation from the incident radiation frequency is a unique characteristic of the scattering molecules and is independent of the incident radiation frequency. This property of the Raman effect affords not only an excellent possibility of distinguishing between different species of a gas mixture, but by measuring the scattered radiation intensity of each component, permits one to determine the concentration of each diatomic or polyatomic specie in the gas mixture, providing the species are Raman active. As is well-known, the Raman frequency displacements agree exactly with the frequencies of the vibrational bands in the near infrared for those gases for which both have been observed. The Raman spectrum can, therefore, be regarded in those cases as the infrared spectrum shifted into the visible or ultraviolet region. It must, however, be noted that not all Raman active molecules are infrared active, and not all infrared active molecules are Raman active.

At this point, it is worthwhile noting that all homonuclear diatomic molecules like H_2, N_2, O_2 are infrared inactive, while they are all Raman active. This is one of the very important properties of the Raman effect and makes the application of the Raman effect to monitoring and measuring of air pollutants feasible, by providing a known reference for calibration purposes as will be discussed later.

The correlation between the incident radiation intensity I_o and the Raman scattered intensity, I, is given according to the polarizability theory of Placzek, including the factors introduced by the optics of the system as a function of direction θ for the Stokes and Anti-Stokes lines by

$$\frac{I_{s,A,\theta}}{I_o} = C_1 \cdot n \frac{(\nu_o \mp \nu)^4 \left\{ (45\alpha'^2 + 7\chi\gamma'^2)(1+\cos^2\theta) + 6\gamma'^2 \chi \sin^2\theta \right\} C_2}{\nu(1-\exp[-\frac{hc\nu}{kT}]) R^2} \qquad (1)$$

Equation (1) reduces to

$$\frac{I_s}{I_o} = C_1 \cdot n \frac{(\nu_o - \nu)^4 \cdot 2(45\alpha'^2 + 7\chi\gamma'^2)}{\nu(1-\exp-\frac{hc\nu}{kT}) R^2} C_2 \qquad (2)$$

for backscattering (i.e., $\theta = 180°$) or for vertically polarized incident radiation and transverse observation.

The ratio of the Stokes and Anti-Stokes line intensity is given, using the Boltzmann distribution for the molecules at the appropriate energy levels by

$$\frac{I_s}{I_A} = \frac{(\nu_o - \nu)^4}{(\nu_o + \nu)^4} \exp\left(\frac{hc\nu}{kT}\right) \qquad (3)$$

and the ratio of intensity of the Stokes and Rayleigh scattered radiation by

$$\frac{I_s}{I_R} = \frac{(\nu_o - \nu)^4}{\nu_o^4} \frac{(45\alpha'^2 + 7\chi\gamma'^2)}{(45\alpha^2 + 7\chi\gamma^2)} \frac{h}{8\pi^2 \mu \nu} \qquad (4)$$

The temperature dependence of the Raman intensity as indicated by the exponential in the denominator of Eq. (1) is shown in Fig. 1.

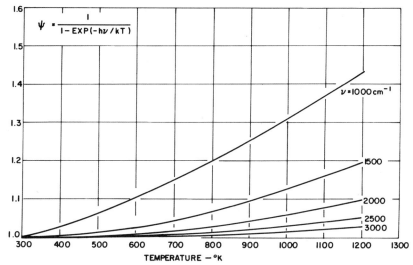

Fig. 1 - Temperature Dependence of the Raman Intensity. Raman Frequency Shift as a Parameter.

The ratio of the Stokes to Anti-Stokes intensity for several species of interest is shown in Fig. 2 and the ratio of the

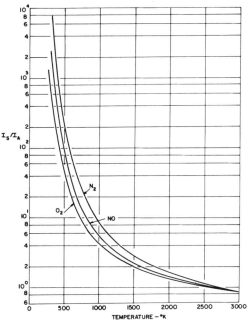

Fig. 2 - Ratio of the Stokes to Anti Stokes Intensity as a Function of Gas Temperature.

Rayleigh to Stokes intensity for a number of species is shown in Table I.

TABLE I

STOKES TO RAYLEIGH INTENSITY

Specie	$I_S/I_R \cdot 10^{-3}$	
	Theory	Experiment
O_2	1.9	1.93
N_2	1.32	1.19
CO_2	1.13	0.57
CH_4	3.29	2.92

From these it is apparent that:

  1. At room or normal atmospheric temperature, the Stokes-Raman lines are almost independent of the temperature.
  2. At normal atmospheric temperatures, the intensity of the Anti-Stokes Raman lines is very much smaller than the Stokes lines.
  3. The Rayleigh line intensity is much higher than the Stokes Raman line. This will require in any application of the Raman scattering technique the utilization of band elimination filters to stop the Rayleigh line.

One of the very important features of the Raman scattering diagnostic techniques is the ability to utilize it for range and depth resolution. The former can be accomplished by utilizing radar techniques, the latter by laser pulses and proper gates following the detector circuitry. The depth resolution ΔD can be presented as

$$\Delta D = \frac{c}{2}(t_o + t_g) \quad (5)$$

Since, as indicated above, the Raman effect is almost instantaneous, the depth resolution appears to be governed by the laser pulse duration t_o and gate width t_g.

Some of the more outstanding features of the Raman effect, making it suitable for the task of monitoring, measuring, and ranging of air pollutants remotely and instantaneously can thus be summarized as follows:

  1. Raman scattering is very specific. Each diatomic or polyatomic Raman active specie has its own characteristic frequency shift.

2. The Raman frequency shift is independent of the illuminating frequency. This permits the choice of a convenient illuminating frequency of the incident light.

3. The Raman spectrometric system can be made single-ended; that is, the transmitter and receiver can be located in close proximity to each other.

4. The Raman diagnostic technique can be utilized for range finding. Here the highly developed radar techniques appropriately modified can be utilized. This, together with the specificity and depth resolution, permits one to determine a three-dimensional distribution of a given specie or number of species simultaneously.

5. The Raman scattered radiation from a given specie is essentially independent of the presence of other species in the volume under diagnosis. The rescattering being predominantly of the Rayleigh kind, which again accounts for a very small fraction of the incident radiation, can, for the most part, be neglected or at most, a small correction factor introduced.

6. Since the Raman effect takes place in a time of the order of 10^{-14}, 10^{-15} sec. for an illuminating frequency in the visible of near u.v, any product of relaxation, chemical reaction or chemical production of a transient nature can be followed utilizing the Raman effect.

III. THE MAJOR AIR POLLUTANTS

It is generally accepted that air pollution in its broader form is a consequence of life itself. Any living creature by its mere existence contributes to the pollution of its environment. Air-breathing creatures, including man, continually add to the pollution of the air. Nature, to sustain life, must be self-cleaning. A substantial part of the pollutants, natural or man-made, are dispersed, dissolved, react among themselves and with others, chemically or physically and are thus removed from the atmosphere, either by reaching a sink such as the oceans or receptors such as plants, animals or man. As the production of pollutants increases with the population expansion and increased industrial and technological activities, the natural process of self-cleaning is not sufficient, resulting in a steadily deteriorating air environment.

As is well known from recent publications, Ref. [6], the content of the major air pollutants is substantially higher than the air which is classified as clean. In some instances, some of these major air pollutants approach on the average the level where they may have adverse effects on life and vegetation. Under special, unfavorable weather conditions, local concentrations may approach the emergency level, that is, the level at which it is likely that sickness or death in sensitive groups of persons may occur. The major contributors to air pollution are, according to

Refs. [6] and [7] the automobile with 60%, industry with 17%, power generation with 14%, and the rest with 9% of the emitted pollutants. The above is of necessity only in a very superficial review of the subject. For more detail, [6] and [7] should be consulted, which provides a substantial list of available literature on the subject.

IV. DESIGN CONSIDERATIONS OF THE APPARATUS

The abundance of components in the air which qualify as pollutants varies over a wide range. The presence of some of these components are counted in terms of hundreds of particles per million (CO_2) and are considered harmless, whereas others may become harmful when their presence amounts to .25 particles per million (NO_2 or NO).

The apparatus must, therefore, be capable of recognizing and measuring large as well as small concentrations of pollutants in the atmosphere. Since the apparatus has to function in the atmospheric environment, the atmospheric propagation properties must be considered. A complete description of the atmospheric effects on laser signal propagation is beyond the scope of this work. As is well-known, atmospheric attenuation can be described by the exponential law of attenuation. Thus the atmospheric transmittivity can be expressed as $\tau_a = \exp(-\alpha_a R)$ where τ can be expressed as the product of the absorption and scattering transmittivities, $\tau_a = \tau_b \cdot \tau_s$. Since the atmospheric absorption is generally due to the molecular constituent such as water, carbon dioxide, ozone, etc., it varies drastically with the wavelength of the illuminating light, in conformity with the molecular absorption of the given constituent. There are certain frequency bands which are completely absorbed and thus unsuitable for atmospheric probing. In selecting the operating wavelength for the apparatus in question, this feature must be considered.

The attenuation due to scattering, although frequency-dependent, can be described analytically and is given by [8] as

$$A = A_o \exp(-R(\beta_o + \beta)) \tag{6}$$

where β, the scattering coefficient, is given by

$$\beta = \frac{32 \pi^3}{3n \lambda^4} (m-1)^2 \tag{7}$$

and n equals the number of scatterers per unit volume; m denotes the index of refraction of the medium.

In Eq. (6), the attenuation of the transmitted beam is given by β_o and the attenuation of the Raman shifted radiation by β. In Fig. 3, a schematic diagram of the transmittivity of the

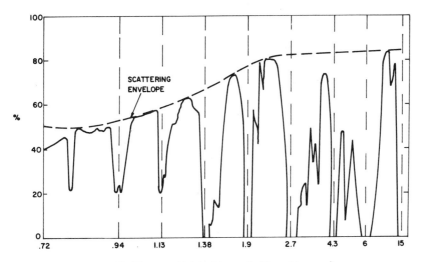

Fig. 3 - Transmittivity of the Atmosphere.

atmosphere as a function of wavelength is shown. Here, several absorption bands as well as the scattering envelope is shown. It was found experimentally that the atmospheric attenuation coefficient due to scattering is also strongly affected by the time of the year as well as the general visibility conditions. This is shown in Fig. 4, where the attenuation coefficient is

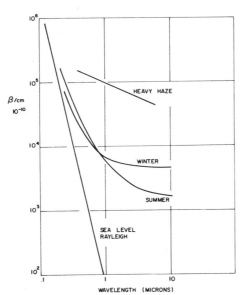

Fig. 4 - Atmospheric Scattering Coefficients for Several Conditions as a Function of Wavelength.

plotted against wavelength for several atmospheric conditions.

The temperature fluctuations in the atmosphere causes also pockets or sheets of warmer, less dense air to mix turbulently with surrounding cooler air. Since the index of refraction of air depends upon its temperature, the interaction of a laser beam with a turbulent atmosphere leads to random amplitude and phase variation of the laser beam. The consequences of atmospheric turbulence on laser propagation may be summarized as follows, [11]:

a) Beam steering-angular deviation of the beam from the line of sight path.
b) Variation in the beam arrival angle.
c) Beam spreading-small angle scattering increasing the beam divergence and decreasing the power density of the target.
d) Beam scintillation-small scale destructive interference.
e) Spatial coherence degradation.
f) Polarization fluctuations.

The above features would seem to present unsurmountable difficulties. This, however, is not so. The utilization of a parallel channel for normalization does provide a solution.

Inspection of Eq. (2) reveals that the scattered signal is proportional to:

1. the number of scatterers in the volume being diagnosed;
2. the 4th power of the shifted frequency of the primary radiation;
3. the intensity of the primary radiation;
4. the inverse square power of the distance from the scatterers to the receiver;
5. the particular constants of the species under investigation;
6. the optics of the apparatus and electronics involved in signal processing as expressed by the constant of the system C_2 which includes also the effect of attenuation of the Raman signal on its way to the receiver.

The desired signal at the end instrument as indicated in Eq.(2) depends, among others, on the constant C_2 which encompasses a variety of factors. It is quite obvious that the signal received will be proportional to the fraction of the scattered photons received, the atmospheric attenuation, the losses of the optics involved, the gain of the phototubes and the electronic processing of the signals received. For a more detailed discussion of these factors, Ref. [12] may be consulted.

As far as the optical losses are concerned, one must design the system such that the product of the Raman intensity, area of receiving mirror and efficiency of the receiving optics is greater than the minimum detectable signal, where the minimum detectable signal is defined as the signal for which the signal-to-noise ratio is equal to 1. Thus, if the noise signal is minimized, the detectable signal can be proportionately decreased.

Since the signal-to-noise ratio can be improved by averaging over a number of pulses, one would tend to design the equipment on a repetitive basis, that is multipulse operation. In that case, the signal-to-noise ratio can be written

$$\frac{S}{N} = \frac{n_p i_s^2}{2e\Delta f(i_s+i_d+i_b)} \qquad (8)$$

This equation would indicate that by increasing the number n of averaged pulses, any signal-to-noise ratio is possible to achieve. This is, of course, limited by the averaging time one can use, and the signal processing electronics. Since, in most modern photomultipliers the dark currents are very small (10^{-14}), the limiting noise is the background current, i_B. With this in mind, the signal power P_s must be

$$P_s \geq \frac{(2e\frac{\Delta f}{n}(\frac{S}{N})_{min} i_B)^{\frac{1}{2}}}{S_{ph}} \qquad (9)$$

for background limited operation, where S_{ph} is the phototube sensitivity. Equation (9) indicates that for daytime operation, the required scattered power or signal power must be larger than for night time operation.

An examination of the frequency dependence of the scattered radiation (Ref. [12]) as a function of the Raman shift for a Ruby and N_2 laser as the primary illuminators reveals that the scattered intensity R not only decreases with the increase in the Raman frequency shift of a given specie, but also that the decrease in intensity is more pronounced for the lower frequency primary radiation source. Whereas the value R for the N_2 laser decreases by a factor of about 15 for a Raman shift between 500 and 3500 cm^{-1}, it decreases by a factor of about 30 for a ruby laser for the same Raman shift interval. This fact makes it desirable to move the primary source of radiation into the u.v. region. There are, however, limitations on the useable frequency imposed not only by the scarcity of high intensity sourses in this region but also by the increase of the scattering coefficient at these frequencies, which enormously increases the atmospheric attenuation of the scattered signals received, Ref. [12].

In Fig. 5, the relative intensity of a Raman signal obtainable using a nitrogen and a ruby laser is shown. An inspection of this figure indicates clearly that the overriding feature here is still the fourth power dependence of the scattered radiation on the frequency. With the increasing distance of the scattering sample,

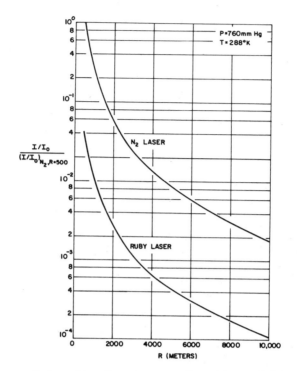

Fig. 5 - Raman Intensity for O_2 as a Function of Distance. Normalized by the Relative Raman Intensity of a N_2 Laser at 500m.

the attenuation at higher frequencies increases and eventually becomes dominant. One must note that the intensity in Fig. 5 was obtained on the assumption that the primary intensities of both lasers were equal. As is well-known, at the present time, ruby lasers capable of delivering several hundred megawatts in a single pulse operation are available as standard off-shelf units, while nitrogen lasers of only several hundred kilowatts can be commercially obtained. For this reason, in some applications it might be advantageous to use a ruby laser with its lower radiation frequency, rather than a nitrogen laser with its higher frequency. The choice of the mode of operation, that is, single pulse or repetitive, would greatly influence the choice of the laser. Whereas a nitrogen laser is capable of operating at a rate of several hundred pulses/sec., a ruby laser is capable of a single pulse or, at most, several pulses/min. operation. Consistent with the requirements for an improved signal resolution as indicated above, the multipulse operation possible with the nitrogen laser is indicated. However, this kind of operation would provide a time average signal.

While it might be advantageous for a number of reasons to operate at a higher frequency in certain applications, other considerations could result in the choice of lower operating frequencies. An inspection of Fig. 5 or Table II and III reveals

TABLE II

EQUIVALENT RAMAN SCATTERING CROSS-SECTION

	Ruby	N_2
O_2	4.72×10^{-30}	1.08×10^{-28}
N_2	2.85×10^{-30}	7.47×10^{-29}
CO_2	6.57×10^{-29}	1.45×10^{-27}
NO	7.2×10^{-31}	3.2×10^{-29}

TABLE III

RAMAN WAVELENGTH OF SOME MOLECULES CORRESPONDING TO RUBY AND N_2 LASERS AS PRIMARY ILLUMINATORS

	NO	NO_2	H_2O	CO_2	CO	N_2
λ_s Å Ruby Laser	7982	7643 7324 7821	9305 7807 9392	7683 7623	8156	8283
λ_s N_2 Laser	3597	3527 3457 3564	3844 3561 3858	3535 3523	3632	3657

that from the point-of-view of signal strength at short distances the N_2 laser is superior, providing a laser of appropriate power is available. If, however, one is interested in field monitoring of some particular species, a size reduction of the apparatus would be desirable. Replacement of the spectrometer with filters would require a better line separation of the Raman scattered radiation. As evident from Table III, a lower operating frequency provides better line separation. While, for example, the line separation of NO from CO is 35Å with a nitrogen laser, it is 174Å for a ruby laser. The wider line separation when using a lower frequency illuminating source (ruby) permits the utilization of filters of reasonable bandpass. This feature in conjunction with the higher pulse power available could, in certain cases, be advantageous.

Finally, the question of eye safety must be considered in choosing a laser for air pollution monitoring. In this respect, the higher frequency N_2 laser might be considered safer than the lower frequency ruby laser, Ref. 12 . It is generally assumed that the high frequency limit of ocular transmission is about 25000 cm^{-1}. This limit, however, is not absolute. There might still be some transmission at this or higher frequency to cause injury, in particular, at high power densities and direct, close observation. The opacity of the eye to high frequency radiation again suggests the superiority of the N_2 laser over the ruby laser. In any case, provisions must be made to avoid direct or indirect exposure to the intense radiation in question.

The choice of the laser will thus have to be made in each case consistent with the given application and with the proper consideration of the previously indicated pro and con features.

From the above considerations, it may be concluded that an air pollution monitoring and measuring apparatus can be constructed which would consist essentially of the following major components:

1. A high powered laser source capable of stimulating Raman scattering at a remote sample.
2. A high powered telescope capable of receiving radiation from a remote source.
3. A spectroscope or narrow bandpass filter capable of sorting and transmitting the different signals received.
4. Signal processing and recording system.

The last component of the pollution monitor consists of a photomultiplier tube power supply and electronic processing and recording equipment. The complexity of the system depends on the applications intended.

Figure 6 shows a photographic view of an experimental apparatus

Figure 6

constructed in our laboratory. This apparatus uses a Q-switched ruby laser capable of delivering a pulse of 100 megawatt peak power. The pulse duration of 10 nanoseconds is sufficiently short to permit a relatively good depth resolution. The receiving telescope consists of a 30 cm diamter parabolic mirror of 3.6m focal length. The system utilizes a Jarrel Ash 0.5m monochromator terminated by a RCA 8853 photomultiplier. The signal can be directly recorded on an oscilloscope or processed by the electronic data processing equipment as seen in the photograph. The whole system is mounted on a movable platform where angle and elevation can be adjusted.

While the ideal mounting of the transmitting and receiving optics would be concentric, this experimental apparatus was constructed utilizing the available equipment and due to the lack of an appropriate Cassegrainian telescope, is not concentric. As indicated above, there are a number of difficulties associated with correlating the received signal and the actual specie concentration at a given range. These may change from day-to-day or hour-to-hour according to the atmospheric conditions, time of day and season of the year. To overcome these difficulties, a parallel channel is generally introduced which is monitoring a known component of the atmosphere, and normalizing the measured specie with respect to this component. This known monitored component may be nitrogen. In case of a single pulse operation, the received signal can be processed using a beam splitter and two separate photon processing channels. In multipulse operation the same procedure may be followed or a single photon processing channel may be employed and the sample can be alternated between N_2 and the specie of interest.

V. CONCLUDING REMARKS

It is quite evident from the above discussions that a pollution monitoring, measuring and detection device utilizing the Raman effect is a practical device and could be constructed. While short distance, remote, single-ended measurements of pollutants of high concentration can be made at present with the available shelf equipment, low concentrations of particular species at greater distances might require further developments not only in lasers, photomultipliers and processing electronics, but also in the field of the so-called "Resonant Raman Effect". The resonant Raman effect provides signals several orders of magnitude higher than the normal Raman effect. These could effectively increasing the sensing distance or decrease the concentration of the pollutant species, without adversely affecting the minimal signal-to-noise ratios required. Preliminary results obtained in our laboratory are very encouraging. However, the data obtained up till now are insufficient to be conclusive.

REFERENCES

[1] G. Herzberg, <u>Infrared and Raman Spectra of Diatomic Molecules</u>. D. Van Nostrand Co., Inc., New York (1945).
[2] G. Herzberg, <u>Infrared and Raman Spectra of Polyatomic Molecules</u>. D. Van Nostrand Co., Inc., New York (1945).
[3] S-I Mizushima, <u>Raman Effect</u>. Handbuch der Physik, <u>XXVI</u>, Springer-Verlag (1958).
[4] G.F. Widhopf and S. Lederman, AIAA Journal, $\underline{9}$, 309 (1971).
[5] S. Lederman, <u>Molecular Spectra and the Raman Effect. A Short Review</u>. Polytechnic Institute of Brooklyn, PIBAL Rept. No. 71-15 (1971).
[6] <u>Cleaning Our Environment</u>. A Report of the American Chemical Society, Washington, D.C. (1969).
[7] Air Conservation Publication No. 80 of the American Association for the Advancement of Science, Washington, D.C. (1965).
[8] L.R. Kohler, <u>Ultraviolet Radiation</u>. John Wiley and Sons, Inc., New York (1965).
[9] E.J. Stansbury, M.F. Crawford, and H.L. Welsh, Canadian J. of Phys., $\underline{31}$ (1954).
[10] H.W. Schrotter and H.J. Bernstein, J. Molecular Spectroscopy, $\underline{12}$, 1, (1964).
[11] W.K. Pratt, <u>Laser Communications Systems</u>. John Wiley and Sons, Inc., New York, (1969).
[12] S. Lederman, <u>The Application of the Raman Effect to Remote Monitoring of Air Pollutants</u>. Polytechnic Institute of Brooklyn, PIBAL Rept. No. 71-29 (1971).

ACKNOWLEDGMENTS

This research was partially supported through the Center for Urban Environment Studies at P.I.B. and partially by the New York State Science and Technology Foundation under Grant No. SSF(70)-4. The authors wish to acknowledge the cooperation of Dr. P. Khosla and Mr. J. Bornstein in preparing this work.

REMOTE SENSING OF POLLUTANTS BY MEANS OF STEREO ANALYSIS

W. Z. Sadeh and J. F. Ruff

Colorado State University

Air pollution levels depend strongly upon the spatial and time variations of various pollutant concentrations. In the vicinity of pollution sources locally high concentrations and significant variation with time of pollutants occur. Surveillance of pollutant concentrations and dispersion requires adequate monitoring over large relevant areas at fixed time intervals. Particularly, strong and stable temperature inversion can cause such circulation conditions where vertical mixing of pollutants is prohibited.

Remote sensing by means of stereo images obtained from flown cameras and scanners provides the potential to monitor the dynamics of pollutant mixing over large areas. Moreover, stereo technology permits the efficient monitoring of the pollutant concentration and mixing in sufficient detail. Consequently, regional standards on air quality can be set forth. Furthermore, methods to detect unpredicted and significant pollution variations can be developed. A method of remote sensing using stereo images is described. Preliminary results based on comparison with ground measurements by alternate methods, e.g., remote hot-wire anemometer techniques, are supporting the feasibility of stereo analysis using aerial cameras.

INTRODUCTION

Broadly, changes in the constitutents of air either by addition to and/or subtraction from them which affect the physical and/or chemical properties of the air, and subsequently cause detectable deterioration of air quality may be defined as air pollution.

Pollutants include any natural and artificial, i.e., man-made, contaminants capable of altering air properties and being airborne. Air pollutants occur in the form of gases, liquid droplets and solid particles, both separately and in mixtures. Gaseous pollutants constitute approximately 90% of the total mass entering the atmosphere. The other 10% are made up of particulates and liquid aerosols [1]. Pollutants emitted directly from identifiable sources are classified as primary pollutants. On the other hand, pollutants produced by interactions among primary pollutants or by reactions with atmospheric constituents are categorized as secondary pollutants [2]. These pollutants are hazardous and particularly health endangering. By and large, the most harmful air pollution is due to the production of acid droplets through the interaction of sulfuric and nitric oxides with humid air. These oxides evolve primarily from combustion of fossil fuel and refuse.

A complete description of polluted air masses is quite difficult since generally the polluting entities do not retain their exact identities after entering the atmosphere. The pollutants undergo thermal and photochemical reactions, and continuous spacial and time variations. In the vicinity of polluting sources locally high concentrations and significant changes with time occur. Subsequent dispersion depends strongly upon the prevailing winds, atmospheric stability and topographic features. Levels of pollutants are determined in terms of areal concentration density, i.e., concentration per unit area. Similar health hazards can arise from short time exposure of order of few minutes to locally high concentrations, or in the case of long time exposure of hours and even days but to low concentrations.

Detection of pollution through direct sampling is rather complex since a great number of sampling stations distributed over large areas are needed. In United States there are over 7,000 sampling stations [1]. Most of these stations are located in the vicinity of major point sources such as high smoke stacks, and in metropolitan and industrial areas, i.e., in area sources. To monitor the spacial and time variations of pollutants in order to obtain contour maps of trace constituents, surveys within relatively large distances downstream of the sources are required. For instance, aerial photographs of power plants reveal that the most dangerous surface pollution concentrations occur within 4 to 8 km downwind of the plants [3]. Surveys of plume rise and dispersion over such large areas using sampling stations are costly and inherently difficult due to the continuous changing meteorological conditions. Moreover, the use of sampling stations to monitor line sources such as highways and runways, mobile sources, and unpredictable downwind precipitation areas is practically precluded due to the extreme range of area involved and concentration changes.

In many cases, furthermore, the set up of sampling stations is prevented by local safety regulations, e.g., set up of tall meteorological towers in vicinity of airports.

Pollutants are deposited into the atmospheric boundary layer where they are conveyed either as active or passive scalars by advection and turbulent diffusion. In the atmosphere the turbulent diffusion depends on both shear flow and vertical temperature distribution. The buoyancy forces, which are determined by the temperature variation, play a prime role in the dynamics of air motion. When stable conditions prevail, i.e., subadiabatic temperature lapse rates, the turbulent diffusion of contaminants is retarded by buoyancy forces. The extreme case of stable conditions is represented by temperature inversions. Strong and stable inversion layers yield extreme circulation conditions where vertical turbulent mixing of pollutants is effectively prohibited by the temperature lid. Dangerous inversion which can persist for several days usually develop within high pressure ranges. Over urban areas temperature inversions are commonly found beyond a certain mixing depth [4,5]. Generally, the urban inversions arise in the form of a dome-shaped envelope due to the uneven heat distribution. Furthermore, on clear mornings at sunrise and often on evenings at twilight with light winds or calm conditions near the surface, a base inversion layer occurs. This layer may extend from the ground up to heights of 150 to 750 m depending upon the local topographic configuration [6]. Simultaneously, a second inversion layer may exist at higher altitudes [6]. Inside the inversion layer the contaminants are uniformly distributed. Usually, highest surface concentrations occur within 5 to 10 km downwind of the sources and can persist for several days.

Air pollution surveys are strongly dependent on detailed knowledge of prevailing turbulent winds and atmospheric stability conditions within relevant large areas. Remote-sensing technology possesses the potential to monitor both meteorological parameters and needed data on the dynamics of mixing of pollutants in sufficient detail and within substantial large areas. The orbital stereo photographs taken during the Gemini missions clearly substantiate the use of stereo techniques to survey contaminated air masses over exceedingly large areas [7]. Furthermore, advance detection of unexpected pollution is feasible. As a result, adequate regional air quality standards and control regulations can be set forth. However, to determine chemical composition of pollutants, size and distribution of particulates, health endangerment levels and so on, direct sampling is needed. Remote sensing by means of stereoscopic observations is capable of yielding a host of information which can augment manifold the knowledge on the dynamics of mixing of pollutants and surface concentrations.

ANALYSIS OF STEREOSCOPIC IMAGES

Stereoscopic viewing of targets essentially reduces to stereo comparison of a pair of images with some overlapping areas. Such two successive photographs can be taken by a precision aerial camera along the flight path of an aircraft. Two pictures taken in succession and the corresponding overlapping areas indicated by shaded regions I and II are shown in Fig. 1(a). Next, two arbitrary objects are selected within the area of overlap. They are denoted by 1 and 2 in the first photograph while their conjugate images in the second picture are designated by 1' and 2', respectively. These objects can be a sharp ground feature and a tenuous but visible polluted air mass as portrayed in Fig. 1(b). Under the assumption of unmovable pollution target, the height and three-dimensional extent of the plume can be determined by stereoscopic comparison of the two images [8]. Thus, the time lapse between two successive exposures must be short enough such that the distance traveled by the pollution target can be disregarded. The human matching of the stereo images is carried out using a stereocomparator. Subsequent attempts to align the ground feature and pollution target will lead to a mismatch for the former. This misalignment is caused by the fact that distances 1-2 and 1'-2' are not equal since the two objects are situated at different heights. The distance difference, i.e., the parallactic distance, can be estimated from the mismatch of the ground reference object as illustrated in Fig. 1(b). Then, the height of the pollution target is computed by triangulation utilizing the parallactic distance and the difference in the view angles from the two images for the pollution target (the parallactic angles). This method can also be used to estimate the features of an inversion layer. When the motion of the polluted mass is not negligible, a third stereo image is needed. Then, two independent parallactic distances are obtained. Similar triangulations yield both plume altitude and its horizontal velocity.

Results using this stereo analysis to deduce the three-dimensional extent of visible plumes are reported in Ref. 9. The three-dimensional extent of plumes were estimated using vertical and oblique stereo image pairs with 60% overlap taken at altitudes of 10,000 and 15,000 ft, respectively. The accuracy of this method decreases drastically with distance from the source. At 3 miles downwind of the polluting source, the error in approximating the plume height is ±100 ft [9]. Beyond this range, evaluation of plume height becomes uncertain. Human photogrammetric stereo comparison is limited even under optimum conditions. For barely visible and invisible plumes such as oxide gases, human stereo interpretation is completely impractical. With increased use of precipitators in high smoke stacks, the visibility problem will become more compound.

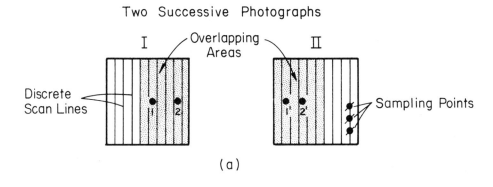

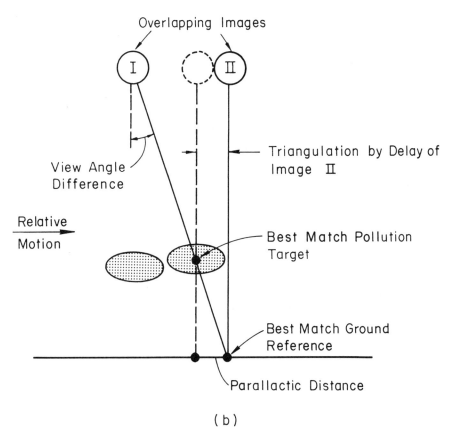

Fig. 1 (a) Stereoscopic imagery. (b) Determination of pollution target height by stereoscopic comparison and correlation techniques.

The stereoscopic analysis of images is highly improved and refined when adequate digital cross-correlation techniques are utilized. Initially, these techniques were developed and successfully employed for remote measurement of sound sources, heat radiation, wind, and turbulence using passive optical crossed beams [10,11,12]. The crossed-beam correlation method was further applied to active sensors. An extensive review of the applications of electromagnetic correlation techniques can be found in Ref. 13. Basically, this method combines remote measurements with adequate data processing and statistical analyses.

The extension of crossed-beam correlation techniques to a stereo-image pair presented hereafter is based on the approach outlined in Ref. 14. To carry out the digital correlation of a stereoscopic image pair, conversion of each picture into discrete scan lines is necessary. These lines, which are depicted in Fig. 1(a), can be obtained by scanning stereo photographs with a microdensitometer [9,15]. Essentially, the discrete scan lines constitute the scan history of the image in the form of a digitized record of the radiation power within a certain field of view. Thus, the scan history consists of a sufficient number of adjacent scan lines which supply an image of the target area in terms of a strip map. It is, further, important to notice that scan histories within wavelength ranges where photographic emulsions are not available can be supplied by utilizing electromechanical stereoscopic scanners.

The scan history can be expressed in terms of a signal

$$x(t) = x(m,n) ,\qquad(1)$$

where m designates the scan line number relative to the position of a ground reference object whose scan line is $m = 0$. The sample point number along a particular scan line is denoted by n. It is determined with respect to the sample point $n = 0$ that intersects the ground reference feature.

The stereo analysis is performed by computing the cross-covariance function of the two overlapping areas I and II. To calculate the cross-covariance, one of the images, e.g., image II, is delayed by k scan lines and ℓ sample points. Then, the sample cross-covariance estimate is

$$C_{xy}(k,\ell) = \frac{1}{4MN} \sum_{-M}^{M} \sum_{-N}^{N} x(m,n)y(m+k,n+\ell),\qquad(2)$$

where $x(m,n)$ is the signal from image I, and $y(m+k,n+\ell)$ is the delayed signal from the overlaying area II. The number of resolution elements per strip is denoted by $4MN$. A direct quantitative measure for the matching is provided by the cross-covariance. The best match of the features related to the polluted mass is indicated by the peak value of the cross-correlation. Similarly, the best match for the ground reference object is also furnished by the cross-covariance. As a result, the parallactic distance is obtained and, then, the plume features (or inversion characteristics) are evaluated by triangulation as portrayed in Fig. 1(b).

The signals from stereo images are random variables of time and, thus, they can be expressed as

$$x(t) = x_1(t) + x_2(t), \quad (3.1)$$

and

$$y(t) = y_1(t) + y_2(t), \quad (3.2)$$

where the subscripts 1 and 2 denote the target signal and transmission noise, respectively. Then, the cross-covariance is

$$\overline{C_{xy}(\tau)} = \overline{x_1(t) y_1(t + \tau)} + \overline{x_1(t) y_2(t + \tau)} + \overline{x_2(t) y_1(t + \tau)} + \overline{x_2(t) y_2(t + \tau)}, \quad (4)$$

where the τ designates the delay time, and the overbar denotes time integration over the entire scan history. The first term on the right-hand side of Eq. (4) is the cross-covariance of the common signals. This product is always positive since both signals are of same sign being generated by the same random process. In other words, both stereo images are affected in a similar way by the polluted air mass target. This is true provided that the pollution target is unmobile. The other three products represent cross-covariances of uncorrelated random variables caused by statistically independent random processes. Consequently, they vanish provided that the integration time is long enough. It is, further, important to notice that the cross-covariance is dominated by the product $\overline{x_1 y_1}$. This term increases linearly with integration time. On the other hand, the other products oscillate with equal likelihood about the average value. Furthermore, their contributions to the time integral increase with the square root of integration time [12]. Adequate averaging methods to cancel the noise by product integration for finite integration time are proposed in Ref. 12.

The mean square error of the cross-covariance estimate due to individual resolution element (or finite integration time) is given by the variance,

$$\text{Var}[C_{xy}(k,\ell)] = \frac{1}{4MN-1} \sum_{m,n} [x(m,n)y(m+k,n+\ell) - \overline{C}_{xy}]^2 , \quad (5)$$

since $C_{xy}(k,\ell)$ is an unbiased estimate of the true value of the cross-covariance $\overline{C}_{xy}$ [16]. This error decreases with increasing integration time. Next, the confidence to obtain a meaningful match is expressed by the ratio of the cross-covariance estimate to its root mean square (rms) error

$$\alpha_{k,\ell} = \frac{C_{xy}(k,\ell)}{\sqrt{\text{Var}[C_{xy}(k,\ell)]}} , \quad (6)$$

which is called the coefficient of confidence.

In practice, additional limitations exist due to uncontrolled variations of the optical environment. Averaging techniques which can efficiently eliminate, to a large extent, these effects were developed [12]. These techniques employ the variations of piecewise mean values to recognize and eliminate this noise. A sample of the instantaneous product xy of two total signals for two different wavelength ranges from a stereo scanner is shown in Fig. 2 [14]. The continuous strongly oscillating curve represents the scan time history. The cumulative average of the instantaneous product

$$C_{xy}(k,\ell,T) = \frac{1}{T} \int_0^T x(t)y(t)dt , \quad (7)$$

is a function of the variable record length (or integration time) since $0 < T < 2MN$. With increasing record length (or integration time) the variations of the cumulative average diminish since the changes in the instantaneous product cancel each other. As mentioned earlier, the contribution of the common signals from the polluted mass to the instantaneous product is positive. Thus, the negative fluctuations of the instantaneous product represent signal components which are unrelated to the target. Moreover, the positive instantaneous products also contain signals which are uncorrelated to the pollution target. These positive products cancel the negative products. The shaded areas in Fig. 2 delineates the power of the unrelevant signals. Their cancellation through the integration process is indicated by the decreasing change in the variations of the cumulative average. Furthermore, with increasing number of instantaneous products the mutual cancellation of the unrelated signals is enhanced. This process of cancellation by product integration is completed when the cumulative average reaches a constant value [12,14]. Practically, this value represents the power of the common signals from the

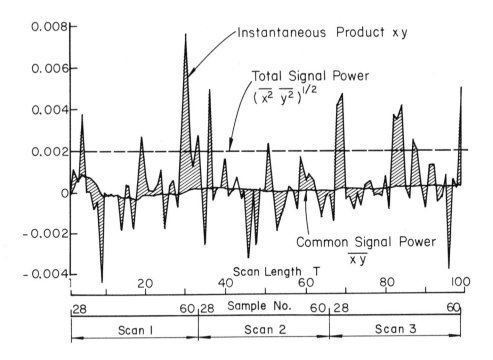

Fig. 2 Cancellation of unrelated signals by the product integration method.

pollution target. Consequently, the product integration process permits recovery of very weak signals provided that the target is sufficiently large and homogeneous. The latter condition is needed to allow complete suppression of the noise [12]. This method can be further utilized for nonhomogeneous polluted masses by subdividing it into homogeneous domains. This can be achieved using recently developed unsupervised classification methods [14]. As a result, even invisible pollution targets are eventually retrievable using this stereo correlation analysis. Furthermore, this technique can be applied to evaluate the water vapor burden within a column of air.

EXPERIMENTAL RESULTS

To demonstrate the feasibility of stereoscopic remote sensing of polluted air masses and of stereo-correlation analysis of stereoscopic images, an extensive field experimental program is currently being conducted. The objectives of the field experiment could not be achieved by simply surveying pollutants randomly emitted into the atmosphere. Particularly, it was imperative to surmount the unpredictable effects of continuous changing meteorological conditions. It was further desired to inject pollutants under controlled circumstances into a known environment.

For these reasons, it was decided to simulate the environmental conditions using the wake flow produced by a 10 ft diameter fan of variable pitch (Hartzell A120-6). This fan constitutes the core of the Colorado State University Environmental Field Station. The fan is driven by an internal combustion engine and, hence, continuously variable air speeds up to 20 mph can be generated. Within the wake large scale turbulence is produced and, consequently, the environmental conditions are adequately simulated for pollutant injection. To reduce undesirable interferences with the ambient winds, all experiments are conducted under calm conditions or light winds up to at most 2 mph. Continuous monitoring of the prevailing winds is performed using cup anemometers.

Contaminants in the form of gases, liquid aerosols and particulates can be released under governed conditions from point sources and/or line sources. The pollutants may be emitted either separately or premixed in any desired ratios. Locations of polluting sources were selected based on velocity distribution and turbulence characteristics within the wake. Moreover, since the sources are mobile they can be located in any configuration and at any position of interest. It is worth pointing out that same sources may be utilized to generate controlled water vapor plumes within the wake.

The Environmental Field Station is located on a flat surface free of immediate natural or artificial obstructions. Almost daily base temperature inversions on clear mornings and on evenings at twilight occur at its site. The station is equipped with an analog Data Acquistion System. In addition, a digital Data Acquisition System is available for on-line data reduction and analysis. Fixed ground reference targets needed for the stereo photographs were set up.

To ascertain the practical applications of the stereoscopic remote sensing, simultaneously with the aerial stereo survey ground measurements are necessary. Furthermore, comparison of the results will lead to development of adequate calibration techniques for the stereo remote sensing. Ground measurements of velocity and turbulence are carried out utilizing a novel three-lead hot-wire anemometer system conceived, designed and built at Colorado State University [17]. This new system permits remote use of hot-wire anemometers since the cables connecting the hot wire, i.e., the sensor, to the bridge can exceed 500 ft in length. In the past, the length of the connecting leads was limited to about 25 ft [17]. The new remote system affords innumerable applications of hot-wire anemometry techniques in atmospheric measurements. Specific hot-wire interpretation methods developed for flow with large fluctuations, such as encountered in the atmosphere, are employed [18].

The aerial stereoscopic photographs are taken employing a Wild RC8 automatic precision mapping camera system with $93°$ coverage (Wild-Heerbrugg Inst.). The camera is equipped with a 6 in Universal Aviogon lens cone (f/5.6) with rotary shutter. An antivignetting filter Wild A.V. 2.2 x (Wild-Heerbrugg Inst.) is utilized to provide uniform lighting across the lens and, thus, uniform exposure on the film is obtained. All pictures were taken using Kodak aerocolor negative estar base film. Included in the camera system are a driving mechanism, a control box, a viewfinder telescope, a clock and an altimeter. A general view of the camera system is shown in Fig. 3. The camera produces 9 x 9 in photographs with remarkably low distortion and high resolution. The former is smaller than 0.01 mm whereas the latter varies from 50 lines per mm in the center to 25 lines per mm in the corners. This high resolution permits adequate conversion of the picture into discrete scan lines. Since the film advance per exposure amounts to 10 in, the readings of the incorporated clock and altimeter can be recorded on each photograph. Continuous adjustment of the exposure time from 1/100 to 1/700 sec is provided by the driving motor. Four overlap areas of 20, 60, 70 and 80%, respectively, are available. The shortest time interval between exposures, i.e., the time lapse, is roughly 3 to 3.5 sec. A plume conveyed by such a velocity that the distance traveled during the time lapse is small compared with its overall horizontal extent, it will appear on two successive photographs as being approximately stationary. At this point, it is important to notice that the plume is "stationary" with respect to stereoscopic photography since it will roughly affect a stereo-image pair in the same manner as a ground reference object. Thus, one can consider the plume of being stereoscopically stationary. Such situations occur particularly within inversion layers, light winds or calm conditions. For aerial operation, the camera system is flown in Colorado State University Aero-Commander 500B research aircraft.

Direct measurement and recording of the space coordinates of any desired image on the stereo-photograph pair is carried out utilizing the Wild STK-1 Stereocomparator/Digitizer/IBM Card Punch (Wild-Heerbrugg Inst.). The coordinates of selected objects can be measured with a precision of $1\mu m$ (1/25,400 in). Algorithms for determination of the space coordinates were put forth. It is important to remark that the evaluation of the vertical coordinates

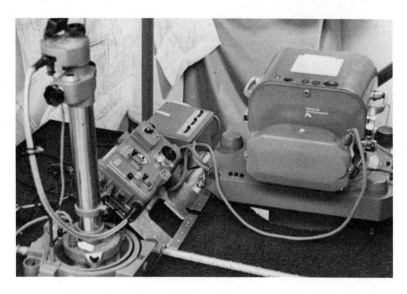

Fig. 3   View of the Wild RC8 camera system.

may be arduous due to the haze of the plume. Scanning by means of a microdensitometer, which measures the transmittance of the photographic emulsion, can undoubtedly supply more reliable estimation of the vertical extent of the pollution target.

Henceforth, the preliminary results of the feasability investigation are presented. To begin with, intensive, albeit not exhaustive, visualization studies of circulation of visible plumes generated by colored smoke were conducted. The smoke point sources were located at various positions along the streamwise direction of the wake flow and at several elevations above the ground. Several frames from a movie showing the smoke plume circulation, entrainment and dispersion without and with temperature inversion are given by Fig. 4. The fan is also shown in Fig. 4(a). Powerful and relatively large scale vortices were clearly discerned when unstable conditions (superadiabatic lapse rates) prevailed. The photographs shown in Figs. 4(a) and 4(b) were taken under such conditions. Without temperature inversion, the vortices disperse quickly the smoke plume at higher altitudes. However, in the neighborhood of the sources, due to the large scale vortices, deposit of pollutants on the ground can occur as indicated by the smoke circulation observed in Fig. 4(b). On the other hand, under inversion conditions small scale vortices were prominent as illustrated by the photographs shown in Fig. 4(c). Relatively high smoke concentrations persisted for prolonged time. Knowledge of the scale of vortices is essentially for study of both dispersion and concentration of pollutants, and temperature distribution. Particularly, within the earth-atmosphere interface, the prevailing vortices play a prime role in the surface concentration of pollutants.

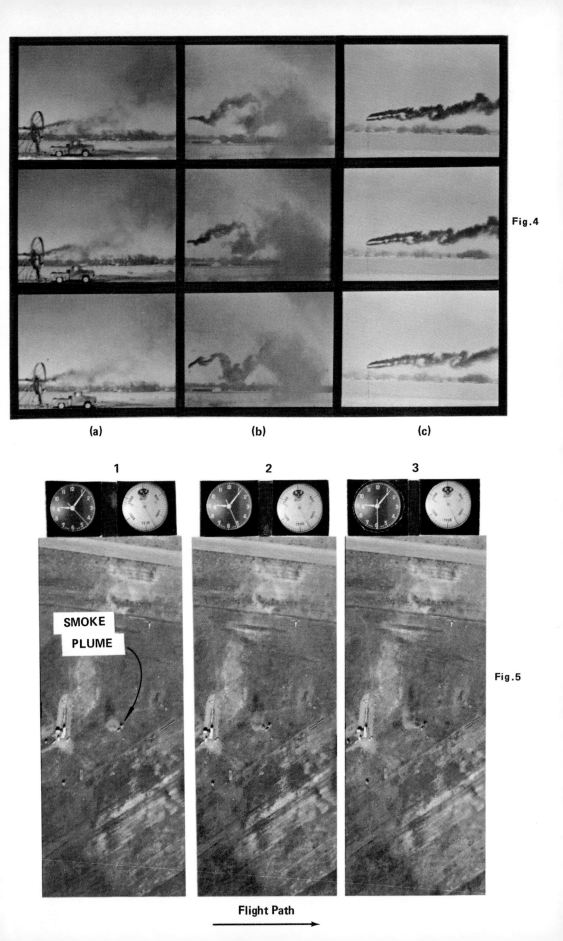

Fig.4

Fig.5

Flight Path

Next, numerous overflight surveys of the smoke plume were carried out. Stereoscopic photographs were taken from several altitudes. All photographs were obtained with 60% overlap and with a time lapse of 3 sec. The aerial photogrammetric surveys were conducted either under calm conditions or light winds up to about 1 mph, and temperature inversion. Reproduction of colored stereo-image triplets of smoke circulation taken from an altitude of 2000 ft is shown in Fig. 5. Analysis of the stereoscopic photographs by means of a stereocomparator revealed that the maximum windwise extent of the plume was about 70 ft. The shape and extent of the smoke plume in the horizontal plane is depicted in Fig. 6. For a wind speed of 1.5 ft/sec (~ 1 mph), the smoke plume traveled a distance of 4.5 ft during the time lapse. Since this distance is about 6.4% of its windwise range, one can conceivably assume that, for any practical purposes, the plume is stereoscopically stationary. The similar extent of the smoke plume obtained by comparison of a series of stereo images taken under same conditions substantiates the foregoing conclusion.

Basically, the smoke plume was produced by injecting smoke into the wake. Subsequently, the smoke was transported by the wake flow. Hence, the wake can be considered of being an invisible plume whose overall spatial extent should coincide with this of the visible smoke plume. To corroborate the results obtained by the remote stereo analysis, hot-wire anemometer ground measurements of the planar extent of a corresponding invisible plume were performed. These measurements were conducted under same wind and inversion conditions as for the remote stereoscopic survey but not simultaneously. The boundary of the invisible plume, i.e., the wake boundary, was defined as the location where the local axial velocity reduces to about 2% of the fan rotor tip velocity [19]. The results of the hot-wire anemometer survey are shown in Fig. 6 together with the horizontal extent of the visual smoke plume. A striking similarity between the extent of the visible plume, deduced by means of remote stereo analysis, and the invisible plume, obtained through ground measurements, is clearly discerned. This consistent general congruence substantiates the feasibility of stereoscopic remote sensing. Furthermore, this remarkable agreement was obtained for a relatively small, tenuous plume.

Fig. 4 Visualization of smoke plume circulation under unstable conditions: (a) point source upstream of the fan; and, (b) point source at 40 ft downstream of the fan; and, within a temperature inversion layer: (c) point source at 150 ft downstream of the fan.

Fig. 5 Reproduction of colored stereo-image triplets of smoke circulation taken from an altitude of 2000 ft.

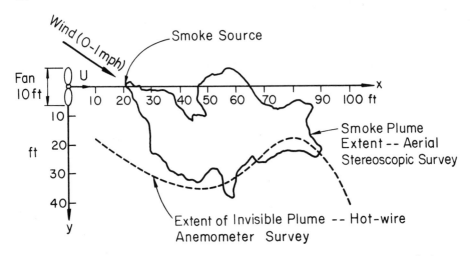

Fig. 6 Planar extent of visible smoke plume and its corresponding invisible plume.

Thus, the stereo remote sensing can be used to survey even barely visible and weak plumes. Currently, the evaluation of the three-dimensional features of the plume is underway.

CONCLUDING REMARKS

The feasibility of stereoscopic remote sensing of polluted air masses by means of flown stereo cameras is indicated by the preliminary results presented in this work. The evaluation of the planar extent of a plume based on both stereo analysis and ground measurements exhibits a striking congruence. Even barely visible and tenuous plumes of relatively limited spacial extent can be monitored by stereo techniques. In this experiment, the windwise extent of the plume was roughly 70 ft.

Stereoscopic analysis of a stereo-image pair can be used to establish the features of a polluted air mass provided that the condition of stereoscopic stationarity is satisfied. Basically, when the distance traveled by a plume during the time lapse between two successive exposures is small compared with its overall horizontal extent, one can assume that the plume is stereoscopically stationary. This situation usually occurs within temperature inversion layer, under light winds or calm conditions. When the condition of stereoscopic stationarity is not fulfilled, e.g., strong winds, stereo-image triplets are necessary. Currently, stereo comparison analysis of such triplets is being carried out.

By and large, standard stereo comparison analysis is limited due to variations in the concentration of pollutants within the plumes, the haze of the plumes, and the continuous changing optical backgrounds. Digital stereoscopic analysis based on correlation techniques possesses the potential to overcome these difficulties. An intensive effort to demonstrate the efficacy of digital stereo correlation is planned. Next, the development of mathematical models for long term pollution distribution and concentration over large areas depends, to a large extent, on detailed knowledge of both meteorological conditions and polluted air masses features. Stereoscopic techniques can be used efficiently to provide contour maps of pollution including isoconcentration density curves in addition to meteorological parameters.

Basically, stereoscopic remote sensing of polluted air masses by flown stereo cameras permits adequate monitoring within large areas at selected time intervals. Furthermore, unpredicted and unexpected pollution can be detected. Consequently, air quality standards and control regulations can be set forth.

ACKNOWLEDGMENT

The assistance of Mr. C. A. Koper Jr., in carrying out the ground measurements is appreciated.

The support of this work by NASA Marshall Space Flight Center is gratefully acknowledged.

REFERENCES

1. Morgan, G. B., Ozolins, G. and Tabor, E. C., "Air Pollution Surveillance Systems," Science, 170, 3955, 289-296 (1970).
2. Chambers, L. A., "Classification and Extent of Air Pollution Problems," Air Pollution, Edit., Stern, A. C., Vol. 1, Air Pollution and Its Effects, 1-22, Academic Press, New York, 2nd ed. (1968).
3. Krause, F. R., Private Communication, NASA, Marshall Space Flight Center, Huntsville, Alabama (1972).
4. Hoffert, M. I., "Atmospheric Transport, Dispersion, and Chemical Reactions in Air Pollution: A Review," AIAA Journal 10, 4, 377-387 (1972).
5. Davidson, B., "A Summary of the New York Urban Air Pollution Dynamic Research Program," J. Air Pollution Control. Assoc., 17, 3, 154-158 (1967).
6. Wanta, R. C., "Meteorology and Air Pollution," ibid. Ref. 2, 187-226.

7. Wobber, F. J., "Orbital Photos Applied to the Environment," J. Photogrammetric Eng., 26, 8, 852-864 (1970).
8. Moffitt, F. H., Photogrammetry, International Textbook Co., Scranton, Penn., 2nd ed. (1967).
9. Veress, S. A., "Air Pollution Research," J. Photogrammetric Eng. 26, 8, 840-848 (1970).
10. Fisher, M.J. and Krause, F. R., "The Crossed-Beam Correlation Technique," J. Fluid Mech., 28, 4, 705-717 (1967).
11. Krause, F. R., Derr, V. E., Abshire, N. L., and Strauch, R. G., "Remote Probing of Wind and Turbulence Through Cross-Correlation of Passive Signals," Proceedings, 6th International Symposium on Remote Sensing of Environment, 13-16 October 1969, Univ. of Michigan, Ann Arbor, Mich., 327-257 (1969).
12. Krause, F. R., Su, M.Y. and Klugman, E. H., "Passive Optical Detection of Meteorological Parameters in Launch Vehicle Environments," J. Applied Optics, 9, 5, 1044-1055 (1970).
13. Krause, F. R., Edit., "Research on Electromagnetic Correlation Techniques," NASA TM X-64505, Marshall Space Flight Center Huntsville, Alabama (1970).
14. Krause, F. R., Betz, H. T. and Lysobey, D. H., "Pollution Detection by Digital Correlation of Multispectral Stereo-Image Pairs," AIAA Paper No. 71-1106, Joint Conference on Sensing of Environmental Pollutants, 8-10 November 1971, Palo Alto, Calif. (1971).
15. Data Corporation Manual, "Man-Data Micro-Analyzer," Data Corp., Dayton, Ohio (1968).
16. Bendat, J. S. and Piersol, A. G., Random Data: Analysis and Measurement Procedures, Wiley-Interscience, New York (1971).
17. Sadeh, W. Z. and Finn, C. L., "A Method for Remote Use of Hot Wire Anemometer," Review Scient. Instrum., 42, 9, 1376-1377 (1971).
18. Sadeh, W. Z., Maeder, P. F. and Sutera, S. P., "A Hot Wire Method for Low Velocity with Large Fluctuations," Review Scient. Instrum., 41, 9, 1295-1298 (1970).
19. Koper, C. A., Jr., and Sadeh, W. Z., "A Preliminary Study of a Fan Wake," RM CEM70-71CAK-WZS23, Dept. of Civil Eng., Colorado State University, Fort Collins, Colorado (1971).

PROFILES OF THE NATURAL CONTAMINANT RADON 222 AS A MEASURE OF THE VERTICAL DIFFUSIVITY

Amiram Roffman

Westinghouse Electric Corporation, Environmental Systems Department, Pittsburgh, Pennsylvania

INTRODUCTION

The concentrations of atmospheric pollutants are affected to a large extent by the intensity of the vertical transport that is governed by the daily variations of the eddy diffusion. This presentation is a report on a study of the vertical eddy diffusion based upon profiles of the natural contaminant Radon 222. The temporal and spacial variations in Radon 222 concentrations are affected by vertical transport and advection, radioactive decay and Radon 222 emanation from the ground. A number of investigations concerning the use of Radon 222 as a tracer in the study of the vertical transport are reported in the literature. These studies were carried out from both analytical and experimental viewpoints. Among the authors who have treated this subject are Israel (1951), Malakhov (1959), Jacobi and Andre (1963), Wilkening (1970), Birot and Andre (1970), and Roffman (1971). Hosler (1969) derived average vertical diffusion coefficients for thermal inversion and unstable convective periods based upon Radon 222 profiles. It is the purpose of this paper to show that averaging of Radon 222 concentrations over a long period of time is insufficient to represent changes in the vertical diffusivity when thermal inversion dissipates and convection due to ground heating is initiated and that the eddy diffusion coefficient changes by a factor of 20 or more during this period.

METHODOLOGY

Vertical profiles of Radon 222 in the first several hundred meters above the ground were obtained using an air sampler attached

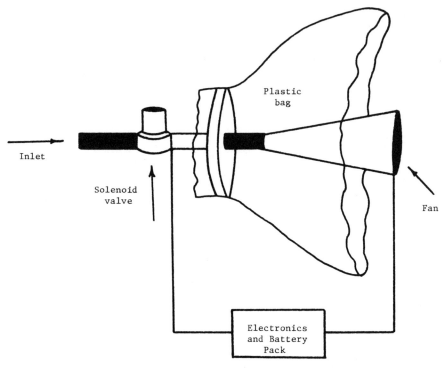

Figure 1. The Air Sampler

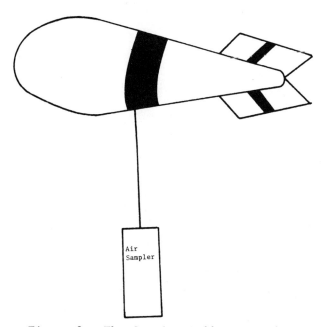

Figure 2. The Captive Balloons Used

to a captive balloon (Roffman, 1972). The air sampler consists of a solenoid valve, a blower, a plastic bag, the electronics and a battery pack (Figure 1). The system is attached to a captive balloon filled with 35 m^3 helium that has a lift of 25-30 kg at an altitude of 1495 m MSL (Figure 2). The balloon is moored by a fiberglass line to a motor driven winch. The weight of the system including the battery pack is about 3.5 kg. At a predetermined height, the instrument is energized by a barowswitch or a mechanical delay timer which closes the electronic circuit. An air sample, 30 liter in volume is collected in the plastic bag, that is collapsed prior to the collection, during a period of 10 min. At the end of the collection the instrument is brought down and the air sample is transferred into an evacuated stainless steel tank for later processing. Experiments in the laboratory reveal no significant loss of Radon 222 by diffusion or absorption to the plastic walls. Simultaneous air collections were made close to the ground using evacuated stainless steel tanks and at a given altitude above the ground using the air sampler attached to the captive balloon. The concentrations of Radon 222 obtained in this way allows the estimation of the vertical diffusivity under different atmospheric conditions.

To estimate quantitatively the changes in the vertical transport the time-dependent one-dimensional diffusion for Radon 222 is solved. The wind term is excluded since the balloon operation can be performed only during calm weather. For periods of time of the order of several hours it is possible to neglect the radioactive decay term for Radon 222 ($T_{1/2}$ = 3.825d). Under these conditions the equation for Radon 222 can be expressed:

$$\frac{\partial n}{\partial t} = \frac{\partial}{\partial z}\left(K_{z,t}\frac{\partial n}{\partial z}\right) \quad (1)$$

where,

n: Radon 222 concentration (number of atoms/cm^3),

$K_{z,t}$: vertical eddy diffusion coefficient (cm^2/sec).

The associated initial and boundary conditions are:

$$n(z, t = o) = f(z)$$
$$n(z = o, t) = \phi(t) \quad (2)$$
$$n(z \, \infty, t) = o$$

where,

$f(z)$: initial concentration of Radon 222,

$\phi(t)$: Radon 222 flux from the ground (number of atoms/cm^2-sec).

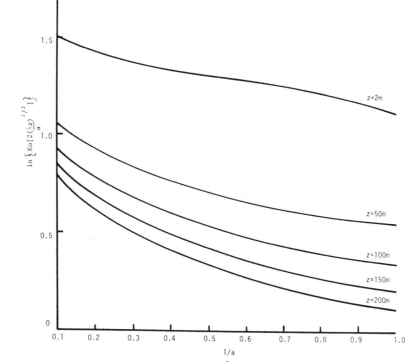

Figure 3. $\ln\{Ko[2(\lambda z)^{\frac{1}{2}}/a]\}$ as Function of 1/a

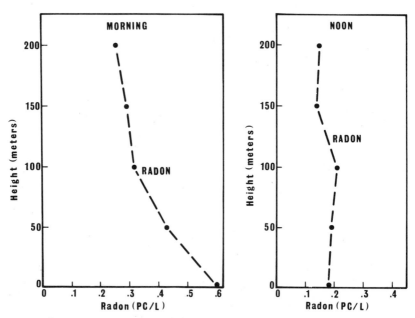

Figure 4. Vertical Profiles of Radon 222 Based upon Averages of Five Morning Flights and Two Noon Flights

Using the method of separation of variables and assuming that:

$$n = n_t n_z$$
$$K_{z,t} = K_z K_t = (az+b)K_t \qquad (3)$$

it is found that the solution of Eq.(1) can be expressed in the form:

$$n = CF(n_t, K_t, \lambda, \phi) K_o \left\{ \frac{2}{a} [\lambda(az+b)]^{1/2} \right\} \qquad (4)$$

The definitions of the symbols used in Eqs.(3) and (4) are:

- a, b: constants with the dimensions cm/sec and cm^2/sec, respectively,
- C: integration constant,
- $F(n_t, K_t, \lambda, \phi)$: a function of the time dependent parts of the concentration and the eddy diffusion coefficient respectively, of the radioactive decay constant of Radon 222 and of the Radon 222 flux.
- K_o: the modified Bessel function of the second kind of order zero.

For practical calculations and for $z \neq 0$ and $b \ll a$ Eq.(4) is expressed as:

$$n = CF(n_t, K_t, \lambda, \phi) K_o \left[2 \frac{(\lambda z)^{1/2}}{a} \right] \qquad (5)$$

Following the method described by Israel et.al. (1967) it is possible to determine the value of "a" for each pair of simultaneous values of n obtained at two different altitudes, assuming $F(n_t, K_t, \lambda, \phi)$ attains the same value at those levels. For $z_i \neq z_j$ Eq.(5) yields:

$$\frac{n_i}{n_j} = \frac{K_o[2\frac{(\lambda z i)^{1/2}}{a}]}{K_o[2\frac{(\lambda z j)^{1/2}}{a}]} \qquad (6)$$

Figure 3 shows typical logarithmic plots of $K_o[2\frac{(\lambda z)^{1/2}}{a}]$ as a function of 1/a for different z values. The distance between two curves corresponding to z_i and z_j yields the value of $\ln[n(z_i) - n(z_j)]$ for which a suitable "a" value is found.

RESULTS

The nature of the vertical transport during thermal inversion conditions, soon after the onset of convection due to the ground heating, and later in the day when mixing conditions prevail is of particular interest. Vertical profiles of Radon 222 based upon averages of five morning flights and two noon flights obtained in the fall of 1970 are shown in Figure 4. Early in the morning the vertical transport is small and the concentration of Radon 222 decreases sharply with altitude. A comparison of the ground level

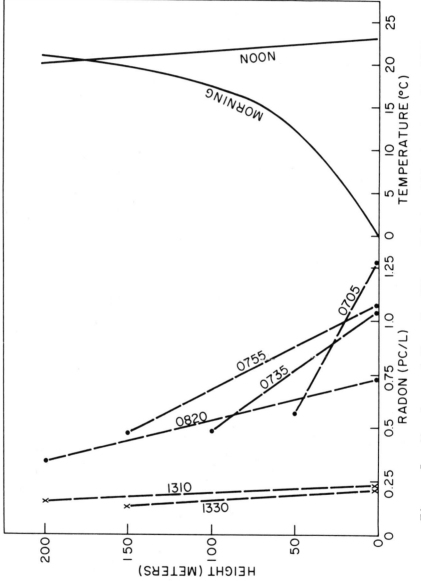

Figure 5. Morning and Noon Profiles of Radon 222, October 30, 1970

Radon 222 concentrations with those at an altitude of 200 m above the ground indicates a decrease by a factor of almost 3. At noon in a well mixed atmosphere the concentration of Radon 222 is almost constant with altitude and is much lower than those observed in the early morning. Typical simultaneous profiles of Radon 222 obtained from a balloon flight on October 30, 1970 early in the morning, and at noon are shown in Figure 5. The average temperature profiles for these two cases are included also. The early morning results demonstrate the relatively fast development in the vertical transport. Similar behavior has been observed on four additional mornings. The values of "a" calculated from the simultaneous Radon 222 concentrations and from the graphs of $\ln \{K_o [2(\frac{\lambda z}{a})^{1/2}]\}$ versus $1/a$ are summarized in Table 1.

The values of "a" in Table 1 increase as time progresses during the onset of convection, and they continue to increase later in the day when the lower troposphere is well-mixed. Soon after sunrise when thermal inversion conditions prevail the vertical diffusivity K_z attains a low value of the order of 4×10^3 cm^2/sec at an altitude of 100m above the ground. The vertical transport due to ground heating causing a graduate decrease in the ground concentrations of Radon 222 and an increase in K_z, which reaches a value of about 10^5 cm^2/sec at an altitude of 100m one hour later.

At noon in a well mixed atmosphere, K_z is of the order of 10^6 cm^2/sec. The net increase in the vertical diffusivity from stable inversion conditions to unstable convective period is of the order of 2×10^2 at an altitude of 100m. The corresponding changes in Radon 222 concentrations as observed during the same period of time are of the order of 10 or more close to the ground and of the order of 2 to 3 at an altitude of 200m.

TABLE 1. Values of the Slope "a" of the Vertical Eddy Diffusion Coefficient Versus Time

LST	a(cm/sec)	K_z(cm^2/sec) at z=100m
0705	0.4	$4 \cdot 10^3$
0718	1.0	10^4
0740	1.8	$1.8 \cdot 10^4$
0755	2.9	$2.9 \cdot 10^4$
0812	9.0	$9 \cdot 10^4$
1200	33.0	$3.3 \cdot 10^5$
1320	100.0	10^6

The average values of the vertical eddy diffusion coefficient obtained in this analysis are in good agreement with those previously reported in the literature. However, early morning results reveal that averaging Radon 222 concentrations over long periods of time when thermal inversion dissipates and convection due to ground heating initiates is insufficient to represent temporal variations of the vertical diffusivity. The results indicate a change by a factor of 25 in K_z over a period of about an hour, and can be applied in the dispersion calculations of atmospheric pollutants emitted from stacks under fumigation situation.

ACKNOWLEDGEMENTS

The author is thankful to Drs. M. H. Wilkening, C. B. Moore and R. D. Stanley for their advice and assistance. This study has been carried out at New Mexico Institute of Mining and Technology, Socorro, New Mexico and was supported in part by the Themis Contract N000 14-68-A-0151. The author is grateful to Westinghouse Electric Corporation, Environmental Systems Department for making it possible for the author to attend this meeting.

REFERENCES

Birot, A., B. Adroguer, and J. Fontan, Vertical Distribution of Radon 222 in the Atmosphere and Its Use for Study of Exchange in the Lower Troposphere, J. Geophys. Res., 75, 2373, 1970.

Hosler, C. H., Vertical Diffusivity from Radon Profiles, J. Geophys. Res., 74, 7018, 1969.

Israel, H., Radioactivity in the Atmosphere, Compendium of Meteorology, American Meteorological Society, Boston, Massachusetts, 155, 1951.

Israel, H., M. Hobert, and C. de La Riva, Measurements of the Radon 222 Concentrations in the Lower Atmosphere in Relation to the Exchange in this Region, Final Technical Report, European Research Office, U. S. Army, 1967.

Jacobi, W., and K. Andre, The Vertical Distribution of Radon 222, Radon 220, and Their Decay Products in the Atmosphere, J. Geophys. Res., 68, 3799, 1963.

Malakhov, S. G., The Vertical Distribution of Radioactive Emanations in the Atmosphere, Izv. Geophys. Ser, 9, 1344, 1959.

Roffman, A., Radon 222 Daughter Ions in Fair Weather and Thunderstorm Environments, Ph.D. Dissertation, New Mexico Institute of Mining and Technology.

Roffman, A., Captive Balloon Measurements of Radon 222 and its Daughter Ions, Pure and Applied Geophys., in press, 1972.

Wilkening, M. H., Radon 222 Concentrations in the Convective Patterns of a Mountain Environment, J. Geophys. Res., 75, 1733, 1970.

SPECTROPHOTOMETRIC DETERMINATION OF INDOLE AND SKATOLE USING LIGNINE EXTRACTIONS

M. Davidson, J. Kindler and A.E. Donagi
Israeli Ministry of Health

Oxidation ponds are one of the common methods for stabilizing domestic and industrial wastes, and are used widely in many countries. They have become one of the most accepted solutions for treatment of wastes in Israel and a large-scale project, the "Recovery of the Dan Region Wastewater", is in operation in this country.

In addition to oxygen, formed throughout the operation of the ponds, algae, bacteria, sulfur oxides, nitrogen oxides, ammonia, methane, hydrocarbons, hydrogen sulfides and also indole and skatole were recognized to be the airborne products emitted from the ponds[1,2].

Therefore, the determination of indole and skatole is one of the routine measurements in the chemical control of odor from oxidation ponds in such countries as Australia, South Africa, etc.

Most of the methods used for the determination of indole and skatole are spectrophotometric[3-12], even though gas-chromatographic methods were also proposed [13-14].

Scott[3] proposed the reaction of indole with p-dimethyl-aminocinnamic aldehyde, for its determination in aqueous solutions. The color formed fades at a rate of 0.7% per minute at a temperature of $30°$ C. and 80% relative humidity.

McEvoy-Bowe(4) has improved the sensitivity of Scott's method, using the reagent p-dimethylaminobenzaldehyde. The color fading remained the drawback of this improved method.

A modification of the well known Ehrilich test(5) (which is specific for indole), has been proposed by Byron and Turnbull(6), which employed the reagent p-dimethylaminobenzaldehyde in trifluoracetic acid. A higher sensitivity, up to 1 μgr/ml indole was obtained, which is superior to other proposed modifications(7,8) of the Ehrlich test.

However, the rapid fading of the color still remained the drawback for this method. Nevertheless, the method had been employed for the determination of indole in tobacco smoke(9).

Russian researchers have described the determination of airborne indole trapped in either dilute acetic acid (10) or 50% alcohol solution(11) and their subsequent spectrophotometric determination after the addition of sodium nitrite reagent. Field tests performed by our laboratory have shown low sensitivities of both the above methods, even when more than 600 liters of air were sampled.

The reagent $NaNO_2$ has been also used by the Japanese investigator Shigaki(12) for the determination of indole in an acid solution. A sensitivity of 1 μgr/ml was claimed for this method.

The gas-chromatographic methods proposed by Daigaku (13) and Hermann and Post(14) are not suitable in this stage for atmospheric measurements of indole and skatole. Problems of efficient sampling and sample concentration previous to the gas-chromatographic measurement, need further investigation.

We have now developed a reliable, simple, rapid and sensitive spectrophotometric method for the determination of atmospheric indole and skatole. The air is trapped in hydrochloric acid solution and the color formed by the addition of vaniline reagent (an extraction product of lignine) is measured spectrophotometrically.

EXPERIMENTAL

Instrumentation

A "Coleman" Junior II Spectrophotometer, Model 6/30 was employed for the spectrophotometric measurements. The pyrex cells, 17 mm. path-length and 105 mm. length had a mean capacity of 15 ml. of solution.

A sampling train, composed of three impingers, a flow-meter and a suction pump was used for the atmospheric samples. The dimensions of the specially designed sinterred glass impinger are given in Fig. 1. The third impinger of the train was a standard bubbling bottle used as a trap to prevent escape of solution drops.

Reagents

All reagents were of analytical grade and were supplied by British Drug Houses Ltd.

0.8% vaniline - was prepared by solving 0.8 gr. of 3-methoxy, 4-hydroxybenzaldehyde (vaniline), in a 100 ml. volumetric flask, and filling to the mark with 25% alcohol solution.

Indole and skatole were purified by sublimation under reduced pressure and gramine. Working solutions were prepared accordingly.

Proposed Method

The air is trapped into the absorbing solution in the two impingers (20 ml. of 2$\underline{N}$ HCl), at a rate of 6-7 liter/min. Minimum 500 liters of air have to be sampled.

The sample is transferred to a 25 ml. volumetric flask, one ml. of vaniline reagent is added and the flask is filled to the mark with 2$\underline{N}$ HCl solution. After allowing a 30 minute interval to elapse, the optical density of the sample is determined spectrophotometrically at a wave length of 505 mμ for indole, 470 mμ for skatole (485 mμ for their mixture) and is compared with a suitable calibration curve.

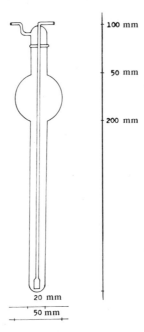

Fig. 1. Impinger of a Special Design for Sampling Indole and Skatole

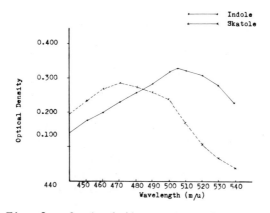

Fig. 2. Optimal Absorption of Indole and Skatole

TABLE 1

Color Stabilization for the Reaction between Indole and Skatole in Acid Medium

Stabilization Time Minutes	Optical Density (O.D.)	
	Indole in 505 mμ	Skatole in 470 mμ
1	0.260	0.245
2	0.270	0.255
10	0.285	0.270
15	0.290	0.270
20	0.310	0.280
25	0.315	0.280
30	0.330	0.285
40	0.330	0.285
60	0.330	0.285

TABLE 2

Color Fading with Time

Hours	Indole - 505 mμ		Skatole - 470 mμ	
	O. D.	Percent Reduction	O. D.	Percent Reduction
0.5	0.330	0.0	0.285	0.0
1.0	0.330	0.0	0.285	0.0
2.0	0.330	0.0	0.285	0.0
6.0	0.315	4.5	0.280	1.8
12.0	0.305	7.6	0.275	3.5
18.0	0.295	10.6	0.265	7.0
24.0	0.290	12.1	0.255	10.5
48.0	0.270	18.2	0.240	15.8

TABLE 3

Sampling Efficiency of Indole and Skatole in Various Absorbing Solutions

Absorbing Solution	Flow Rate LPM	Volume of Air (liters)	Percent Impinger Efficiency		
			First	Second	Third
Ethyl Alcohol 30%	2	180	87.6	9.4	3.0
	4	360	77.0	16.6	6.4
	6	360	75.0	18.5	6.5
Propyl Alcohol 30%	2	180	86.3	10.5	3.2
	4	360	81.2	13.9	4.9
	6	360	79.8	12.2	8.0
1N HCl	2	180	90.2	9.8	-
	4	360	88.0	12.0	-
	6	360	87.2	12.3	0.5
2N HCl	2	180	97.3	2.7	-
	4	360	96.4	3.3	0.3
	6	360	95.8	3.9	0.3

RESULTS

Indole gives a stable yellowish-orange color with a relatively well-defined maximum at 505 mμ. Skatole's maximum appears at 470 mμ. The intersection point of the curves in the extinction vs wavelength given in Fig. 2, is at 485 mμ.

The calibration curves, in Fig. 3 show that the Beer-Lambert Law is obeyed up to a concentration 1.6 μgr/ml, and demonstrates the utility of the method for trace levels.

The sensitivity of the method was found to be up to 0.1 μgr/ml and this sensitivity may be further increased up to 0.02 μgr/ml by buthyl or propyl alcohol extraction. The calculated molar extinction coefficients are 28,100; 27,200; and 24,400 for indole, skatole and their mixture respectively.

The yellowish orange color appears immediately with the addition of the vaniline reagent to the indole or skatole sample, but the color is stable only after about 30 minutes, as may be seen from Table 1. The color is stable for about two hours (Table 2), which can be conveniently utilized for appropriate analysis.

Table 3 shows that the color fading is relatively slow, as compared with other methods related in the relevant literature for indole and skatole determination. After 6 hours the fading reaches a reduction of about 15% in the color of indole and less than 2% for skatole respectively.

Preliminary experiments for the determination of the sampling efficiency have shown that only alcoholic or hydrochloric acid solutions, are suitable for this purpose. Table 3 summarizes the sampling efficiency of indole and skatole in the above media. It may be seen from Table 3 that the best absorbing medium is hydrochloric acid probably because of the formation of the chlorohydrate which contributes to the stability of the final reaction product.

Aliphatic amines, diamines and aminic acids do not interfere with the reaction, whereas aromatic aminocompounds do.

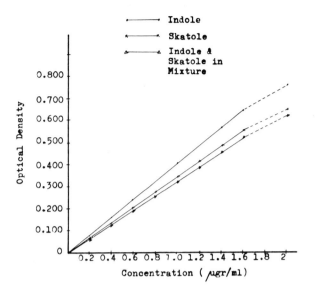

Fig. 3. Calibration Curve for Indole, Skatole and Their Mixture

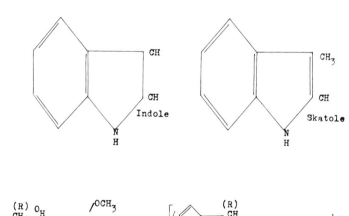

Fig. 4. The Reaction of Vaniline with Indole

Phenols aromatic acids, aldehydes and inorganic salts do not interfere even in high concentrations.

DISCUSSION

The basic idea leading toward the utilization of vaniline was the empiric test mentioned by Nenitescu(15). By this test a stick of pinetree soaked in concentrated hydrochloric acid produces a red color in the presence of indole. Preliminary tests have shown that lignine extracts from pine trees react in the same way. We found vaniline (3-methoxy, 4-hydroxybenzaldehyde), the oxidation product of lignine(15), to give the best results following the same line.

Direct spectrophotometric determination of indole and skatole in the presence of substances which absorb in the same region of the spectrum is difficult. In these circumstances, particularly where high sensitivity is required, as in atmospheric samples, a specific colorimetric method is desirable. A color reaction, which is virtually specific for indoles, having vacant 2- or 3- position is the well-known Ehrlich test(5). This consists of condensation of the indole with electrophile p-dimethylaminobenzaldehyde in hydrochloric acid solution, to give a violet color. But this test as well as its proposed modifications are lacking sensitivity, and in practice cannot be used for the accurate determination of airborne indole and skatole because the color tends to fade.

The proposed method using the vaniline reagent is overcoming most of the above drawbacks. The reaction mechanism is almost similar to the Ehrlich test. The reaction product is probably a mesomeric form in which thechinoidal form is the strongest, thus affecting the color stability. In indole, the active 3- position will react (Fig. 4), whereas in skatole the 3- position is blocked by a methyl group and the condensation occurs at the (less reactive) 2- position.

The reaction sensitivity which is up to $0.1\ \mu gr/ml$. is at least 10 times higher as compared with the other colorimetric methods proposed in the relevant literature. This may be further improved by buthyl or propyl alcohol extraction of the color.

The field sampling of the ambient air for the indole and skatole determination is performed using a 2N HCl absorbing solution. The chlorohydrate formed is stable for quite a long period of time. For the subsequent laboratory analysis, the color of the reaction with vaniline reagent is stable for at least two hours, which is enough time for all practical purposes of spectrophotometric measurement. When fading of the color starts it is less than 5% for indole and less than 2% for skatole after 6 hours.

This method may be easily adopted for the determination of indole and skatole in water pollution control.

Bibliography

1) Amrami, A., Tavruah (Sanitation), 2,17 (1953)(in Heb.)
2) Burnett, W.E., Environm. Sci. & Technol., 3,744(1969).
3) Scott, T.A., Biochem. J., 80,462 (1961).
4) McEvoy-Howe, E., Analyst, 88,893 (1963).
5) Kolsek, J., Novak, N. and Prepor, M., Z. anal. Chem., 159,113.
6) Byrom, P. and Turnbull, J.H., Talanta, 10,273 (1940).
7) Allsop, C.B., Biochem, J., 35,965 (1941).
8) Chernoff, L.H., Ind. Eng. Chem., Analyt. Ed., 12,273 (1940).
9) Sugawara, W., Nippon Sembai Kosha Chuo Kenkuyasho Kenkyu Hokoku, No. III,169 (1969).
10) Paletaev, M.I., Hygiene and Sanitation, 30,76 (1965).
11) Mokhov, L.A. and Mareeva, N.S., Lab., Delo, 6,29 (1962).
12) Shigaki, H., Nippon Kagaku, Zassi 79,225 (1958).
13) Daigaku, H., Suissan Gakub Kenyu Iho, 15,42 (1964).
14) Hermann, T.S. and Post, A.A., Anal. Chem., 40,1573 (1968).
15) Nenitescu, C.D., "Elementary Treatise of Organic Chemistry"(in Rumanian) Vol. II Ed. Techn. Bucurest (1958)

16) Bodea, C. et al., "Treatise of Vegetable Biochemistry" Vol. II, Editure of the Academy of the Roumanian Socialist Republic, Bucarest (1965).

THE STEADY-STATE DEMAND-OUTPUT-WASTE ECONOMY

H. MELVIN LIEBERSTEIN

*University of New Castle,
New South Wales, 2308, Australia*

1. INTRODUCTION

On the whole, man inherited an economic universe that was essentially in steady-state. Until rather modern times animals that he might need were largely kept from multiplying to a point of endangering their own food supply by predator animals, by malnutrition, by disease, and by other natural forces. Similarly, the oxygen level man needed to breath was maintained in the air by a carbon dioxide and food energy chain through a green plant-animal symbiosis on a global scale. Nature was said to be in balance, and modern man found it, including his cities and farms, almost so or simply it and he would not have come to be.

But lately he has come to take over the management of resources and to contribute his wastes (as well as his depredation) so significantly in this universe that he has cause to wonder if he may not be upsetting this balance and thus destroying the universe he is adjusted to survive in. Unless he wishes to create an unnatural environment in which he and his progeny can forever be encased, or otherwise, to accept alterations in his own makeup so that he can survive in conditions he could not now tolerate, he has no alternative but to seek to construct an economic consumer-producer-pollution universe which is somehow in steady-state. We hear a great deal about recycling, but what else is the purpose of this to be? All round us in the newspapers, magazines, the media, lectures, books and just private conversations, we hear quite serious concerns expressed about detailed aspects of using wastes for positive purposes or of managing production in such a way that wastes are easily degradable into nature's own balance. From this massive movement of thought we now synthesize a common, though perhaps largely unspoken, desire to move toward a steady-state economic universe, and while

the discourse on details of how this is to be achieved are the
matters we think should be of primary concern for the present, there
may also be value in forthrightly looking to our overall goals.

To whatever extent, certainly limited here, we can bring mathematical precision to the definition and examination of our goals we can begin to agree to an understanding of them. Eventually, within a given model framework we will state here precisely what reasonable meaning we can think to give to the phrase "in a steady-state economic universe", completely unaware at the moment of any earlier attempts to clarify this all important concept or even expressly to state it. If others find our framework to be incomplete or our interpretation of goals inept, then they may say so, of course, if they suggest another framework and goal that is equally precise. It is thus hoped that we can enter into a meaningful discourse on global goals. Likewise, certain economic structures will be found to be needed in order for a universe to be in steady-state in accordance with our framework and definition, and once these properties are precisely delineated they may be found so undesirable that a new framework and definition will be sought. Of course, mathematical formulation has an advantage over verbal discussion in allowing formal deductions, and thus it has the capacity of a faultless examination of the implications of clearly stated assumptions.

The problem of creating a growing economic steady-state is not really new in microcosm. In 1850 Paris was torn apart and completely rebuilt. It had several times built over its medieval confines without being renewed and was filled with horse traffic over tortuously twisted and badly connected routes. Cholera, inhuman degradation, crime, insurrection, bad water, and the smell of human excreta infested many quarters. The rebuilding project opened beautiful vistas and gave fresh air and water to the inhabitants. In so rebuilding the French were deliberately mimicking the ideals of the Romans in rebuilding ancient Rome, and, strangely enough, some of the 1850 Parisian plans were (probably unknowingly) used in a recent rebuilding of the riverfront district in St.Louis. Most American cities either are being rebuilt or need to be, and it certainly must have occurred to many that we ought to rebuild in the future with a planned obsoletion schedule to keep the cities in renewed steady-state. Big ships and bridges are being continually repainted and repaired. Most large, well managed properties are so treated. No industry or army can long operate without planned obsolescence.

But when one somehow intends to include wastes, however tenuously, in an account of an economic universe for the purposes we have announced, he is forced to think globally. Certainly, no more provincialism can be allowed, but no more gross statistical averages over huge and highly diverse nations can be tolerated either. It matters a great deal whether harmful products are released downtown in Philadelphia or in the lonely stretches of wind-swept Kansas, but the very concerns that bring us to this discourse makes a pollution in either locale a concern of the citizens of Israel or any other land in the world. The atmospheres of whole countries and even of entire oceans have been reported as seriously polluted. The scope of our discussion is thus forced to be so broad that variation with distance cannot be ignored; i.e., distance must be brought in as an independent variable. As long as electrical circuits were only built in relatively small laboratories or within small mounting chassis, only time had to be considered an independent variable, but when transmission lines were being designed in mid nineteenth century to span oceans, and propagation rates could no longer be considered infinite, then distance as well as time had to be considered an independent variable.

Here a similar situation has arisen. Former studies of economics were only crudely national or perhaps multinational; but because of the unavoidable global nature of the present discourse, it becomes mandatory that economic variables (here demands, outputs, wastes) be treated as dependent both on space and time. Dynamic (i.e., time-dependent) systems using demands and outputs in essentially the same manner as here have been treated in extenso by Leontief [1] and others. In recent years there has been a dramatic realization of the importance of spacial considerations in economic planning by Tinbergen and others [2] for example; the economic variables are simply described for each of a number of locations as well as for each of a number of production and transportation sectors, and then essentially static linear programming models are utilized. In this manner the importance of different locations of production are acknowledged, but so far as we know no time-dependent models have been suggested that account for spacial variations even in this sense; in any case the dynamics of space-time relationships are not acknowledged. We now envisage the inclusion of these things.

2. OUTPUTS VS DEMANDS EXPRESSED AS PARTIAL DIFFERENTIAL EQUATIONS

As a novice in economics, I will not try at this time to derive the detailed mechanisms of economic exchange which the detailed specification of a differential equation would demand. Rather, I will indicate a framework for possible economic discussions, describe simply the variables involved, and discuss the properties of the framework we would expect as well as the properties we believe solutions should have in order to describe a steady-state. Thus it is the forms of equations, relating to the variables and operators involved and those properties of solutions we think should be imposed, that will be discussed here. In spite of our almost urgent need, none would seem to be ready to draw up a set of equations completely describing the global economy, but we hope that economists will want to try their hands at further describing the nature of such equations. Some questions, as we will see, can be answered from just this knowledge.

It would be clearly absurd for us to believe that the demand for a given commodity is the same, or at all similar, in all the many quarters of the globe. Some societies are not sophisticated enough to demand some items in nearly the same per capita amounts as other societies demand them. For example, light bulbs are far more in demand in Los Angeles than on the Navaho reservation in neighboring Arizona, and, in fact, they are more in demand in Los Angeles than anywhere else in the world. Also, the level of this given demand at any one location changes with time, possibly being dependent on the level of sophistication in that location, the current popularity of that commodity in that locale, and other less subtle factors such as level of income and competing demands. Adopting the Leontief notation [1] we let Y_i represent demand (value) for commodity i, but we put

$$(2.1) \qquad Y_i = Y_i(z,t),$$

where Leontief only had $Y_i = Y_i(t)$. Here $z = (z_1, z_2)$ represents a given point on the globe, and t is time from a reference zero. Of course, the globe is a two dimensional manifold in 3-space, but for the purpose of simplicity we ignore its curvature (though not its basic "connectedness" properties) and represent it as though in 2-space. This is a feature of convenience in discussion only.

But we model here an economically open society, and thus we regard the demand as driving production. The output X_i of commodity i is thus denoted by

$$(2.2) \qquad X_i = X_i(z,t),$$

where Leontief has it only as $X_i = X_i(t)$. But for this purpose our outputs must be considered to be produced at the point they are eventually transported to; in fact, at the point of consumption. Cost of transportation and cost of production thus become twin contributors to output value as we now suggest it to be used. For purposes of the <u>present</u> discussion we feel we must assume that X_i has two partial derivatives with respect to components z_1, z_2 of z. This implies (at least) that the space variables are a continuum, for simplicity here to be thought of as based on Euclidean length in physical space. Economists may want to think of space in terms of some economic scales, not just in crude terms of length, but some economic scales may be, for all practical purposes, indivisible, so such scales may present some difficulty here. We take the attitude that when we are more certain of the nature of our models there will be time to re-examine this point and to formulate a model in terms of discrete variables z_1, z_2. It is not our intention to prescribe even a complete framework for all economic models. We stick here with Euclidean lengths for the scales of z_1, z_2 though we do not mind if they are proportionately reduced in some parts of the globe where, say transportations and thus outputs, are cheaper and demands more uniform than in other locales. It should be kept in mind that the use of ordinary length as a space scale is tied to the highly unconventional definition of outputs as including costs of transportations.

In the notation of Leontief we now propose demand-output equations that are partial differential equations but are comparable to his dynamic equations [1, p.57, equation(3.3)] which constitute first order ordinary differential equations. Where, as already emphasized, we now redefine output to include transportation costs to a point $z = (z_1, z_2)$ on the globe, we write our partial differential equations as

(2.3) $F_i(z,t; X_1,\ldots,X_n; \partial X_1/\partial t,\ldots,\partial X_n/\partial t; \partial X_1/\partial z_1,\ldots,\partial X_n/\partial z_1;$
$\partial X_1/\partial z_2,\ldots,\partial X_n/\partial z_2; \partial^2 X_1/\partial z_1^2,\ldots,\partial^2 X_n/\partial z_1^2;$
$\partial^2 X_1/\partial z_2^2,\ldots,\partial^2 X_n/\partial z_2^2; \ldots; Y_1,\ldots,Y_n) = 0$

for $i = 1,\ldots,n$, being noncommittal for the present as to how many partial derivatives of X_i with respect to components of z and t should be included. Perhaps operators other than derivatives, such as integrals or differences, should be included, but to do so can immensely complicate the questions we wish to address ourselves to now, and the analysis should be kept as simple as possible. If integrals must be introduced later, it is possible that they can be eliminated at the expense of introducing higher derivatives or perhaps of introducing more variables. Of course, the demands $Y_i(z,t)$, $i = 1,\ldots,n$, are taken to be given. If we are to study the global economy, it is evident that very large economic sectors (such as light industrial, heavy industrial, grain agricultural, animal agricultural, etc.) must be chosen for any detailed practical analysis, although for our present limited purpose of examining our overall goals this restriction may not be felt to be compelling. Of course, if large sectors are accepted, and most economists would recommend this for the study of the global economy, then the number of sectors n in (2.3) would be small.

We can do nothing with the form (2.3) without being willing to be much more specific. Yet, lacking any detailed economic knowledge and being unwilling at this time to commit ourselves to a set of beliefs on global economic mechanisms, we can rely only on matching the known properties of the solutions of partial differential equations with what we presume from the literature are the expected properties of the global economy. This is perhaps a grossly unsatisfactory guessing at forms, but it will be seen to allow for the deduction from one set of widely accepted properties some other properties that perhaps have not been quite anticipated, and thus it can be seen to supply somewhat new knowledge.

We believe that such a guessing at forms then can represent a
start at serious space-time output-demand studies.

Economists (and even anthropologists) often speak of the diffusion of goods throughout the world, and our definition of output to include the cost of transportation leads us to believe that we can expect the system (2.3) to have properties like those describing physical diffusion processes; that is, mathematically speaking, we would expect (2.3) to be of parabolic type. One would not expect to see any clear wave propagation phenomena except possibly in highly dynamic or perhaps terrifying explosive situations (see note 1), so hyperbolic type is not expected in the general situation. The properties of solutions of elliptic type are usually thought of as quite static (though this can be misleading), and these properties certainly seem inappropriate here. This is the usual classification of one equation of second order, and if (2.3) were one such equation, we could be rather certain that it was of parabolic type (see note 2).

A system like (2.3) can be of parabolic type, of course, without each equation being so, but since the Leontief dynamic system (which has no space variations) can be written so as to give each equation in (2.3) as involving one derivative only of output X_i with respect to time t, let us boldly assume (tentatively, of course) that the output of each commodity diffuses in a manner much like heat, but affected possibly by the values (not the rates) of the other outputs. That is, let us assume that (2.3) can be written as

(2.4) $\partial X_i/\partial t = f_i(z,t; X_1,\ldots,X_n; \partial^2 X_i/\partial z_1^2, \partial^2 X_i/\partial z_2^2) + g(Y_1,\ldots,Y_n)$,

$i = 1,\ldots,n$, where $g(0,\ldots,0) = 0$. Derivatives $\partial X_i/\partial z_1, \partial X_i/\partial z_2$ are deleted because they would seem to give forced diffussion about a moving centre, and this is a phenomenon we do not expect to see except possibly over extremely long time periods*. Assuming linearity of the f_i's in the second order derivatives appearing in (2.4) (i.e., assuming quasilinearity), we now write that

(2.5) $\partial X_i/\partial t - (\mu_{1i}(z,t;X_1,\ldots,X_n)\partial^2 X_i/\partial z_1^2 + \mu_{2i}(z,t;X_1,\ldots,X_n)\partial^2 X_i/\partial z_2^2)$

$= \sum_{j=1}^{n} m_{ji}(z,t;X_1,\ldots,X_n)X_j + g_i(Y_1(z,t),\ldots,Y_n(z,t))$,

---

* Eventually, we will want to include the effect of derivatives $\partial^2 X_j/\partial z_1^2$ and $\partial^2 X_j/\partial z_2^2$ for all $j = 1,\ldots,n$ (not just for $j = i$)

$i = 1,\ldots,n$, where μ_{1i}, $\mu_{2i} > 0$. As we have emphasized, this is only our conjecture as to the form of a space-time output-demand model. It represents our first attempt to put in some precise terms what economists only vaguely discuss when they get out of the realm of static models or of models that treat of gross statistical averages over large regions of space. We would certainly not like to be held responsible for determining the detailed properties of functions μ_{i1}, μ_{12} without a derivation of (2.5) being expounded starting from basic principles, and modern economic thought would seem to regard even the determination of any values for the m_{ji}'s for a global economy to be specious unless very large industrial sectors (small n) are chosen.

3. WASTES AND RECYCLING

It seems evident that the masses of wastes W_{ij} ($j = 1,\ldots,N$) are functions of the output X_i. Because of our redefinition of (the value of) outputs to include costs of transportation to the point of consumption, they are certainly not simple functions. The outputs X_i will be functions of z and t, but the wastes will be distributed all the way from the point of production along the transportation route (exhaust fumes of the transport carriers are now part of the waste output) and at the point of consumption. Moreover, they may diffuse (physical) widely over the globe. We write

(3.1) $$W_{ij} = W_{ij}(X_i(z,t); z,t).$$

The waste functions W_{ij} will in general be very complicated, and their determination will take a considerable portion of the effort in any treatment of this model. Moreover, a major effort should be expended to make them minimal in some norm (or at least optimal in some sense that is found desirable). All the recycling or positive use of wastes studies should be oriented to this end. No matter how efficient we become in recycling or reusing wastes, it would seem that there will be an irreducible minimum. At least, this is the position taken in modern economic thought, and environmental efforts should be exerted to reduce this to a minimum possible at any given time. The thought that there is an irreducible minimum comes from a thermodynamic analogy, but thermodynamics does not account for the evolution of more complex systems. If society can be regarded as an organism, then its existence, as ours, represents a local violation of thermodynamic principles, but, of course, an organism is not a closed system. Still, rising levels of technology find ways to use materials once thought to be wastes, as numerous examples can show. Whether or not, as it would seem, in the end there is still an irreducible minimum is of no importance, for the efforts to reduce wastes must be made over a relatively short time period.

The waste functions (3.1) may be extremely complicated, involving physical diffusion in air, water and soil and involving mathematically some intricate functionals (say integrals) or even solutions of partial differential equations of far greater complexity than those describing our demand-output model, but they are to be treated as a side show, so-to-speak, and do not <u>directly</u> complicate the mathematical model of demand-output or our concerns over steady-state. Their complication has already been introduced by forcing consideration of a global space-time output-demand model. The possibly practical restriction to large economic sectors caused by the necessity to acquire a global view would also require an analysis in terms of large categories of wastes and thus a treatment of them in gross terms.

Perhaps we might <u>talk about</u> a model which is closed with respect to demands and which recycles wastes, but the analysis of such considerations is beyond us at present, and talk is all we could do; we would be quite unable to undertake anything significant at present, even with respect to stating our meaning of a steady-state. From (3.1), of course, waste is in an economic steady-state whenever the output is, unless some very peculiar waste functions are involved. Any more precise statement of the properties of the waste functions and the influence of these functions on this point will have to await further study of them. We can only see that at any given level of technology they should be strictly increasing functions of the outputs, and processes must be avoided which do not provide either for inherent constraints or policy constraints of the mass of a given waste (even assuming large associated outputs).

4. THE STEADY-STATE; MEAN-SQUARE ASYMPTOTIC STABILITY OF OUTPUTS AND WASTES

The equation (2.5), or possibly the more complicated (2.3), may be read as saying simply that a given time-dependent and spacially varying demand drives a resulting output of commodities in a global community by mechanisms that are somewhat analogous to heat diffusion. But we believe it should be <u>possible to state that the output is entirely determined by the demand, except possibly for some rather quickly decaying transients</u>. That is, we believe that in a well-designed economy the outputs will not depend on the state of the outputs at a very appreciable time in the past but almost wholly on the current or rather recent pattern of time changing demands alone.

The usual boundary value problem for one equation of parabolic type like those in (2.5) is the initial-boundary value problem in which the value of the dependent variable (here an output X_i) is given for an interval of time on the boundary of a curve enclosing the region where the effects of the equation are to be studied, and then the value of the output is given initially (at $t = 0$) at all points within this region. However, in a healthy economy we believe the effects of the initial values should wash out quickly. In fact, initial values would appear to us to be highly impractical quantities to use in dynamic economic analysis because the system is at no time static nor is there a beginning moment, and the necessary output values may change radically before they can all be measured. Moreover, when we think globally, there is no exterior region for the region we study, and, therefore, we do not expect to need the prescription of any boundary values. Then, indeed, except for transients the outputs should be determined by demands.

Let us examine what properties equations (2.5) must have in order that the effects of initial values wash out quickly. Let U_i and V_i be solutions of (2.5) that correspond to the same set of demand functions. We ask that

$$(4.1) \qquad \lim_{t \to \infty} I(t) = \lim_{t \to \infty} \sum_{i=1}^{n} \int_{G_z} (U_i - V_i)^2 dA_z = 0,$$

or even, if possible, that $I(t)$ be bounded by an exponential in time with a large and negative exponential rate so that solutions corresponding to different initial values go quickly together in mean-square.

This is the condition we mean to describe as steady-state. Let us quickly point out that it is not stationary (time-independent or static) because the demands depend on time. If they are say uniform over the globe and oscillatory in time, one would expect outputs that are uniform over the globe and oscillatory. Of course, no such simple pattern is ever to be realized. Since any time can be considered time zero, the property (4.1) gives economic stability with respect to changes in outputs, and it is not, therefore, going to run wild in the face of deliberate attempts at manipulation like market flooding (providing outputs far beyond demand) or commodity withholding (as in strikes). It is likewise stable against the errors of producers who simply err in keeping close communications with demand. The maintainance of such communications, including the anticipation of demands, is perhaps the most difficult task a producer faces and the one where he is most likely to err. Wastes are tied to output so the output stability we speak of will give a similar waste stability provided the waste functions are properly controlled. An economy with the property (4.1) is responsive to demands, and neither its production of outputs, nor its wastes are grossly subject to the whims of an individual manager or group of managers. In fact, we believe that the magnitude of the anticipated negative exponential rate will be a valuable measure of the health of the global economy and may well deserve the name "resilience" since it measures its ability to recover from displacements.

We have said that (4.1), which by way of (3.1) implies a similar result with respect to wastes, is our delineation of a steady-state economic universe. Yet such a definition will cause some distress these days because it does not necessarily limit the growth of wastes. In fact, we believe the waste functions (3.1) must be strictly increasing functions of outputs*, so in order to limit the growth of irreducible wastes we would have to limit the growth of output, and the world would not yet seem ready to do that. Probably we will have to undertake it someday, either that or reduce wastes to such a small percentage of output that we can live indefinitely on this globe without taking serious note of its accumulation. But limiting production would seem to be a difficult matter, and we would not like to have to wait for the time when the world is ready for such a limit to start working toward a steady-state economy.

Once we have determined what properties of (2.5) guarantee a steady-state of outputs and wastes, then in this model these both depend (except for transients) only on demands, and thus we can study what demand functions give what properties of the steady-state outputs. If wastes are to be bounded, or their growth restricted, then one must investigate what properties the demand functions must have in order to achieve this. Unfortunately, demand functions will be hard to control, and undoubtably the easiest methods available will be those of population control, but before extreme measures are proposed, we might investigate if some of the desired effect could be accomplished by altering the economic structures as reflected in functions μ_{i1}, μ_{i2}, and $m_{i1},\ldots,m_{in}$. We are not proposing such production or demand management but only showing their relation to the problem of achieving a steady-state universe. It is a very direct relation because we must first achieve a steady-state before outputs can be seen as essentially determined by demands and not in large measure by the history of the economy.

5. RELATION TO THE LEONTIEF MODEL

Our equation (2.5) reduces to the Leontief model [1, p.57] if for every $i = 1,\ldots,n$, $j = 1,2$,

$$\partial Y_i/\partial z_j = \partial X_i/\partial z_j \equiv 0$$

and all coefficients are constant. Then (2.5) becomes a system of ordinary differential equations with constant coefficients but with forcing terms (demand functions) appearing as functions of time. Solutions of such systems, of course, have the property that they can be expressed as the sum of the so-called complementary solutions (of the homogeneous equations; with $Y_i(t) \equiv 0$ for every i) and one particular solution involving the given functions $Y_i(t)$. Moreover, if the eigenvalues of the Leontief coefficient matrix M,

---

* In the absence of evolving systems of higher complexity.

where*
$$dX/dt = MX + Y(t),$$
have negative real part, each term of the complementary solution is multiplied by a decaying exponential, so it is considered the transient, and the particular solution is the steady-state. In mechanics and electricity this is very often the only term of any importance.

The complementary or transient portion of the solution contains the only part that depends on the initial values $X_i(0)$, and the effects of these wash out if the real part of the eigenvalues of M are negative. Leontief says that the outputs at any time t depend on the outputs at some previous time (regarded as initial), which is quite true strictly speaking, but if the properties of M are appropriately adjusted (i.e., in a stable economy) the contributions of past performance will be negligible after a rather short time.

Without being forced, as now, to take a global view of the economy, it appears that it was justifiable to treat of statistical averages of outputs X_i over small regions in order to eliminate the necessity to account for spacial variations. Practical limitations demanded it. Also, presumably practicality could be argued as a justification for assuming constant coefficients; we are not ready to specify and analyse an economy in terms we can only now propose in (2.5). But we believe a mistake was made in former economic dynamic analyses in not recognizing that it is only the steady-state (which may well be time-dependent) that is important, that a stable economy should be operating somewhat independently of its history (except as history perhaps affects the structural coefficients). If an economic analysis does not show a steady-state, then either an error has been made in the quantification of the model (determining the components of M in the Leontief model) or the economy must be made stable before we can have much success in studying its time variations. At least, stability ought to be a primary goal of analysis.

6. A STEADY-STATE WHEN THE STRUCTURE IS INDEPENDENT OF THE LEVELS AND LOCATIONS OF OUTPUTS

Of course, partial differential equations, even those with constant coefficients, have no such simple properties that the solution can be written as a sum of transient and steady-state terms.

---

*In matrix notation, the Leontief dynamic equation (3.3) of [1,p.57] reads
$$(I-A)X - B\dot{X} = Y(t),$$
where $\dot{X} = dX/dt$, so in this notation
$$M = B^{-1}(I-A),$$
where B is a stocks matrix and A is a structural matrix.

The conditions providing that a steady-state (4.1) occurs must be indirectly inferred and cannot be expected to arise from a delineation of all solutions. But the ferreting out of the conditions on the general equations (2.5) (i.e., on the coefficients μ_{i1}, μ_{i2}, and $m_{i1},\ldots,m_{in}$) so that mean-square asymptotic stability occurs is more of a task than we care to undertake until some detailed considerations of the mechanisms gives us a form of equations that we can feel confident do describe the global economy. Otherwise, since the eventual equations to be considered may be either much less complicated or perhaps more so (in the sense of containing other operators), we would be perhaps performing an extensive mathematical analysis without economic content.

However, for an understanding of (4.1) it will be well to prove it is valid under some understandable and suggestive, if highly artificial, conditions. For the purpose of this demonstration, we examine an economy described by (2.5) where the structural coefficients μ_{i1}, μ_{i2} and $m_{i1},\ldots,m_{in}$ are independent of the outputs or the locations; that is,

(6.1) and
$$\mu_{i1,2} = \mu_{i1,2}(t) > 0$$
(6.2)
$$m_{i1},\ldots,m_{in} = m_{i1},\ldots,m_{in}(t),$$

where the positivity of the coefficients $\mu_{i1,2}$ have to do with specification of the positive direction of time, and we will find a condition on the coefficients $m_{i1},\ldots,m_{in}$ that has to do with making the steady-state system (2.5) analogous to the Leontief dynamic system if it is in steady-state. Assumptions (6.1)(6.2), of course, make (2.5) linear and the coefficients independent of spacial variables without going so far as to assume constant coefficients.

Let $U_i(z,t)$ and $V_i(z,t)$ be any solution of (2.5), let $W_i(z,t) = U_i - V_i$, and let

(6.3)
$$I(t) = \tfrac{1}{2} \int_{G_z} \sum_{i=1}^{n} W_i \, dA_z ,$$

where the region of integration is the whole globe. We have

$$I'(t) = dI/dt = \int_{G_z} \sum_{i=1}^{n} W_i \partial W_i/\partial t \, dA_z .$$

But since U_i and V_i satisfy all equations (2.5) (with the same demand function terms $g_i(Y_1,\ldots,Y_n)$ given in the right member), we have by subtracting equations for U_i, V_i and multiplying by W_i,

$$W_i \partial W_i/\partial t = \mu_{i1}(t) W_i \partial^2 W_i/\partial z_1^2 + \mu_{i2}(t) W_i \partial^2 W_i/\partial z_2^2 + \sum_{j=1}^{n} m_{ji}(t) W_i W_j$$

$$= \mu_{i1} \partial/\partial z_1 (W_i \partial W_i/\partial z_1) + \mu_{i2} \partial/\partial z_2 (W_i \partial W_i/\partial z_2) - \mu_{i1}(\partial W_i/\partial z_1)^2$$

$$-\mu_{i2}(\partial W_i/\partial z_2)^2 + \sum_{j=1}^{n} m_{ji} W_i W_j .$$

Since $\mu_{i1,2} > 0$ (see (6.1)) we temporarily adopt new space coordinates
$$Z_1 = z_1/\mu_{i1}^{\frac{1}{2}}; \quad Z_2 = z_2/\mu_{i2}^{\frac{1}{2}}$$
in each ith equation, from which
$$W_i \partial W_i/\partial t = \sum_{j=1}^{2} \partial/\partial z_j(W_i \partial W_i/\partial Z_j) - \sum_{j=1}^{2}(\partial W_i/\partial Z_j)^2 + \sum_{j=1}^{n} m_{ji} W_i W_j .$$
Then
$$I'(t) = \sum_{i=1}^{n} (\mu_{i1}\mu_{i2})^{\frac{1}{2}} \int_{G_Z} \sum_{i=1}^{n} W_i \partial W_i/\partial t \, dA_Z$$
$$= \sum_{i=1}^{n} (\mu_{i1}\mu_{i2})^{\frac{1}{2}} [\int_{G_Z} \sum_{j=1}^{2} \partial/\partial Z_j(W_i \partial W_i/\partial z_j) dA_Z - \int_{G_Z} \sum_{j=1}^{2} (\partial W_i/\partial z_j)^2 dA_Z]$$
$$+ \int_{G_Z} (\sum_{i=1}^{n} \sum_{j=1}^{n} m_{ji} W_i W_j) \, dA_Z .$$

To simplify an argument that the first integral is zero, we simply delete a neighborhood T (the larger the better) of some economically isolated point, and with perhaps some slight inaccuracy, assume that on the boundary δG of sufficiently small neighborhoods T no output is found. We have then

$$I'(t) = \sum_{i=1}^{n} (\mu_{i1}\mu_{i2})^{\frac{1}{2}} [\int_{\delta G_Z} (W_i \partial W_i/\partial n) dS_Z - \int_{G_Z - T_Z} \sum_{j=1}^{2} (\partial W_i/\partial z_j)^2 dA_Z]$$
$$+ \int_{G_Z - T_Z} (\sum_{i=1}^{n} \sum_{j=1}^{n} m_{ji} W_i W_j) dA_Z$$

or

(6.4) $$I'(t) = - \sum_{i=1}^{n} (\mu_{i1}\mu_{i2})^{\frac{1}{2}} \int_{G_Z - T_Z} (\sum_{j=1}^{2} (\partial W_i/\partial z_j)^2) dA_Z$$
$$+ \int_{G_Z - T_Z} (\sum_{i=1}^{n} \sum_{j=1}^{n} m_{ji} W_i W_j) \, dA_Z$$

* * *

Assumption. Suppose now that the matrix $M = (m_{ji})$ is negative definite; i.e., suppose now that
(6.5) \qquad for every $W \neq 0$, $W^T M W < 0$.

* * *

Under this condition we see from (6.4) that $I'(t) < 0$ except when $W_i = U_i - V_i = 0$ for all $i = 1,...,n$. Moreover, under this condition we have that
$$I'(t) \leq - \sum_{i=1}^{n} (\mu_{i1}\mu_{i2})^{\frac{1}{2}} \int_{G_Z - T_Z} (\sum_{j=1}^{2} (\partial W_i/\partial z_j)^2) \, dA_Z .$$

Reverting back to physical coordinates ($Z_j = z_j/\mu_{ij}$, $j = 1, 2$).

$$I'(t) \leq - \sum_{i=1}^{n} 1/(\mu_{i2}\mu_{i1}) \int_{G_z-T_z} \sum_{j=1}^{2} (\partial W_i/\partial z_j)^2 dA_z.$$

By a well-known argument, there exists a positive real number λ^2 such that for all i,

$$-\int_{G_z-T_z} (\sum_{j=1}^{2} (\partial W_i/\partial z_j)^2 dA_z \leq -\lambda^2 \int_{G_z-T_z} W_i^2 dA_z$$

so that

$$I'(t) \leq -\lambda^2 \sum_{i=1}^{n} 1/(\mu_{i1}\mu_{i2}) \int_{G_z} W_i^2 dA_z ,$$

since W_i is zero in T for every $i = 1,\ldots,n$.

Letting
(6.6)
$$D = \max_{i=1,\ldots,n} (\mu_{i1}\mu_{i2}) ,$$
we have

$$I'(t) \leq -(\lambda^2/D) \int_{G_z} \sum_{i=1}^{n} W_i^2 dA_z = -(2\lambda^2/D)I(t),$$

and, finally, we have our bounding exponential

(6.7) $I(t) \leq I(0)e^{-(2\lambda^2/D)t} \cdot I(t).$

The quantity λ^2 depends on properties that are independent of μ_{i1}, μ_{i2} or $m_{i1},\ldots,m_{in}$ so it is independent of the economic structure unless length scales for z_1, z_2 are adopted that depend on the economic structure. Because of the importance we attach to (6.7) and to the magnitude of the stability parameter (the exponential rate of distortion decay we have called resilience),
(6.8) $-2\lambda^2/D$,
this independence is thought to be an important result. Probably a similar result can be derived in some nonlinear cases, but they will be much more difficult. The numerical determination of λ^2, of course, is a well-known problem and is available to modern investigation. In this computation, the larger an isolated region T can be taken the sharper a value of λ can be found. The resilience (6.8) shows a relation to economic structure through its dependence on D.

The condition (6.5) which we find we need in order to arrive to a steady-state economy is consistent with the one we found to be necessary in order to arrive to the same state for the Leontief dynamic model, and, in fact, the condition (6.5) can be replaced with the condition that the real parts of the eigenvalues of M be negative. Since eigenvalues of large matrices are sometimes hard to assess, the condition (6.5) may be preferred. When one knows by what method the components of M have been assessed, it is sometimes rather easy to verify a condition like (6.5) directly.

7. SUMMARY

In summary we have pointed to several features that should be noted in a model of the kind of global waste economy we propose to formulate.

(1) A full account must be made of the dependence of economic variables on space and time. If output values are to be utilized as dependent variables, they must then include the costs of transportations to the points of consumption.

(2) Wastes must be treated as functions of outputs both at the point of production and along the route of transport as well as at the point of consumption.

(3) The steady-state economy may be studied as a dynamic system without reference to initial values, but certain properties of economic structure will have to prevail in order for a steady-state to prevail.

(4) The steady-state economy will have some desirable features of stability, with respect to perturbations of outputs and wastes, which seem essential to the continued existence of any economy.

(5) It would seem that there is an irreducible level of waste residuals that can only be limited if we are willing to limit outputs.

8. NOTES

Note 1. We have mentioned that in the presence of a highly dynamic, explosive or expanding output, the equations might become locally hyperbolic with an accompanying high rate of propagation of outputs and wastes. Such a circumstance would obviate our analysis of a steady-state, which is based on an assumption of a form that gives parabolic type. Local transition to hyperbolic type seems to occur in a number of situations usually thought to be governed by parabolic type [3,5]; for example, in flame fronts, where large gradients of heat maintained by a chemical release of energy represents an explosive driving force, hyperbolic type, characterized by a constant rate of propagation, pertains. A similar circumstance occurs in the passage of an action potential on a nerve axon. (Impact loading on a rod which under normal loads is elastic but which flows under impact, can cause local transition of an equation of hyperbolic type to another one of hyperbolic type, both associated with constant rates of propagation, but with different rates.) What sort of extreme economic event might cause such a transition, or if such a transition is possible, we do not know. Probably war would tend to stop production in a given location, and if it also stopped demand by decreasing population, it would seem to create a collapse or contraction not an expansion which we would visualize as associated with transition to hyperbolic type. Perhaps such a transition could occur if some very prolific country suddenly undertook an explosive unloading of its goods or outputs. Actually, the possibility of such a transition seems to us remote under any circumstances, and we are confident that hyperbolic type does not govern economic dynamics.

Note 2. The classification (elliptic, hyperbolic, parabolic type) for second order equations as we mentioned in section 2 is not complete for equations with variable coefficients because mixed type is not considered. It is possible that a valid output-demand economy will be described where each one of a system is of elliptic-parabolic type [4].

Note 3. Models spoken of here are intended for a discourse on gross properties, principally for one on framework and definition. They emphasize the relation of our pollution problems to global economics, but are not presented here as being available for immediate use in global economic analysis. Even the nature of the equations used are presumed from gross properties, not derived from economic mechanisms.

Note 4. We often like to view the world with its flow of goods in society as an organism with much the same capacities as biological organisms. But this is a dangerous thesis to propose since it has the capacity to cause reliance on reasoning by analogy. Also, of course, organisms are not closed systems at all whereas the global society is essentially so. Analogy, of course, should only be used as a guide to new ideas. In this case we have used a mathematical technique which we developed earlier for a study of blood flow [3] in the study of an economy. In blood flow, it is the point velocity of flow which is governed by an equation of parabolic type and which is mean-square asymptotically stable against perturbations (though no condition like $W^T MW < 0$ is needed there). Asymptotic stability is needed in blood flow to replace the property of uniqueness because it gives uniqueness up to decaying transients; i.e., it gives uniqueness of the steady-state. We do not choose to stop the blood flow just in order to have a strictly unique solution. Also, there is really no adequate way to measure an initial velocity profile, and, as in the economy, the situation is perhaps too dynamic for this to be a practical prospect.

Note 5. In our model of blood flow (see Note 4) the driving force is the pressure gradient just as demand is for our model of a demand-output economy. However, we have carefully considered the role of the pressure gradient as the primary data to be used in blood flow [3, Chapter 1]. It was chosen because it is the only basic entity which is really feasible to consider being accurately measured in the body. Velocity, regarded as determined in the steady-state by the pressure gradient and independently of the initial values, is not really feasible to measure. For the economy, on the other hand, we have chosen demand to be the primary driving force in our model because this seems to be the prevailing attitude in the literature, but we have little confidence that it can be easily measured or that it can be rigorously derived from rather easily measured quantities like population and cultural style. Of course, if demand is the logical driving force, we can come to some conclusions without knowing the demand functions; it can, for example, be noted in the case analyzed above (structures depending only on time) that the property of mean-square asymptotic uniqueness does not depend on the demand functions but only on the nature of the structural coefficients.

It is quite possible that outputs are more easily measured than demands, and from some points of view of modern science it would then be more suitable as primary data. In moving to models involving differential equations, as dynamic problems would seem to demand, it becomes quite important to distinguish properly between the forcing functions (here demands) and the dependent variable functions (here outputs), but we cannot know if we have made the correct choice until some equations like (2.5) are derived out of considerations of the underlying mechanisms of economics. The dynamic system of Leontief [1, p.57] does seem to be so derived, and the forcing functions there are demands. After studying the Leontief model, the philosophy behind using demands as the driving forces in an economy begins to seem natural, and we have, therefore, adopted it. Probably also, it is natural to treat demands as given and not pretend to derive them from other economic factors just because their determination arises partly from non-economic factors or, at least, from factors that are difficult to relate to an economic framework.

References

[1] Wassily W. Leontief, The Structure of the American Economy, 1919-1939, an Empirical Application of Equilibrium Analysis, 2nd ed., Oxford Univ. Press, New York, 1951.

[2] L.B.M. Mennes, Jan Tinbergen, J. George Waardenburg, The Element of Space in Development Planning, North-Holland, 1969.

[3] H.M. Lieberstein, Mathematical Physiology, blood flow and the electrical activities of cells, American Elsevier, 1972

[4] H.M. Lieberstein, A Course in Numerical Analysis, Harper & Row New York, 1968.

[5] H.M. Lieberstein, A possible four mode operation of neurons and chains of fibroblasts; transmission mechanisms for an early warning alert system. Math. Biosci., 12(1971), 7-22.

SIMULATION OF POPULATIONS, WITH PARTICULAR REFERENCE TO THE GRAIN BEETLE, TRIBOLIUM

N. W. Taylor

*University of New England, Armidale, N.S.W.
Australia*

THE MODEL

Suppose a population consists of a species whose life cycle goes through a number of stages. Let

$N(j;t)$ = the number in the jth stage at time t

$\Delta N(j;s,t)$ = the number aged between $s-\Delta t$ and s at time t, s being measured from the time of entering the stage j

$a(j)$ = the age span of the jth stage

$b(s)$ = the birth rate per adult of age s

$p(j;s,t)$ = the probability of surviving in the jth stage from age s to $s+\Delta t$ at time t.

Since $p(j;s,t)$ is the proportion of those of age s in stage j which survive for a time Δt, then

(1) $\Delta N(j; s+\Delta t, t+\Delta t) = \Delta N(j;s,t) p(j;s,t)$

for $s = \Delta t$ to $a(j)-\Delta t$, and for all stages j.

Since the number entering stage j at age zero depends on the number leaving the previous stage j-1 at age $a(j-1)$, then

(2) $\Delta N(j;\Delta t, t+\Delta t) = \Delta N(j-1; a(j-1, t) p(j;0,t)$

for $j = 2, 3, 4, \ldots$.

The number entering the initial stage, $j = 1$, depends on the birth rate. If J denotes the mature adult stage, then

(3) $\Delta N(1;\Delta t, t+\Delta t) = \sum_{s=\Delta t}^{a(J)} \Delta N(J;s,t) b(s) \Delta t.$

The total numbers in each stage are given by summing these:

(4) $N(j; t+\Delta t) = \sum_{s=0}^{a(j)-\Delta t} \Delta N(j; s+\Delta t, t+\Delta t)$

where the equations (1) and (3) are used for $j = 1$, and (1) and (2) for $j = 2, 3, 4, \ldots$.

These equations give the population at time $t+\Delta t$ in terms of that at time t, so that if the initial population is known, its whole development can be traced. A population model in this form is easily programmed for a digital computer.

In populations consisting of a number of species, as in the cases of competition, predator-prey, and host-parasite systems, there will be a set of equations like (1) - (4) for each species.

Tribolium, a grain beetle that infests flour, is a convenient animal for laboratory experiments on population studies, and an extensive literature exists. The populations are self-regulatory and always tend to a steady state provided the food supply and environment are maintained. The cause of this population regulation is cannibalism, the live forms destroying the inert and perhaps also the weaker live forms. For the model, six main stages are taken: eggs ($j = 1$), small larvae (2), large larvae (3), pupae (4), callows (5), and mature adults ($J = 6$).

If $k(j)$ = the proportion of eggs destroyed by one member of stage j in unit time,

then, to a first approximation,

(5) $p(1;s,t) = 1 - \{k(2)N(2;t)+k(3)N(3;t)+k(6)N(6;t)\} \Delta t.$

If $n(j)$ = the proportion of pupae destroyed by one member of stage j in unit time, then, similarly,

(6) $\quad p(4;s,t) = 1 - \{n(2)N(2;t)+n(3)N(3;t)+n(6)N(6;t)\} \Delta t.$

The other $p(j;s,t)$ depend mainly on accidental and, particularly in the case of mature adults, on the natural mortality. Losses due to accident are usually negligible compared with those due to cannibalism.

For species of <u>Tribolium</u> in competition the live forms of each species attack the inert forms of all species, so that there will be a set of equations like (5) and (6) for each species, with the added complication that each equation will contain a set of terms like those in the braces { } for each predator species.

For predator-prey and host-parasite systems the probabilities of survival depend again on the numbers present, but the form of the dependence is different since the presence of some species assists the survival of others.

If the environment changes, the probabilities are functions also of the time-dependent variables describing the environment.

USES OF THE MODEL

The population characteristics $a(j)$, $b(s)$, $k(j)$, $n(j)$, and those $p(j;s,t)$ which do not depend on cannibalism can be determined from separate experiments. When the values for these are inserted into the equations (1) - (6) a computer simulation of the population can be run, and the results compared with experiments on whole populations. In this way, assumptions concerning the behavior of the insects can be checked.

It can be shown, by making selective alterations to the model, which factors are the most effective in determining the population and hence where further experimental data are required. The model can also be used to predict the results of various treatments such as changing the parameters or imposing external effects.

In both the model and the experiments the populations tend to a steady state (or to a steady state with a small oscillation superimposed, depending on the relative magnitudes of the parameters). This is due to the manner in which the $p(j;s,t)$ depend on the $N(j;t)$. The model is in fact an elaboration of the Pearl-Verhulst model given by the equation

$$dN/dt = bN(1 - N/N_o), \quad b > 0$$

which tends to the steady state solution, $N = N_o$.

Changes can be incorporated easily into a computer model of this type. A start can be made with a very simple model, and extensions and alterations made as additional information becomes available.

THE COMPETITION MODEL

In the case of competition between *Tribolium* <u>confusum</u> and *Tribolium* <u>castaneum</u>, one species dies out and the other takes over and reaches a steady state. The outcome depends on the strains used, and the model gave results in general agreement with the experiments of Park, Leslie, and Mertz (1964).

For *T. confusum* (strain bIV) and *T. castaneum* (strain cII) the outcome of the experiment was uncertain, cII being eliminated in 6 out of 10 replicates. The model reflected this uncertainty in the very long time taken for bIV to be eliminated. A small change in the parameters could have made cII the winner.

THE EFFECT OF CHANGING THE PARAMETERS

Young (1970) describes experiments to determine the relative effects of the various cannibalism processes on the adult population. An investigation such as this is very easily performed on the model. Again, the results for the model agree with the experiments. For both *T. confusum* and *T. castaneum* the process most effectively limiting the adult population is the destruction of pupae by adults. Eggs are usually more susceptible to destruction than pupae ($k(j) > n(j)$), but this is more than offset by the greater reproductive value of the pupae, the stage closely approaching that in which reproduction occurs (Fisher 1929).

A SELECTION OF THE LITERATURE

Among the classic works on population modeling, before high-speed computers entered the field, are those by Gause (1934), Lotka (1925), and Volterra (1926). More recently, Watt (1968) and Patten (1971) have shown how a variety of problems in population studies may be attacked by the use of computers. Early mathematical models of *Tribolium* populations have been given by Stanley (1932a, b) and Landahl (1955a, b). There are more details of the model described above in papers by the author

(Taylor 1968, 1972). Some examples of the various ways of making the Tribolium and other models more life-like by introducing an element of randomness have been given by Leslie (1962), Niven (1967), and Coulman, Reice, and Tummala (1972). Some papers containing the experimental results from which the parameters for the Tribolium populations were calculated are by Park, Mertz, and Petrusewicz (1961), Brereton (1962), Park, Leslie, and Mertz (1964), Mertz, Park, and Youden (1965), Park, Mertz, Grodzinski, and Prus (1965), Mertz (1969), and Young (1970). More extensive bibliographies on these topics can be found in the works cited.

REFERENCES

Brereton, J. Le G. 1962. A laboratory study of population regulation in Tribolium confusum. Ecology 43: 63-69.

Coulman, G. A., S. R. Reice, and R. L. Tummala. 1972. Population modeling: a systems approach. Science 175: 518-521.

Fisher, R. A. 1929. "The Genetical Theory of Natural Selection." Ch. 2. Dover, New York (Reprinted, revised 1958).

Gause, G. F. 1934. "The Struggle for Existence." Williams and Wilkins, Baltimore, Maryland.

Landahl, H. D. 1955a. A mathematical model for the temporal pattern of a population structure, with particular reference to the flour beetle. Bull. Math. Biophys. 17: 63-77.

Landahl, H. D. 1955b. A mathematical model for the temporal pattern of a population structure, with particular reference to the flour beetle. II. Competition between species. Bull. Math. Biophys. 17: 131-140.

Leslie, P. H. 1962. A stochastic model for two competing species of Tribolium and its application to some experimental data. Biometrica 49: 1-25.

Lotka, A. J. 1925. "Elements of Mathematical Biology." Dover, New York (Reprinted 1956).

Mertz, D. B. 1969. Age-distribution and abundance in populations of flour beetles. I. Experimental studies. Ecol. Monogr. 39: 1-31.

Mertz, D. B., T. Park, and W. J. Youden. 1965. Mortality patterns in eight strains of flour beetles. Biometrics 21: 99-114.

Niven, B. S. 1967. The stochastic simulation of Tribolium populations. Physiol. Zool. 40: 67-82.

Park, T., D. B. Mertz, and K. Petrusewicz. 1961. Genetic strains of Tribolium: their primary characteristics. Physiol. Zool. 34: 62-80.

Park, T., D. B. Mertz, W. Grodzinski, and T. Prus. 1965. Cannibalistic predation in populations of flour beetles. Physiol. Zool. 38: 289-321.

Park, T., P. H. Leslie, and D. B. Mertz. 1964. Genetic strains and competition in populations of Tribolium. Physiol. Zool. 37: 97-162.

Patten, B. C. (Ed.). 1971. "Systems Analysis and Simulation in Ecology." Volume I. Academic Press, New York and London.

Stanley, J. 1932a. A mathematical theory of the growth of populations of the flour beetle, Tribolium confusum Duv. Can. J. Research 6: 632-671.

Stanley, J. 1932b. A mathematical theory of the growth of populations of the flour beetle, Tribolium confusum Duv. II. The distribution by ages in the early stages of population growth. Can. J. Research 7: 426-433.

Taylor, N. W. 1968. A mathematical model for two Tribolium populations in competition. Ecology 49: 843-848.

Taylor, N. W. 1972. Simulation of Tribolium populations. Proc. Ecol. Soc. Australia 6: (At press).

Volterra, V. 1926. Variazioni e flutuazioni del numero d'indivui in specie animali conviventi. Mem. Acad. Lincei 2: 31-113.

Watt, K. E. F. 1968. "Ecology and Resource Management." McGraw-Hill, New York.

Young, A. M. 1970. Predation and abundance in populations of flour beetles. Ecology 51: 602-619.

POLLUTION BY DIFFUSIVE PROCESSES

H. S. Green
University of Adelaide

The Role of Diffusion in Pollution

The most serious types of pollution are those following the release of the pollutant from points within or on the boundary of a gaseous or liquid medium. The problems to be considered in this paper relate to the determination of the distribution of the pollutant within the medium and along the boundary of the medium.

In most instances, the propagation of the pollutant is the combined effect of three natural processes: transport by currents in the fluid, diffusion by macroscopic processes of a stochastic character, and molecular diffusion.

The effect of a current is simply to carry the pollutant from one place to another in a manner which can, in principle, be predicted exactly by hydrodynamical or rheological methods. This is true whether the current is steady or (like tidal currents in the ocean and estuaries, or macrometeorological currents in the atmosphere) varying with time in a known, determinate way. Because of its predictable character, such transport offers relatively simple problems to both the applied mathematician and the pollution engineer. But, as the meteorologist, for instance, knows only too well, the predictions usually have a range of validity which is severely limited in time and volume, partly because of incomplete data and partly because of non-linear dynamics leading to the rapid growth of small disturbances under favourable conditions. Some of the resulting uncertainties can no doubt be reduced. But, very likely, the details of turbulent motion, local convection currents and other instabilities are not predictable except over quite short periods of time, with surprising consequences for the transport of pollutants in the fluid medium. Thus, beyond a certain point, the refinement of the input data will not noticeably improve the quality of the forecast, and something more than the usual hydrodynamical and rheological modelling is needed.

As an indication of the type of analysis required, it may be noticed that the existing theories of turbulence[1], for instance, are of a statistical character, and that consequently only statistical predictions can be expected. The "best" theories of turbulence are, however, not yet well adapted to practical problems concerning the transport of pollutants, even if there were no other processes to take into account. A much more flexible tool is provided by the theory of the Brownian motion. Though it was not originally intended for such purposes, it is in fact, ideally suited to the study of transport by macroscopic processes of a stochastic character. Such diffusive processes include not only turbulence, convection, and the like, but surface phenomena such as breaking waves in fluids and the gusts and eddies associated with boundary irregularities in gases, and quite generally effects arising from a multitude of small environmental features which it is not possible or practicable to take into account exactly. Naturally the random motions of the pollutant which result must be superposed on the predictable motion associated with the current, where both are present. Although the original theory of the Brownian motion, as formulated by Einstein, did not provide for this possibility, subsequent generalizations by Fokker and Planck, Chandrasekhar and others[2] do so in a very natural way. Of course the Brownian motion of tiny foreign bodies in fluids is due to fluctuations in the resultant force exerted by the molecules of the fluid, and is thus quite different in origin from the random motion of the typical pollutants, but it is nevertheless of the same general character and the same mathematical techniques may be applied. It is perhaps worth mentioning that there are quite a number of other applications, to processes as different as the passage of vesicles across endothelial cells[3] and the diffusion of electrons and ions in a plasma[4].

Another important consideration is that, if the pollutant is itself a fluid or a sufficiently fine suspension, the effects of molecular diffusion must be taken into account, especially over long periods of time. Without such diffusion, a small volume of pollutant released at a particular place and time would still be located in a well defined region at any subsequent time. Although molecular diffusion is normally a slow process whose effect on transport is small compared with those already considered, from the point of view of pollution it is much more insidious, since it is, in the strict sense of thermodynamics, irreversible. It confronts us with the second law and with the foreboding that the state of absolute chaos in which pollution has done its worst can only be somewhat postponed! Actually it is hard to distinguish between the effects of molecular and macroscopic diffusion, and there is good reason to believe[5] that some of the macroscopic phenomena, like turbulence, have a molecular origin. Anyway, as Kirkwood was the first to demonstrate in detail[6], diffusion and other irreversible processes can be accounted for very well by the theory of the Brownian motion. In the transport of particles in fine suspension, of course, this theory is only one and is manifestly appropriate.

We may conclude that the theory of the Brownian motion is able to take into account all three processes which we have implicated in the pollution of and by a fluid medium, and although the mechanisms in macroscopic and molecular diffusion are quite different, it is not necessary to distinguish them in the subsequent mathematical analysis. It is reassuring to find that we do not need to enter into a detailed study of molecular physics to grapple with the practical problems to be considered. But we should be as cautious not to oversimplify the analysis as to avoid unnecessary complications, and it may serve a useful purpose to point out, in a more general context, some of the pitfalls into which a too naive approach may lead.

First it is worth asking whether, since diffusion is involved, we could not dispense with the stochastic theory (of the Brownian motion) altogether and make use of the better known laws of diffusion and conservation which have proved adequate in wide areas of macroscopic physics and chemistry. Indeed, in simple problems where the diffusion is isotropic, it is known that the elementary theory of diffusion gives results which are asymptotically identical with the stochastic theory, so that if one is interested in distributions a long time after the entry of the pollutant, there is no valid reason why the elementary theory should not be used. The "long time" referred to is the relaxation time necessary for steady state diffusion to be established, and is a parameter of the problem which should be determined at the outset. Within this relaxation time the elementary theory is inapplicable and may give quite misleading estimates of the rate of accumulation of the pollutant in the neighbourhood of the source. Some sources of pollution, such as exhaust gases, are periodic in intensity, and if the period is not long compared with the relaxation time, again the elementary theory may lead to incorrect conclusions.

The relaxation time, according to the theory of the Brownian motion, is related to the effective viscosity of the medium, and, more directly, to the effective coefficient of diffusion of the pollutant in the medium. The precise relations will be discussed later, and the point of mentioning them here is to underline the distinction between the effective transport coefficients involved and those determined by ideal experiments, which can be very important if the pollutant consists of solid particles, or even of molecules which are much larger than those of the fluid medium. It is precisely here that the complexities of molecular theory have a bearing on the problems we wish to consider. Nevertheless these complexities can be avoided by the direct experimental determination of the parameters involved, and there seems to be an immediate need for experimental work of this kind, since the data available at present are quite scarce.

A cautionary remark appears to be needed concerning the applicability of certain well known results of the theory of Brownian motion. Einstein's formula

$$<\Delta x^2> = 2Dt \qquad (1)$$

for the mean square displacement of a Brownian particle in a given direction, in terms of the coefficient of diffusion D, is of course only correct when the time t is long compared with the relaxation time. More importantly, it is only correct in parts of the fluid remote from any boundary, and is practically useless in the estimation of the distance traveled by pollutants from their source, where this is located on or near a boundary. An incorrect application of this formula can easily lead to errors in order of magnitude. Thus, a careful study of the boundary value problems associated with the diffusion equation is essential to the problems confronting us. A proper understanding of the boundary conditions cannot be dispensed with in this connection. Most pollutants are neither completely absorbed nor ideally reflected from a boundary of the fluid with which they make contact. Then an important parameter is the attachment coefficient, which measures the probability of attachment to the boundary if such contact occurs. In some instances it is possible for absorption to be followed by re-emission, and the mean time of attachment is a further parameter required for a complete mathematical solution of the diffusion problem.

A further limitation of both the elementary theory of diffusion, and the stochastic theory in its simplest form, is that in many situations where turbulence, convection and similar phenomena are dominant, diffusion is no longer isotropic. In consequence, the diffusion process itself can no longer be described in terms of a single parameter (the coefficient of diffusion, or the friction constant) and a tensorial formulation is necessary. The nature of the required generalization will be indicated subsequently.

What has been said so far was mostly independent of the nature of the pollutant, whether fluid, fine solid suspension, or consisting of solid objects of larger size. The final parameters of the problem are the characteristic length, or diamater, and the mass, of the independently moving particles or pieces of the pollutant. The characteristic length, and the attachment coefficient already referred to, are important in determining whether the pollutant will escape into the fluid medium, or be returned to a point of the boundary not far from the point of origin. The mass, of course, determines the effect of gravity on the diffusion process, but also influences other parameters, such as the relaxation time and the coefficient of diffusion, which affect the transport process in a fundamental way.

In spite of all these complications, the stochastic theory, and the elementary theory of diffusion, are both sufficiently well developed to allow us to predict the distribution at any time of a pollutant from a known source. The inverse problem, that of determining the source or sources of pollution from its observed distribution, is of equal, or perhaps greater interest, and can be analysed by the same methods. Examples of the qualitative and quantitative results already available for systems with simple geometries will be given later in this paper. We shall, however, next proceed to consider in more detail the principles which have been used in these analyses.

Characteristics of the Diffusive Process

In what follows, we shall generally assume that the concentration of the pollutant has not reached the level where interactions between its particles play an important role. We thus avoid consideration of certain types of non-linear phenomena beyond those already essential to our problem which, though of great intrinsic interest, appear only in the final stages of pollution. The density of the pollutant in the fluid (mean number of particles per unit volume) is a function $P(\underline{x},t)$ of position and time. Under the present hypothesis it is linearly related to the source density $S(\underline{x},t)$, which represents the number of particles per unit volume released from the point $\underline{x}$ at the time t:

$$P(\underline{x},t) = \int n(\underline{x},t\,;\,\underline{x}',t')\,S(\underline{x}',t')d^3x' \,. \qquad (2)$$

Here $n(\underline{x},t,\underline{x}',t')$ is the probability, per unit volume of finding a particle, which was released from the point $\underline{x}'$ at time t', in the neighbourhood of the point $\underline{x}$ at time t. The reabsorption of the pollutant at the boundary of the fluid is determined by the boundary conditions.

According to the elementary theory of diffusion, the Green's function $n(\underline{x},t;\underline{x}',t')$ which provides the master solution to our problem satisfies the diffusion equation

$$\frac{\partial n}{\partial t} + \frac{\partial}{\partial \underline{x}} \cdot \left(\underline{u}n - D\frac{\partial n}{\partial \underline{x}}\right) = 0 \qquad (3)$$

where $\underline{u}$ is the local velocity of the current in the fluid, D is the effective coefficient of diffusion and $\partial/\partial \underline{x} = \nabla$ is the three-dimensional differential operator often written as div or grad. The limitations of this theory have already been noted. Where necessary, it should be replaced by the corresponding stochastic theory, in which the fundamental function is not n, but the velocity distribution function $f(\underline{v},\underline{x},t)$, a density in velocity space as well as actual space, from which n is readily determined:

$$n(\underline{x},t \; ; \; \underline{x}',t') = \int f(\underline{v},\underline{x},t) d^3 v \; . \qquad (4)$$

This velocity distribution function satisfies Liouville's equation

$$\frac{\partial f}{\partial t} + \underline{v} \cdot \frac{\partial f}{\partial \underline{x}} + \frac{\partial}{\partial \underline{v}} \cdot (\underline{a}f) = 0 \; , \qquad (5)$$

in which $\underline{a}$ represents the mean acceleration of a particle with velocity $\underline{v}$ at the point $\underline{x}$ in the fluid, at time t. The stochastic theory is completed by writing[7]

$$\underline{a} = \underline{F} - \xi(\underline{v} - \underline{u}) - \frac{\xi kT}{mf}\frac{\partial f}{\partial \underline{v}} \qquad (6)$$

where k is Boltzmann's constant, T is the absolute temperature, m is mass of the particle, $\underline{F}$ is the resultant force per unit mass acting on the particle, not of course including the velocity-dependent force $-m\xi(\underline{v} - \underline{u})$ exerted by the fluid. The constant ξ is called the friction constant, and is related to the effective coefficient of diffusion, when the elementary theory is applicable:

$$\xi = kT/(mD) \; . \qquad (7)$$

Another relation connects the friction constant with the effective viscosity of the fluid. For a large particle, Stokes' formula

$$\xi = 6\pi b\eta/m$$

(where b is the mean radius of the particle) is applicable, but it follows from Kirkwood's theory[6,7] that this is only an approximate form of the relation

$$\eta = \frac{m\xi}{6\pi b} + \frac{\nu kT}{2\xi} \; , \qquad (8)$$

in which ν is the molecular number density of the fluid medium. We are obliged to use the terms "effective viscosity" and "effective coefficient of diffusion" because of the influence of turbulence and other phenomena of the type already discussed.

The friction constant is the reciprocal of the relaxation time required for a steady state of diffusion to be established, when no variable forces are present. Under these conditions, therefore, the relaxation time is

$$t_r = 1/\xi = mD/(kT) \ . \tag{9}$$

It has already been noticed that in certain applications, particularly those in which turbulence, convection and the like are important, the diffusion may be anisotropic. In such instances, the formula (6) must be generalized by replacing the friction constant ξ by a tensor:

$$\underline{a} = \underline{F} - \underline{\underline{\xi}} \cdot (\underline{v} - \underline{u}) - \frac{kT}{mf} \underline{\underline{\xi}} \cdot \frac{\partial f}{\partial \underline{v}} \ . \tag{10}$$

The coefficient of diffusion, D, must also be replaced by a tensor $\underline{\underline{D}}$, and the relation (7) becomes one between tensors.

We next consider the boundary conditions which are required to integrate either the diffusion equation (4), or the stochastic equation resulting from the substitution of (6) into (5). If every particle reaching the boundary became attached to it, the condition P = 0 would apply. If this condition were even approximately satisfied, the effect of the random motions would be to return almost every particle to the boundary not far from its point of release, just as a bottle thrown into breaking waves on the sea shore is almost inevitably washed up again. On the other hand, the condition for ideal reflection from the boundary is $\nabla P = 0$ at the boundary, and in the absence of forced diffusion, the pollutant will tend to approach a uniform distribution throughout the fluid, at a sufficiently long time after its release. This is more likely to happen if the pollutant is in molecular form, though even then adsorption at the boundary may be significant. In general, the correct boundary condition is of the type

$$\underline{v} \cdot \frac{\partial P}{\partial x} + \lambda P = 0 \tag{11}$$

(at the boundary), where $\underline{v}$ is the outward normal to the boundary of the fluid and λ is a coefficient whose value obviously has a critical effect on the distribution of pollutant not only at the boundary but within the fluid. It is related to the attachment coefficient α, whose importance has already been noticed, by

$$\lambda = \alpha/b \ . \tag{12}$$

The imposition of the boundary condition is the thing which makes the solution of the diffusion equation, and of course also the stochastic equation, mathematically challenging. Probably numerical techniques will be needed for systems with more complicated geometries, but since these do not generally provide much insight into the problems which they purport to solve, we shall prefer to consider here some of the more powerful analytical techniques. A good introduction to the latter is provided by Sneddon's well known treatise[8]. Perhaps the most useful is a generalization of the method of images which is commonly used in obtaining analogous solutions of Laplace's equation. The efficacy of such a generalization is suggested by the fact that the static solutions approached after a long time by the solutions of the diffusion equation must in fact satisfy Laplace's equation. The method in question allows us to ignore the presence of the boundary of the fluid, provided that it is replaced by an equivalent distribution of sources and sinks outside the boundary. Thus, the formula (2) is replaced in actual calculation by

$$p(\underline{x},t) = \int n(\underline{x} - \underline{x}', t - t') s(\underline{x}',t') d^3x' , \qquad (13)$$

where $p(\underline{x},t)$, $n(\underline{x} - \underline{x}', t - t')$ and $s(\underline{x}',t')$ have the same values as $P(\underline{x},t)$, $n(\underline{x},t;\underline{x}',t')$ and $S(\underline{x}',t')$ within the fluid, but are defined also outside the fluid, and the integration also extends outside the fluid. It will be noticed that n is now a function of the displacement $\underline{x} - \underline{x}'$ from the source, and the time $t - t'$ elapsed following release from the source. The solution of the diffusion equation, and also of the stochastic equation, is now a relatively simple matter, and the remaining problem, which is not so simple in general, is to choose the source distribution $s(\underline{x},t)$ outside the fluid in such a way that the condition is still satisfied at the boundary. The relation between $s(\underline{x},t)$ and $S(\underline{x},t)$ is a linear one, but as it depends on the geometry no general formula can be stated. However, the method by which $s(\underline{x},t)$ may be determined in principle, will be illustrated in the final section of this paper.

Applications and Results

For the purposes of this section, we shall suppose that the hydrodynamical or rheological problem of determining the current $\underline{u}$ in the fluid, as a function of position and time, has been solved completely. In fact, as we have already acknowledged, this task is not without its difficulties; but the means of overcoming them have been the subject of numerical and analytical studies for a long time, with not unsatisfactory results.

When there is no current, the solution of the diffusion equation (3) required for substitution into (13) is well known:

$$n(\underline{x},t) = (4\pi Dt)^{-3/2} \exp\left[- \underline{x}^2/(4Dt)\right] . \qquad (14)$$

More generally, when $\underline{u}$ no longer vanishes, the substitution

$$n(\underline{x},t) = (4\pi Dt)^{-3/2} \exp\left[- (\underline{x} - \underline{r})^2/(4Dt)\right] \tag{15}$$

leads to the transparent requirement $d\underline{r}/dt = \underline{u}$, so that

$$\underline{r} = \int_0^t \underline{u}(t' + \tau) d\tau , \tag{16}$$

where the integration takes in values of $\underline{u}$ along the fluid trajectory passing through the point $\underline{x}'$ of origin of the pollutant, at the time t' of release.

The appropriate solutions of the stochastic equation have also been found, at least for the simpler types of current and force field experienced by the pollutant. It is even possible to state solutions of a generalized equation, which is in fact needed when diffusion is anisotropic, so that the friction constant ξ has to be replaced by a friction tensor $\underline{\underline{\xi}}$, and the formula (10), rather than (6), is applicable. In general, the solution takes the form

$$f = \sqrt{\Delta} \exp\left[-\tfrac{1}{2}(\underline{w} \cdot \underline{\underline{\alpha}} \cdot \underline{w} + 2\underline{z} \cdot \underline{\underline{\beta}} \cdot \underline{w} + \underline{z} \cdot \underline{\underline{\gamma}} \cdot \underline{z})\right] ,$$

$$\underline{w} = \underline{v} - \underline{c}, \quad \underline{z} = \underline{x} - \underline{r} , \tag{17}$$

where $\underline{\underline{\alpha}}$, $\underline{\underline{\beta}}$, $\underline{\underline{\gamma}}$, Δ, $\underline{r}$ and $\underline{c}$ are all functions of the time. The vectors $\underline{r}$ and $\underline{c}$ can be simply interpreted as the displacement, relative to its point of origin, and the velocity, which the pollutant would have, at time t after its release, if it were moving under the influence of the force $\underline{F} - \underline{\underline{\xi}} \cdot (\underline{c} - \underline{u})$ alone:

$$\frac{d\underline{r}}{dt} = \underline{c} , \quad \frac{d\underline{c}}{dt} = \underline{F} - \underline{\underline{\xi}} \cdot (\underline{c} - \underline{u}) . \tag{18}$$

The normalization of f requires that Δ should be proportional to the determinant of the six-dimensional matrix formed by the tensors $\underline{\underline{\alpha}}$, $\underline{\underline{\beta}}$ and $\underline{\underline{\gamma}}$:

$$(2\pi)^3 \Delta = \det \begin{pmatrix} \underline{\underline{\alpha}} & \underline{\underline{\beta}} \\ \underline{\underline{\beta}} & \underline{\underline{\gamma}} \end{pmatrix} . \tag{19}$$

These tensors satisfy the non-linear differential equations

$$\frac{d}{dt} \underline{\underline{\alpha}} + 2\underline{\underline{\beta}} = 2\underline{\underline{\alpha}} \cdot \underline{\underline{\xi}} - \frac{2kT}{m} \underline{\underline{\alpha}} \cdot \underline{\underline{\xi}} \cdot \underline{\underline{\alpha}} ,$$

$$\frac{d}{dt} \underline{\underline{\beta}} + \underline{\underline{\gamma}} = \underline{\underline{\beta}} \cdot \underline{\underline{\xi}} - \frac{2kT}{m} \underline{\underline{\beta}} \cdot \underline{\underline{\xi}} \cdot \underline{\underline{\alpha}} ,$$

$$\frac{d}{dt} \underline{\underline{\gamma}} = - \frac{2kT}{m} \underline{\underline{\beta}} \cdot \underline{\underline{\xi}} \cdot \underline{\underline{\beta}} . \tag{20}$$

If the tensor $\underline{\xi}$ is diagonal, $\underline{\alpha}$, $\underline{\beta}$ and $\underline{\gamma}$ will also become diagonal, and the above equations can be solved in the following order of variables:

$$\gamma_i/(\alpha_i\gamma_i - \beta_i^2) \quad , \quad \beta_i/\gamma_i \quad , \quad 1/\gamma_i$$

(where the subscript i distinguishes the three diagonal elements). Thus, the solution is a simple generalization of that which applies under isotropic conditions:

$$\frac{\gamma}{\alpha\gamma - \beta^2} = \frac{kT}{m} \quad , \quad \frac{\beta}{\gamma} = -\frac{(1 - e^{-\xi t})}{\xi} \quad ,$$

$$\frac{1}{\gamma} = \frac{2kT}{m\xi^2}(\xi t - 1\tfrac{1}{2} + 2e^{-\xi t} - \tfrac{1}{2}e^{-2\xi t}) \quad , \tag{21}$$

assuming thermal release of the pollutant. The density distribution, which is of most interest, is given by

$$n = (4\pi D_1 t)^{-3/2} \exp\left[-(\underline{x} - \underline{r})^2/(4D_1 t)\right] \tag{22}$$

where

$$D_1 = \frac{1}{2\gamma t} + \frac{kT\beta^2}{mt\gamma^2} \tag{23}$$

is now a function of time, which however approaches the constant value D when the time t elapsed after release becomes long compared with the relaxation time $1/\xi$. Although the solution (22) resembles (15), the stochastic theory differs from elementary diffusion theory also in the formula for $\underline{r}$, which must be obtained by integration of the differential equations (18). This difference is quite important in taking account of gravity and other forces.

The above results show that the determination of the Green's function required for the evaluation of the pollutant $p(\underline{x},t)$ by means of (13) is not difficult, and hardly more troublesome in the stochastic theory than in the lementary diffusion theory. We may therefore turn to the more difficult problem of finding the extended source distribution which corresponds to the prescribed boundary conditions. To illustrate this problem, we consider a fluid medium between two parallel boundaries, such as a river between parallel banks. Analogous problems have been considered previously in other contexts[3,8].

Suppose that the distance between the two banks is d, and that x measures the distance from one of the banks. The boundary conditions require that the extended source distribution should satisfy

$$\int_{-\infty}^{\infty} n(x,t)\left(\frac{\partial s(x)}{\partial x} - \lambda s(x)\right)dx = 0$$

$$\int_{-\infty}^{\infty} n(d - x, t) \left(\frac{\partial s(x)}{\partial x} + \lambda s(x)\right) dx = 0 \qquad (24)$$

and are clearly fulfilled, provided that

$$\frac{\partial s}{\partial x} - \lambda s$$

is an even function of both x and $d - x$. Thus, if

$$s_1(x) = s(d - x),$$

we must have

$$s(x + d) = s_1(x) - 2\lambda \int_0^x e^{\lambda(x-y)} s_1(y) dy$$

$$s_1(x + d) = s(x) - 2\lambda \int_0^x e^{\lambda(x-y)} s(y) dy . \qquad (25)$$

As $s(x) = S(x)$, a known function, for $0 < x < d$, this pair of integral equations allows us to compute successively values for $d < x < 2d$, $-d < x < 0$, $2d < x < 3d$, $-2d < x < -d$, etc., and thus determine the whole extended distribution. Contributions from regions far inside the boundaries of the fluid may usually be safely neglected.

A variety of other geometries can be treated by similar methods, for instance those with boundaries intersecting at angles of 60° or 90°. For more complicated geometries, the same approach may be used, but a numerical technique based on a variational method is needed to make the appropriate integrals vanish.

We shall consider, finally, how to use the results of such calculations to make predictions or inferences of practical importance. Firstly, the distribution of the pollutant in the fluid medium is represented by the function $p(x,t)$ and is obtained by performing the three-dimensional integration required by (13). Another important function is the superficial density of pullutant on the boundary of the fluid. This is, of course, the integrated flux across the boundary and is given by

$$\int_0^t D_1 \underline{\nu} \cdot \nabla p \, dt$$

at time t following the release of the pollutant from its source. The values obtained in this way take no account of the possibility that absorption may occur at the boundary, followed by partial or total re-emission after a period of time. In problems where such a possibility exists, it is necessary to use the boundary density given above as an additional source.

Another useful procedure which can be carried out with the help of the above analysis is the determination of the source or sources of a pollutant from its observed distribution. In this application, the function $P(\underline{x},t)$ is known, and we are required to solve the integral equation (13) to determine the source function $S(\underline{x},t)$. The first step is to obtain the extended pollutant distribution $p(\underline{x},t)$, by the use of formulas similar to (25). The integral equation to be solved is thereby reduced to the form (13), which is amenable to standard techniques, i.e., Fourier transformation with respect to the spatial variables, and Laplace transformation with respect to the time variable.

These examples provide some indication of the utility of the theories of diffusion and the Brownian motion in the study of problems of pollution.

Work described in this paper was performed in part while the author was visiting Michigan State University and the University of Florida. He is indebted to Dr T. Triffet and Dr R.B. Leipnik for useful comments; also to Dr J. Casley-Smith, who suggested the biological application.

BIBLIOGRAPHY

[1] G. K. Batchelor, Theory of Homogeneous Turbulence (Camb. Univ. Press, Cambridge, 1953); J. O. Hinze, Turbulence (McGraw Hill, New York, 1959); J. L. Lumley, Stochastic Tools in Turbulence (Academic, New York, 1970).

[2] See selected papers, edited by N. Wax, Noise and Stochastic Processes (Dover, New York, 1954), or S. Chandrasekhar, Rev. Mod. Phys. 15, 1 (1943).

[3] H. S. Green & J. R. Casley-Smith, J. Theor. Biol. 35, (1972).

[4] See, e.g., B. Kursunoghu, Phys. Rev. 132, 21 (1963); J. H. Williamson, J. Phys. A (Proc. Phys. Soc.) 1, 629 (1968). Other uses of the theory are given in reference 2.

[5] See H. S. Green, Int. J. Engng. Sci. 1, 5 (1963).

[6] J. G. Kirkwood, J. Chem. Phys. 14, 180 (1946).

[7] H. S. Green, Molecular Theory of Fluids, p.160 (North-Holland, Amsterdam, 1952, and Dover, New York, 1969); Handbuch der Physik, 10,

[8] I. N. Sneddon, Mixed Boundary Value Problems in Potential Theory (North-Holland, Amsterdam 1966).

COMPUTER CONTROL OF PHYSICAL-CHEMICAL WASTEWATER TREATMENT

Dolloff F. Bishop,[1] Walter W. Schuk,[1]
Robert B. Samworth,[2] Ralph Bernstein[3]
and Elliott D. Fein[3]

[1]*National Environmental Research Center,
Cincinati, Ohio 45268*
[2]*Water Resources Managemement Administration,
Department of Environmental Services, District
of Columbia Government, Washington D.C. 2004,*
[3]*Federal Systems Division, International
Business Machines Corporation, Gaithersburg,
Maryland 20760*

INTRODUCTION

The District of Columbia Department of Environmental Services (DES) and the United States Environmental Protection Agency's (EPA) National Environmental Research Center, Cincinnati, Ohio, are engaged at the EPA-DC Pilot Plant in Washington, D.C. in a major pilot study of several advanced wastewater treatment systems[1]. The studies are providing design data while demonstrating the feasibility and reliability of the processes. The development of process control and data acquisition methods is a critical part of the pilot effort. The stringent objectives of product quality sought from the advanced systems mandate close process monitoring and control.

The DES and EPA engaged International Business Machines Corporation (IBM) to provide a computer system with flexible hardware and software as a research tool for the pilot plant. The broad objective of the computer research is to develop models and programs for data acquisition, alarm monitoring, and process control for the various treatment processes in the pilot plant.

The primary goal and the subject of this study are to develop hardware and software configurations for digital process control, alarm monitoring, and data acquisition of physical-chemical treatment of raw wastewater.

The secondary goal is to provide data logging and alarm monitoring for data from a second treatment system, the three-stage activated sludge system. The computer hardware for these goals must be suitable for the broad objective of future research in process control on other advanced wastewater treatment systems.

In the current study, the physical-checmical treatment of raw wastewater includes two-stage lime precipitation to remove solids and phosphorus, filtration to remove residual turbidity, chlorination with pH control to remove ammonia, and granular carbon adsorption to remove soluble organics. The three-stage activated sludge treatment consists of five processes: primary sedimentation for solids removal, modified activated sludge with mineral addition for BOD and phosphorous removal, nitrification for biological oxidation of ammonia to nitrate, denitrification with alum addition for removal of nitrate and residual phosphorous, and filtration for removal of residual turbidity.

The control work included analyzing the pilot plant processes and analog control systems; developing the dynamic response data; defining the control algorithms; and performing a simulation of the controlled plant.

PHYSICAL-CHEMICAL TREATMENT

The pilot system (see Figure 1) selected for digital process control consists of hydro screening, two-stage (high pH) lime precipitation with intermediate recarbonation, dual-media filtration, pH control with chlorine and with CO_2 stripping, breakpoint chlorination, and downflow granular carbon adsorption. The process is designed for a nominal capacity of 50,000 gallons per day. A flow controller may impress a diurnal variation up to 4:1 maximum to minimum flow across the clarification process (2:1 maximum to average flow).

In the first stage of the precipitation process, raw wastewater, dry powdered CaO at a dosage of approximately 350 mg/l, and recycled solids are rapid mixed at 250 rpm for 7.5 minutes (average) and then flocculated for 31 minutes in a turbine mixed (34 rpm) flocculator. The lime increases the wastewater pH to above 11.5, and precipitates and removes bicarbonate, phosphate, and magnesium ions from the water (equations 1-3).

$$Ca^{++} + HCO_3^- + OH^- \rightarrow CaCO_3\downarrow + H_2O \qquad (1)$$

$$5Ca^{++} + 3H_2PO_4^- + 7OH^- \rightarrow Ca_5OH(PO_4)_3\downarrow + 6H_2O \qquad (2)$$

$$Mg^{++} + 2OH^- \rightarrow Mg(OH)_2\downarrow \qquad (3)$$

The magnesium hydroxide is an efficient flocculant which aids in the removal of the particulate organics.

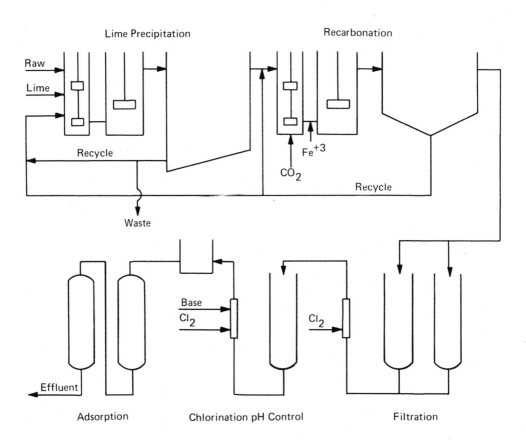

Figure 1. Physical-Chemical Wastewater Treatment

The precipitated and flocculated solids are settled in a rectangular settler with 130 minutes average detention time at an average overflow rate of 1000 gpd/ft^2. Sludge from the slurry pool at the bottom of the settler, recycled at a rate equal to 10 percent of the average influent flow, produces solids concentrations within the reactor of 3000 to 5000 mg/l. Waste rates of approximately 1.5 percent of the influent flow with solids concentrations of 30,000 to 50,000 mg/l maintain the solids balance of the slurry pool.

The limed water, after sedimentation, flows through an open channel to recarbonation. Carbon dioxide, added at an approximate average dosage of 145 mg/l below a turbine operating at 171 rpm in a 17-minute recarbonation tank, reduces the wastewater pH from 11.6 to 9.5 and precipitates the excess calcium ions added in the liming (first) stage (equation 4).

$$Ca^{++} + CO_2 + 2OH^- \rightarrow CaCO_3\downarrow + H_2O \tag{4}$$

$$Fe^{+++} + 3OH^- \rightarrow Fe(OH)_3\downarrow \tag{5}$$

Five mg/l of ferric ions are also added in the recarbonation tank to form $Fe(OH)_3$ (equation 5) which flocculates the precipitated calcium carbonate. The water is then flocculated in a 31-minute flocculation basin which is turbine mixed at 34 rpm.

After flocculation, the water flows to a center circular settler with an average overflow rate of 1000 gpd/ft^2. Settled slurry recycled from the bottom of the settler to recarbonation at a flow rate equal to 10 percent of the average influent flow provides nuclei for the calcium carbonate precipitation in the recarbonation tank. Wasting at flows equal to 1.5 percent of the influent maintains the solids balance of the slurry pool in the settler.

The overflow from the second stage settler is pumped to a distribution box ahead of two filters. It then flows by gravity through two dual-media filters to remove residual particulates. Each filter is packed with 24 inches of 0.9 mm coal and 6 inches of 0.45 mm sand. When pressure drop across a filter exceeds 9 feet of H_2O, air scrubbing for 5 minutes at 5 SCFM of air and backwashing at 15 gpm/ft^2 for 10 minutes automatically cleans the filter and returns it to service.

After filtration, chlorine reduces the pH of the filter water to 6 to 7 in the prechlorination reactor (static mixer with 1.6

seconds average retention time). The water then is pumped to the CO_2 stripper to remove carbon dioxide prior to "breakpoint" chlorination. The carbon dioxide stripping reduces the base required in breakpoint. To remove ammonia, the wastewater is breakpointed with chlorine and base (equations 6, 7, and 8) under careful pH control, usually 7.0 (equation 9) in a second static mixer with a 1.6 second average retention time followed by a 1-minute holding tank.

$$Cl_2 + H_2O \rightarrow HOCl + HCl \tag{6}$$

$$NH_4^+ + HOCl \rightarrow NH_2Cl + H_2O + H^+ \tag{7}$$

$$2NH_2Cl + HOCl \rightarrow N_2\uparrow + 3HCl + H_2O \tag{8}$$

$$NaOH + HCl \rightarrow NaCl + H_2O \tag{9}$$

The breakpoint occurs at the point where NH_3-N is reduced to zero, free available chlorine is detected, and the total residual chlorine is minimized. Several undesirable side reactions also occur.

$$NH_2Cl + HOCl \rightarrow NHCl_2 + H_2O \tag{10}$$

$$NHCl_2 + HOCl \rightarrow NCl_3 + H_2O \tag{11}$$

$$NH_4^+ + 4Cl_2 + 3H_2O \rightarrow HNO_2 + 8HCl + H^+ \tag{12}$$

The amount of chlorine required depends upon the ammonia and nonammonia chlorine demand and the amount of residual free chlorine in the wastewater. In the D.C. lime clarified raw wastewater, the $Cl:NH_3$-N dosage weight ratio usually is approximately 9:1. The reaction conditions are manipulated by rapid and complete mixing, pH control, and free residual chlorine dosage control to favor oxidation to N_2 and thus minimize the products of equations 10, 11, and 12.

The flow from the breakpoint reactor and holding tank is pumped through four downflow carbon columns to remove most of the soluble residual organics. Each column is packed with 840 pounds of 8 by 30 mesh granular carbon. The carbon is supported by a 50 mesh stainless steel screen on a gravel bed. The first column is backwashed and surface-washed on a daily basis with a quantity of water equal to about 4 percent of the product water. To simulate two-stage carbon treatment, the spent carbon is replaced with virgin carbon in two columns at a time. The carbon is not regenerated as part of the study but would be in full-scale operation.

The sludges from the treatment systems are thickened and classified in a centrifuge to separate the calcium carbonate from the noncarbonate sludges. The calcium carbonate in the centrifuge cake is recalcined to recover most of the lime and in full-scale operation to produce CO_2 needed in recarbonation.

$$CaCO_3 \xrightarrow{\Delta} CaO + CO_2\uparrow \qquad (13)$$

Research is currently being conducted to evaluate the best handling method for the noncarbonate sludges in the centrate.

The plant's process control may be achieved with the plants sensors and actuators (see Table 1) by either analog or digital systems which provide control of CaO, CO_2, $FeCl_3$, Cl_2, and NaOH chemical feeds and solids wasting (see Figure 2). The control system also includes an electropneumatic logic network based upon time, level, and differential pressure to automatically control the filter's backwashing cycle.

Table 1. Physical Chemical Control System

Stage	Control Objective	Control Variable	Sensors	Actuators
Lime clarification	pH setpoint range: 11.3-12.0	CaO Feed range: 0-36 lb/hr	pH assembly Magnetic flow meter	Gravimetric feeder
	Sludge wasting range: 0.5-2.5% of flow	Volume	Magnetic flow meter Level switch	Electropneumatic ball valve
Recarbonation	pH setpoint range: 9-9.8	CO_2 Feed range: 0-200 lb/hr	pH assembly Magnetic flow meter	Equal percentage valve
	Sludge wasting range: 0.5-2.5% of flow	Volume	Magnetic flow meter Level switch	Electropneumatic ball valve
	$FeCl_3$ dosage range: 0-10 mg/l	$FeCl_3$ Feed	Magnetic flow meter	Peristaltic pump
pH Control (prechlorination)	pH setpoint range: 6.0-7.5	Cl_2 Feed range: 0-150 lb/day	pH assembly Magnetic flow meter	V notch chlorinator
Breakpoint chlorination	Free Cl_2 residual range: 0.5-8 mg/l	Cl_2 Feed range: 0-150 lb/day	Magnetic flow meter AutoAnalyzers for Cl_2 and NH_3	V notch chlorinator
	pH setpoint range: 7-8.5	NaOH Feed range: 0-3.7 liter/min. (8.5%) NaOH by weight)	pH assembly Magnetic flow meter	Positive displacement pump

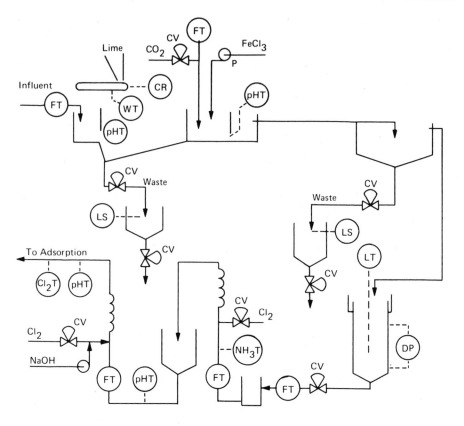

Figure 2. Physical Chemical Control System

The control system indirectly controls the water quality by minimizing error in the pH at four points, the dosage of $FeCl_3$ and the concentration of Cl_2 in breakpoint. The complete treatment system under analog control typically removes (see Table 2) approximately 95 percent of the BOD and COD, 98 percent of the total phosphorus, and 88 percent of the total nitrogen from the District of Columbia wastewater.

Table 2. Typical Physical-Chemical Treatment of D.C. Wastewater

Effluent	Pollutants, mg/l			
	BOD	COD	P	N
Raw	129	307	8.4	22.0
Clarification	24	55	0.3	15.0
Filtration	20	48	0.2	15.0
Chlorination	*	48	0.2	2.6
Adsorption	5	15	0.15	2.6
Removal %	95	95	98.0	88.0

*Chlorine interferes in BOD analysis

CONTROL SYSTEMS ANALYSIS

The solids wasting control loops in lime clarification and recarbonation are simple feedforward systems where periodic pulses proportional to flow produce a discharge into a fixed volume. The discharge is controlled by a level switch sensing the fixed volume. The control loop for $FeCl_3$ feed in recarbonation is a feedforward system in which the duty cycle of a peristaltic pump is controlled based on the $FeCl_3$ concentration, wastewater flow rate, and desired dosage. In each case, the computer observes the process flow, computes the control signal, and applies the signal to the process actuator. Control analysis and development of controller algorithms are not required.

The CaO, CO_2, and pH (with Cl_2) control loops are systems in which the chemical dosage is controlled by both a feedforward signal proportional to flow and a feedback signal based upon pH error. In the generalized digital approach (see Figure 3) to pH control, an actuator provides a chemical feed into the process, which causes a change in the process effluent pH that is sensed by a pH meter. A digital computer, consisting of an analog-to-digital converter (sampler), comparator, controller, and digital-to-analog converter (hold) performs the function of periodically sampling the process pH, comparing the pH with an operator set point value, computing a chemical feed control signal, and applying this signal to the chemical feed actuator.

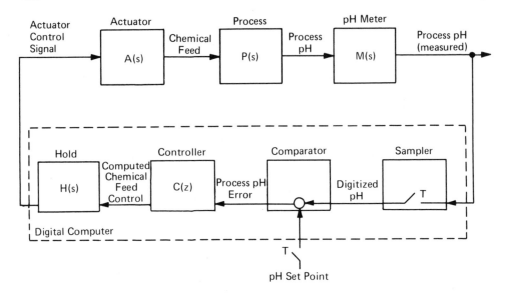

Figure 3. Generalized Form of Chemical Feed-pH Control System

In experiments to determine the plant characteristics of these three chemical feed loops, the process dynamics completely dominated the dynamics of the sensors and actuators. For example, the transient response of the first-stage process (CaO feed) has a time constant of about 30 minutes, compared with a response time of several seconds for the actuators and sensors. The experiments also reveal that the process dynamics are nearly first order but exhibit some nonlinearity.

In the chemical feed control systems where CaO, CO_2, and Cl_2 are the manipulated variables, two candidate control algorithms were evaluated; dead beat control and proportional-integral control with adaptive gain tuning and feed forward flow rate compensation. The process control systems were analyzed mathematically using LaPlace transform[2] and z-transform[3] techniques. The mathematical definition of the variables and coefficients used in the analysis are provided in Table 3. Since the three chemical feed loops are similar, the control analyses in generalized form based upon the CaO feed loop are applicable to the other pH control loops.

Table 3. Mathematical Definitions

Symbol	Definition
Controlled plant components	
A(s)	LaPlace transform of actuator transfer function
P(s)	LaPlace transform of process transfer function
M(s)	LaPlace transform of pH meter
τ	Component time constant
Computer components	
C(z)	z–transform of control algorithm
H(s)	LaPlace transform of zero-order hold
T	Sample period of A/D converter
System variables	
E(z)	z–transform of pH error
R(z)	z–transform of pH set point
G(z)	z–transform of G(s) = Z[G(s)] = Z[A(s)P(s)M(s)H(s)]
s	LaPlace transform variable
z	z–transform variable
n	Integer = 0, 1, 2, . . .

The pH response of the pilot plant lime process to a step change in CaO feed rate exhibits a time constant of about 30 minutes and is approximated by a first-order transfer function. The mathematical model of the stage is

$$A(s) = \text{Constant actuator gain} = A \qquad (14)$$

$$M(s) = \text{Constant pH meter gain} = M \qquad (15)$$

$$P(s) = \frac{P}{\tau s + 1} \qquad (16)$$

P = Process gain = f (flow, alkalinity)

τ = Process time constant

It should be noted that the process gain P is dependent upon variables other than just wastewater flow and alkalinity. However, wastewater flow is the major disturbance of the process. Because the flow rate is easily measured and the process is slow, steady-state feedforward compensation is incorporated to diminish the effects on the process due to flow changes, while feedback control is employed to compensate for all other disturbances.

The computer compares the sampled pH signal with the pH set value producing a difference or error signal which is operated on by the controller algorithm. The algorithm is a difference equation which computes the chemical feed rate every T minutes. The general expression for the controller transfer function is:

$$C(z) = \frac{a_n z^n + a_{n-1} z^{n-1} + \ldots + a_0}{b_m z^m + b_{m-1} z^{m-1} + \ldots + b_0} \tag{17}$$

The computed output value is maintained between sampling events by a digital-to-analog converter or hold circuit. The hold is characterized as

$$H(s) = \frac{1-e^{-sT}}{s} \tag{18}$$

The control system (see Figure 3) may be redrawn as shown in Figure 4. The control input-to-output relation in z - transform notation is

$$\frac{\varepsilon(z)}{R(z)} = \frac{1}{1 + C(z)G(z)} \tag{19}$$

The control objective is to reduce the error $\varepsilon(z)$ to zero in the shortest time possible. The fastest controller, dead beat control, theoretically reduces the process pH error to zero in one sampling interval or control cycle. This type of control causes $\varepsilon(z)$ to go from the value R at t = 0 to zero value at t = T and remain there.

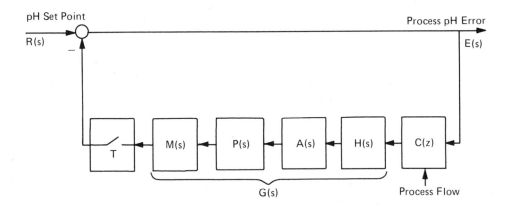

Figure 4. Transfer Function Characterization of Controlled System

For a step change Q in pH set value at t = 0 (the most severe case)

$$R(z) = \frac{Qz}{z-1} \tag{20}$$

and for the process pH error to go to zero at t = T, 2T, ...

$$\varepsilon(z) = Q \tag{21}$$

Solving for the controller transfer function results in

$$C(z) = \frac{R(z) - \varepsilon(z)}{\varepsilon(z)G(z)} \tag{22}$$

and substituting values of $R(z)$, $\varepsilon(z)$, and $G(z)$ produces

$$C(z) = \frac{z - e^{-T/\tau}}{(z-1)(1-e^{-T/\tau})APM} \tag{23}$$

The difference equation that is implemented in the computer for the dead beat controller transfer function is

$$0_1(nT) = 0_1[(n-1)T] + \frac{\{I(nT) - e^{-T/\tau} I[(n-1)T]\}}{APM(1-e^{-T/\tau})} \tag{24}$$

where

0_1 (nT) = Computer output at t = nT, n = 0,1,2, ---

I (nT) = Computer input at t = nT, n = 0,1,2, ---

For a linear system, proportional and integral control provides equivalent process response to dead beat control if

$$K_P = \text{Proportional Constant} = \frac{e^{-T/\tau}}{APM(1-e^{-T/\tau})} \tag{25}$$

$$K_I = \text{Integral Constant} = \frac{1}{APM} \tag{26}$$

The difference equation for proportional integral control is

$$O_2(nT) = K_P I(nT) + K_I \Sigma I(nT)$$

$$= O_2[(n-1)T] + (K_P + K_I)I(nT) + (-K_P)I[(n-1)T] \quad (27)$$

The difference equations $O_1(nT)$ and $O_2(nT)$ contain the plant gain terms A, P, M. Thus, if the plant components change characteristics, or if the wastewater flow rate changes, the controller is designed to accommodate and compensate for the change. The advantage of formulating the controller with variable proportional and integral coefficients is that it will provide for a range of selectable performance, varying from dead beat to any reduced speed of response.

The breakpoint chlorination stage is a complex, two variable process (see Figure 5) that is difficult to control. Proper control requires good pH control (NaOH addition) and good chlorine dosage control (proper minimum residual free chlorine). The loop controlling the Cl_2 dosage in breakpoint employs a feedforward signal proportional to the amount (mass) of ammonia in the process influent and a feedback signal based on a free residual chlorine concentration error. The feedforward signal is calculated from the wastewater flow, influent ammonia concentration, a preselected $Cl:NH_3-N$ dosage ratio and from the chlorine used in pH control by prechlorination. The feedback signal is produced from the free residual chlorine concentration after breakpoint and the free residual chlorine set point. Finally, the control loop for base (NaOH) addition in the breakpoint process employs a feedforward signal proportional to the wastewater flow rate and a feedback signal based on the pH error.

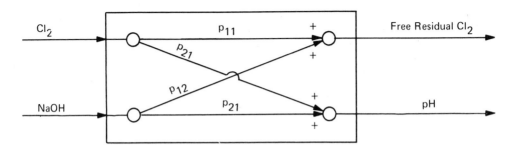

Figure 5. Breakpoint Chlorination Process Model

The Cl_2 and NaOH both affect the free residual chlorine and the pH in an interactive manner. The digital approach is to control the breakpoint process in a noninteractive manner (see Figure 6).

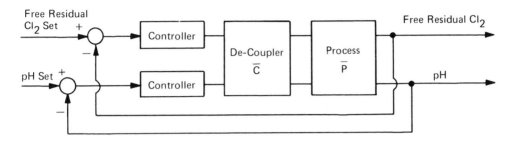

Figure 6. Breakpoint Chlorination with Noninteractive Control Applied

Here the controller provides simultaneous but noninteractive Cl_2 and NaOH control when a set point change is made in either the desired value of free residual Cl_2 or pH; that is, each output variable is controlled independently of the other output variable.

The noninteractive control is achieved by developing the combination of the decoupling matrix $\bar{C}$ and the process matrix $\bar{P}$ to be a matrix $\bar{D}$ with zero nondiagonal elements, that is

$$\bar{P}\bar{C} = \bar{D} \tag{28}$$

or

$$\bar{P}\bar{C} = \begin{bmatrix} p_{11} & p_{12} \\ p_{21} & p_{22} \end{bmatrix} \begin{bmatrix} c_{11} & c_{12} \\ c_{21} & c_{22} \end{bmatrix} = \begin{bmatrix} d_{11} & 0 \\ 0 & d_{22} \end{bmatrix} \tag{29}$$

The combination requires that

$$p_{21} c_{11} + p_{22} c_{21} = 0 \tag{30}$$

and

$$p_{11} c_{12} + p_{12} c_{22} = 0 \tag{31}$$

letting $c_{11} = c_{22} = 1$,

$$c_{12} = -\frac{P_{12}}{P_{11}} \qquad (32)$$

$$c_{21} = -\frac{P_{21}}{P_{22}} \qquad (33)$$

If the controller $\bar{C}$ satisfies these conditions, the computer achieves direct, noninteractive control of the output variable with the set point variable.

Clearly, the process transfer function $\bar{P}$ must be known, and a practical controller $\bar{C}$ must be designed. Data is presently being evaluated in order to complete this control approach. While ideal noninteraction cannot be achieved, sufficient noninteraction should improve existing analog control methods.

SIMULATION

In the physical chemical wastewater treatment system, it is essential to test the candidate algorithms prior to implementation on the actual pilot plant. The need for simulation arises because the treatment processes possess significant nonlinearities which cause difficulty in obtaining analytic solutions in closed form. In addition, simulation is needed for performing sensitivity analyses; that is, predicting controlled process response in the presence of process uncertainties.

The simulation technique used is CSMP (Continuous System Modeling Program)[4]. CSMP is a continuous system simulator that combines a functional block modeling capability with a powerful algebraic and logical modeling capability. The input language enables a user to prepare structure statements describing a physical system, starting from either a block diagram or a differential equation representation of that system. The program provides a basic set of 34 functional blocks. Included in the basic set are such conventional analog computer components as integrators and relays plus many special purpose functions like delay time, zero-order hold, dead space, and limiter functions. This is complemented by the FORTRAN library functions, and a macro capability that permits individual existing functions to be combined into a larger functional block. Data output options include printing of variables in standard tabular format, print-plotting in graphic form, and preparation of a data set for user-prepared plotting programs. A system to be simulated is described to the program by a series of structure, data, and control statements. Structure statements describe the functional relationships between the variables of the model, and, taken together, define the network to be simulated. Data statements assign numeric values to the parameters, constants, initial conditions, and table entries associated with the problem. Control statements specify options relating to the translation, execution, and output phases of the System/360 CSMP program, such as run time, integration interval, and type of output.

The simulation modeling approach for chemical feed control (CaO, CO_2, pH with Cl_2) is illustrated in terms of the first-stage (lime clarification) process. The block diagram for the lime clarification process is as follows:

where:

    CS   = control signal (dimensionless)

    A   = activator gain (CaO dosage/CS unit)

    UACS   = unconstrained activated control signal (CaO dosage)

    INSW   = System/360 CSMP input switch block (explained below)

    ACS   = constrained activated control signal (CaO dosage)

    DELAY = System/360 CSMP dead time block (explained below)

    DACS   = delayed ACS (CaO dosage)

    INTGRL = System/360 CSMP integrator block (explained below)

    ECAO   = effective CaO dosage (CaO dosage)

    P   = ECAO-to-pH conversion factor (pH/CaO dosage)

    PH   = pH

    M   = pH meter constant (dimensionless)

    PHM   = pH metered (pH)

The unconstrained activated control signal is given by

UACS = (A)(CS) (34)

Since the lime feeder imposes a physical limitation on the maximum ACS (MACS) that can be introduced per unit time:

ACS = INSW (UACS - MACS, UACS, MACS) (35)

which literally means

$$ACS = \begin{cases} UACS & \text{if } UACS - MACS < 0 \\ MACS & \text{if } UACS - MACS \geq 0 \end{cases} \quad (36)$$

The ACS either goes directly into the lime process or first suffers a "dead-time" delay (caused by the lime feeder belt and/or by a transportation delay in the tank). DELAY is a System/360 CSMP block to model this assumption; thus, DACS is ACS delayed a given number of minutes. The INTGRL block defines the process itself as a first-order system:

ECAO = INTGRL (IECAO,(DACS-ECAO)/TAU) (37)

which corresponds to

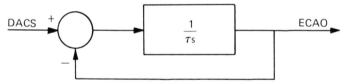

and is equivalent to

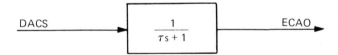

The yield of the first-order process model is ECAO, the effective CaO dosage. By "effective dosage" is meant the level of dosage actually acting on the pH of the wastewater, according to the following relationship:

PH = (P)(ECAO) (38)

P, the ECAO to pH conversion factor, is itself a nonlinear function of ECAO for given levels of flow and alkalinity. The modeling of this process nonlinearity is conveniently handled as a table function in the simulation program.

The closed-loop simulation diagram for the digital controllers is shown in Figure 7.

In the simulation runs, the objective was to determine the relative performance of the candidate controllers and their sensitivity to uncertainties in the knowledge of the process transfer function. The controlled plant response to a pH error resulting from a set point change applied to an assumed first-order process and second-order process, is shown in Figure 8. The dead-beat controller, designed for a first-order process, achieves very rapid reduction of pH error. However, if a second-order process is assumed, some oscillation in the response occurs. This oscillation can be reduced by using smaller proportional and integral coefficients as shown in Figures 8c and 8d. Dynamic compensation of plant non-linearity has been used to achieve rapid and improved control of the first-stage clarification process. By experimentation with the simulated plant, the candidate controller has been verified and predictive response data generated.

Other simulation runs conducted to predict performance in the presence of noise and uncertainty in knowledge of process parameters and to assess other sampling periods reveal that the controllers yield a stable response. The dead beat controller is superior if the process is first-order; the proportional integral controller with non-dead beat parameters if the process is a second-order. The System/7 computer, when installed, will allow experimentation and data collection that will provide better plant characterization for modeling and control purposes. More sophisticated control algorithms could be implemented in the computer if necessary to achieve various control objectives.

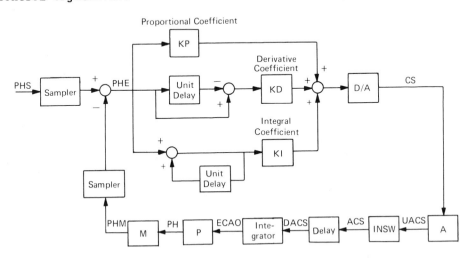

Figure 7. Simplified Characterization of CSMP Simulation of First Stage Clarification with Proportional Integral Derivative Control

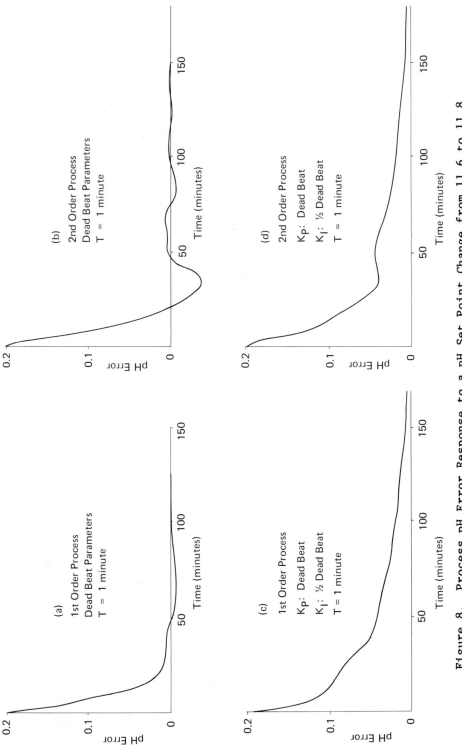

Figure 8. Process pH Error Response to a pH Set Point Change from 11.6 to 11.8

COMPUTER SYSTEM HARDWARE

The IBM System/7[5] selected for the pilot plant is a modular, sensor-based, interrupt-driven, digital computer which consists of a processor module, two input/output modules, and an operator station (see Table 4). The processor module has 10,240 words of monolithic memory, with a storage cycle time of 400 nanoseconds. The input/output modules are multifunction modules which provide points for analog input, analog output, digital input, and digital output.

Sixty analog input points are used to bring information from pH meters, flow meters, etc. One input signal containing information from six temperature sensors is multiplexed under control of a stepping motor driven by a digital output. Four signals from Technicon* AutoAnalyzers* (which measure phosphate, nitrate, ammonia, and free chlorine concentration) are each multiplexed to permit the operator to switch to different streams for calibration.

One of the modules provides in three groups 35 digital inputs from switches; the switches indicate various operating conditions such as pressure in receiver tanks, hydraulic level, etc. A "process interrupt" feature is included for one of the groups of 16 input lines. This feature permits the status of the group to be compared on a bit-by-bit basis to a 16-bit reference register. An interrupt is initiated on the basis of an equal or unequal compare, where choice of compare and the contents of the reference register are selected by programming.

A single analog output is multiplexed through an analog hold device to drive four process control actuators: the carbon dioxide valve in the recarbonation stage, the two chlorine valves in the breakpoint chlorination, and the sodium hydroxide pump.

Digital output points are used to control the first-stage lime belt and sludge blowdown valve, the recarbonation stage ferric chloride pump and blowdown valve, the operator alarm (visual and/or audio), and the watchdog timer. (The latter is a device designed to indicate an abnormal system condition such as endless looping or loss of power. It is powered externally to the computer, and times out to give an alarm unless reset periodically by the computer). Also, one of the digital outputs is used to control a stepping motor to advance the temperature multiplexer to the next point.

Finally, an operator's station equipped with a paper tape reader and punch, printer, and keyboard is attached to the processor module and is used for program preparation, initial program loading, paper tape input and output, and operator communication with the system.

---

*Technicon and AutoAnalyzer are trademarks registered by Technicon Instruments Corporation, Ardsley, N.Y.

Table 4. Control Computer Characteristics

Component	Characteristic
Processor	
Cycle time	400 nanoseconds
Word size	16 data bits plus 2 parity bits
Storage size	10,240 words
Priority interrupt levels	four
No. of instructions	40
No. of input/output commands	68
Interval timers	two, 50-μsec resolution
Input/Output	
Analog input	
Scan speed	200 points/second/module
No. of modules	two
ADC resolution	14 bits plus sign
No. of points	60
Sensors scanned	pH (8)
	wastewater flow (14)
	water pressure (4)
	sludge density (2)
	air pressure (1)
	dissolved oxygen (11)
	chemical flow (7)
	temperature (6) *
	chemical concentration (19) **
	airflow (8)
Digital Input	
Scan speed	250,000 Hz
No. of points	35 total, 10 interrupting
Noninterrupting points	chemical concentration type (19)
	temperature point (6)
Interrupt points	chemical concentration (4)
	temperature (1)
	pressure (3)
	hydraulic level (2)
Analog Output	
Resolution	10 bits
No. of points	one
Actuators controlled	gas valves (3)
	chemical pump (1)
Digital Output	
Rating	100 volt-amperes
Speed	250 Hz
Actuators controlled	lime belt (1)
	sludge valve (2)
	chemical pump (1)
	operator alarm (1)
	watchdog timer (1)
	stepping motor (1)

* multiplexed to 1 point
** multiplexed to 4 points

COMPUTER SYSTEM SOFTWARE

The software system is designed to satisfy the following functional requirements: process control, data acquisition, alarm monitoring, and system operation.

The process control function was described earlier. The data acquisition task consists of periodically scanning the 60 analog and 35 digital input points, converting the values to engineering units, and punching the values on paper tape. Certain computer process control values are also punched on paper tape. Finally, the operator may manually enter text or data to be punched on paper tape.

The alarm monitoring task is derived from the process control and data acquisition tasks. Important variables are monitored and the operator notified if they are out of range or otherwise invalid. Several alarms are provided for the lime addition control loop. First, the operator is notified if the controlled pH varies by more than 0.2 units from the set point. A second alarm occurs if the actual dosage of lime is not within the expected range of dosage for the selected set point. The lime belt duty cycle produces another alarm if the duty cycle exceeds 95 percent or is less than 5 percent. Finally, the weight signal from the lime feeder belt causes an alarm if it indicates no lime on the belt. These alarms in the lime process may be produced by blockage of the lime hopper, empty hopper, malfunction of the pH sensor, failure of the mixer, or others.

The sludge blowdown processes are monitored for alarm conditions. The constant volume tank must fill within two minutes, but must not be full when a blowdown is to begin or an alarm will be given. In the recarbonation process, pH, CO_2 feed, and ferric chloride pump duty cycle are monitored. In a similar fashion, the prechlorination and breakpoint chlorination processes are monitored for pH control, free chlorine control, and dosage vs set point. Finally, alarms are given for low air pressure in a receiver tank and out-of-range analog signals.

Operational flexibility is provided by the use of system parameters which are external to the programs and can be modified by input through the IBM 5028 Operator Station. The operator can (a) logically disconnect any control loop; (b) manually alter the actuator settings; (c) inform the computer of a change in the operating range of any sensor, solution concentration, tank volume, pump or valve capacity; (d) change the alarm limits on a controlled variable; and (e) change a set point or frequency of data acquisition.

Using "Modular System Programs for System/7" (MSP/7), which is available programming support, as a base, the system was designed as shown in Figure 9. MSP/7 program names start with "$".

Programs to satisfy operator requests are accessed as follows. The operator makes a request by depressing the REQUEST button on the operator station. This is sensed by the MSP/7 program $OPR which unlocks the keyboard. The operator types in a three-character code indicating the program desired, followed by additional information required by the program. The MSP/7 program $OPTR gives control to the program designated and passes on any typed-in data. For example, the operator may have all outstanding alarms printed by typing in "DAL." He may acknowledge a specific alarm (e.g., *A12) by typing "AAL *A12" which would cause the audio and visual alarm signals to cease, unless other outstanding unacknowledged alarms were present. System parameters may be displayed via DSP and modified via ECP or OCP. Last read or written analog or digital values may be displayed by calling DLV. Selection of ELV permits manual changing of the actuator setting for a loop in HOLD status. WIP is called to punch a paper tape containing all current values of system parameters. This allows faster reinitialization, if required by power outage, after which RIP would be called to read in the system parameter values from the paper tape.

The analog values required for process control (pH, flow, etc., in each loop) are developed as follows: The program RPCAI is entered once every second by the MSP/7 routine $SCHD. RPCAI requests that the analog sensors be read by passing a list of sensors to the routine $AI. RPCAI then returns control to $SCHD. $AI initiates analog-to-digital conversion for the first sensor, accepts the interrupt when conversion is complete (meanwhile allowing other processing to go on), and then initiates conversion on the second sensor, etc. When all of the requested analog sensors have been read, $AI passes control to ZPCAI. This routine converts the values into engineering units with conversion information from the System Parameter Table, applies smoothing, and stores the smoothed values.

The various loop control routines are entered once every minute by $SCHD. These programs use the values in the Smoothed Value Table with the appropriate algorithms to develop the required control signals for the actuators. For actuators driven by an analog signal, the loop control routines pass control to AOCTL which sends the signal through the analog hold device to the actuator. This is accomplished by putting the analog signal on the line, waiting for the output to settle to within 0.1 percent of its final value, then turning on the digital output bit that selects the particular memory amplifier desired. The bit is left on long enough for the memory amplifier to reach the desired value and then the digital output bit is turned off.

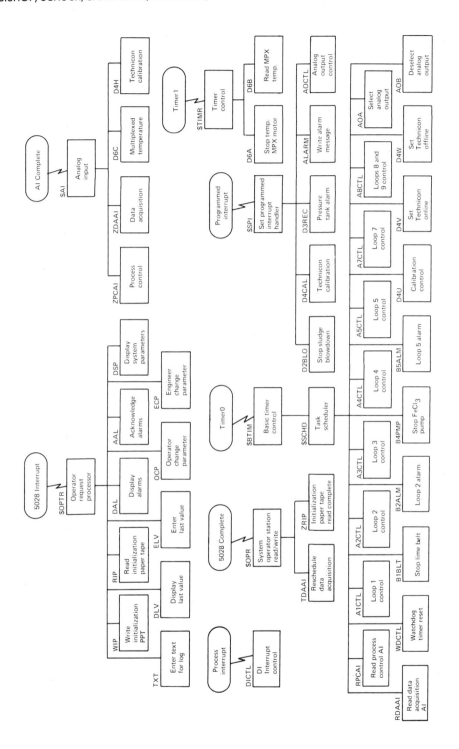

Figure 9. Programming System Design

Data acquisition analog values are developed in a manner similar to that of process control. RDAAI is entered once every 15 minutes and passes a list of devices to $AI. When all devices have been read, control is passed to ZDAAI which converts the values into engineering units and punches them onto paper tape.

Alarm monitoring is performed where the information is developed. For example, pH limit checks are made in ZPCAI where the pH values are converted from analog voltages to pH units. Dosage checks are made in the loop control routines A1CTL, A2CTL, etc., where the dosages are computed.

When an alarm condition is detected, or a previous alarm condition becomes normal, the information is passed via a set programmed interrupt to the routine ALARM which notifies the operator.

SUMMARY

The analog automated physical-chemical wastewater treatment system that consists of two-stage lime clarification with recarbonation, filtration, pH control with Cl_2, breakpoint chlorination, and carbon adsorption typically reduces residual pollutants to 5 mg/l of BOD, 15 mg/l of COD, 0.15 mg/l of total phosphorus and 2.6 mg/l of total nitrogen. The control system indirectly controls the water quality and minimizes the chemical dosages by controlling and minimizing: the pH error in the wastewater for chemical feeds, CaO, CO_2, Cl_2, (for pH control), and NaOH; the dosage for $FeCl_2$ feed; and the residual free Cl_2 concentration error for the Cl_2 feed in breakpoint. The system also controls solids wasting from lime clarification.

Direct digital control has been developed for chemical feed control. Control systems analysis reveals plant process dynamics that are nearly first order but exhibit nonlinearities. In the control analyses and simulation, control algorithms were developed for chemical feed control.

The digital control also promises to reduce the complex interaction in breakpoint chlorination to essentially noninteractive control of the process. Additional analysis is required to complete the development of the noninteractive control approach and evaluate results.

The digital computer (IBM System/7) to be installed will also perform data acquisition and alarm monitoring for physical-chemical wastewater treatment and the three-stage activated sludge process in the EPA-DC Pilot Plant. The software is designed to implement process control, data acquisition and logging alarm monitoring, and overall system operation.

ACKNOWLEDGMENTS

James E. Bowers and Luis Gutierrez of IBM and Thomas A. Pressley of EPA contributed to the data collection, algorithm development, and process simulation.

REFERENCES

1. Bishop, D.F., O'Farrel, T.P., Stamberg, J.B., and Porter, J.W., "Advanced Waste Treatment Systems at EPA-DC Pilot Plant," presented at the 68th National Meeting of the AIChE, Houston, March 1971.

2. Truxal, J.G., "Automatic Feedback Control System Synthesis," McGraw-Hill Book Co., Inc., New York, New York, 1955.

3. Ragazzini, J.R., and Franklin, G.F., "Sampled-Data Control Systems," McGraw-Hill Book Co., Inc., New York, New York, 1958.

4. IBM Corporation, "System/360 Continuous System Modeling Program," User's Manual Publication H20-0367-2, White Plains, New York, IBM 1968.

5. IBM Corporation, "IBM System/7 System Summary, Form GA34-0002-1," Second Edition, White Plains, New York, September 1971.

COMPUTATION AND MAPPING OF THE DISPERSION AND HERBAGE UPTAKE OF GASEOUS EFFLUENTS FROM INDUSTRIAL PLANTS

A. J. H. Goddard, R. E. Holmes, H. Apsimon

Imperial College of Science and Technology, Pollution Prevention (Consultants) Ltd.

There are many industrial processes which release substances known to be injurious to man or damaging to his environment, where the extent of damage is a complex function of a large number of variables. The task of Management in monitoring and controlling the releases to the atmosphere can thus be formidable. Decisions may have to be made on widely divergent topics such as: the number and type of sampling devices to install on the plant; the extent to which specific operational or maintenance procedures which require a short duration release above the 'acceptable' average release can be tolerated; whether off-site sampling is required and if so what type, where, how and at what frequency. Often decisions of this type are required at short notice and are subject to considerable commercial pressure. To assist in this situation the Authors have written a suite of computer routines which will enable rapid computation of air contamination downwind from complex release patterns. Routines exist whereby isopleths of ground level air concentration averaged over varying time intervals are automatically plotted for rapid appraisal. In addition integrated air concentrations can be predicted for specified locations, such as air sampling stations, and these results printed out. Further routines compute the levels of contamination within plant and animal species as a function of time during the exposure.

As an example of this application of some of their Routines the Authors have selected the release of flourides from Aluminium Smelting.

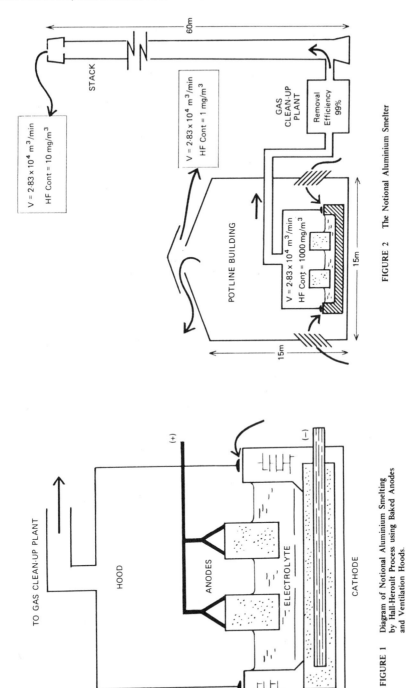

FIGURE 1 Diagram of Notional Aluminium Smelting by Hall-Heroult Process using Baked Anodes and Ventilation Hoods.

FIGURE 2 The Notional Aluminium Smelter

2. THE SIGNIFICANCE OF ALUMINIUM SMELTING IN THE U.K. IN 1972 - THE NOTIONAL SMELTER

The chief sources of atmospheric fluoride pollution in the U.K. are steel, glassmaking and phosphate fertiliser factories, cement works, potteries, brickworks and aluminium smelters. The extent of this pollution was reviewed by the Agricultural Research Council in 1967. At that date aluminium production in the U.K. was small but subsequently decisions were made to expand production and there has been a large investment in aluminium smelters aimed at increasing output of primary aluminium by a factor of ten. These smelters are gradually becoming operational. The Management of these plants will face operational decisions on the extent of public pollution during the difficult phase of commissioning and setting to work these new plants; the public in the vicinity of these plants, newly conditioned by international publicity on pollution to voice their concern, will be demanding more rigorous and informed evaluation of their new risks.

Thus for the purposes of illustrating the computational program we have taken as our notional source of fluoride air contamination an aluminium smelter operating the Hall-Heroult process to yield 100,000 tons of Aluminium per annum. This is an electrolytic process with alumina (aluminium oxide Al_2O_3) dissolved in molten cryolite (Na_3AlF_6) to form the electrolyte which is maintained at a temperature between 800 and 1000°C in normal operation. The electrolyte is contained within steel trays lined with refractory bricks and an inner carbon shell with steel inclusions acting as the cathode. For simplicity of illustration we have assumed a baked carbon block anode although the program could be applied to the Soderberg system if required. Typically additional fluorides as aluminium fluoride (AlF_3) and fluospar (CaF_2) are added to the electrolyte to increase the overall efficiency of the process. Ball and Dawson (1971) estimate that fluorine is consumed during production of aluminium at a rate of about 30 kg/t of aluminium and that about 50% of this is discharged with the waste gases from the smelting pots. Measurement by Waddington and Less (1971) measured that with this type of plant about 50% of the fluorine emission is in the form of hydrogen fluoride. Typically fluoride concentrations of between 500 and 2500 mg/M^3 are measured in the untreated gas above this type of furnace, and for the purpose of illustration we have assumed a mean concentration level of 1000 mg/M^3 in the untreated furnace gases. With some designs of smelter furnace hoods are provided to limit the exposure of the Pot Rooms Operatives to fluorides, in other designs the exposure is limited by providing a large volume of clean ventilation air to limit the air concentration by dilution. There are operational advantages and disadvantages of both design approaches and a compromise is sometimes adopted whereby both hoods and potroom ventilation are used to limit the exposure of the operatives. Under these conditions there is typically a release of fluoride materials at certain routine stages of operation direct to atmosphere through roof louvres. The fluoride contamination level of such air would not be expected to exceed 50% of the recommended TLV for continuous 8 hour day exposure and an average European TLV of 2 mg/M^3 for HF has been assumed for illustration purposes in this paper. This release would be directly to atmosphere through roof louvres running the length of the potrooms at roof level. Our notional plant assumes a Pot-Line building approximately 60 metres in length by 15 metres wide and 15 metres high, with two ranks of roof louvres running the length of the building. The discharge volume is assumed to be 2.83×10^4 m^3/min (10^6 cfm) and is not filtered, whereas the highly contaminated air from the furnaces, which it is assumed will be renewed at a rate of 2.83×10^4 m^3/min, is conventionally treated either by wet scrubbers or bag filters. For the purposes of illustration a removal efficiency of 99% for HF is attributed to our notional gas cleansing system which results in an effluent contaminated with HF to a level of 10mg/M^3. It is further assumed that the air would be released through a 60 metre stack. The notional smelter model is illustrated in Figure 2.

3. THE RELATIVE SIGNIFICANCE OF AIR AND PASTURE
CONTAMINATION WITH FLUORIDES

Although fluorides are ubiquitous in nature and have been attributed at low concentrations with beneficial effects, they are in fact cumulative poisons. Damage has been demonstrated to a wide range of plants; farm animals have suffered injury through eating fluoride contaminated forage and chronic fluorosis has been identified as an important industrial disease. In order to determine the scope of the program and the limits of individual routines it was necessary for the Authors to establish both the absolute and relative significance of air and pasture contamination and to determine, as far as practicable, the relationship between them.

3.1 Human Toxicity

Considerable evidence exists that continuous exposure to small concentrations of fluorides can result in the eventual development of severe crippling fluorosis, an industrial disease in which the victim is slowly debilitated until he can no longer carry out simple physical tasks. The storage or utilisation of fluorides in the body is usually negligible and the normal adult will excrete almost all of his daily uptake of fluorides in the urine and faeces. The excretion of absorbed fluorides has a limit imposed by the limited excretory capacity of the kidneys, normally about 20 mgs of fluoride per litre of urine. To allow for metabolic differences and variations in sensitivity a control level of 5 mgs per litre of urine is often taken as a level at which significant accumulation and therefore chronic fluorosis is highly unlikely. A major technical problem is that of relating these levels of urine contamination to air contamination so as to derive standards of acceptable air quality. The currently acceptable relationship in the U.K. is implied by a TLV (Threshold Limiting Value) of 2 mg/M^3 HF of air when inhaled continuously for 8 hours per day. For occupational workers it has been recognised that short duration exposure to levels in excess of the TLV may be inevitable and Emergency Exposure Limits of 8 ppm HF/m^3 of air for 1 hour exposure and 10 ppm HF/m^3 of air for 30 minutes exposure have been applied.

3.2 Animal Toxicity

Fluorosis has been diagnosed in a wide variety of animals.(Shupe 1969). Fluorosis in cattle is of particular relevance to this study and has been extensively studied because of the direct economic consequences of effects such as loss of weight, reduction in milk yield, lameness and general debility. Uptake is almost entirely controlled by the level of direct fluoride contamination of the animals' diet. Fluorine tolerance levels for foodstuffs is given by Shupe (1969) as 30-40 ppm for breeding or lactating animals and 100 ppm for 'finishing' (presumably meat) animals. Harris et al (1963) has calculated only slightly higher levels for sheep i.e. 50 ppm for breeding and 160 ppm for 'finishing' animals. In the vicinity of the Aluminium smelters the controlling dietary factor is the level of fluoride contamination of grassland for animals reared on natural pasture.

3.3 Plant Toxicity

In plants fluorides enter the leaves through the stomata without causing damage at the point of entry. They are translocated to the tips and margins of the leaves causing plasmolysis and cell collapse. Sensitivity to fluorides varies widely from species to species. McCune (1969) suggests a maximum limiting 24 hour mean for Conifers of 3 µg F/m^3; for corn of 10.5 µg F/M^3 and for tomatoes of 12 µg F/m^3. The devastating effects on conifers from the release of high concentrations of fluorides from aluminium smelting in Norway has been reported by Robak (1969). McCune (1969) reviewing the factors influencing plant toxicity draws attention to the complexity of the situation. He suggests for example that the phytotoxicity of HF is considerably increased when the leaves are wet, but can be partially removed by washing; that uptake of HF is significantly greater by day than by night; and that although soluble fluorides are accessible to vegetation there seems to be no evidence that fluorides from the air build up in the soil in sufficient quantities to adversely affect crops. In addition there is some evidence of variation of sensitivity within species. The complexity of these factors makes it difficult to relate a variable air contamination level to specific plant damage.

It is even more difficult to predict plant contamination levels from exposure to variable air pollution levels, and yet it is precisely this relationship which is required for predictive studies of fluorosis in farm animals from the use of contaminated forage.

4. MODELLING OF DISPERSION AND HERBAGE UPTAKE

4.1 Calculation of Dispersion

In order to calculate space and time dependent gaseous fluoride levels at herbage level over a basic grid of 900 points covering a 15 km square region having the plant at its centre, the Gaussian plume diffusion scheme of Gifford (1961) and the related stability categories of Pasquill (1961) have been used. In formulating the model we have been guided by methods developed in the safety assessment of nuclear power stations, Slade (1968) and Clarke R.H. (1970). In order to examine both average and time dependent concentrations over a period of a month the variable meteorological record is converted to a series of quasi steady-state periods. This division is made primarily on the basis of significant changes in wind direction with secondary subdivisions based on change of weather category. Within each period mean wind speed and weather category are assigned and the horizontal standard deviation of the Gaussian concentration distribution is derived from analysis of wind direction records. The data is obtained from a single meteorological station within the area covered by the 15 km square grid.

The expression evaluated for each period for both stack and pot-line releases to yield ground level concentration at distance x metres downwind from the source and y metres crosswind from the plume axis is:

$$\chi(x,y) = \frac{Q \exp\left\{-\frac{1}{2}\left[\frac{y^2}{\sigma_y(x,p)^2} + \frac{z(x,y)^2}{\sigma_z(x,p)^2}\right]\right\}}{\pi u \sigma_y(x,p) \sigma_z(x,p)}$$

z = height of source above ground $x \geqslant 0$ at x,y
u = a representative wind speed
p = weather category

$\sigma_y(x,p)$ = horizontal dispersion parameters

$\sigma_z(x,p)$ = vertical dispersion parameters

Reflection of the plume from the ground is assumed. Within the framework of the Gaussian plume model a number of additional features are incorporated. To allow for the extended nature of the pot-line an additional lateral dispersion is added which is proportional to the standard deviation of the building emission points about the line of mean wind direction through the centre of the building. Allowance for an elevated inversion layer is made by assuming a reflection model in which the cloud is reflected alternately at the inversion layer and at the ground. First order corrections for small deviations in ground level are made by allowing z to vary with position. Features are available in the program for treating washout and ground deposition. Washout introduces an exponential depletion of the plume with distance downwind depending on wind speed. Deposition is difficult to apply in practise for rates will vary considerably with the type of ground.

The momentum and excess temperature of the pot-line effluent introduce negligible effects but for the stack both momentum and temperature lead to considerable rise of the effluent plume. Momentum rise and temperature rise are combined to yield an effective source height for the diffusion model.

4.2 Computer Mapping of Concentrations

The program is intended to be used initially in conjunction with a program of measurements of gaseous concentrations and herbage sampling measurements. As an aid to assessment of sampling position disposition, and as an aid in routine reporting, the program generates a contour map of monthly average ground level fluoride concentrations, together with sampling position locations. Contours are only generated outside a radial limit of 1 km from the plant; this circle is presumed to enclose land owned by the plant operators. To aid management decisions on plant operation, total concentration and stack contribution are contoured separately, using different line-styles. Printed lists of monthly average concentrations at the individual sampling positions are also produced.

4.3 Calculation of Time-Dependent Herbage Levels

The importance of herbage sampling in checking fluoride levels to which cattle may be exposed has already been noted. However, interpreting herbage levels in terms of plant emission is virtually impossible unless meteorological variations, the effect of growth rate and the effect of grazing are taken into account. With the aid of published data a model has been prepared which takes the time-dependent ground level fluoride concentration at one sampling position and computes time-dependent herbage levels. The model relates principally to rye-grass.

It is assumed that a particular leaf has a finite life and that upon decay no translocation of fluoride occurs. Anslow (1966) while quoting seasonal rates of leaf appearance points out the difficulty of defining leaf decomposition. In this model leaf lives ranging from 28 days in the summer to 60 days in the winter have been assumed and, together with growth dilution and accompanying grazing, are the removal mechanisms for fluoride. No allowance has been made for loss by weathering; Hitchcock et al (1971) found no significant loss of fluoride from Orchard grass (Dactylis glomerata) during periods of up to 20 days subsequent to fumigation. Orchard grass (or Coxfoot grass) is a closely related member of the same tribe of grasses as Ryegrass (Lolium perenne).

Hydrogen fluoride gas is absorbed through stomata with a possible minor contribution from absorption at the leaf surface (McCune and Weinstein 1971). Leaf uptake rates, taking account of possible stomatal closure at night, have been fitted to the linear regression model of Hitchcock et al (1971) based on relatively short-term experimental exposures of Orchard grass. Equal uptake rates per kg dry weight have been assumed for leaves of all ages although there is evidence that this rate decreases in old leaves.

While controlled fumigation experiments may usefully represent an integral over the whole day, in the variable conditions of exposure produced by meteorological conditions diurnal effects such as stomatal closure should be taken into account. In the absence of conclusive published data a reduction of uptake rate to 20% at night has been assumed. Published evidence on the effect of humidity upon uptake is inconclusive and it has not been included in the model, nor has the effect of direct wetting of the leaf. It may be speculated that wetting of leaves may cause an initial increase in uptake, but that prolonged rain may cause some loss from the leaf.

Translocation of soluble fluoride in the transpirational stream to tips and margins of leaves is observed in a wide variety of plant species. Benedict et al (1964) have reported measurements of the distribution along Orchard grass leaves and show that absorbed fluoride is located almost entirely in the end 5%. In the present model the grass crop is described in terms of the dry weight yield per hectare. Although means exist in the program for assigning to different ages of leaf different effective lengths, growth and grazing rates and fluoride uptake rates, in the calculations reported here these factors are assumed independent of age. To take some account of the vertical distribution of leaf tips in the sward we have taken a distribution of absorbed fluoride varying as the fourth power of the effective leaf length.

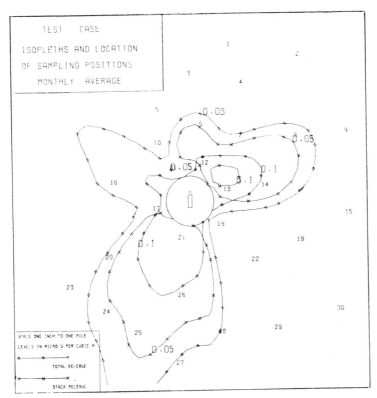

Figure 3. Computer Generated Map showing Plant, Sampling Postiions and Isopleths

MONTHLY AVERAGE GROUND LEVEL CONCENTRATIONS.

SAMPLING POSITION	TOTAL RELEASE (MICR.G/CUB.M)	STACK RELEASE (MICR.G/CUB.M)	BUILDING REL. (MICR.G/CUB.M)
1.	.0228	.0196	.0032
2.	.0171	.0143	.0027
3.	.0288	.0249	.0039
4.	.0276	.0226	.0050
5.	.0225	.0186	.0039
6.	.0594	.0481	.0112
7.	.0575	.0435	.0140
8.	.0569	.0471	.0098
9.	.0386	.0327	.0059
10.	.0312	.0249	.0063
11.	.0531	.0322	.0210
12.	.1126	.0645	.0481
13.	.1876	.1021	.0854
14.	.0919	.0684	.0235
15.	.0241	.0201	.0040
16.	.0465	.0359	.0106
17.	.1098	.0574	.0524
18.	.0460	.0225	.0235
19.	.0138	.0103	.0035
20.	.0629	.0490	.0139
21.	.2168	.0849	.1320
22.	.0044	.0036	.0008
23.	.0419	.0342	.0078
24.	.0611	.0472	.0139
25.	.0661	.0504	.0157
26.	.0813	.0596	.0217
27.	.0451	.0364	.0088
28.	.0573	.0460	.0113
29.	.0044	.0037	.0006
30.	.0002	.0002	.0001

Figure 4. Computed Average Ground Level Concentrations of HF at Sampling Stations 1 to 30.

Seasonal growth rate variations for ryegrass in the United Kingdom have been published (Anslow and Green 1968) and the variation used in this model is a simplification of this. Growth is assumed to begin in mid-April for the northern part of the U.K. and to achieve a maximum rate of 80 kg dry weight per hectare per day. Dilution by growth alone may be examined or, alternatively any growth above a typical basic 1000 kg per hectare residual herbage (not stubble) may be removed by continuous grazing. Obviously, in view of the assumptions relating to fluoride distribution, this grazed herbage will have a fluoride concentration substantially above the average.

5. RESULTS OF TEST CASE CALCULATION

The methods described in Section 4 have been applied to the notional smelting plant described in Section 2. For this calculation a typical meteorological record together with herbage growth and life data appropriate to the month of April has been used.

The computer generated map giving plant location, 1 km radius sampling positions and average isopleths for the month is shown in Figure 3. Isopleths for both total release and the stack contribution are contoured at levels of 0.1 µg m^{-3} and 0.05 µg m^{-3}. Figure 4 shows the corresponding output from the program at the individual sampling positions. It will be seen that although the stack release (greater by a factor of 10 at source) dominates at large distances as shown by, say, sampling position 29 at 6.5 km from the source, the pot-line release has the greater effect at short distances, for example at position 13, 1.6 km from the plant.

Figure 5 summarises the results of two calculations to obtain the variation of herbage fluoride levels with time at a selected sampling position, number 14, which is 3.2 km to the east of the plant. It will be seen that after 14 April wind directions are such as to blow fluoride away from this position for the remainder of the month. In these calculations a natural fluoride concentration of 5 mg kg^{-1} dry weight is assumed to be uniformly distributed, together with an initial absorbed level of 5 mg kg^{-1} dry weight due to earlier plant emissions. In the first calculation, with no grazing, after the onset of growth the average concentration is seen to decrease slowly due to both growth dilution and leaf decay. If however continuous grazing is assumed there is an initial high level in the grazed crop which falls subsequently due to the removal of the most highly concentrated portion of herbage, which in this case is not immediately replenished. The average concentration in the residual crop falls rapidly to the naturally occurring level. Continuous grazing must yield an upper limit to the grazed herbage level; development of the method to yield more realistic grazed herbage concentrations will depend upon the availability of data on grazing patterns and leaf size distribution.

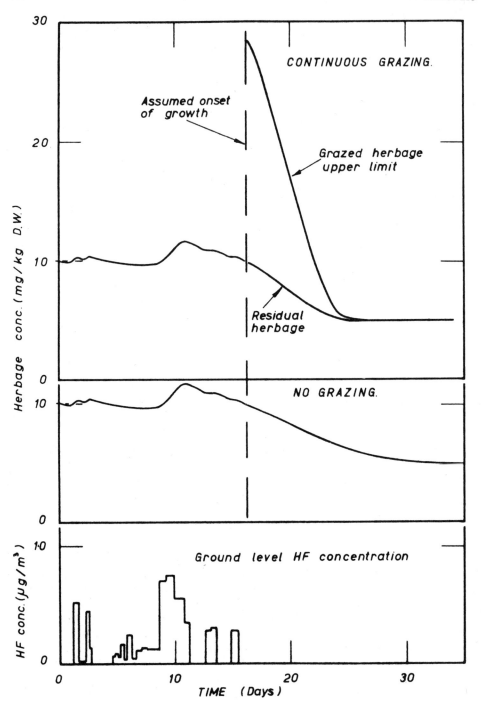

Figure 5. Computed Fluoride Levels and Herbage Concentrations at Sampling Station 14

6. CONCLUSIONS

The Authors' primary object in developing the ecosystem pollution program which is applied in this paper to fluoride emission from a notional Aluminium Smelter, was to provide a management tool whereby commissioning, routine operation and maintenance procedures could be planned against a predetermined, quantified and readily comprehensible level of risk of environmental damage.

The numerical results computed and plotted for this paper illustrate the application of this program to the initial analytical stages of this type of environmental study. The program provides rapid and readily comprehensible predictions of air and herbage contamination. By virtue of the low cost of operation computations can be made to relate herbage monitoring and district air sampling results to specific variables such as significant changes in weather, seasonal changes in herbage growth rates, and changes in the release pattern of fluorides from the plant.

The accumulation of analytical experience of a particular plant within its own specific environment will enable the program (possibly modified as a result of the empirical data acquired) to be used with confidence in a predictive manner. This will allow predictions to be made of pollution levels in unusual operational and maintenance conditions, such as the increased release of fluorides from overheating a smelting pot in specified conditions of climate and season. It will further allow the cost optimisation of the environmental monitoring programme of routine air and herbage analysis since the number and location of sampling stations could be selected for optimum effect.

In addition since computation and plotting of environmental contamination is significantly less costly than sampling and analysis, there are appreciable financial benefits in proving the accuracy of prediction for a specified plant in order to reduce the overall number of measurements to the minimum.

The limitations of the program arise primarily from the paucity of basic data. It is this lack of basic data which leads to difficulty in the establishment of specific tolerance levels for air and pasture. The Authors would like to see their program applied to the establishment of the significance of the parameters influencing damage functions, to direct research along the shortest iterative path to the establishment of adequate parametric data.

REFERENCES

Agricultural Research Council Report
The Effects of Air Pollution on Plants and Soil.
H.M.S.O.(1967).

Anslow R.C.(1966) The Rate of Appearance of Leaves on Tillers of the Graminae, Herbage Abstracts,
36 (3): 149 - 155.

Anslow R.C. and Green J.O. (1967) The Seasonal Growth of Pasture Grasses, J. Agric. Sci. Camb. 68, 109 - 122.

Ball D.F. and Dawson P.R. (1969) Air Pollution from Aluminium Smelters. Chem. and Process Eng. June 1971
49 - 54.

Benedict H.M., Ross M.R. and Wade W.W. (1964)
The Disposition of Atmospheric Fluorides by Vegetation.
Int.J.Air Wat. Poll.,8 (5):279 - 289.

Clarke R.H. (1970) A Method for Assessing the Radiological Effects of Radioactive Effluents from Nuclear Installations. Proc. Second International Congress of the International Radiation Protection Association.
Brighton. U.K.

Gifford F.A. (1961) Use of Routine Meteorological Observations for Estimating Atmospheric Dispersion.
Nuclear Safety,2 (4): 47 - 51.

Harris et al (1963) J. of Animal Science 20: 51 - 55.

Hitchcock A.C., McCune D.C., Weinstein L.H.,MacLean D.C., Jacobson J.S. and Mandl R.H. (1971) Effects of Hydrogen Fluoride Fumigation on Alfalfa and Orchard Grass: A Summary of Experiments from 1952 through 1965, Contributions from Boyce Thompson Institute, 24 (14)
363 - 385.

McCune D.C. (1969) Fluoride Criteria for Vegetation.
Environmental Science and Technology 3 (8).

McCune D.C. and Weinstein L.H. (1971) The Metabolic Effects of Atmospheric Fluorides on Plants. Environmental Pollution 1 (3): 169 - 174.

Pasquill F.(1961) The Estimation of the Dispersion of Windborne Material, Meteorol. Mag., 90 (1063);33 -49.

Shupe J.L. (1969) Levels of Toxicity to Animals. Environmental Science and Technology 3 (8).

Slade H.S. (Ed.) (1968) Meteorology and Atomic Energy, U.S. Atomic Energy Commission.

Robak H. (1969) Aluminium Plants and Conifers in Norway. Proceedings of the First European Congress on the Influence of Air Pollution on Plants and Animals. Wageningen 27 - 31.

Waddington J., and Less L.N. (1971)
A.I.M.E. 100th Annual Meeting, New York, March 1971.

PARTICLE COLLECTION EFFICIENCIES ON CYLINDRICAL WIRES*

George A. Sehmel

Battelle, Pacific Northwest Laboratories Richland, Washington

ABSTRACT

Particle mass transport and deposition mechanisms of eddy diffusion, particle inertia, and lift forces are described by "effective" eddy diffusion coefficients, ε, directed toward the deposition surface. Of principal concern is the eddy diffusion coefficient in the region very close to a surface in which the turbulence scale and intensity decrease rapidly as the surface is approached. These eddy diffusion coefficients are predicted from particle deposition velocity measurements through the use of a model which describes particle deposition in terms of eddy diffusion, Brownian diffusion, and the terminal settling velocity.

INTRODUCTION

The turbulent deposition of aerosol particles from an air stream onto a surface is one means by which airborne particulates are removed from the polluted atmosphere. Aerosol particles will deposit on a surface due to the action of the turbulent air motions of the carrying stream. The particle mass transfer at the surface has been characterized by a deposition velocity, K, defined as

$$K = \frac{\text{amount deposited/cm}^2 \text{ of surface/sec}}{\text{airborne particle concentration above the surface}} \quad (1)$$

---

* This paper is based on work performed under United States Atomic Energy Commission Contract No. AT(45-1)-1830.

which has units of cm/sec. This deposition velocity is the combined effect of all processes contributing to particle deposition. In laboratory experiments, K has been defined in terms of a particle concentration within a cm of the surface. For these experiments, K has been directed toward measuring the controlling mass transfer resistance at the deposition surface. In contrast for field experiments, K has been defined in terms of particle concentrations 1 to 2 m above the vegetative cover. Consequently, the K for field experiments includes both the controlling mass transfer resistance at the surface as well as the mass transfer resistance to about the 2 m height.

Particle deposition velocities were predicted in the pioneering work of Friedlander and Johnstone (4) by eddy diffusion transport followed by a final free flight. Subsequently, others (1, 2, 3, 5, 11) have attempted to improve upon the basic assumption for the particle velocity at the start of the free flight. This model approach has subsequently been shown to be erroneous by both Sehmel (8) and Rouhiainen and Stachiewicz (6) by using two different approaches. The conclusion is that the equality of particle eddy diffusivities to the eddy diffusivities of air momentum is untenable adjacent to a surface. In the present paper, the effective eddy diffusivity approach of Sehmel will be used to include a larger data base for calculation of eddy diffusivities near a floor surface and will be extended to include eddy diffusivities near a ceiling surface.

MODEL DEVELOPMENT

The purpose of using effective eddy diffusivities for particles is to establish a procedure for predicting deposition velocities to smooth surfaces. In reality, this eddy diffusivity procedure first predicts the overall mass transfer resistance and subsequently predicts the deposition velocities. Although only the overall mass transfer resistance can be predicted, reasonable estimates of particle eddy diffusivities and comparison with air eddy diffusivities can be made.

Particle deposition near a surface is assumed in the model to occur by diffusion and gravitation effects. The basic model assumptions are that there is a constant flux of particles from a uniform concentration source C_1 at 1 cm, that particle eddy diffusivities can be predicted, that the gravitational force can be described by the terminal settling velocity, and that the particles are completely retained by the surface after contact. Based upon these assumptions, the deposition flux to a floor surface is

$$N = -(\varepsilon+D)\frac{dC}{dy} - v_t C. \qquad (2)$$

Similar equations (10) can be set up for deposition on wall and ceiling surfaces. Prediction of the deposition velocity

$$K_1 = -N/C_1 \qquad (3)$$

is from a dimensionless integral form of Equation (1) in which

$$-\int_{C_1}^{0} \frac{u_* dC}{N + v_t C} = \int_{y^+ \text{ at 1 cm}}^{r^+} \frac{dy^+}{\varepsilon/\nu + \mathcal{D}/\nu}. \qquad (4)$$

The right-hand integral is termed the resistance integral Int. Once Int is predicted, the deposition velocity can be predicted from

$$K_1 = \frac{y_t}{1 - (1/\alpha)} \qquad (5)$$

where

$$\alpha = \exp(-v_t \, \text{Int}/u_*). \qquad (6)$$

EXPERIMENTAL

The experimental techniques for measuring deposition velocities have been previously described (9). The essential features are that monodispersed uranine particles [density of 1.5 g/cm^3 (7)] were introduced into a 61 × 61 cm cross section wind tunnel. Downstream, the particle deposition flux and airborne concentration profiles were measured. Particle concentration was measured by impaction on cylindrical wires or by collection on filters. The closest concentration measured was 1 cm from the surface. The mass balance of the uranine was determined by dissolving the uranine in water and fluorometric analysis of the wash solutions.

DATA ANALYSIS

Particle effective eddy diffusivities were calculated to predict the observed deposition velocities. The constants in the final eddy diffusivity correlations were selected to minimize the sum of squares of the logarithms of the ratios of the predicted to experimental deposition velocities. The final correlations for eddy diffusivities are shown in Fig. 1.

The eddy diffusivities are a function of the dimensionless distance y^+ from the surface, the particle relaxation time, τ^+,

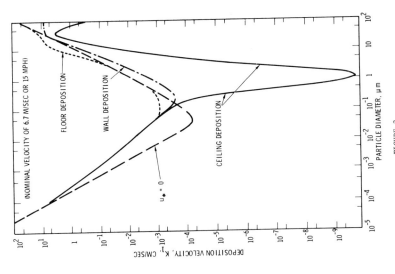

FIGURE 2

COMPARISON OF DEPOSITION VELOCITIES TO SMOOTH FLOOR, WALL, AND CEILING SURFACES FOR A $u_* = 34.1$ CM/SEC AND $y_0 = 0.004$ CM

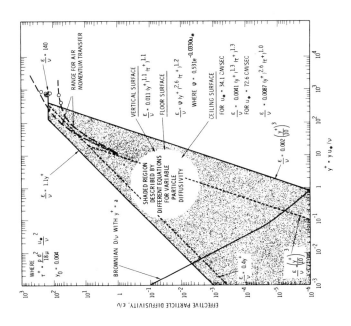

FIGURE 1

PARTICLE DIFFUSIVITY DISTRIBUTIONS NEAR A SMOOTH SURFACE

and are also different for vertical surfaces (8), floor surfaces, and ceiling surfaces. The different diffusivity equations are applied to the shaded region to calculate the diffusional resistance Int. For instance, typical calculations for floor deposition might start from either

$$\text{Int} = \int_{y^+ \text{ at 1 cm}} \frac{dy^+}{140 + D/\nu} + \int \frac{dy^+}{0.002\left(\frac{y^+}{10}\right)^3 + D/\nu}$$

$$+ \int^{r^+} \frac{dy^+}{\psi(y^+)^{2.6}(\tau^+)^{1.2} + D/\nu}, \quad (7)$$

or

$$\text{Int} = \int_{y^+ \text{ at 1 cm}} \frac{dy^+}{140 + D/\nu} + \int \frac{dy^+}{1.1y^+ + D/\nu}$$

$$+ \int^{r^+} \frac{dy^+}{\psi(y^+)^{2.6}(\tau^+)^{1.2} + D/\nu}. \quad (8)$$

The limits of integration would depend upon where the line representing the eddy diffusivities intercepted the upper and lower boundary of the shaded area. Also shown in the shaded area are dashed lines which were used by Sehmel in the original development of effective eddy diffusivities.

DEPOSITION VELOCITY COMPARISON

Deposition velocities to floor, wall, and ceiling surfaces were calculated from Equation (5) using the appropriate form of Equation (7). The resulting deposition velocities are shown in Fig. 2 for a nominal velocity of 6.7 m/sec. As expected, the ceiling deposition is least, wall deposition intermediate, and floor deposition the greatest. For particles smaller than about 0.8 μm, the predicted deposition velocities are all the same. The reason for this identical curve is that for each geometry the eddy diffusivities were small and were represented by the lower boundary limit of the shaded region of Fig. 1. Even though small, the eddy diffusivities were large enough to predict a greater deposition velocity than for particle transport caused by only Brownian diffusion and gravity settling to a floor surface which is shown by the $u_* = 0$ curve.

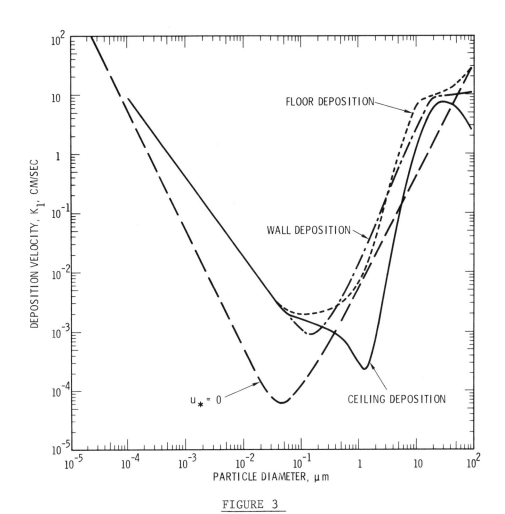

FIGURE 3

COMPARISON OF DEPOSITION VELOCITIES TO SMOOTH FLOOR, WALL AND CEILING SURFACES FOR A u_* = 72.6 CM/SEC AND y_0 = 0.004 CM

(Nominal Velocity of 13.4 m/Sec or 30 MPH)

A further comparison of deposition velocities is made for a larger nominal air velocity of 13.4 m/sec in Fig. 3. The greater turbulence has caused the three curves to be closer together. However, improvement in the eddy diffusivity correlations appears to be needed since a portion of the wall deposition velocity curve is greater than for the floor deposition velocity curve.

CONCLUSIONS

Much remains to be learned about particle deposition. Nevertheless, a model is developed to estimate particle sampling losses in smooth straight lines as well as the lower limit of deposition velocities expected for atmospheric surfaces. The real practical problem now is determining methods for predicting particle deposition to environmental surfaces in which factors such as surface roughness, particle agglomeration, impaction, interception, electrical, and phoretic effects are more important in the eddy deposition process.

NOMENCLATURE

C = local particle concentration, particles/cm^3
C_1 = average particle concentration at reference height of 1 cm, particles/cm^3
d = particle diameter, cm
$\mathcal{D}$ = Brownian diffusivity, cm^2/sec, see Sehmel (1971)
Int = diffusional resistance integral, see Equation (4)
k = von Kármán's constant equal to 0.4
K_1 = deposition velocity, cm/sec (number of particles depositing/(cm^2 sec)/average number concentration/cm^3 of air at reference height of 1 cm)
N = particle flux, particles/(cm^{2}sec)
r = particle radius, cm
r^+ = ru_*/ν, dimensionless
u = local horizontal air velocity, cm/sec
u_* = friction velocity, $u = (u_*/k) \ln[(y+y_0)/y_0]$
v_t = settling velocity, $\rho d^2 g/(18\ \mu)$, cm/sec
y = height above surface, cm
y^+ = yu_*/ν, dimensionless
y_0 = roughness height, cm
α = $\exp(-v_t\ \text{Int}/u_*)$, see Equation (6)
ε/ν = ratio of [(inertial + diffusivity + lift)/viscous] forces for particles
μ = gas viscosity, gm/(cm sec)
μm = micrometer, which is 10^{-4} cm
ν = kinematic viscosity, μ/ρ_g, cm^2/sec
ρ = particle density, gm/cm^3

ρ_g = gas density, gm/cm^3
τ = particle relaxation time, $\rho d^2/(18\mu)$
$\tau^+ = \tau u_*^2/\nu$, dimensionless

REFERENCES

1. Beal, S.K. Deposition of Particles in Turbulent Flow on Channel or Pipe Walls. Nucl. Sci. Eng., 40, 1-11 (1970).

2. Davies, C.N. Deposition of Aerosols from Turbulent Pipe Flow. Proc. Roy. Soc. Ser. A, 289, 235-246 (1966).

3. Davies, C.N. Aerosol Science, New York, Academic 408, 425-440 (1966).

4. Friedlander, S.K. and Johnstone, H.F. Deposition of Suspended Particles from Turbulent Gas Streams. Industr. Eng. Chem., 49, 1151-1156 (1957).

5. Owen, P.R. Dust Deposition from a Turbulent Airstream. Int. J. Air-Water Pollution, 3, 8-25, 50-53 (1960).

6. Rouhiainen, P.O. and Stachiewicz, J.W. On the Deposition of Small Particles from Turbulent Streams. J. Heat Transfer, 29, Ser. C, 169-177 (1970).

7. Sehmel, G.A. The Density of Uranine Particles Produced by a Spinning Disc Aerosol Generator. Amer. Industr. Hyg. Assoc. J., 28, 491-492 (1967).

8. Sehmel, G.A. Particle Deposition from Turbulent Air Flow. J. Geophys. Res., 75(9), 1766-1781 (1970).

9. Sehmel, G.A. Particle Diffusivities and Deposition Velocities over a Horizontal Smooth Surface. J. of Colloid and Interface Sci., 37, 891-906 (1971).

10. Sehmel, G.A. Particle Eddy Diffusivities and Deposition Velocities for Smooth Surfaces, BNWL-SA-4324, Battelle, Pacific Northwest Laboratories, June 1972. Submitted for publication.

11. Wells, A.C., and Chamberlain, A.C. Transport of small Particles to Vertical Surfaces. Brit. J. App. Phys., 18, 1793-1799 (1967).

INDUSTRIAL NOISE POLLUTION

L.S. Goodfriend and F.M. Kessler

L.S. Goodfriend & Associates, Cedar Knolls, New Jersey

INTRODUCTION

Of all the pollutants, noise is the only one that does not leave a residue. To determine how much noise has been made at any location, it must be measured as it is being made, or at least recorded precisely for measurement and analysis at a later time. It is also difficult to study the adverse effects of noise because there are no directly observable tangible effects of noise on people when the levels of noise are below those which will cause temporary loss of hearing; and these levels are well above those that will cause interference with speech communication and distraction from creative activities.

The subject of this paper is industrial noise pollution. This pollution occurs both in plant and in the adjacent community due to a multiplicity of noise sources located either within the plant building or exterior to the building on the roof, at its walls, and within the plant grounds.

In-plant industrial noise, depending upon its frequency range and amplitude, may annoy, interfere with speech/hearing relationship, or cause irreversible hearing damage.

Each of these effects upon human beings have been the object of investigations for many years. The amount of hearing damage among factory workers has reached a sufficient level to prompt the United States Congress to pass legislation setting maximum acceptable noise levels and duration of occupational exposure. There is presently before Congress, proposed legislation to limit industrial plant and other sources of noise which affects a

community. This law, it is hoped, would provide on a national
level, the limitation on industrial noise which now exists only
at local levels through zoning ordinances and municipal laws.
A difficulty exists in the community in separating that portion
of the noise ambient due to industrial noise from that portion
due to a multiplicity of other sources, such as aircraft, surface
transportation, construction, and other local effects. A considerable number of descriptors of community noise have been proposed, but each are lacking in effectiveness in describing the
community response relative to specific noise sources. Presently,
the U. S. Environmental Protection Agency and the Department of
Transportation are sponsoring research to develop the community
descriptor needed for the evaluation of noise from many sources
and its impact upon the community.

IN-PLANT HAZARDOUS NOISE EXPOSURE

The criteria established by the Federal Occupational Safety
and Health Act of 1970[1], based upon the Walsh-Healey Public Contracts Act of 1969[2], relates to results reported in 1965 by the
National Academy of Sciences-National Research Council on Hearing,
Bioacoustics, and Biomechanics[3]. This committee has been mandated to arrive at a damage risk criteria for hazardous exposure
to intermittant and steady-state noise.

They found that hearing loss in humans results from two major
sources besides accidents and congenital effects:

A. Presbycusis, a decline in hearing acuity due to old age, and
B. Long duration exposure to high levels of noise.

The basic hazardous noise criterion which the committee
adopted, provides that a noise environment is unacceptable if,
after 10 years of eight hours per day exposure, the average
employee has suffered a permanent work-induced hearing loss of
10 decibels at 1000 hertz, 15 decibels at 2000 hertz, or 20 decibels at 3000 hertz or above.

The Walsh-Healey damage risk criteria are presented in Table
1. The Occupational Safety and Health Act - 1970, requires the
employer to provide..."Protection against the effects of noise
exposure...if the sound levels exceeds those...(Table 1)...
when measured on the A-scale of a standard sound level meter at
slow response..." The Act also specifies that when employees are
subject to sound levels exceeding those in the Table, feasible
administrative or engineering controls must be utilized.

TABLE 1
PERMISSIBLE NOISE EXPOSURES

Duration per Day, Hours	Sound Level dB(A) Slow
8	90
6	92
4	95
3	97
2	100
1-1/2	102
1	105
1/2	110
1/4 or less	115

Interestingly, the Department of Labor takes the following approach with regard to personal protective equipment in lieu of engineering or administrative controls. They state in their Bulletin 334[4], that "the use of personal protective equipment is considered by the Department to be an interim measure while engineering and administrative controls are being perfected. There will be very few cases in which the use of this equipment will be acceptable as a permanent solution to the noise problem."

The American Academy of Ophthalmology and Otolaryngology[5], has available data which indicate the percentage of industrial workers which may have noise-induced hearing loss due to hazardous exposure. These data are presented in Table 2. You will note, that a significantly large number of employees (6.6 percent) will suffer noise-induced hearing loss if exposed for eight hours per day to 90 decibels A-weighted for a 10 year period of time, suggesting that the eight hours per day exposure level be reduced to 85 dB(A).

TABLE 2
PERCENTAGE RISK OF DEVELOPING A HEARING HANDICAP

Noise Level dB(A)	Exposure - Years						
	5	10	15	20	25	30	35
85	1.0	2.6	4.0	5.0	6.1	6.5	8.0
90	3.0	6.6	10.0	11.9	13.4	15.6	17.5
95	5.7	12.3	18.2	21.4	24.1	26.7	28.3
100	9.0	20.7	30.0	35.9	38.1	40.8	41.5
105	13.2	31.7	44.0	49.9	54.1	57.8	57.5
110	19.0	46.2	61.0	68.4	73.1	73.8	71.5
115	26.0	61.2	79.0	83.9	86.1	84.3	89.5

MEASUREMENT AND ANALYSIS OF IN-PLANT NOISE

In-plant noise may be readily measured, using a sound level meter set on slow response A-weighting network. The noise level can be compared to the values presented previously in Table 1, to indicate the levels of exposure a worker may be subjected to before the compliance limits of the Occupational Safety and Health Act are exceeded. The noise level thus obtained, does not provide the knowledge, for engineering controls, of the major source noise spectrum, nor which component of the major noise source is the principal radiator of the excessive noise. The acoustical consultant or in-plant noise specialist, desires the data in a spectral format.

Tape recording the noise data has the advantage of minimizing the field measurement time, allowing more time to be spent in analyzing the data for the funds expended.

A measurement system used for acquiring the acoustical data might consist of a Bruel & Kjaer Type 4145 Condenser Microphone, used in conjunction with a Bruel & Kjaer Type 2203 Precision Sound Level Meter, which functions as a linear amplifier and step attenuator. Its output, in turn, is fed to a Kudelski-Nagra IV Magnetic Tape Recorder. The recordings are made using Scotch Type 175 Magnetic Recording Tape (1.5 mil TENZAR), at a speed of 7-1/2 ips. The noise levels are monitored during recording using headphones.

At the beginning and end of each reel of tape, the entire system is calibrated by means of a Bruel & Kjaer Type 4220 Pistonphone. The calibration consists of introducing a sound pressure level of 124 decibels into the system through the microphone and recording this signal on the magnetic tape. As well as providing a reference tone, this method insures that no change in calibration of the system occurs during the measurements.

The analysis of rms sound pressure levels in one-third octave bands, octave bands, and A-, B-, and C-weighted noise levels is performed in the laboratory from the magnetic tape. This analysis might incorporate the use of a General Radio Type 1921 Real-Time Analyzer and a Digital PDP-8/I Computer. The magnetic tape is played back using a Crown Tape Recorder. The output signal from the recorder is fed to a Real-Time Analyzer, which consists of a One-Third Octave Multifilter, and a Multichannel RMS Detector. The digital output of the real-time analyzer is fed to a PDP-8/I computer, which computs the desired noise spectral analysis information.

The data may be sampled for durations as short as one-eighth second to durations as long as 32 seconds.

TABLE 3
TYPICAL EQUIPMENT NOISE LEVELS (A-WEIGHTED)

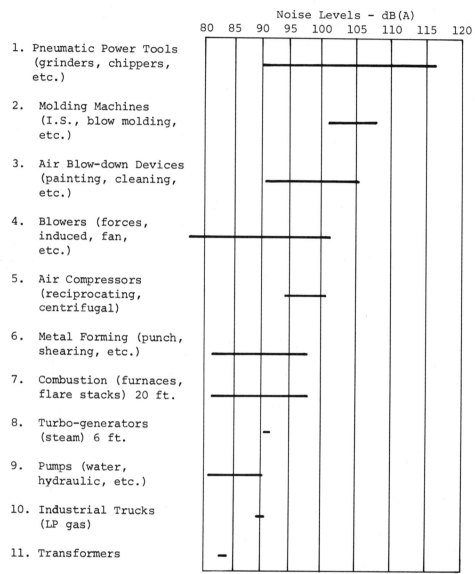

NOISE CONTROL

Two examples of noise control, of the source and along its transmission path are presented. One may eliminate the source of noise by impeding or absorbing the noise along its transmission path. An enclosure is an example of this approach.

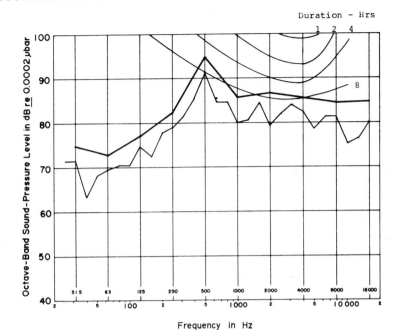

FIGURE 1   CAN FILLING OPERATION
—— OCTAVE   —— ONE-THIRD OCTAVE

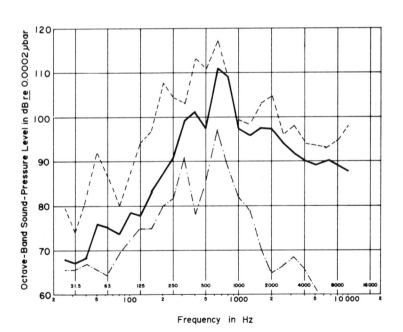

FIGURE 2   LARGE POLISHING WHEEL OPERATION
—— BEFORE ENCLOSURE   ---- INSIDE ENCLOSURE   —·—·— OPERATOR'S POSITION

INDUSTRIAL EQUIPMENT NOISE

The general approach to noise control in industrial plants is well established. However, because of the multiplicity and complexity of industrial plant noise sources and their associated environments, solutions to industrial noise problems have been obtained more or less on an empirical basis. In other words, an analytical solution to every industrial noise problem does not exist. Experimental investigations of the noise source should form part of a noise control development program. Excessive noise in existing industrial plants can be reduced (to conform to established criteria for hearing damage, annoyance, or speech communication), by applying current state-of-the-art noise abatement technology.

A recent study[6] conducted for the Federal Environmental Protection Agency at a number of typical industrial plants, allowed us to develop Table 3, which presents the noise level, A-weighted, of plant equipment. Many plants contain numerous areas which are not in compliance with the Federal regulations. Often, equipment such as compressors indicated as item 5 in Table 3, produce noise levels in excess of 90 dB(A), but are usually at unattended areas where they do not impact upon the general plant environment. Other industrial plants often locate their noise making equipment in the general production area with no concern for its hazardous impact.

The process of filling cans can be quite noisy. At a typical oil refinery where one quart cans are filled in a semi-automatic line, the noise level at the operator's position was recorded. Figure 1 presents the spectral data resulting from the analysis of this recording. The heavy black line is the octave band sound pressure level, while the lighter line is the one-third octave band spectra. Superimposed on this figure are the Walsh-Healey contours. Note the additional information one can obtain from the one-third octave presentation. At high frequencies the contribution in narrower bands may be readily noted.

The next figure, Figure 2, presents the one-third octave band sound pressure levels measured at a 120 inch polishing wheel operation prior to noise control measures. This noise has considerable tonal content as indicated by the difference in sound pressure levels between third octaves.

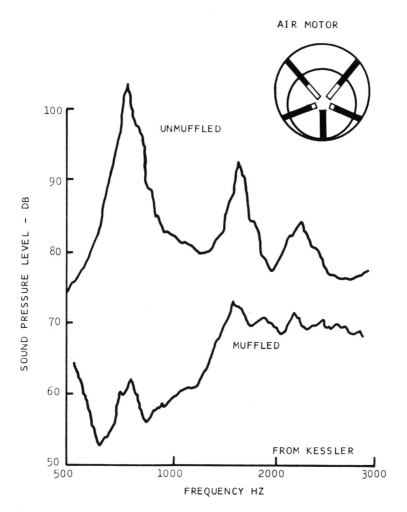

FIGURE 3 NOISE SPECTRA FOR A MUFFLED AND UNMUFFLED FIVE VANE AIR MOTOR UNDER FULL LOAD

Referring again to the large polishing wheel operation, acoustical enclosures provided the best solution to the reduction of the high noise levels at the operator's position for this polishing process. The criteria considered to be of primary concern in the enclosure design were:

1. Acoustical effectiveness.
2. Accessibility for operators.
3. Accessibility for maintenance.
4. Compliance with fire regulations and safety codes.
5. Adequate visibility of process by operators.
6. Durability.
7. Economy.

Careful construction techniques are of extreme importance in achieving maximum noise reduction from an enclosure. Sound leaks through small openings in an enclosure greatly reduce the potential transmission loss. All voids should be filled with resilient sealant. Doors and access ports should be fully gasketed on all four sides. If ventilation is required, a sound trap in association with forced air ventilation, must be employed.

An enclosure was constructed for the polishing operation. Figure 2 presents the one-third octave band spectra of the original noise levels, the noise levels inside of the enclosure, and the noise levels observed outside of the enclosure at the operator's position. Note the abatement achieved. The noise level, while increasing within the enclosure due to the absence of perfect absorption, is reduced considerably by the glass fiber and plywood or sheet steel exterior.

An alternate approach is to modify the equipment generating the noise. An example of the reduction of the source noise of an air powered hoist is presented.

It also allows the presentation as an example of spectra in a narrow band format. Here, the filters used for the analysis are 10 percent of an octave wide. Figure 3 presents the noise of a hoist in its unmuffled configuration. Also presented in the figure, is a schematic of the vane-type air motor, which is the principal source of noise. The hoist air motor generates noise at the vane passing frequency and harmonics. Note how clearly this noise at the vane passing frequency and its harmonics are evident by using the narrow band analysis. The noise control approach for this equipment, is to reduce the noise at the principal frequencies by muffling prior to the exhaust air entering the environment.

A muffler was designed to be incorporated within the body of the hoist. This muffler was a double expansion chamber unit with an inter-connecting tube. The double expansion chamber was selected

Figure 4   Internal Muffler Section

Figure 5   Hoist Air Motor Section Containing Internal Muffler Section

due to its insertion loss characteristics and the fact that a
chamber which already existed in the hoist casting could be
utilized for one of its sections. The second expansion chamber
unit was added to the hoist. This added section is shown in
Figure 4 as fabricated, and in Figure 5 as installed in the hoist.
Note the reduction in sound pressure level at 700 hertz, the
vane passing frequency, Figure 3. A sound pressure level reduction in the order of 40 decibels was achieved.

IN-PLANT NOISE CONTROL

Industrial plant noise, anticipated at the early phase of
plant development, can readily be controlled. Table 4 presents
a summary of noise reduction methods which include plant planning
as an important section.

TABLE 4

I. Plant Planning
    a) Selection of Equipment
    b) Location of Equipment Within the Plant
    c) Location of Plant with Respect to the Community

II. Control at the Source
    a) Maintain Dynamic Balance
    b) Minimize Rotational Speed
    c) Decouple the Driving Force
    d) Reduce Velocity of Fluid Flow
    e) Reduce Turbulence
    f) Use Directionality of Source
    g) Use Reactive or Dissipative Mufflers

III. Control of the Transmitted Noise
    a) Vibration Isolate the Source
    b) Enclose the Source
    c) Absorb Sound Within the Room

IV. Control of Radiated Noise
    a) Increase Mass
    b) Increase Stiffness
    c) Shift Resonant Frequencies
    d) Add Damping
    e) Reduce Surface Area
    f) Perforate the Surface

The noise as indicated earlier may be controlled at the source,
may be controlled along its transmission path, and may be controlled at the surface which re-radiates the noise into the environment. The latter can also be considered a transmission path.
Noise reduction programs for plants already in operation are

usually directed at reducing noise along its transmission path with enclosures as the approach which is often used.

Many corporations are developing noise specifications for new equipment. When used by their purchasing agent, these specifications should aid in the noise abatement effort as obsolete noise equipment are replaced.

INDUSTRIAL PLANT EXTERIOR NOISE

The noise of an industrial plant, or plant noise plus surface transportation noise, contributes to the residual noise level in its community. Industrial noise is a local problem with each plant presenting individual intrusive characteristics which may not be comparable on a nation wide basis. The plant location, community residual noise levels, and other sources such as major highways, airports, or construction activities contribute to the community climate.

It is usually difficult to assess the impact of plant noise on the community by simply observing the A-weighted ambient noise levels at various locations in the community during the workday, worknight or weekend. To better understand the effects of the noise and to obtain some quantitative measure of these effects, various rating systems have been devised. Rating systems most commonly used today are the Composite Noise Rating, the Noise Exposure Forecast, and Robinson's Noise Pollution Level. These descriptors require complex computations.

Community noise may be assessed using data in A-weighted noise level format by using a new rating system which has recently been introduced by Wyle Laboratories and reported in their contractor's report[7] to the Environmental Protection Agency NTID 300.3. Its name is "Community Noise Equivalent Level" (CNEL).

Histograms of the data observed at numerous locations provide an intrusive noise level L_{10} value, that is the noise level exceeded 10 percent of the time during the measurement period. This statistical noise level for each location for day, evening, and nighttime periods are tabulated and energy averaged. The CNEL is computed from the following equation:

$$CNEL = 10 \, Log \, (\frac{1}{m+n+l}) \left[\sum_{i=1}^{m} Antilog \, (DL_{10}/10)_i \right.$$
$$\left. + 3 \sum_{i=1}^{n} Antilog \, (EL_{10}/10) + 10 \sum_{i=1}^{l} Antilog \, (NL_{10}/10)_i \right]$$

Where m, n, and l are the number of intrusive noise level values for day, evening, and nighttime sampling periods, respectively.

DL_{10}, EL_{10}, NL_{10} are intrusive noise levels A-weighted for day, evening, and nighttime sampling periods, respectively. The CNEL values thus computed are corrected for seasonal factors, for the background levels, and for history of previous noise exposure and community attitudes. If pure tonal or impulsive noise are evident, additional corrections are applied. These corrected CNEL values are known as normalized community noise equivalent noise levels in dB and are presented in Figure 6 for five typical industrial plants. Information regarding complaint levels were obtained by interviewing various township police and board of health officials in the industrial plant community.

The heavy line is the average of data of many industrial plants and aircraft community NCNEL values which Wyle Laboratories had obtained. Note a satisfactory agreement at higher NCNELs, where sporadic complaints or wide-spread complaints are indicated. Some error is evident where the noise is noticeable and the NCNEL values are fairly low.

INDUSTRIAL PLANT-COMMUNITY NOISE ABATEMENT

It appears that complaints or a lack of complaints, may not be a satisfactory indicator of the impact of plant noise on its neighbors.

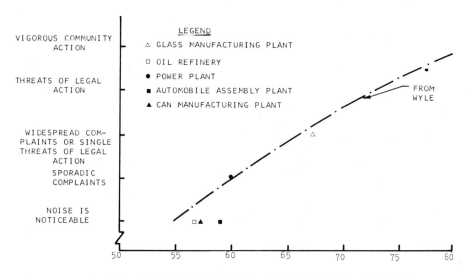

FIGURE 6   NORMALIZED COMMUNITY NOISE EQUIVALENT LEVEL IN dB

Industrial plant neighbors in a community may not object to plant noise even at fairly high levels

   a) if it is continuous,
   b) if it does not interfere with speech communication,
   c) if it does not include pure tones or impacts,
   d) if it does not vary rapidly,
   e) if it does not interfere with getting to sleep, and
   f) if it does not contain fear-producing elements.

Some political, social, or economic situations develop where noise which is normally objectionable causes no complaints. Often single indivuduals or families may be annoyed by an industrial noise which does not annoy other plant neighbors. This, in many cases, may be traced to unusual exposure conditions, or to interpersonal situations involving plant management personnel.

In the past, the primary motivation for reduction of plant noise reaching the community was the desire to be good neighbors and to maintain good community relations. It was found through discussions with industrial plant management, that the large corporations of national stature are particularly sensitive to public relations. Funds and personnel are quickly made available to solve noise problems which the plants are made aware of by community complaints.

One of the more important considerations for industrial plant planning for noise control lies in the initial design of new plants and the modernization of existing ones. Architectural noise control concepts have been successfully applied to this field for the past two decades.

The site selection and industrial plant design processes, together with the local government control of industrial zoning, provide the motivation and the early opportunity to institute noise abatement efforts. It is known that this early phase of industrial plant development provides the most economic period for application of noise reduction techniques. Local municipal pressures in the form of noise nuisance ordinance and, more recently, realistic zoning regulations, have produced legal pressures to reduce plant noise.

It is anticipated that, in general, industrial plant noise reaching the community will not increase in the near future, but may in fact, decrease, as noise abatement efforts required by the Occupational Safety and Health Act of 1970 become effective. But it must be pointed out, that at specific locations where interior plant noise is reduced by simply locating the noise sources outdoors, the impact upon the community may increase.

It appears that noise due to construction job sites, surface transportation, and aircraft exceeds in importance the contribution of industrial plants to community annoyance. At some future date, when noise abatement efforts applied to the above primary sources successfully reduce their levels, the contribution of industrial plant noise to the community residual levels will rise in importance. One might conclude though, that using the present state-of-the-art in noise abatement, it is possible to control industrial noise and thus, provide satisfactory in-plant and community environments.

ACKNOWLEDGEMENT

This paper was essentially based upon the results of an L. S. Goodfriend & Associates' study of industrial plant noise pollution sponsored by the United States Environmental Protection Agency's Office of Noise Abatement and Control.

REFERENCES

1. Williams-Steiger Occupational Safety and Health Act of 1970.

2. Walsh-Healey Public Contracts Act, Part 50-24, Safety and Health Standards for Federal Supply Contracts, Federal Register (May 1970).

3. Kryter, K. D., et al., "Hazardous Exposure to Intermittant and Steady Noise," Journal of the Acoustical Society of America, Vol. 39, No. 3 (1966).

4. U.S. Department of Labor "Guidelines to the Department of Labor's Occupational Noise Standards," Bulletin 334 (revised - 1971).

5. House, H.P., et. al., "Guide for Conservation of Hearing in Noise," A Supplement to the Transactions of the American Academy of Ophthalmology and Otolaryngology (1969).

6. Environmental Protection Agency "Noise from Industrial Plant," NTID 300.2 (1971).

7. Environmental Protection Agency "Community Noise," NTID 300.3 (1971).

MOTOR VEHICLE NOISE

William J. Galloway

Bolt Beranek & Newman, Inc., Canoga Park, California

Noise from motor vehicles is the most pervasive source of urban noise. Except for locations near to airports where aircraft noise is excessive or near large industrial centers, motor vehicle noise is the controlling factor in setting the background noise levels of our environment. In those few situations where extensive community noise surveys have been made, vehicular traffic noise controlled the noise environment in more than 85 percent of the locations. In this paper we will briefly examine the overall noise characteristics of automobiles, motorcycles, and heavy trucks, and then consider how they aggregate into what we generally describe as "traffic noise."

Automobiles, through their total numbers, are the largest total source of urban noise. However, individually they are not the most intense. At typical street speeds, the modern passenger car is basically, a quiet device. It is normally equipped with very adequate intake and exhaust mufflers and engine noise is controlled by the car body. However, as speeds are increased, time/roadway noise becomes the controlling factor in automotive noise. A typical variation of sound pressure levels with speed is shown in Figure 1. Note that the noise levels with engine power on and with it off are little different at 60 mph.

One can ask how passenger car noise levels vary from one make to the next, or what happens with age. The California Highway Patrol (1) has made an extensive set

588   10. NOISE

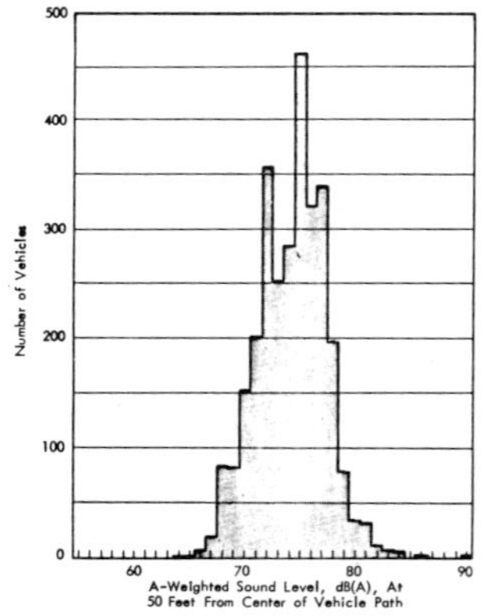

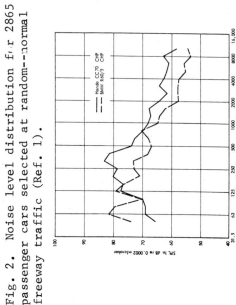

Fig. 2. Noise level distribution for 2865 passenger cars selected at random--normal freeway traffic (Ref. 1).

Fig. 4. Noise spectra for two motorcycles--50 feet--CHP method.

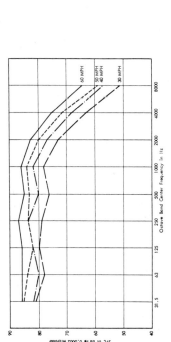

Fig. 1. Noise spectra for passenger car at different speeds--25 feet.

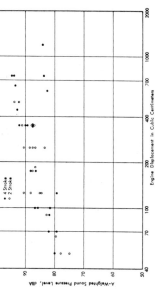

Fig. 3. Noise levels for 48 motorcycles--50 feet driveby, wide open throttle--SAE proposed test method.

of measurements from which one can draw a general picture of the passenger car population. As an example, the distribution of observed noise levels measured in sound level A (dBA) under freeway conditions is shown in Figure 2. The analysis made by the Highway Patrol indicated no significant differences between makes of vehicles, and no significant difference in vehicles more than five years old.

The automobile noise described above is with reference to concrete roadways and constant speed conditions. These levels will be exceeded by about 5 dB during acceleration. Very smooth asphalt roadways will result in noise levels as much as 10 dB below those indicated above.

Motorcycles are quickly becoming an important part of the total vehicular noise picture. In recent years, the registration of new motorcycles has been growing at an ever increasing rate. Not only are they used increasingly as a means of transportation among certain age groups, but the mini-cycle used off the street on trails and in parks is becoming an important source of high intensity noise. We recently had the opportunity to measure the noise levels of 48 different motorcycles, covering essentially all the popular makes and models being sold in this country for street use. Noise measurements under several different operational test procedures were performed. One of the most informative was that using a proposed Society of Automotive Engineers test procedure. This test essentially calls for a wide open throttle driveby such that the engine achieves its highest rated rpm when passing the test microphone.

The results of these tests are shown in Figure 3 where the maximum sound level-A is plotted as a function of engine displacement. It is interesting to note the general trend toward higher noise with increasing size. The fact that this is not necessary is shown by the low noise levels produced by three of the largest motorcycles tested, indicating that better muffling is available than most manufacturers supply. The 10 dB variation in noise output between the noisiest and quietest motorcycles in the 125 cc class is further indication that quieter motorcycles can be produced.

Comparison of the sound level spectra for a 70 cc and a 600 cc motorcycle having almost the same sound level-A values is shown in Figure 4. The difference in muffling effectiveness for the two machines is apparent

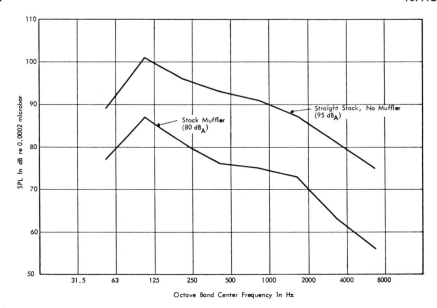

Fig. 5. Noise spectra for Cummins NH-250 engine in tractor on dynamometer--1900 RPM--50 feet.

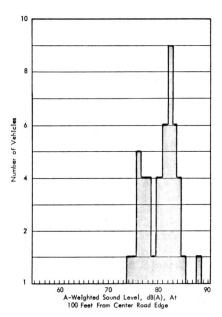

Fig. 6. Noise level distribution for 48 diesel trucks selected at random--normal highway cruising speed.

when one considers that the larger engine is almost 10 times larger than the smaller.

It is worth noting that, in a recent social survey we performed to estimate the public acceptability of motor vehicle noise from 1200 respondents, about 25% of the traffic noises eliciting objectionable reactions were identified with motorcycles. This result is quite remarkable when one realizes that, in total, motorcycles represent only a very small percentage of the total vehicle population

Diesel trucks are clearly the noisiest vehicles on our streets and highways. They are considerably larger in size and more complex as noise sources than passenger cars. Present diesel tractor/trailer designs are not optimized in any way from a noise control point of view. With the exception of exhaust noise, no attempt is made to reduce other sources of noise such as gear noise or intake noise. Current practice, good quality mufflers can reduce exhaust noise by 10 to 15 dB when compared to an unmuffled, straight stack diesel. At this level of exhaust noise control, little further reduction in noise can be achieved without attacking the other noise sources on the vehicle.

Typical noise reduction obtained by exhaust muffling is shown in Figure 5. These data were taken on a dynamometer test stand and do not properly reflect the effects of tire/roadway noise. Operating the same vehicle on the road at the same engine speed may raise the noise level by approximately 5 dB at low frequencies and approximately 10 dB at the higher frequencies where tire noise predominates.

In both passenger car and truck data it is clear that tire generated noise, at least for highway speeds, is often a controlling factor in vehicle noise. Some data available show that tread design, tire loading, and wear can produce as much as 10 to 15 dB difference in the noise produced by motor vehicles. Thus, this is clearly an area deserving much more attention than it has received to date. An example of the noise level distribution from a randomly selected group of trucks operating at highway speeds is shown in Figure 6.

Traffic noise is the result of the combined effect of many individual vehicles having the noise characteristics outlined previously. Instead of the relatively short duration, single event noise signal produced by a passing vehicle, we are now concerned with a random noise produced by the superposition of the sounds from many vehicles. As with most time series phenomena, a statistical description of the noise signal is appropriate. Combining elements of traffic flow theory and knowledge of the noise characteristics of individual vehicles allows us to generate models of traffic noise, for use in urban design problems. Using traffic flow volumes, vehicle speeds, and passenger/truck mixes as input parameters, we can generate the mean and variance of noise levels, and their spectral distribution, for many practical traffic situations.

A number of examples of the use of simulation models are given in Reference 2. An example of a noise level distribution predicted by the model and a comparison with observed values is shown in Figure 7. For simplified, quick estimates of average noise levels, at vehicle volume flows greater than 1000 per hour, the following equation can be used in the case of passenger cars:

$$\overline{L}_A = 10 \log_{10} \frac{q}{d} + 20 \log_{10} \overline{v} + 20$$

where $\overline{L}_A$ = time average noise level in dBA

q = vehicles per hour

$\overline{v}$ = average speed in miles per hour

d = distance in feet from geometric mean centerline of road to observer

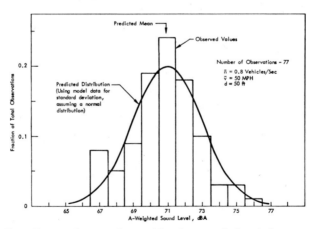

Fig. 7. Comparison of prediction model with measured noise level for passenger cars.

The effect of adding trucks to the above can be estimated from the following table.

Effect of Trucks on Average Traffic Noise

% of trucks in traffic	incremental increase in $\bar{L}_A$
0	0
2.5	1
5	2
10	4
20	8

In this paper we have presented some of the basic components of traffic noise: a) motor vehicles are the predominant source of noise pollution in urban environments; b) passenger vehicles are usually reasonably quiet, individually, but collectively produce substantial noise at higher speeds and traffic flow volumes; c) motorcycles are unacceptably loud to a high proportion of the population--they can be made very much quieter if public pressure demands this be done; d) diesel trucks are the loudest noise sources in most urban areas--they can be made 10 to 15 dB less noisy if major redesign is performed--public pressure, on a proper time table, can cause this redesign to take place; e) tire noise can well become the major controlling source in vehicle noise--the basic mechanisms of tire noise generation have received very little attention and should be investigated; f) practical models for predicting traffic noise are available for planning use.

This paper was also presented and published in the Proceedings of the 17th Annual Meeting & Equipment Exposition, Institute of Environmental Sciences, Los Angeles, Ca., April 27, 1971.

References:

1) "Passenger Car Noise Survey," California Highway Patrol, January 1970.

2) W. J. Galloway, W. E. Clark, J. S. Kerrick, "Highway Noise: Measurement, Simulation, and Mixed Reactions," NCHRP Report 78, Highway Research Board, 1969.

ENGINEERING AND SCIENTIFIC IMPLICATIONS OF NOISE CONTROL LEGISLATION

ALVIN F. MEYER, JR.

Director, Office of Noise Abatement and Control, U.S. Environmental Protection Agency, Washington, D.C.

There has been much discussion regarding the needs for even greater engineering and scientific action for environmental management. However, as pointed out in a report of the Subcommittee on Science Research and Development of the Committee on Science and Astronautics, U.S. House of Representatives, "the human race is in fact managing the environment today." The multitude of subjects discussed at the UN Conference on Environment in Stockholm, Sweden, in the past few days clearly are indicative of the many aspects of environmental management or "mismanagement" (depending upon the particular situations and circumstances) which are involved. Of major concern are, in fact, the mistakes in environmental management which require the use of science, technology and engineering to minimize the impact on society and mankind throughout the world.

There has developed an ever increasing awareness of the needs for and in fact the practice of multi-discipline efforts to deal with problems at first glance, arise from failures of science and technology. Environmental quality is an abstract term, difficult to qualify, and involves a complex of technological, economics, social institutions, legal and political problems. As pointed out further in the previously referenced U.S. Congressional Report, all too often the material upon which decision making must be made is in the form of data from the natural sciences and engineering.

The regulation of noise through legislation presents some very special problems of environmental quality management because of the very nature of the environmental attribute involved. Acoustic noise, being defined as unwanted sound, has a wide range of effects on

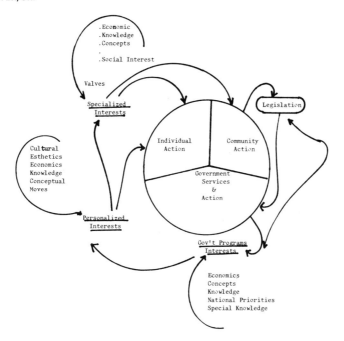

FIGURE 1

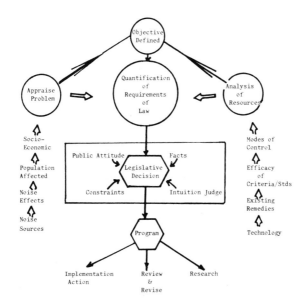

FIGURE 2

humans. These extend from the extreme of actual hearing impairment to the very human subjective reactions of annoyance and irritation. Taking into account the effects of interference with communication, appreciation of music or other desired "listening," disturbance of sleep and other possible psychological or pysiological impacts of a non-auditory nature, but difficult to measure or define, very real problems are presented in attempting to devise regulatory schemes which lend themselves to enforcement.

As part of an extensive action taken by the U.S. Environmental Protection Agency in investigating noise and possible means of environmental action, as required by the United States Congress in Title IV, PL. 91-604, considerable attention was given to the matter of laws and regulatory schemes for noise abatement. The material contained in the EPA Report to the President and Congress on Noise is based on a comprehensive publication on this subject (EPA NTID 300.4, "Laws and Regulatory Schemes for Noise Abatement").

It is not the purpose of this paper to abstract the voluminous work. Some of the more important observations and findings will be highlighted, in the context of how they relate to the multi-discipline environmental management concept. There have been many efforts to reduce the complex array of issues in considerations of environmental management to some sort of schematic visualization, and one such is presented in Figure 1. It is evident from this illustration that the need for "community action" leading to legislation and regulation arises from a coming together of individual or personal value systems with those of specialized interest (such as industry or labor) and those of government. These value systems often may be compatible and work together. Quite frequently, they are in conflict and may, depending upon the circumstances, vary in terms of a social-economic-political reaction within the same individual or group, depending upon the particular circumstance or situation. Again, for the reasons already stated, in the case of noise, these attitudinal changes in the same person or group of persons may be quite pronounced. For example, an individual or group of individuals may enjoy participating in some noise producing activities (such as motorcycle riding) but object vigorously to the sound of aircraft interfering with their sleep. Among some of the very real conflicts, which exist in regard to noise regulations, are those arising not only from the foregoing but from honest differences and opinions as to the effects of noise and their implications from a health and welfare viewpoint. Perhaps the most serious aspects of this is that the courts in many nations have taken a view that environmental noise is a nuisance which well may have to be borne as part of the cost of other socially desired values such as transportation, technological progress, or economic expansion. What has happened is that in the main, the social cost of noise in terms of detrimental effects in relation to environmental quality has been externalized by the creator of the acoustical energy involved, be it industry or an individual operator. That trend seems to be changing worldwide as evidenced in the results of the EPA Study on Noise Concerns in Other Nations.

The expert in the field of noise control is well aware that there are classic approaches to noise abatement. These in their most simplistic form are measures aimed at the source of noise, those modifying or acting upon paths of transmission, and those concerned with personal protection of the receiver. There are obviously many alternatives available within these general areas of engineering and scientific application. Many of these are also associated with other areas of environmental concern such as water or air pollution and involve land use planning, zoning and architectural arrangements and design.

In the absence of law or regulations, criteria for noise control are selected by those sponsoring the control measure, be they an individual concern, manufacturing a potentially noisy product; the developer of a housing complex; or the purchaser of specially designed potentially noisy equipment, and so on. As stated by Beranek, the cost of noise control measures are obviously directly related to the criterion selected and, conversely, the criterion selected may depend upon an assessment of the cost (criterion here used not in the classic environmental sense). As already indicated in arriving at whatever cost versus benefit estimations are conducted in the absence of legislative requirements, the levels of controls, if any selected, may be of such a nature that, in effect, the noise cost is externalized; namely, the burden is placed upon those exposed thereto rather than being included in the basic design.

The criteria usely available are related to scales, or procedures for evaluating noise in relation to human effects; translated into some measure useful for engineering design or planning. From the statutory enforcement viewpoint, unfortunately, many of the measures of value currently in use present difficulties in application to regulatory enforcement procedures involving court action. James Hilderbrand, in his book, <u>Noise Pollution and and the Law</u>, succinctly discusses some of the problems which arise in the absence of specific noise control statutes.

Once statutory requirements for noise control are enacted, the foregoing situation rapidly changes. Instead of "market place" decisions as to "costs" to be borne for some level of "benefits" acceptable to the purchaser of the item involved, the level of abatement depends upon a governmentally imposed "noise control standard." Such standards or regulations, however, must if to be practical, reasonable, and effective, take into account a wide array of scientific, sociological, economical, and political considerations, as alluded to already in this paper.

Dr. Louis Mayo and his associates at the George Washington University, Washington, D.C., in preparing the EPA publication on regulatory schemes mentioned above, prepared an "Environmental Control Matrix" which is a most comprehensive analysis of virtually all of the elements of the subject. It highlights the many multi-discipline considerations associated with noise control by statutory regulations. An effort to present a simplified block diagram of this most extensive matrix is presented in Figure 2. The principal steps, involving scientists and engineers, along with their other professional colleagues, the general public, and the sources and makers of noise are as follows:

- ... assessment of the nature and extent of the noise problem.

- ... determination of the knowns and unknowns and discernment of needs for both near term and long term research.

- ... consideration of a possible near term and longer term of abatement and control possibilities (including physical modification, control of the sources, transmission past barriers, protection of receivers through the use of sound protectors, operational modifications of use controls.)

- ... review of available existing criteria and standards including an appraisal of the effectiveness of existing remedies and procedures.

- ... assess the implications of abatement alternatives by determining the economic, sociological and political effects on elements of society involved.

- ... through some socio-political mechanism define objectives, based upon a balanced appraisal of the problem and of the resource situation, and then quantify them into legislative proposals.

Throughout the steps described above, and outlined in Figure 2, it is evident that scientists and engineers with acoustical expertise will be deeply involved in providing an input into the multi-discipline studies and reviews. One guiding principal is that there must be maximum involvement of all of the various interests in this process. A group of engineers dealing with a problem which involves biology, medicine, economics, sociology, and law can only come up with solutions based on the best level of knowledge on those subjects within the group of engineers. Likewise, a group of lawyers, social scientists, or political scientists, dealing with the same problem, will only come up with the recommendation or solution based on their understanding of the engineering and life-science principles involved. On the other hand a group of lawyers, physical, life and social scientists, engineers and representatives of those affected will arrive at far more meaningful recommendations for needed programs. It, of course, would be extremely useful if the education, training and experience of the multi-disciplined specialists participating had included some central core of knowledge and activity relating to the fundamentals of human physiology and of environmental quality management. Perhaps one of the most urgent needs with regard to the role of the engineer and scientists, in these procedures and in those described immediately following, is the providing of such a central core and the educational process for professional workers. All engineers regardless of their particular specialty should have, in the curricular at the undergraduate level, some fundamental courses in engineering, physiology, and in principals of environmental control and management. Likewise, common core courses ought to be available for students of law, medicine, sociology, political science, any arts and humanities. This is no new concern of this author, and many of his colleagues. The time has come to accelerate action in this regard.

Once legislative proposals have been developed, the need for ultimate decisions by the appropriate legislative authority then arises. The degree of effectiveness of the legislative proposal will be subject to question by those affected in the industrial sector, by those whose private activities may be regulated and by those who are advocates of improved environmental quality. In the case of the noise control, this may appear to result in a division of views to be presented to the legislature on the part of the advocates of noise versus the proponents of quiet. The ultimate decision as to the nature of the authority to be conferred upon government, takes into account a variety of apparently, at times, conflicting influences. These, as shown in Figure 2, include four principal elements. These are:

>... First, there is the role of facts developed in accordance with the procedures already discussed above. Once again, the importance of a multi-

discipline participation in the assemblage of the basic information and in the definition of the objectives is emphasized. Politicians have no more special expertise in the fields of science and engineering than do the general public. Engineers and scientists per se also have no special qualifications with respect to political processes; thus the need for the greatest effort to produce the best information possible arises. The real challenge is to provide the requisite information in readily understandable language. The array of facts must include what's known, what's not known, and certainly should include various alternatives.

... Second, against the array of facts, careful consideration must be given to the constraints which may effect the utility and effectiveness of the law. These constraints involve those relating to economics and the number of attributes of the various value systems of the society involved. These may include cultural patterns, social mores, the degree of knowledge of health, environment, ecology, and the conceptualization of environmental attributes, esthetics, overall national goals and priorities including profit motives of industry, and certainly, the degree of professional and technical skill available to deal with noise and noise problems.

... The public attitude, which is also reflected in the constraints, is the third major influence. There are a number of interfaces between citizens, groups of citizens, the community, government and industry. The noise problem is perhaps even more difficult than most environmental problems for reasons already stated. The public attitude results in a number of conflicting pressures, included among these will be the pressure by industry to maintain the status quo against which there may be pressure of individuals and groups for reform in action. These in turn may be reflected in demands visible to the legislative body to achieve quiet and control noise through political aims. At the same time within government, there may be pressures to provide services consistent with economic constraints, which tend to resist the collective community action. On the other hand, government may discern

> that the need for effective noise control is greater than the public perceives.
>
> ... Finally, the legislature must then exercise, in the decision-making process, its intuition and judgment out of which then arises the legislative authority.

Obviously, the final product is an authority for a regulatory program which will include implementing actions, such as standards and regulations, abatement procedures and enforcement, and research. A major requirement is, of course, for continuing review of the effectiveness of the program in producing abatement of noise. Once again, the role of the engineer and scientist in providing his technical skills and expertise again is self evident. In the review and revision process the real challenge is to come up with meaningful and understandable evaluations of program effectiveness, and a clear translation to government and to the public of the results of such studies and evaluations.

The United States Environmental Protection Agency in the last twelve months has been deeply involved in activities relating to the pending United States Noise Control Legislation. The foregoing has been a general review of the actual scheme of activity undertaken. One of the more useful results of this has been the evolution of the model discussed above. It is hoped that these experiences and observations will be useful to others concerned with this important aspect of noise control.

NOISE POLLUTION IN DEVELOPING COUNTRIES AND IN ISRAEL IN PARTICULAR

F. MICHAEL STRUMPF

*MAMASH, Applied Science Laboratories, Ltd.
Ramat Aviv, Israel*

Noise can now, all too justifiably, be added to the list of major pollutants. Much has been said about the noise with which we are forced to be acquainted day to day. However, little is known about the psychological and physiological aspects of our exposure. In addition, even in industry where noise is defined as a problem and treated as such, little understanding has been achieved.

Although it is not necessary to acquaint this forum with the problems of noise, I want to read to you the following article which was published in Stockholm with regard to the U. N. Environment Conference:

Man is fast driving himself mad with noise. Of all forms of pollution, noise is the most insistent yet the easiest to control, the World Health Organization said in a report to the U. N. Environment Conference.

Noise is measured in decibels and 140 decibels produces insanity. Pneumatic drills produce 120 decibels--the pain threshhold--and a motorcycle, 110 decibels.

The report described noise as "a curse of modern times," and "a major environment problem" today.

The hearing of people is becoming severely affected, it said.

A recent survey in Sweden showed that 79.5 per cent of all persons between 15 and 20 applying for jobs had hearing troubles caused by noise.[1] This is double the figure in 1956.

The report said that doctors now attribute one in every three cases of neurosis to noise and four of every five headaches.[2]

"It has now become urgent to set permissible limits," the report said.

The increase in environmental noise is a by-product of technological advance. It is posing an increasing source of annoyance, and what is more important, a rapidly growing damage. If we follow the literature on the subject, we will find that environmental noise is growing rather steadily. Our protection by means of reducing noise and enforcing standards so far has not eliminated the problem. It has rather categorized it into varying grades of suffering. The fact is, that the increase in environmental noise causes our increased exposure to noise--noise here meaning not only that generated by industry but also day to day sources such as traffic, sirens, horns, etc.

The advance of Developing Countries is accompanied by industrialization. We witness not only a higher noise level in industry and traffic, but also a population concentration on the one hand and higher building on the other. To this add household noises, and we soon see that we are surrounded by noise sources from which we cannot tear ourselves away. Industrial, municipal, and regional planning are indeed the basics that in the future will establish Environmental Noise.

Today we work with higher noise levels than ever before. The ear, one of our most sensitive organs, protects itself in its own way; however, the results are often less than desirable. As said before, research shows the increasing number of noise-impaired people in their various forms and degrees. Industrial Medicine unfortunately does not keep pace with noise problems. The real degree of damage is still unknown, especially its long range aspects.

Damage by noise ranges from annoyance to insanity and death. Where are the limits? We live in the midst of incessant background noise. This "minimal quietness" may be reacted to differently by different people. Its definition may in fact be fatal for certain population groups. I personally doubt the famous statement often made: "Noise is a thing you can get used to." The psychological stress and its influence are expressed statistically in various works. Noises have been graded and listed as well as the "human society reaction" that was directly manifested.

It was found that Environmental Noise in the dwelling house was the most annoying. Its annoyance factor was so great that in densely populated areas and in high store buildings the rate of clinic cases and suicides was highest compared with all other groups of the population.[3,4] These facts call for a revision of our building designs.

In a dwelling we have vertical and horizontal noise propagation We have staircases, banging doors, and yards filled with children. It seems that in this instance we experience environmental noise pollution in its most serious stages. The design of a quiet home is not necessarily that expensive. Moreover, the advantages achieved are far more extensive than the dry physical figures of annoyance rates. In hot climate countries like Israel, we live with open windows. Noise intrudes easily. Here, therefore, the education towards the importance of quiet should be emphasized.

The school is one of the places where this education should start. However, does the school building receive our attention? Long noisey corridors, shouting children, and noisey classrooms necessitate voice raising. Noise is rapidly increased. Is this the foundation on which to build a quiet atmosphere? The design of classrooms is one step toward controlling the environment. Rooms are silenced by acoustic ceilings and carpets. The reflected energy is reduced and intelligibility is improved by raising the signal to noise ratio.

In developing countries standards must be created and/or accepted. A society aware of the noise hazard is that much closer to improved living. Manufacturers will start to reduce noise in all fields and this reduction will even more emphasize and define those noisey products which will then be avoided.

Standards can be divided into four major groups:

1. Standards that establish the maximum permissable noise levels for workers in high intensity noise areas.

2. Standards that define the "minimum noise levels"--levels that state what "quiet" is.

3. Standards for buildings.

4. Standards to control the environment: traffic noise, home appliances, etc.

I shall briefly comment on each of the standards:

1. The criterion for maximum noise level was stated as 90 dB(A), a level at which man can work for 8 hours. An increased level reduces permissable time of stay. For example: 15 minutes in a 115 dB(A) noise.

2. Environmental quiet should protect rich as well as poor in their free time.

3. Standards in buildings provide a control of the dwelling industry. This is not true for the developed countries and obviously not for the developing countries where the special needs of this market make it possible to sell everything.

4. Standards to control environmental noise pollution need, more than other, governmental support. Often it may come through the back door, due to claims and lawsuits.

Let us come back to protection from noise when its sources can not be reduced. Research has been done on extended use of ear protectors. Still, we speak only about the ear protection, not including the physiological aspects. It should be noted that noise annoyance is a source of increased accident rates, fatigue, lack of ability to concentrate, hearing difficulties, and reduction in efficiency.

The aspects of noise pollution have been discussed, and in view of all alarms, its scope should be further investigated in hearing conservation programs. One of the immediate fruits of such a program will be the exchange of working places for "noise sensitive people." This will enable them to save their hearing before the stage in which it is too late. When this turning point has already been past, damage must be treated.

Israel is in a sense more than a developing country. Being a population melting pot, it is of utmost importance to know the extent to which noise pollution has penetrated. A statistical survey can state the scope as well as the severity of the problem.

Let me summarize by saying that the hazardous influences of noise are more far reaching, and, as noise pollution will grow, we must start protective activity. And the earlier, the better.

REFERENCES

1. Jerusalem Post, daily paper, Article on U.N. Environment Conference in Stockholm, 1972, edited by W. Dixon and J.E. Fricke.

2. Noise as a Public Health Hazard, Asha report no. 4, 1969.

3. Noise in Buildings, SBI report no. 58, 1962.

4. Occupational Hearing Loss, edited by D.W. Robinson, Academic Press, London, 1971.

ENVIRONMENTAL PROTECTION IN THE U.S.

Stanley M. Greenfield
Environmental Protection Agency

Every kind of environmental deterioration known to man can be found in the United States of America. A nation which has long taken pride in its technological advancement now finds itself troubled by the by-products of that advancement: polluted water and air, wasteful use of materials and energy, urban crowding, noise, and a host of other threats to the public health and welfare.

Indeed, some of us are beginning to think that the deterioration may not be the by-product, however unintentional, but instead the main product of technical advancement. As Irving Kristol put it, in a more general indictment of human folly: "The unanticipated consequences of social action are always more important, and usually less agreeable, than the intended consequences." <u>On the Democratic Idea in America</u>, Harper and Row, quoted in New York Times Review, 10 May 1972.

To list the particular environmental problems about which we are most concerned in the United States is useful only if we realize that such a listing reflects our ignorance as well as our knowledge. The processes by which Nature absorbs polluting substances and renders them harmless are still very imperfectly understood. We do not know enough about the natural environment to take control actions with confidence that they will be effective. We must continue to study the relationship between man and the environment and at the same time do the best we can with our imperfect knowledge and techniques to reduce pollution wherever it seems possible to do so.

The most urgent environmental problems in the United States are those which pose immediate threats to human health: impure water, polluted air, solid waste disposal, pesticides and other toxic substances, and ionizing radiation.

During the last two decades the U.S. Congress has been increasingly concerned with these problems. It has passed legislation to support research, to develop and demonstrate control methods, and to aid state and local governments in pollution control work.

Such legislation has usually dealt with a particular kind of pollution or polluted medium. It has tended to assert and establish a federal responsibility, supplementing and in some ways superseding the local and state laws that were based on long-standing local powers to protect public health and to prevent nuisances. The imposition of federal controls has accelerated over the last decade, and federal legislation is now dominant in all five pollution areas, with the exception of some aspects of solid waste disposal.

Toward the close of the 1960s public concern increased, and it became apparent that the divided approach - controlling pollution by the type of pollutant or the polluted medium -- was likely to be wasteful and sure to be ineffective. The environment is complex and interconnected; its protection would require a unified, coordinated effort.

This policy was officially recognized in the National Environmental Policy Act, signed by President Nixon on January 1, 1970, which declared it a national policy to "encourage productive and enjoyable harmony between man and his environment" and set forth the principles for coordination of the federal responsibility.

In December, 1970, the Environmental Protection Agency was established to bring together in one central unit - reporting independently to the President -the various anti-pollution programs already established in more than a dozen different agencies.

The existing environmental programs had each been mandated by separate legislation, and they were administered by separate agencies (in some cases by more than one). Moreover, the programs were handicapped by their attachment to traditional government departments. The agency responsible for an environmental program had been established for, and long concerned with, other purposes. Environmental considerations were latecomers in the agency's raison d'etre, and often conflicted with other purposes of longer duration and higher priority.

The result was the familiar beaucratic schizophrenia, or split personality, within the agency. For instance:

* The Department of Agriculture's prime purpose is promoting the production of food and fiber. It's interest in pesticides was mainly to increase agricultural production; it was only secondarily interested in environmental effects. moreover, some aspects of pesticide regulation were the responsibility of an entirely different agency, the Food and Drug Administration.

* Water pollution control was assigned by law to the Department of the Interior, where it sometimes took a back seat to Interior's traditional tasks of developing new dams and reservoirs. The same department's National Park Service and Bureau fo Sport Fisheries and Wildlife have both been positive forces in studying, understanding, and preserving the natural environment; they also sometimes found themselves in conflict with the department's developmental aims.

* The Atomic Energy Commission, established to control and foster nuclear power production, also had the primary responsibility for policing itself in setting safety standards for radiation in the environment.

The Department of Transportation did not supply any component of the new environmental agency, but a potential conflict exists in the regulation of noise from aircraft. The department's present regulations are being followed by the industry with some reluctance and delay, and many people feel they are not stringent enough. A congressionally authorized study has proposed that EPA be given partial control of aircraft noise standards and full control of noise from vehicles and land-based machinery. All of the transporation functions which the department oversees have important effects on the environment as regards land use, urban and industrial growth, and air and water pollution.

In addition to such internal conflicts, the former division of environmental programs among various agencies made it difficult for them to keep in touch with each other. It does little good to make an industry stop discharging a pollutant into the air if it can liquefy the offending substance and pour it into a river, or solidify it and dump it on the land.

EPA brought the five principal anti-pollution programs together for unified action standard setting, research, and monitoring, under one organizational roof. It put the enforcement function outside the government components that have been long associated with certain sectors of the economy and are usually committed - by law or by custom - to promoting the very activities they are supposed to regulate in the public interest.

As Administrator William D. Ruckelshaus expressed it: "EPA is not obligated to promote anything but a better environment."

If this obligation conflicts with another department or agency, the conflict must be resolved by one of three higher authorities:

* The President himself, using the expertise of his Council on Environmental Quality, and his scientific and economic advisers;

* New legislation from the Congress; or

* New judicial interpretation of existing law.

It is exceedingly difficult to conceal such conflicts from the press and the public. Indeed, we in EPA feel they should be publicized, debated openly, and decided on the merits of each case. The National Environmental Policy Act provides a very potent procedure for this purpose: the environmental impact statement.

All federal agencies, before taking any action that may affect the environment, must first draft a statement setting forth such expected effects, listing other possible actions, and giving reasons why the particular proposed action was chosen over the options.

These impact statements are submitted in draft to the Council on Environmental Quality. They also are sent to EPA and to any other agency having a legal interest in the proposed action or technical expertise in the kind or quality of the environmental effects involved. Impact statements are also circulated to state and local governments when they are affected. The reviewing bodies' comments - often highly critical - furnish the originating agency with a basis for revising its final statement, or they are included with it as attachments.

The drafts, comments, and final statements are all public records, available to any interested party and to the press. Public hearings are frequently held to help an agency assess the environmental impact and to revise its statement.

Last year EPA reviewed more than 2,400 impact statements.

In rough order of frequency they concerned road building, watershed protection projects, dams, canals, harbors, airports, parks, and power plants. Since most large construction by private industry or municipalities involve federal licensing of some sort, the environmental effects of many non-governmental actions are also subject to this kind of public review.

The impact statement procedure has helped to convert project planners, both public and private, to greater environmental awareness. Government agencies are becoming more competent in environmental assessment, and they are coming to realize that such assessment is an integral part of their function. A new purpose has been added to the host of organic laws under which government agencies operate.

Naturally there are conflicts over technical matters, over legalisms and interpretation, and especially over the estimating of benefits and costs. These open conflicts help to stimulate and give a focus to environmental protection.

Although EPA is the federal government's principal anti-pollution agency, setting and enforcing standards, the agency does not try to be an all-powerful, central policeman of pollution. No single group can watch every mile of waterway, every industrial stack and outfall, every automobile tailpipe. The agency uses legal action vigorously but selectively. It tries to work through state and local governments wherever possible, to create pressures, to encourage as well as threaten.

Last year EPA took legal action against industrial and municipal polluters on an average of 15 times per month. The threat of a law suit frequently produces voluntary compliance of the offender, or a specific plan for achieving compliance by a certain date.

Nationwide standards for air quality were set by EPA last summer, as required by law. They define for the first time the quantities of six major types of air pollutant that are dangerous to health and well-being. These are very specific standards - parts per million, measured in certain ways, and persisting over cetain times. They apply throughout the nation, in rural areas as well as in cities.

Declaring a standard does no good unless it is implemented, that is, unless definite actions are taken to achieve the standard and these actions are enforced. Fifty states plus the District of Columbia, Guam, Puerto Rico, Virgin Islands, and American Samoa submitted implementation plans to EPA. Two weeks ago (May 31) all of them were formally approved with or without exceptions by

the EPA administrator, but only ten of the plans were approved
without exceptions specified by the EPA administrator (Alabama,
Connecticut, Florida, Mississippi, New Hampshire, Oregon, West
Virginia, Guam, Puerto Rico, and American Samoa).

Each state plan sets limits to pollutant emissions from
all major sources. The limits are based on a careful inventory
of sources: industrial and power plants, automotive vehicles,
and so on. Each plan sets forth the control regulations to
be established, enforcement provisions and penalties, and detailed
procedures for coping with air pollution emergencies.

Direct Federal regulations of industrial air pollution
have been established for new or substantially modified installations
in five types of industry, all notorious air polluters: fuel-
burning steam generators, large incinerators, and plants producing
cement, sulfuric acid, and nitric acid. The regulations specify
maximum amounts of pollutants in weight per unit of plant output.
The limits are calculated to require these industries to use
the best available control technology.

A permit system to control the dumping of waste materials
in waterways is being established under the authority of a
little-used provision of the River and Harbor Act of 1899.
The provision, called the Refuse Act, forbids the discharge
of any waste into navigable waters or their tributaries.

The actual issuance of permits has been delayed by a ruling,
at the lowest level of the federal courts, that an environmental
impact statement should have been submitted with each permit
application. We regard this as an obstructing tactic, and
we are seeking to get the ruling reversed, either by appealing
to a higher court, or getting the law changed, or both.

Until this obstruction is overcome, the permit program
is stalled, but we are already making use of the information
contained in the 19,000 permit applications. These constitute
an extremely valuable inventory of the sources of liquid waste
in our waterways, information that is unique and unprecedented.
When combined with information on other effluent sources (e.g.,
municipal sanitary and storm sewers, not covered by the 1899
law), we will have a comprehensive factual base for all future
efforts to control water pollution.

The agency has established strong regional organizations
in the ten Federal administrative regions throughout the country.
Each now has a regional EPA administrator and a complete staff
of local experts in the programs areas and in law enforcement.
The object is to bring more of the decision-making power close

to the areas concerned, so that EPA's actions will be prompt, effective, and sensitive to local conditions. On-the-scene decisions will be particularly important in the administration of the Refuse Act permit program.

EPA has reorganized the research and development programs of its predecessor bureaus. Three national environmental research centers have been established: in Cincinnati, Ohio, for control technology, solid waste, and radiation health effects; in Research Triangle Park, North Carolina, for air pollution studies; and in Corvallis, Oregon, for water pollution studies. A fourth center, in Las Vegas, Nevada, specializes in environmental monitoring. The programs of these four centers and their satellite laboratories - 31 altogether, in 19 states from Alaska to Florida and from Washington state to Rhode Island - are designed to enable the agency to predict and anticipate environmental hazards as well as reacting to them. This entire program is organized as a single directed effort under my office and as such represents a somewhat unique prominence of research in a federal regulatory agency.

Research and monitoring are vital to environmental improvement. Effective standard-setting and enforcement requires sound data on what is being introduced into the environment, its impact on ecological stability, on human health, and on other factors important to human life. In closely coordinated research programs EPA strives to develop a synthesis of knowledge from the biological, physical and social sciences which can be interpreted in terms of total human and environmental needs.

Major aims of the Agency's research efforts at this time include:

> * Expansion and improvement of environmental monitoring and surveillance to provide baselines of environmental quality.
>
> * Better understanding of long-term exposures to contaminants, of sub-acute or delayed effects on human and other organisms, and of the combined and synergistic actions of chemical, biological, and physical stresses.
>
> * Acceleration of applied research into the control of pollutants, the recycling of wastes, and the development of non-polluting production processes.
>
> * Improved assessment of trends of technical and social change and potential effect - first, second,

and even third-order effects - on evnironmental quality.

* Improved understanding of the transport of pollutants through the environment; their passage through air; water, and land; their ability to cross the various interfaces; and the various changes that can make them innocuous at one point and hazardous at another.

In addition to performing research in its own laboratories, EPA, through grants and contracts, supports the studies of scientists in universities and other research institutions. The agency also consolidates and evaluates information as it is developed throughout the scientific community to develop the best possible scientific base for environmental action.

EPA provides training both in its own facilities and in universities and other educational institutions to develop the skilled manpower needed to combat environmental problems. The largest portion of EPA's manpower training activities has been in the water area; some 10,000 waste treatment plants operators have been trained. EPA is assessing national manpower needs in solid waste activities. In addition to laborers, solid waste programs will require supervisors, middle management officials, engineers and others in solid waste management. Studies also are under way to determine the number of trained personnel required for air pollution control. In pesticides, EPA conducts formal and informal training to produce pesticide residue chemists. Effective pesticides control will require trained and licensed applicators; our training needs in this area are just beginning to be felt.

Government can only do so much. The rest is up to the private sector. Our system leaves a lot to its citizens, encouraging them to seek reform directly instead of always turning to government. Even the most stringent regulation of industry will not alone succeed in moving it to meet its social and legal obligations.

Industrial innovation in manufacturing and marketing has made the United States a world leader in technology and material well-being. But industry must now use its innovative capabilities to solve pollution problems. A number of technological innovations by industry have already contributed dramatically to our present ability to cope with pollution or soon may do so.

Although control of some pollutants awaits new technology, many pollutants could be curbed drastically by more widespread use of current technology. We estimate that if the best presently available technology were applied to all existing sources of particulate emissions, they would be reduced by 95 percent.

The American public has become aware of its environment as never before. The environment has become a major national issue - a concern of Americans in all walks of life. It is impossible to predict what the staying power of citizen interest in the environment will be, but there are some hints.

Public opinion polls rate environmental quality as a top domestic concern. Bond issues for environmental improvement did extremely well at the polls in recent elections. The nation's environmental and conservation organizations are growing at rapid rates. Local groups, often dealing with a single issue, spring up daily. Environmental groups have entered political campaigns - endorsing candidates with good environmental credentials - and have influenced ballot box decisions. And citizens increasingly are turning to the courts for redress of environmental degradation. "Public interest" law firms are challenging in the courts major decisions of government and forcing industries to take cleanup actions.

After only a year and a half of operation, the Environmental Protection Agency cannot yet be judged a success or a failure. We think we have made a good start. We know that any success we have will constitute only a small part of total effort needed to establish and maintain a rational, balanced relationship between man and the total environment.

The reduction of pollution is a first step - necessary but not sufficient - to achieve such a balance.

It has long been recognized that the unchecked growth of the world's population and the ever increasing use of limited natural resources are logically impossible to maintain, regardless of how we do in reducing pollution.

These larger questions were eloquently set forth in the recent study, "Limits to Growth," by Dennis Meadows and others, under the sponsorship of the Club of Rome. In this study the M.I.T. scholars attempted to model the global ecosystem.

Meadows and his colleagues attempted to predict, for the whole world, the future course of half a dozen very generalized factors, including population growth, food production, resource depletion, use of fuels and energy, and pollution levels. Their projections invariably lead to a global breakdown at some early, but not well defined, point in time.

Whether or not you agree with their basic data, their assumptions, and their admittedly simplified relationships among these factors, the technique suggested by them and many others demands consideration. We view this type of approach as one of growing importants in EPA's research program.

A century and a half ago, Thomas Malthus propounded a similar mathematical model that also shook people up. Mathus proved to be wrong on his time scale, but his main conclusion that population tends to outstrip the food supply has not yet been finally refuted.

Vast changes will have to take place in our society if Malthus and Meadows are to be disproved and man is to achieve environmental equilibrium.

Meanwhile, we have challenge enough in the immediate task of cleaning up the environmental damage already wrought. Only by controlling pollution with the means at hand or soon to be developed can we stay alive and healthy and gain time to face and solve the greater problems that lie ahead.

POLLUTION PROBLEMS IN AUSTRALIA - A LARGE, SPARSELY POPULATED, RAPIDLY DEVELOPING COUNTRY

D.O. Jordan
Department of Physical and Inorganic Chemistry, University of Adelaide, South Australia

Introduction

Australia is a large continental island of almost 3 million square miles in area; it is equivalent in size to the conterminous size of U.S.A. It has control of its coastline of 12,000 miles and its river systems and so no pollution is received from any neighbouring country. It is the driest continent in the world, about one third is desert and the average rainfall is 16.5 inches. Coupled with this low rainfall is a high rate of evaporation. Of the present population of 12.8 million, about 7 million live in the six State capital cities. The cities have, over the last 30 to 40 years, become heavily industrialised and with an Air Pollution Potential in each one equivalent to that of New York they face similar air pollution problems. The cities, apart from the commercial and business areas, are large urban areas composed of single storey houses at a density of four or five to the acre.

The author has recently been concerned in a study of the environment in South Australia, one of the States of Australia. Since the problems of South Australia are similar to those of Australia as a whole, we will concentrate mainly on this State.

Population of South Australia

The population of South Australia is at present 1.2 million. South Australia was founded in 1836 and took 86 years to attain 0.5 million (1922), another 41 years to reach 1 million (1963), and will reach 1.5 million in 1985 after only a further 22 years.

It is predicted that the population will reach 2.0 million in 2003, 3 million in 2025 and 4 million in 2043. These figures are small by any standards. However, the doubling rate of 40 years is only a little greater than the world average. There are two ways in which Australia could limit its population, by fertility control and by immigration control. The latter is determined by Federal policy and the number of financially assisted migrants, although small, makes a significant increase in the population over the years. South Australia's main problem is water; it has two sources at present, reservoirs in the catchment areas around Adelaide and the River Murray. On the basis of these two sources aided by private reservoirs and some underground reserves, it would appear that with present water usage (120 gallons per day per person) the State cannot support a population greater than 3 million. Whether distillation using nuclear power could allow a significant increase in this total has not yet been estimated.

Water Pollution in South Australia

South Australia is the driest State in the driest continent. Its water pollution problems are not great. There is some eutrophication occurring in the reservoirs due to excessive nitrate run-off from hills farms. This is controlled by copper sulphate addition, but is being tackled at source by improved methods of farming, by moving farms away from catchment areas and in one case by moving a whole village. The main problem with the River Murray water is salinity. This arises through ground water, inflow from salt water aquifers and through irrigation, only the latter can be easily controlled.

The whole of Adelaide is served by a deep sewerage system with four main sewage treatment works. In country areas sewage is disposed of either by septic tanks with or without a common drain leading to evaporation ponds or a sewage system with an Imhoff tank. In only two places is raw sewage pumped to the sea. These are two relatively small towns. The reuse of water is important. At present about 40 million gallons a day of effluent can be discharged to the sea. The effluent is saline, but a great deal is capable of use, and is used, for irrigation. However, at present this is only for sports fields and the airport and not for crops. The main problem is the almost certain presence of the hepatitis virus and so chlorination (or irradiation, or both) is necessary before safe use for crop irrigation can be contemplated. However, satisfactory experiments with ground irrigation, as distinct from overhead sprays, have been carried out. No reuse for human consumption is at present contemplated. At some northern, desert towns, a second reticulation system has been installed and sewage effluent is used to irrigate parks, gardens and street trees.

Of industrial pollution there is little direct evidence except for one area where the discharge from a pulp and paper mill went direct into a fresh water lagoon.

Air Pollution in South Australia

The main fuels in South Australia are oil, natural gas and electricity. One power station, situated in a small desert town, burns a low grade coal. Coupled with a low density of industry, there is no great air pollution problem. Fall-out ranges in Adelaide from 11 to 24 tons/square mile/month, the former in residential areas, the latter in industrial. However, Adelaide does have a smog problem, due to motor car exhaust emission, garden incinerators and industry. The problem is being tackled by a Clean Air Act controlling emissions and the banning of dark smoke and Air Pollution Potential alerts on days where temperature inversions are likely. At present the A.P.P. alert is not compulsory. No legislation exists at present on car exhaust emissions and all that is to come into force is a control of carbon monoxide, to 4.5 per cent for an idling engine.

Unfortunately the industrial areas of Adelaide are all upwind from the residential areas, an example of bad initial planning or lack of planning.

The best solution for garden incinerators, which are a major source of pollution, is to have a much more extensive garbage collection system than exists at present.

Refuse Disposal

The main method of refuse disposal is tipping and sanitary landfill. At present there are no incinerators and no refuse is dumped at sea. There are within the city many old clay pits of considerable size which are being filled. Some industrial wastes enter these pits and this is unsatisfactory; these include oil (notably sump oil), cyanide waste, acids, etc. This problem is being tackled at source by neutralisation of acids, oxidation of cyanide and re-refining of sump oil. Liquid waste will probably be treated, in future, at one site with extensive evaporation lagoons.

Land Use in South Australia

The cities of South Australia tend to sprawl with 4 to 5 houses per acre. Since the city exists on a coastal plain between a low range of hills (2300 feet) and the sea, it is long and narrow, 15 to 20 miles by 6. Urbanisation in South Australia is acute. Of the total population of 1.2 million over 800,000 live in Adelaide. A new major city is urgently needed and planning has already commenced. It is important that Adelaide should not grow beyond about 1 million. By keeping the population reasonably low, it should be possible to avoid a freeway system which increases air pollution and noise pollution problems and probably does little to ease the traffic problem.

The flora and fauna of South Australia are unique and varied. It is important to preserve areas as national parks. Various forms of park are contemplated: National Parks, Conservation Parks, Recreation Parks and Game Reserves. In addition the preservation of whole regions as "Scenic Preserves" is to be attained. Two regions, Kangaroo Island and the Flinders Ranges, have been termed planning areas for this purpose.

Mining and quarrying can cause great damage to the land. The new Mining Act demands rehabilitation and a Rehabilitation Fund has been created by placing a tax on mineral produced. However, in the opal mines of the far north, heavy machinery has ripped up the country and created a 'moon-like' landscape.

There has also been some overgrazing by sheep on some properties. At present there is a swing from sheep to cattle. Not enough thought is being given to the ecology of the regions involved in this change.

Conclusion

By reasonable controls, it appears that pollution in South Australia is being kept to a reasonable level. Controls will ensure that this level is maintained or improved. There is still time in South Australia and most probably in other parts of Australia to place controls on pollution before the situation becomes too bad. To visitors from overseas we appear to have no pollution problems, but potentially they are there. Reasonable controls now should be able to maintain a clean environment for all time. The main problem is population. Do Australia and South Australia have to increase their populations? Many believe it should be kept small but other factors, other than environmental ones, e.g. economic and defence, are involved.

THE CANADIAN POLLUTION PROBLEM - A NEED FOR BROADER PERCEPTIONS

P. M. HIGGINS

*Water Pollution Control Directorate
Environmental Protection Society, Environment
Canada*

Canada, as it happens, is the name of a pretty big corner of the biosphere. And within our borders we face complex manifestations of the environmental crisis. Some of the details differ - but the basic problems are the same as yours.

Before giving you a quick and necessarily condensed review of the Canadian situation - first let me outline some of the basic thinking behind the Canadian Government's response to the crisis.

In common with other nations we believe that the key to success is a broadening of vision. We believe the environmental crisis to be a crisis of perception. There are gaps in the data.

Whenever you start thinking about how the world got into the present mess, you keep running up against this matter of narrow mandates - inadequate inputs. Segregated attention - tunnel vision - consideration of only some of the data got us into this situation. We concentrated on our own narrow objectives. This fondness for a tight focus has given us some benefits, to be sure. Indeed it has had a lot to do with the construction of what we fondly call civilization. For when we set out to go from point "A" to point "Z" we learned to keep our gaze fixed on the destination - on point "Z" and forgot other preoccupations. "Keep your eye on the ball" was the slogan, and it has been followed in everything from industrial production to the linking-up of populated centers with roads or railways.

What we didn't do was keep our eye on the interests of that other ball - the small blue one we live on. They say that hindsight is always 20/20. And it is very easy to see today just what

we left out of our equations. We ignored environmental factors.
We were narrow in our concepts. Not only about what a given
activity might do to our vital natural life support systems; but
also narrow in the sense that we did not put man in perspective
in the biosphere. We did not consider his place in the big mosaic.

In Canada, as in other countries, we think that responses to
the environmental crisis must be characterized by <u>width</u> of perception.
They must be free of the same fatal constriction that caused the
trouble to begin with. Which means that corrective measures must
be based on complete analyses. We must paint on a wide canvas.

We believe, that the number one priority is to restore
balance by building environmental factors into the equation. We
also believe that from here on we must keep our vision wide in the
sense of viewing man's total environment.

And the total environment includes the environmental and the
economic aspects.

Canada's intention is to consider both aspects. We have a
two-pronged attack on environmental problems.

Our number one environmental goal is to prevent further
deterioration. (By ensuring that the highest degree of waste
control practicable be built into all new developments). We seek
protection and maintenance of our air, our water, our land, our
natural resources. In addition we must roll back the present threat
to the biosphere.

The second thrust of our attack is in the quest for better
environmental data, on which sound comprehensive forward environmental
planning can be based. Sound economic planning involves
environmental statesmanship of the highest order. A good
engineer understands the forces of nature; a good economist,
because he takes the long view, must also be concerned with
nature's biological scheme of things. So the biologist must be
more precise.

In addition, we are committed to maintaining the economic
well-being and vigour of Canada. We are fully aware of the perils
of unchecked exponential growth - of expansion run wild. But we
do not believe that the only option open to us is to sacrifice
the gains made by a modern technological society. Breadth of
vision demands that we recognize that there have been such gains.
We do not visualize giving ground to disease, poverty, physical
isolation and other basic foes associated with the past. We are
committed to seeking a healthy protected environment not as a
setting for a non-technological society, but as the foundation
for an advancing civilization.

This does not mean that we believe in business as usual. We think, in Canada that industrial growth of the old narrowly-focussed kind - expansion indifferent to environmental havoc - must go forever. We believe that industrial growth, undertaken with a full consideration of environmental needs is possible and desirable. We think, moreover that this kind of growth, along with its accompanying advances in science and technology is our best hope for solving the problems of pollution and want.

We believe too that we must broaden our concepts not only of the physical environment, but of our political and social arrangements.

The extent to which we can learn to work together, across the lines of jurisdiction, and national interest will determine the day. Divisions of power and territory have their uses no doubt, and in many cases, good effects. But these do not include a fragmenting of the world's response to the environmental crisis. Canadians have agreed that these divisions should never be allowed to stand in the way of protecting the environment.

Before going on to the subject of how these basic principles are applied in our programs, let me introduce you briefly to the setting in which they are carried out. To Canada itself.

We are talking about a very big country - in land area, the second largest in the world. We have a relatively small population about 22 million people. We have the longest coastline in the world - 117,000 miles,* and we face on three oceans. Our country is rich in lakes and rivers. Our fresh water expanses cover 291,000 square miles.** We have an abundance of natural resources. Our continental shelf is huge - up to 40 per cent of our total land area.

Given statistics like these - given this combination of natural riches, plenty of room to move around in, and a relatively small population, one might think that Canada doesn't have much to worry about. But of course we do. Not simply because our situation is affected by what happens elsewhere in the world, but because of the specific economic forces at play in our country.

I refer to the fact that Canada is a growth country. There was a time when a nationalistic statesman would make such a statement swelling up with pride. Nowadays, at least among

*187,000 Kilometers
** 75,000,000 Hectares

the environmentally informed, there is apprehension mixed with the pride.

Here are some of the aspects of economic growth in Canada. Our population is growing - and in ways that growth presents possibilities for problems. In one century, from 1850 to 1950, Canada's population increased from 2.5 to 13.7 million. It took just 20 years to add a further 7.5 million. We have both high and low projections about where we will be by the year 2000. If you go by the low projections, we'll have 34 million people. If you accept the high figure, 41. All of which might seem to pose no particular problem in a nation the size of Canada.

But the fact is that most of this will be urban growth. Canada, in spite of its international tourist circuit image of wilderness - mounties, singing woodsmen and rosemarie - has a higher rate of urban growth than any other developed nation in the world. The cities swallowed up almost all of our total population increase of six million in the 15 years between 1951 and 1966. By 2000, we believe that 70 per cent of our people will be living in 37 cities having populations over 100,000. One half of our population -- 17 million by the low estimates -- will be living in just nine cities in the million-and-up population category. So you see, it isn't all wide open spaces, clear streams and blue skies.

It is within these urban areas of Canada, that most of the work will go on, most of the money will be earned, most of the goods and services produced and consumed. This will be the setting for a staggering increase in consumption of raw materials, in transportation of goods, in energy demands. Here we can expect, a massive demand for renewable resources, a massive thirst for water, for domestic as well as other purposes. And here, unless we act to prevent it, we can expect massive increases in air, soil, water and noise pollution.

Because the long-range forecast is for explosive growth. (By 2000 we expect that the Canadian GNP (and hopefully not our GGA - gross garbage accumulation index) will rise sixfold, compared with 1967. If it grows at the fastest rate projected, it will double every 13 years. Even at the slowest projected rate, it will double every 16. There's also good reason to believe that our per capita GNP will double by 1985 and triple by 2000).

We have no illusions about the inexhaustibility of our natural resources. For example: We've watched the harvest of fish from our continental shelves fall 800,000 tons per year in three years. This in spite of, - yes and because of -- the best efforts and most advanced technology employed by the fishing fleets of the world.

We have no illusions about our area being too vast to pollute. We are a land criss-crossed with flowing waters. For a long period of our history they were our continental highways. Yet today eight of our major river basins are in real pollution trouble. (Canada has or shares with the United States 11 of the world's 76 longest rivers, and 9 of 20 of the world's largest lakes).

We have suffered some of the more obvious predictable symptoms of the environmental sickness, and a few unexpected, offbeat ones. In company with the U.S., we completed the St. Lawrence Seaway a few years ago - opening up the continental heartland to world shipping. Along with ocean-going ships unloading at inland terminals we encountered some less welcome visitors. Sea Lamprey. Canals begun early in this century allowed them to circumvent Niagara Falls with the result that they cut the annual commercial take of lake trout from 5.5 million pounds to 400 pounds. Spelling doom for a once highly profitable fresh water fishery. For those of you not familiar with the Sea Lamprey - it is an eel-like parasitic animal (growing up to 18" in length) which attaches itself to the side of its prey and with its bloodsucker mouth sustains itself from its host. The Lamprey have been observed hitchhiking to the Upper Great Lakes by attaching themselves to the undersides of ocean going ships. Fortunately our control program has finally brought the Lamprey under control.

We are living evidence that environmental problems are not curable without international cooperation. Recently Prime Minister Trudeau and President Nixon sat down in Ottawa and signed an Agreement to clean up the Great Lakes water-system. (Pollution has reached crisis proportions in parts of these lakes, notably Lake Erie - often referred to as North America's "Dead Sea" which is really a misnomer - Lake Erie is really too much alive. It is the abundance of life which is causing problems). As you know the Great Lakes constitute the largest integral body of fresh water in the world. Pollution flows across the international border, in both directions, with total undiplomatic contempt for demarcation lines. So now our nations are pledged to roll back pollution in these waters within the next ten years, and to set up joint machinery to keep it protected thereafter.

We are concerned about many other potential problems of this kind - for instance about the terrible impact of oil-spills on our coast-lines, including very specifically the fragile Canadian Arctic, where such catastrophes would have unfortunate problems soon become international problems. Indeed someone has estimated that a full

20 per cent of Canada's environmental problems are as a result of extra territorial activities and which are controllable only by international effort. Problems that will require efforts not by Canada alone, but by Canada and at least one other nation.

And so to our responses.

I said at the outset that Canadians were sold on the need for wide terms of reference - on the need to consider the total environment, not just a part of it.

The total environment includes not only physical aspects but political.

Ours is a Federal rather than a Unitary State. Back 104 years ago when the political blueprints for Canada were being drawn up, no one was worried very much about the environment. Consequently, it shouldn't come as any surprise to learn that our political arrangements, although they were entirely appropriate to our diverse cultures and geography, are not what you would call environment-oriented. Canada is a honeycomb of jurisdictions and responsibilities.

I don't want to get into a detailed description. Just let me move to the bottom line. For purposes of environmental control policy the jurisdictional questions in our country, fall into two major areas. The first concerns the proprietary rights of each Government. Who owns what? The second concerns the right to legislate. Who's in charge here? If jurisdiction over environmental management were based solely on ownership of natural resources, there wouldn't be much difficulty defining them. But Provincial and Federal Governments have legislative powers which overlap. For instance, the Provincial Government has primary legislative jurisdiction over water through its proprietary rights. This enables it to make laws on such matters as domestic and industrial water supply, power development, irrigation, reclamation and recreation. The Federal Government, however, is responsible for protecting and managing Canada's fisheries - both sea-coast and inland. This, as we will see, allows the Federal Government a powerful tool to indirectly control water pollution.

So part of the Canadian response - a high priority part - has been to develop ways for own Governments to work together against pollution.

One visible organizational response. One that reflects the determination of Canadians to work together as a nation in this field, was formation of a Federal Department of the Environment. This organization - Environment Canada we call it - was created in 1971. It brought together under one Ministerial roof, a multitude of agencies that had been involved in activities impinging on various aspects of the problem. These included management of renewable resources and protection of the environment. The Department includes components and services working in such fields as Meteorology and Atmospheric Research, Fisheries, Forestry, Wildlife Water Management, Environmental Protection, and Policy Planning and Research Co-ordination. Special Advisory Councils were also created to advise the Government on environmental matters. These include representatives from industry, from the universities and from the public at large.

Within the last several years all 10 provinces of Canada have created new Departments or Institutions responsible for environmental management. Responsibilities once the exclusive domain of health authorities which recognized man as the single most important life form have been broadened to include the total environment. This is not to say that we consider man any less important, but rather we are beginning to better understand a little more of the complex ecosystem of which we are only one part.

One new component of Environment Canada is the organization known as Environmental Protection Service. This group, of which I am a member is responsible for carrying out the Department's Environmental Protection responsibilities. It is Environment Canada's action arm - or its cutting edge - depending on your taste in metaphors.

Environmental Protection Service - E.P.S. - is responsible for both preventatives and cures. We are charged with implementating various remedial measures that fall within the Federal jurisdiction and for monitoring and evaluating the success of these measures.

E.P.S. is also the Federal door on which Canadians in all sectors can knock when they want to talk about environmental problems.

It is also the Federal Government's own house cleaner - responsible for cleaning up Federal installations across the country.

Here are some of the things we are doing now. A high priority task is the cleanup of industrial water pollution.

At this point I would like to have you meet a law which has played a big part in this activity. The Fisheries Act of Canada. This piece of legislation has been on the books for over 100 years. Parliament put new muscle into it a couple of years ago, specifically to permit us to move to prevent pollution of Canadian waterways. Prior to the amendments, the law provided for punitive and corrective action only after damage had been done. I will not go into exhaustive detail about the Fisheries Act pollution control provisions here because we'll be covering much of the material separately at this Conference. But here are some basic features.

National regulations are being developed under the Fisheries Act which strive to achieve the dual goals of the total environment. We have tried to build concern for both ECO's - Economy and Ecology - into the equation. Regulations require that pollution losses be held to a minimum based on best practicable control technology applied both to industrial practices and waste treatment.

We are thus proceeding, industry by industry, to develop regulations governing the quality of the waste-waters permitted to be discharged into Canadian waterways.

The way we have gone about framing these regulations offers a good example of how we are implementing basic policy decisions of reconciling environmental and economic goals.

Regulations for two industries - pulp and paper and chlor-alkali producers - have already been announced. In both cases they were the result of deliberations by task force representing both the public and private sectors. These teams comprised both Federal and Provincial Government officials and top-level representatives of the industries concerned, with consultants lending a hand as well. Environment Canada, however, was the final arbiter to ensure that the regulations indeed will enhance environmental quality.

The regulations, as drafted, reflect an interest both in clean water and a busy economy. For instance, they are based squarely on the possible. They demand compliance with standards which are achievable, given the current state of pollution abatement art. We don't order industries to do what cannot be done. On the other hand, the regulations do allow for more stringent controls where local conditions require it, or when the state of the abatement art advances. Local or provincial control requirements if more stringent are considered to take precedence over the national minimum standards.

The national regulations also distinguish between the problems of new plants and old ones. They require immediate compliance by new plants. Older mills built under different rules will have to measure up too, but the schedules of compliance are negotiated individually. - And will be based on local priorities and the minimum lead time technically to implement.

We do however weigh heavily the possible economic impact of possible plant closures particularly in one-industry towns.

Another decision was to stop pollution at the water's edge as opposed to setting pollution limits for the receiving body. The reason is simple: We don't really know what level of pollution is "safe" and we don't plan to find out the hard way. This principle applies also to our air pollution control programs.

Similar regulations are being drafted for a list of major industries in Canada including petroleum refining, mining and food processing.

Being completely candid - we are, I think, being very successful with new developments - new Canadian industry ranks with the cleanest in the world. However, our successes with existing industry occur with frustrating slowness.

Other major new environmental legislations recently passed by Parliament includes the Canada Water Act and the Clean Air Act. Also a new Wildlife Act has been submitted to the current session of Parliament.

When the Parliament of Canada passed the Clean Air Act in 1971 it gave our country several badly-needed new weapons against pollution.

First and foremost - it will allow us to get some uniformity into our responses to the air pollution problems of Canada.

Prior to the Clean Air Act, and still for that matter, there has been a great diversity both in expertise and in relevant legislation among the ten provinces. Wherever this kind of unevenness exists, there is opportunity for the existence (not by design) of pollution havens. One fairly obvious national need was to set up national air quality objectives - air quality targets which would apply across Canada. The Act has allowed us to develop these aiming points. Objectives were proposed last year for five major pollutants - sulphur dioxide, particulate matter, carbon monoxide, photochemical oxidants and hydrocarbons. A sixth pollutant, nitrogen oxide is under consideration.

We also wanted to ensure that what needs to be done to control air pollution does indeed get done. These capabilities are built into the Clean Air Act. We are also able to provide some Federal leadership and coordination in key areas. For instance by compiling a national inventory of source emission data, establishment of national air quality objectives, prescription of national emission standards and guidelines, and control of air pollution from the many installations and undertakings run by the Federal Government. We also have the authority under the Act to control the composition of fuels that may be produced in Canada, or imported into the country.

Here again, the approach has been to call for application of the best practicable pollution abatement methods available - and to stop pollution at the factory fence.

At the international level, Canada is a concerned participant. We have contributed several air quality monitoring stations to a global network sponsored by the world meteorological organization.

We are working actively against air pollution caused by motor vehicles - a very live issue in our country. There are approximately seven million automobiles on the road now in Canada. We expect to have nearly 15 million by the end of the century. Over the next 30 years interurban passenger travel by road is expected to double.

The control of motor vehicle pollution from new vehicles is exercised by the Canadian Ministry of Transport, with the help and advice of Environment Canada. We recognize that we need to keep Canadian standards similar to those in the U.S. Not only because we are neighbours, and we visit each other in massive numbers, and in automobiles, but also because traffic in Canadian cities differs little from comparable U.S. centers.

Canadian control is based, essentially, on self certification by the manufacturer. Environment Canada has set up a test facility to monitor a sampling of new cars offered for sale.

Canada, like everyone else, has solid waste problems. The amount of municipal garbage, to name just one source, presently generated in Canada is about 4.5 pounds per head per day. This is expected to go up to 7.5 pounds by the end of the century - if we do nothing about it. We have recently conducted our first national solid wastes survey. The information obtained is presently being analyzed.

Environment Canada has also embarked on an ambitious research and innovation program in search of new methods and technology. We are working with universities and the business sector. The Environmental Protection Service will initiate and help fund research into solid waste management problem areas such as collection and disposal, technology and economics of recovery, and studies of recycling possibilities and other environmental problems.

Technology development and demonstration programs for air and water are also under way - with those in water being the most advanced.

Prevention is always better than cure. But accidents happen, and we are taking formal steps to prepare for them. Comprehensive contingency plans are being developed for all kinds of environmental accidents, relating to land, air and water. One of the things we are currently doing is putting together a central operations center in Ottawa for control of Federal responses to accidents of this kind. In all our work in this area, we are in cooperation with other Governments, with industry and with the private sector.

In more general terms we are working upon an array of policy instruments, mechanisms and programs either to reduce the total production of wastes or to render them harmless to the environment. We are encouraging recycling of wastes and the development of new products from them. We are working on better legal means to hold the polluter liable for the damage he causes. We are working on a program to inform the general public regularly on the state of the environment. We are seeking better criteria for defining the tolerance of flora and fauna to contaminants. We've made a good beginning at cutting back or eliminating certain of the better understood forms of pollution at their source. And we have come up with regulations to control levels of phosphorus in laundry detergents.

We are carrying out environmental impact studies - building into the planning of major developments the kind of environmental questions that - if they had been asked in the past - might have saved us from our present situation. Along with the provinces involved we are studying the impact of a proposed oil transshipment port at Lorneville, New Brunswick, on our East Coast. We have mounted studies of the proposed MacKenzie Valley Pipeline Corridor, the oil refinery at Come-by-Chance, Newfoundland. And we are studying the whole question of marine oil exploration and transportation in terms of environmental impact.

One useful discipline we have imposed on ourselves was the setting of goals - aiming points for the future. By doing this we feel we've been able to establish definite checkpoints in time which will tell us not only about effort expended, but about how much progress we're making towards our targets.

Here are some of our major aspirations:

- By 1980 we aim, in broad terms, to have rolled back the effects of gross pollution and to have done so without sacrificing a high standard of living.

- We hope to have achieved our acceptable national air quality objectives. Also to have achieved a 30 per cent reduction in motor vehicle pollution. To have regulations on the books and complied with, dealing with fuel additives and compositions.

- By 1980 we aim to have compliance with minimum national effluent standards by all major water pollution sources.

- We aim at instituting water management plans, including water quality and pollution abatement programs in our nine major water systems.

- By 1980 we aim to have our own Federal Government housekeeping done - our own pollution cleaned up in all installations and facilities run by the Government of Canada.

- We plan to have legislation and regulations in force to control environmentally toxic substances that affect our living resources.

- We plan to have a system functioning in which the environmental factors, pro as well as con, benefits as well as costs - routinely defined and considered in the course of decision-making about major developments, public or private.

Another thing we are doing is asking questions.

We, like you, have some pretty basic ones to ask. We want a proper balance between economic growth and a clean environment. But what is the proper balance? How clean is clean? Growth creates wastes. How much waste can the environment handle? What wastes should be recycled?

Who's right? Between the optimist and the pessimist, which is the realist? What about priorities? How much should we spend in research on toxicology? How much on environmental impact studies? How much on monitoring systems and national inventories? How much information will political decision makers need in order to make their decisions? And so on.

To get at a certain important category of answers - we have strong research programs. One of the major thrusts is to discover the best approach to reaching an understanding of the nature of ecosystems. How can they be modelled, expressed in quantitative terms.

We need answers on the macro problems of possible world-wide pollution of the atmosphere, ocean pollution, the wide dispersal of certain chemicals. We need to learn more about the influence of oil spills and other man-made changes in our ecosystems, chemicals, transportation routes, etc. We are involved in research into these questions. The basis of our science policy can be simply expressed - we want to make effective and efficient use of scientists toward the achievement of Canada's environmental objectives.

We need, in short, data to fill the gaps - to add the necessary breadth to our perceptions. We need maps to steer by.

We believe that the single most important task facing the nations of the world is to gain knowledge. Knowledge about the environment - about ecosystems, about the balance of life forces - about sources of energy; about pollution. And knowledge about our fellow nations - their questions and their answers. A widening of our perceptions - an improvement of our peripheral vision - is an essential element in the formula for survival. And it cannot be fulfilled in isolation.

A CO-OPERATIVE APPROACH TO POLLUTION PROBLEMS IN CANADA

C. G. Morley

Policy, Planning Research Service of the Canadian Federal Department of the Environment, Ottawa

INTRODUCTION

Canada is a nation with social, economic, legal and political institutions which were conceived and have evolved within a federal, not a unitary framework. This Canadian federalism has generated a number of unique advantages such as a considerable decentralization of executive, legislative, administrative and judicial functions. In a country the size of Canada such decentralization is essential if human needs are to be sensitively and adequately responded to. This same federalism which encourages a more effective political-legal system is not without its disadvantages, and as is often the case, many of these perceived disadvantages are integrally related to the advantages themselves. The disadvantage most often identified, particularly by persons with a federal government bias, is the fact that our Constitution divides legislative jurisdiction amongst the provincial and federal governments. No one of these governments can be indifferent to the limitations imposed on their legislative powers by this constitutional allocation of responsibility.

Of the eleven senior governments in Canada 10 of them are provincial and one is federal. The ten provincial governments are limited in what they can subject to laws originating with them by two principle factors. First, a provincial government can legislate only in respects of those classes of subject matter which have been specified in our Constitution as being within provincial legislative competence. Second, one province cannot presume to give extra-territorial, that is, extra provincial, effect to its laws. It will be appreciated that each of these limitations incorporate a principle of law which is more easy to state then to apply. Neither of these principles is absolute. There are instances where provincial initiatives have been accepted or interpreted as exceptions to the general rule.

While each of the provincial and federal governments is senior, the federal government is more senior then any other. Being the senior of the senior governments it is restricted in its legislative activity only by the constitutional allocation of legislative subject matters and by those rules of the international community by which it feels constrained. The federal government is not constitutionally constrained by any consideration of extra-territoriality. It can legislate so as to affect rights within any of the 10 provinces so long as the legislation originates in one of the subject matter categories constitutionally allocated to the federal government. In some instances the federal and provincial governments have a concurrent legislative competence. The consequences of this is that both the federal and provincial governments may legislate in respect of the same matter but should federal legislation occupy the field it renders provincial law invalid to the extent that the two laws are in conflict or otherwise incompatible. Legislation affecting water is a good example of this. Since the provinces own most of the public lands located within provincial boundaries they are the ones who can impose controls over the use of all waters which flow over these lands. However, the federal government has a constitutional right to legislate in respect of, inter alia, inland fisheries and navigation, two matters which clearly relate to water. Federal legislation for both of these subject matters presently exists and within it there are some federally imposed controls over provincial waters. To the extent that these controls are consistent with and necessary for both the preservation of fisheries and the protection of navigation the controls are validly imposed. To the extent that any provincial legislation is inconsistent with the federal legislation the provincial legislation must fall.

In any case there are very real limitations on the kind of legislation which any Canadian government can enact. In many instances comprehensive laws which are consistent across the country are difficult if not impossible to realize. Comprehensive laws to deal with environmental degradation fall within this category. Provincial governments have no power at all to develop comprehensive controls which are national in scope. The federal government is not quite so constrained. While it cannot provide any comprehensive national controls to deal with certain of our environmental problems, it can provide legislative frameworks within which comprehensive and national initiatives may be taken. The Canada Water Act and the Clean Air Act are two recent federal statutes having this framework character. Because of this legislation federal-provincial co-operation is encouraged and in some instances is required. In each case there is a potential for developing controls which transcend both the territorial and subject matter limitations to which brief reference has been made.

Water Problems

Canada is experiencing an increasing demand for its apparently abundant but undeniably finite fresh water resources. The most immediate and obvious pressures originate in Canada itself. Both industries and municipalities need an increasing amount of water for consumption and waste assimilation purposes. Increased leisure time for many of our citizens results in increasing demands for recreational uses of water in both wilderness and urban areas. Agricultural consumption continues to rise. Demands for electrical energy are increasing sharply and frequently it is our rivers which are dammed and their waters impounded in response to this demand. And of course contributing to the problems' complexity are at least three other factors: the traditional belief that water is a 'free' good; a growing population; and persistent demands by those increasing number of people for a higher standard of living.

Some of the less immediate demands for Canada's water originate outside of Canada. From the United States comes unofficial pressure in the form of repeated suggestions that Canada share this and other of its resources. Preliminary studies have already been completed and proposals made for dams, lakes, diversions and canals, intended to sustain existing and anticipated transportation, energy and consumption demands. This kind of external pressure does and will continue. The quantity and quality of Canadian water is such that the United States (confronted as it is with the increasing consumption, decreasing supplies, and rising costs of water) presently has few significant alternative sources of this essential resource.

THE CANADA WATER ACT

The Canada Water Act is one of the more recent Canadian federal legislative initiatives intended to respond to at least some of these problems. At the time the Act was being debated in the Canadian House of Commons the government considered it the legal framework within which comprehensive planning could be initiated with a view to optimizing water uses and minimizing or avoiding conflict over competing uses. During the debate the government Minister then responsible for seeing that the Act became law stated in the House:
"Water must be so used as to ensure a maximum stream of benefits to society. This optimization can only occur

if we have comprehensive planning to achieve our goal of
multi-purpose use. We must look at each basin as an
integrated whole. We must examine all the uses which
can be made of each basin. We must plan for the future
so as to achieve the greatest long-term net social
benefit of our water resources."

The preamble to the Act endorses these objectives. In it the
rapidly increasing demands for Canada's water resources is acknowledged. So also is the need for information respecting "... the
nature, extent and distribution..." of Canadian waters, the present and future demands for them and the means, existing or required,
to satisfy such demands.

The preamble goes further and identifies water pollution in
Canada as a separate kind of demand and characterizes it as a "...
significant and rapidly increasing threat to the health, well-being and prosperity ..." of Canadians as well as a threat to the
Canadian environment at large.

Although water pollution has been isolated as a problem of
particular concern, the Act was not conceived and developed with
an emphasis on water pollution control. The preservation and enhancement of Canadian water quality is but a part of a more comprehensive concern. Research, planning and comprehensive programmes for the conservation, development and utilization of water
resources is a principle concern of the Act. So also is federal-provincial co-operation whenever these or related activities are
initiated within the Act's legislative framework. Those parts of
the Act which encourage or require federal-provincial co-operation
are there because of the political and constitutional facts of
Canadian life. The preamble acknowledges this in part when it
states that the Act's framework for governmental co-operation is
predicated on the respective federal and provincial responsibilities for water resources.

Water Resource Management

The Act envisages two principle areas of co-operative effort.
In Part 1 the concern is for comprehensive water resource management. While 'comprehensive' is no where defined, 'water resources
management' is. Efforts directed towards "... the conservation,
development and utilization of water resources..." including activities ranging from data collection to program implementation to
water quantity and quality control are part of a water resource
management exercise. Presumably the greater the range of activity
the more 'comprehensive' the exercise is. In any case within this
part two areas of possible co-operation are identified. The Act
provides that the Minister of the Environment, who is the Minister

presently responsible for the administration of this Act, may at his discretion enter into agreements with one or more of the provincial governments to establish inter-governmental committees or other bodies on any or all of the following lines: national, provincial, regional, lake or river basins. The raison d'être for the creation of such bodies is (1) to consult on water resource matters, (2) to advise on research, planning, conservation, development, and utilization of water policies and programmes and (3) to facilitate the co-ordination and implementation of any water policies and programs developed and proposed. Federal-provincial consultative committees for each of the ten provinces have already been established. These committees provide both a necessary forum for discussing and agreeing on water resource priorities and a body capable of co-ordinating water oriented activities of both the federal and provincial governments.

In addition to those provisions which encourage consultative arrangements there is provision for federal-provincial agreements for the management of water resources. A condition precedent to such co-operative involvement is that the water constituting the subject matter of the agreement be one in which the provincial governments have an 'interest' and in which the federal government has a "... significant national interest ...". The emphasis on the federal governments significant national interest reflects both the practical need to establish priorities and the jurisdictional sensitivity which discourages federal intrusion into areas which are primarily of provincial concern. In any case where these interests do exist and both governments agree, the written agreement may provide for (1) the establishment and maintenance of water inventories, (2) the collection and processing of data on water quality, quantity and distribution, (3) any kind of water research, (4) the formulation of 'comprehensive' water resource management plans including the cost of implementing such plans, (5) the design of projects for the efficient conservation, development and utilization of the waters under consideration, (6) the implementation of any plans or projects formulated and designed and finally, (7) the establishment of joint bodies with the responsibility of directing, supervising and co-ordinating the comprehensive water resource management programs which have been jointly developed and approved. The sharing of both costs and responsibility for developing and subsequently implementing such programs is required to be a term of any federal-provincial agreement which is negotiated. This requirement to negotiate and agree on the basic issues of cost sharing and the division of responsibility guarantees serious pre-consideration of the general objectives of the agreement and minimizes the possibility of subsequent disagreement over the proper allocation of responsibility for giving effect to any programs developed. Further, it is a characteristic of Canadian political life that agreement on common and general objectives frequently takes less time than agreement on cost sharing. Tying the two together assures long term co-operation and more effective agreements even if the negotiations take longer.

While co-operation is encouraged a provincial government cannot veto an initiative in water resource management simply by refusing to co-operate. If the comprehensive water resources programme is to be developed for federal waters, that is for waters under "... the exclusive legislative jurisdiction..." of the federal government, provincial co-operation is not necessary and need not even be requested. For waters such as these, which include the Canadian marine belt and most of Canada's waters north of 60° of latitude, the federal government may unilaterally conduct research, collect data, formulate comprehensive water resource management programs, design projects, establish boards and implement any programs. Where, however, the water of 'significant national interest' to be studied is inter-jurisdictional (that is within the real or perceived jurisdiction of both the federal and provincial governments) provincial co-operation must be sought. Should the provincial government refuse to co-operate the federal government may then do for these inter-jurisdictional waters everything it can do in federal waters except that it may not implement the plans and projects which the federally consituted study group developes and recommends for those waters. So far no provincial government has provoked unilateral federal action by refusing to co-operate. However, should such co-operation ever be withheld and unilateral action is then taken by the federal government it is possible that the end which was frustrated by the initial withholding of provincial co-operation will still be achieved. It is difficult to believe that any provincial government could withstand the public pressure which is practically guaranteed by such an exercise. If the federal government studies a water area, identifies the water resource problems which are peculiar to that area and then proceeds to develop and recommend a means for resolving or managing those problems, the pressure upon the recalcitrant provincial government to implement the federal recommendations might be substantial. It is not likely that an informed and aroused public will be impressed by constitutional niceties if they see their water resources in need of protection.

In the case of international and boundary waters the Act requires that the co-operation of interested provincial governments be sought. However, should this requested co-operation be withheld the Act provides that the federal government can unilaterally do whatever it can do in respect of inter-jurisdictional waters and may in addition implement any programs, plans and projects considered necessary for the conservation, development and utilization of those waters. International and boundary waters are included in the Act's definition of inter-jurisdictional waters but because of the international responsibility and potential liability of the federal government in respect of these kinds of inter-jurisdictional waters this additional power to implement plans and projects is important. It is more important here than it is in the case of other inter-jurisdictional waters where the consequences of failing to implement needed programs for them can at worst generate a domestic controversy capable of being settled in any available Canadian political or legal forum.

Water Quality Management

In addition to providing a framework for co-operative development and implementation of comprehensive water resource management programs the Act encourages federal-provincial co-operation in water quality management. Water quality management is a concept which includes but is greater than water pollution control and is less than water resource management. It is concerned with conflicting water uses and the restoration, maintenance and improvement of water quality with a view to resolving such conflicts.

A water quality management agency is the corporate entity created or designated for a specified water quality management area. Once named, the Agency is vested with the responsibility of planning, initiating and implementing programs to restore, preserve and enhance the water quality level in the water quality management area which has been subjected to the agency's jurisdiction. In the case of federal waters, negotiation with and the co-operation of provincial governments is not essential. For these waters the federal government can unilaterally establish an agency. In the case of inter-jurisdictional waters, including international and boundary waters, "... the water quality management of which has become of urgent national concern..." the federal government must negotiate with all interested provincial governments. For waters falling within this category provincial government co-operation and agreement is needed in order to either create a new or name an existing corporation to act as the water quality management agency for any specified inter-jurisdictional waters.

While all reasonable efforts must be made to get provincial government co-operation, this requirement for provincial co-operation does not give a provincial government any power to veto attempts to establish needed agencies. Federal unilateral action is contemplated once two conditions precedent have been met. The first condition is that the quality of those waters intended to be subjected to agency management must be of "urgent national concern...". This requirement is in part a constitutionally imposed condition precedent to unilateral federal action over inter-jurisdictional waters. Constitutional support for federal controls over boundary waters (one category of inter-jurisdictional waters) is strong. Such support for federal legislative action in respect of international and inter-provincial waters (another category of inter-jurisdictional waters) is probably available although there is little authoritative case law. The most controversial initiative, however, would be any federal attempts to subject to its control waters which are exclusively intra-provincial. Such an attempt would raise serious jurisdictional disputes.

It is no doubt because of this spectrum of constitutional support for federal controls over different kinds of waters (and in particular the different categories of inter-jurisdictional waters) that the Act has built in to it the requirement that the federal government involve itself only in those inter-jursditional waters which are of urgent national concern. Within the Canadian constitutional context this is indeed an ingenious drafting accomplishment. There is strong support amongst constitutional jurists and in Canadian case law for federal intervention in areas which have become of 'urgent national concern' even if the subject matter affected is otherwise within provincial legislative competence. The governing constitutional principle is that once a matter has acquired this characteristic it is, by its very nature, beyond the effective control and exclusive concern of any one provincial government. Being a problem, it demands a solution. Having this aspect of 'urgent national concern' it necessarily (one can argue) comes within the jurisdiction of the federal government. With the Act drafted this way the result is that while the federal government can undoubtedly be challenged if it unilaterally creates an agency, the principle issue then to be resolved is whether or not the designated waters are of 'urgent national concern'. If the issue is litigated, the court will have presented to it evidence which both substantiates and denies that the waters have this character. A judicial finding that the evidence supports the federal conclusion would go far in sustaining the federal initiative. A contrary finding would say little more than that the federal government erred in assuming that the necessary 'urgent national concern' existed in respect of those waters. But the result is that any challenge of the federal government will, if successful at all, result only in the disallowance of that particular federal initiative and not in an 'ultra vires' ruling for all or even part of the Act itself.

The second necessary condition precedent to unilateral action in inter-jurisdictional waters is the refusal of interested provincial governments to co-operate either in designating a water body and creating an agency or in accepting the recommendations of any agency which had been earlier established by agreement. Where these conditions, that is the 'urgent national concern' and the absence of provincial agreement, are satisfied then the federal government may itself designate the specified inter-jurisdictional waters and create a water quality management agency for those waters.

Every agency established, no matter how it comes into existence, has the same statutorily determined objectives "... to plan, initiate and carry out programs to restore, preserve and enhance..." the quality of the water in the designated area. To achieve this the agency may for its designated area do any or all of the following: (1) identify kinds and quantities of wastes, (2) ascertain water quality levels, (3) forecast future kinds and quantities of waste and (4) develop and recommend a water quality management plan. Included in these plans are to be recommended water quality standards and dates for achieving them, recommendations for kinds and quantities of tolerable waste discharges and conditions for such discharge, recommendations for the degree and type of waste treatment, an estimate of the costs of implementing any proposed plan and finally, recommendations for both effluent discharge fees and waste treatment charges - two of the unique features of the Act.

Both the waste discharge fee and the treatment charge are legislative attempts to discourage the traditional attitude that water is a free good which may be freely used as a waste receptacle. The discharge fee is a sum of money required to be paid by any person who lawfully deposits a waste into any designated waters. The theory is that such fees encourage both good waste management practices and the internalizing of waste disposal costs. In practice this is to be accomplished by fixing the amount of the fee at a level which reflects the traditionally externalized cost absorbed by the acquatic environment when the specified untreated waste is deposited. All fees paid are available to be used to remedy or mitigate the pollution damage caused by the untreated waste discharge. Furthermore, if fees are sufficiently high there is an economic inducement to treat the waste to a pre-determined and environmentally tolerable standard. In either case the costs of remedying or avoiding pollution damage becomes a calculable cost of production which is absorbed initially by the producer and perhaps ultimately by the consumer of the product. In either case the environment is no longer required to bear a significant part of the production cost.

Waste treatment charges serve a similar function in that these pollution costs also become a cost of production. Here also it is the polluter who must, at least initially, pay the true costs of his production. These charges, however, are payable only for whatever waste treatment service is provided by the agency. In order to assure that the necessary and adequate waste treatment facilities are available, the agency itself may operate and maintain these facilities. The need for them is to be identified in any proposed agency plan.

Once the proposed agency plan has been accepted by the federal government in the case of federal agencies or by the federal and provincial governments in the case of inter-governmental agencies, the agency becomes operational. This means that the agency is then empowered to construct, operate and inspect treatment facilities, collect waste treatment charges and effluent fees, monitor water quality, analyse wastes, publish information and generally do anything necessary for effective water quality management.

Regulations

A significant legislative technique common to most federal anti-pollution legislation is the reliance on regulations as an integral part of any effective program. Regulations supplement a statute and provide the detail for what would otherwise be a general law. They are of but not in a statute. No regulation will be found within a statute but no regulation can be promulgated unless it is of a kind contemplated in specific and enabling legislation. The Canada Water Act is one such statute which requires an extensive use of regulations to support all of its water quality management activities.

Water quality standards are little more than objectives to be met in order that pre-determined levels of water quality may be achieved. Unless they are fixed by regulation no sanctions are available to encourage or compel compliance. However, once standards are fixed by regulation they become legal norms and sanctions exist if the norm is violated. Again, it is the prescription of specified wastes in regulatory form that more readily assures the effective control of the discharge of those wastes. And it is by regulation that the criteria for setting effluent discharge fees and waste treatment charges are prescribed.

Since the prescription of specified substances as a waste and the determination of standards, fees and charges are essential if there is to be effective water quality management, and since these factors acquire legal significance only through regulations, the source of these regulations is important. Their source and the source of all ancillary but supportive regulations under the Act is the federal government. What this means is that ultimately the effectiveness of any water quality management agency is dependent upon federal government regulations, not provincial nor inter-governmental regulations. Even where inter-governmental co-operation has existed in all other respects, the fact that the federal government must finally give legal effect to essential regulatory controls and that it is ultimately federal law which is to be administered and enforced qualifies the co-operative character of the Act. Even when the administration and enforcement of governing regulations is the responsibility of an agency created by inter-governmental agreement these regulations must originate at the federal government level. Provincial concern over this has already been expressed in some federal-provincial negotiations. No doubt this requirement will invite further questioning as the number of inter-governmental negotiations increases.

CLEAN AIR ACT

Canadian constitutional law respecting the jurisdictional responsibility for ambient air and air pollution control is effectively non existent. This may change as the federal government takes steps to implement its recently proclaimed Clean Air Act.

This Act, unlike the Water Act, was not conceived for the purpose of managing air as a resource. Since the Act provides a less comprehensive legislative framework there is less emphasis in it on inter-governmental co-operation. However, there are circumstances where co-operation is encouraged and others where it is essential.

The Act provides a basis for accomplishing several objectives, two of which are the prevention and control of undesirable air pollution originating in all federal works, undertakings and businesses. Where federal works and undertakings constitute the source of undesirable air pollution steps may be taken under this Act to impose needed controls. Provincial co-operation or agreement is not a condition precedent to federal initiatives in such cases for the federal government has strong constitutional support for such action.

The federal actions which may be taken in respect of federal works are any one or a combination of the following approaches. Specific emission standards can be determined and prescribed by regulation for any air contaminant. Once regulated standards have been established each federal work, undertaking or business is thereafter legally required to meet these standards. Another available and potentially effective control is through Ministerial insistence that all plans and specification for proposed federal works and undertakings be submitted and reviewed with a view to either approving, requiring modification of or rejecting the construction as proposed. Finally, standards may be prescribed for both the composition of and the maximum concentration of additives in all fuels used in federal works.

In addition to the quite extensive provisions allowing for the clean up of federal works, the Act provides for the setting of national emission standards and their prescription by regulation. National emission standards applicable to federal works may be unilaterally fixed by the federal government. However, the constitutional constraints require that national standards for any specified air contaminant be unilaterally prescribed only where there is 'significant' danger to human health or where Canada would be in breach of her international obligations and responsibilities in their absence. While Canadian constitutional law offers more support for the first than it does for the second of these unilateral federal initiatives some constitutional support is available for both.

Prescribing national standards for the concentration of both elements and additives in fuel imported into Canada or produced for use or sale in Canada is also anticipated and equally open to challenge. However, the federal government control over imports has significant jurisprudential support while the controls over fuels produced for use and sale in Canada is less certainly justifiable. Nevertheless there are substantial arguments which can be made in both cases.

The kind of federal controls here identified are either narrowly directed (federal works), or are dependent upon their being 'significant' danger to human health or require that their be outstanding international commitments. Unfortunately not all Canadian air pollution problems fit one of these categories. There are others which cause considerable concern but are beyond any substantive federal response unless federal-provincial agreements have been negotiated. It is for these kinds of problems that the Act provides a framework for co-operation.

An important non-substantive exercise which can be undertaken without inter-governmental agreement is the formulation of ambient air quality objectives for specific air contaminants. Such objectives may be determined for an air contaminant either alone or in combination with others. In any case, where objectives are set there are three possible ranges which may be determined: tolerable, acceptable, and desirable. Federal-provincial agreement on objectives and their ranges is not necessary. However, the responsible federal Minister has a statutorily identified discretion to seek the advice of and consult with, industry, labour, provincial and municipal authorities and any other persons with an interest in ambient air quality and pollution control. While this is a discretion which the Minister has even without it being specifically mentioned in the Act its presence may encourage the government to actively pursue outside advice and direction from knowledgeable, interested and concerned persons and consult with them prior to formulating objectives and determining standards. This same quasi-public dialogue is further encouraged and the more effective administration of the statute assured with the provision allowing the establishment of Ministerial advisory committees to assist the Minister in the performance of his air pollution control and abatement duties and in the co-ordination of related activities of governments, institutions and person.

Indirectly dependant upon there being such objectives is a substantive initiative which is beyond federal competence in the absence of provincial co-operation. This initiative is the prescription of specific emission standards for works, undertakings or business which are constitutionally beyond federal control. Since much of Canada's air contamination is generated by provincially controllable stationery sources, the Act has provided a statutory basis for co-operative federal involvement in this area. Because these sources are within the provincial governments legislative competence provincial co-operation is essential. This provision for federal-provincial co-operation has yet to be utilized. When it is the procedure to be followed has been specified.

Before applicable specific emission standards can be prescribed there must first be negotiated a federal-provincial agreement for formulate, co-ordinate and implement policies and programs designed for the control and abatement of air pollution including the formulation and adoption of national ambient air quality objectives. Once such an agreement exists and specifically when there is federal-provincial agrement on air quality objectives, specific emission standards may be set by joint governmental consultation and thereafter federally regulated. These particular specific emission standards will thereafter apply to those non-federal works, businesses and undertaking for which the standards were developed.

CONCLUSION

This paper has ignored completely the role of the municipal government. Within our legal-political system this level of government is a creature of the province with its powers and responsibilities defined for it by provincial legislation. Many, perhaps most, of our air and water pollution problems originate within municipal boundries and many proposed solutions to these problems must somewhere involve these governments if the solutions are to be successful. Developing a legislative framework for the co-operative involvement of municipal governments is a task which we have yet to meet.

Other of our Canadian air and water pollution problems are international in character. We have a long common frontier with the United States, much of it consisting of boundry waters through which the border runs. There are rivers which traverse this same border, some originating in Canada and running south, others originating in the United States and running north, with each one having the potential of transporting wastes across the frontier. And of course the ambient air likewise refuses to confine itself to the territory of one or the other of the two countries. For many of these problems there are existing Canadian-

United States agreements which encourage a co-operative approach at the international level. These agreements which exist and are used have not been considered here even though they exemplify how two countries have approached common pollution problems.

The legislation here considered was selected because it is new and because its co-operative character reflects the federal system it is intended to serve. These are not the only Canadian laws which govern the Canadian response to air and water pollution but they, more than any other, were consciously designed for the purposes of both encouraging and requiring federal-provincial co-operation.

There is no doubt that these laws are insufficiently comprehensive to provide a response to all of the air and water problems which have arisen, are existing or will emerge in Canada. But they do provide a tool which, if properly used, will correct many of the problems causing us concern. Like any other law their true potential will be realized more in their use than in their conception.

SOME POLLUTION PROBLEMS IN NIGERIA

Alaba Akinesete
University of Lagos, Lagos

Nigeria covers an area of about 353,000 square miles and has a population of some 60,000,000. By a lot of standards, she is still sparsely populated. Furthermore, by these same standards, she can hardly be called an industrialized nation. Consequently, she cannot have many pollution problems. On the contrary, our pollution problems are numerous and the sooner we mastered them the better, otherwise we would find ourselves in the same predicament that the more highly industrialized nations now find themselves.

Most of the pollution problems in Nigeria center around the larger cities. The poplulation, which used to be overwhelmingly agrarian and therefore rural, is now becoming progressively more urban. This is due, in the main, to the siting of most industries in and around the larger cities such as Lagos, Ibadan, Kaduna, Kano, Port Harcourt and Enugu. This population movement has brought in its wake the attendant pollution problems. In order to stave off this movement, the government now advises that the industires be taken to the people. Luckily, most of the new industries will be agriculture-based. Hence the farmers will tend to stay on on their farms, while the industries will attract the necessary technical and clerical expertise from nearby areas.

While this government policy may be seen as a partial solution to the city problems, it may in fact spell doom to the otherwise clean environment of the countryside. It is therefore essential that the government makes sure that all new industries satisfy certain pollution standards before being granted their licenses.

While the above are problems of the future, we should have a look at some of the present day problems, especailly as regards the cities. I shall attempt to enumerate these problems with particular reference to Lagos, although these can be identified in most of the other cities.

SANITARY WASTE DISPOSAL

Lagos in all its glory, and with its nearly 2,000,000 inhabitants cannot boast of a central sewage system. The reason is quite simple. We left it till too late. The cost of such a system is now so prohibitive that we have had to seek outside help. In the meantime, and until some aid is found, most homes have to have individual sewage systems utilizing the traditional septic tanks and soakage pits. While these structures take away a good part of very valuable land (the average plot is about 50 ft. x 100 ft.), the problem that most householders encounter is not in siting them but in desludging them after some period of use. The City Engineers Department, whose responsibility it is to desludge sewage systems, is hardly able to cope with a job of this magnitude. As a result, a lot of sewage systems are left uncleared for months after the due date. (In anticipation of this, architects and engineers frequently oversize septic tanks.) To add to this, large buildings, new community schemes, the larger institutions and some factories are equipped with some form of sewage treatment plant - usually packaged systems working on the extended aeration principle. In this case, too, desludging must take place at some time. When it does take place, the question arises as to what to do with the sludge. The practice in Lagos now is to discharge it into the lagoon for onward transmission to the sea. This can be unsightly and can be a health hazard. Consequently, a better disposal system should be found. Land disposal offers a hope.

TRADE EFFLUENT DISPOSAL

Nigeria is becoming an industrialized nation. The pace of industrialization has quickened tremendously since our own war. Most of these industries are still small-scale but there are some respectable-size ones, such as the petroleum refinery, various textile mills, plywood and veneer industries, hides and skins factories, cement works, flour mills, soap and detergent factories, etc. Many others, predominantly petro-chemical agro-allied and iron and steel are planned within the four-year development plan. All these industries exude effluents - gas, liquid and solid - which are dangerous to the environment. While the government has laid down several guidelines regarding the quality of trade effluents, it does not have enough staff who will ensure that these guidelines are adhered to. As a result, the industrialists take advantage of the situation. Unfortunately, or perhaps fortunately, most of these industrialists come from the highly-industrialized nations. I can only ask you to please pass on the mesage. One particularly depressing example is that of a cement works near Lagos. This works is situated in an area that used to be good farmland. The land is now stark and naked. Leaves which used to be green have now taken on a greyish hue. The land that used to be black is also the same color. Houses are the same color. People who have to stay outdoors are not left out. Even motorists need to drive fast and with the glass up, otherwise they will assume grey color. Such a situation must not be allowed to continue.

SOLID WASTE DISPOSAL

There are two methods of solid wast removal in use in Lagos. The first one is the traditional house-to-house collection. The second method involves the siting by the City Engineers Department of "dumpers" or portable garbage containers at strategically located collection points. Nearby householders empty their garbage cans into these containers which are removed and replaced almost daily. Notwithstanding the method of collection, the garbage is normally used as sanitary landfills. Incineration is now hardly used. The only drawback in our landfill method is that there is no compaction prior to filling. At the best a roller is put over the fill after dumping. Consequently, consolidation takes place over a much longer period. However, it is gratifying to note that many former landfills are now fashionable residential areas in Lagos. The chances are that this method will continue since there is still a lot of shoreline and marshland to reclaim.

CONCLUDING REMARKS

I have attempted to enumerate some Nigerian land and other pollution problems especially with particular reference to Lagos. These are by no means our only pollution problems. The object of the paper has been to highlight some of the problems in the hope that some of you may have some common ground. Pollution research and control is still in its infancy in Nigeria. With the introduction of Schools of Environmental Studies in some of your universities and with the keen participation of Nigeria at the United Nations conference in Stockholm, I have no doubt whatsoever that Nigeria will contribute its own quota in achieving a much cleaner world.

LEGAL CONTROL OF INDUSTRIAL AIR POLLUTION IN ISRAEL

Gerald M. Adler

Faculties of Architecture and Civil Engineering, Technion-Israel Institute of Technology, Haifa

Unlike Canada, Sweden, the United Kingdom, the United States and others, Israel has no comprehensive legislative code for environmental protection generally, and air pollution control in particular. It has rather over twenty statutes and other enactments of which about half relate to a greater or lesser extent to air quality. Responsibility for setting norms, coordinating programmes, supervision of standards and their enforcement is distributed among various government bodies, both central and local, on a pragmatic principle. Accordingly it is perhaps more useful to analyse the situation from a functional viewpoint, namely:
1) the planning of industrial location, 2) the implementation of policy in the establishment and construction of industrial plants, 3) the management of industrial operations, and 4) private and governmental supervision of air quality, external to the industry.

1. PLANNING

Under the Planning and Building Law, 1965, the responsibility for the planning of industrial location is divided among three institutions - the National Board, the District Planning and Building Commission, and the Local Commission. The National Board is composed of some twenty persons representing central and local government interests and professional and public bodies. It is responsible for the preparation of a National Scheme which should, inter alia, provide for the setting aside of industrial zones, areas for the production of minerals and the siting of power stations. The National Scheme, like other schemes prepared under this law, is declared to be binding both on the public and private sectors. At present there is no formally binding national scheme which lays down the areas of industrial concentration, although a plan for the location of a power station network is in preparation.

A District Commission, of which there are six, is composed of some fifteen members representing primarily central and local government interests. It is responsible both for the preparation of a District Outline Scheme and for appraising and approving the planning decisions of the Local Commission. In addition its objective is the creation of appropriate conditions in regard to employment. By inference that would entitle the Commission to plan specifically for the location of industrial and commercial capacity; indeed, it is permitted, after consultation with the relevant Local Commission, to make provision for different types of industrial zones within the framework of the District Outline Scheme. This power, to the best of the writer's knowledge, has not been exercised, there being at present no such schemes in effect, although some are in an advanced stage of preparation. Partly because of the lack of a National and District Scheme the Knesset, after considerable public pressure, was impelled to enact in 1967 a special law for the location of the Tel Aviv Power Station Law. This statute, in effect, allowed The Israel Electric Company to circumvent those planning and licensing statutes which would have otherwise applied, and rendered it subject to direct central government control regarding location, construction and operation of the station. The Planning and Building Law requires that all local outline and detailed schemes both privately and publicly sponsored, as well as nonconforming uses and relaxations, receive District Commission approval before coming into force. With one exception, no regulatory guides are provided for the proper exercise of this function. Section 61, setting the outside limits and objectives for legitimate local commission policy, reads in part:
 The objects of a Local Outline Scheme are ...
> (2) to ensure appropriate conditions from the point of view of health, sanitation, cleanliness ... and to abate nuisances by the planning and use of land,including the setting aside of zones for residential,industrial and commercial purposes.

Judicial interpretation of this section is almost non-existent in regard to air pollution. In <u>Megidovitch vs Northern District Planning and Building Commission</u>, 11 P.D. (2) 287, the Court held a local commission to be justified in limiting the number of abattoirs within an area so as to enable their efficient supervision by the health authorities. Speaking for the Court, Landau J. suggested that for appropriate planning control there was no difference between controlling the distribution of unhealthy meat in a particular location and the unsanitary nuisances emanating therefrom; both matters were appropriate for town planning regulation. The Court thus implied that it was quite legitimate for the planning authorities to control, directly through a land use allocation policy an industrial activity which may create a nuisance, despite the availability of more appropriate regulation under the Licensing of Business Law.

There is no reason to suppose that the dicta in this case need be restricted to abattoirs. Arguably any industrial undertaking which by its activities is likely to create a health hazard may be legitimately controlled at the planning stage without having to wait until it goes into operation.

This interpretation would justify and explain the employment of informal administrative guides, called buffer standards, by District Commissions in approving local planning decisions. These standards, designed by an interministerial committee, comprehend a list of some 450 industries divided into six categories distinguished according to their potential nuisance effect on the nearest designated residential land. Minimum distances ranging from 0 to 2000 metres, have been set. Power stations, for example are required to locate 2000 metres distant, while quarrying for cement may be conducted only 500 metres away. The scale has several weaknesses. It fails to take into account the magnitude of the industrial operation involved. Only residential interests are protected; nature reserves which also may be affected seem to go unprotected. No information is available as to when, in what circumstances, and how these standards are employed by the District Commission in its decision making, and in particular the degree of flexibility employed in regard to residential areas coming within the outer limits of an industrial zone. Again it is not known how the standards are applied where two or more industrial plants, each located at its appropriate distance and on its own creating no injury, produce together a synergistic effect. These matters remain in doubt, since the deliberations of District Commissions take place in private and their minutes are not available for public inspection.

The preparation and submission to the District Commission of local schemes is the responsibility of the Local Commission. This body in a large number of cases is constituted by the local municipal authority. It is at this level, therefore, that claims for local development and employment and the need to expand the local tax base on the one hand, are pitted against the demands for environmental quality in air pollution control, on the other. The Planning and Building Law, fails to take this aspect adequately into consideration. It overlooks the conflicting responsibilities and priorities of a local council when questions of attracting industry to its area are raised. In such circumstances, a local authority may perhaps be less thoughtful of the environmental implications of the new industrial developments it wishes to support and encourage for a variety of fiscal and other reasons . Thus, under the Capital Investments Law, 1959, the governmental Investment Authority may grant recognition to an industry if it fulfils the objectives of the Law, i.e. the encouragement of local and foreign capital investment with a view to the development of the productive capacity of the national economy, the effective utilization of its natural resources, the full utilization of existing productive capacity, the absorption of immigrants, the planned distribution of population and the creation of new sources of employment.

Clearly in a developing country like Israel investment for the full exploitation of natural resources and industrial capacity is likely to take a higher place on the scale of priorities than investment in air pollution control, which is seen as providing at best a zero return.

The problem which arises is however not the investment itself but the effect of recognition on industrial location. The Law sets up three zones of preference, based not necessarily on topographical and meteorological suitability, but on a series of non-functional map coordinates, the criterion for whose exact placement is difficult to determine. An examination of the areas discloses that of the approximately 30 new towns, built since 1948, 28 of them are in these zones. Clearly population dispersal and immigrant absorption is the dominant policy. What is of concern is the intense competition among the towns for potential industry and the extent to which local leadership is prepared to give a relatively free hand to industry should it decide to locate in such a town. Some degree of control might theoretically be exercisable by the Investment Authority in the imposition of conditions relating to air pollution control. The Ministry of Commerce and Industry, to whom the Authority is responsible, is not represented on the District Commission, and therefore, apart from the various ad hoc interministerial committees set up from time to time, its views on both private and public industrial location are not coordinated with those of the bodies formally responsible for physical planning.

There has, however, been some advance in planning coordination of those governmental decisions which are likely to have an adverse environmental effect. The Government, in 1971, appointed a committee of chief executive officers (Directors General) of fourteen ministries and government bodies for the protection of environmental quality. These include Defence, Health, Agriculture, Commerce and Industry, Labour, Interior, Development, Housing, Transport, and Tourism Ministries, the Prime Minister's office, the Budget Director, and the Water Commissioner. The Committee's general responsibility is coordination of government activity as it affects the environment and the initiation of action for its improvement. In particular the Committee is intended to be the highest administrative authority for reaching decisions which affect the environment and, government ministries are directed not to implement any action which may have a detrimental influence, without the Committee's prior authorization. What this means in practice remains to be seen. Thus far the Committee has been active only in initiating research on the state of the Lake of Galilee - freezing development in its immediate vicinity, and the appointment of a team of experts to investigate the environmental, economic and administrative implications of power station locations on the sea coast.

Whether the Committee will exercise powers similar to those of the U.S. Environmental Protection Agency or play even an active coordinating role is doubtful, in view of the fact that the Committee has a negligible secretariat and no technical staff directly responsible to it. Furthermore the Committee's terms of reference restrict it from engaging in "environment" planning in spheres which are the exclusive responsibility of a particular government ministry.

In summary, it may be said that the planning functions authorized by legislation do enable decision makers to take into consideration the potential effect of industrial air pollution from new development, but the lack of prexisting plans and publicised norms in this regard inhibit a positive response, especially when and where investment objectives tend in the contrary direction. Decision making in the public sector is, it is true, subject to review by a high level ministerial committee, but what this may leave cannot, for the moment, be determined.

2. PLANT CONSTRUCTION

In regulating construction, the Planning and Building Law provides an extensive code, largely devoted to specification standards for ensuring stability of structure and safety of occupants. Before any construction is permitted, a building permit must be obtained from the Local Planning Commission. A permit will only be issued upon conformity to a detailed or outline planning scheme, and in this manner, the implementation of a development plan may be controlled in the interests of environmental protection generally, and air pollution in particular.

In deciding upon an application for a building permit, a Local Commission is authorized to impose conditions in a variety of matters connected with building construction. Conspicuous by its absence is any direct reference to heating systems which are likely to cause air pollution or to incinerators or furnaces, unless such are impliedly included in "garbage disposal." The issue of a permit is prohibited unless the application also conforms to the provisions regarding sanitary installation. An examination of these extensive provisions discloses no reference at all to pollution abatement devices. The only reference in the whole body of regulatory material is an indirect one which prohibits the installation of incinerators "in the residential, commercial or industrial areas of a building", whatever this means. Absent is all reference to what, if anything, may be incinerated, when it may be done and under what conditions.

The Licensing of Business Law, 1968, which regulates more the manner in which an undertaking is to be conducted, rather than the construction of the building, makes but one brief reference to air pollution. Where a chimney is required for a particular type of industry, the stack shall be at least 2 metres higher than the roofs of the buildings in the immediate vicinity; satisfaction of this condition, however, does not justify the emission of vapours or smoke in such quantity as to cause nuisance; the furnace must be built so that combustion is completed as far as possible. Regretfully, this requirement provides no effective solution which takes into consideration the type of emission, the topographical and meteorological conditions, and the nature of the surrounding physical environment.

The position generally, therefore, is that there are no construction standards either for residential or industrial uses, which ensure the installation of pollution control devices. The one exception to this proposition relates to the Tel Aviv power station which, as already mentioned, is governed by a special statute. Section 3 thereof imposes an obligation on the government, inter alia, to issue regulations prescribing the modes of construction and implementation of the power station plan. These regulations limit the capacity of the station and the number of its units, set a minimum stack height, require the installation of special reservoirs for low sulphur fuel and inspection stations for checking the quality of the emissions. The regulations do not mention other installations, but the Electric Corporation is required to take whatever technical steps the Operations Inspector, appointed under the Law, may order to prevent pollution.

3. OPERATIONAL CONTROL

Operational control of an industrial undertaking which causes pollution may be considered from various points of view; the factors of, or inputs to the production process, the systems of working and operating the plant, and the effects thereof on air quality standards external to the plant.

Israeli law starts with the proposition that under the Licensing of Business Law, the Ministers of Health and Interior may, by order, designate and define businesses and industries which require licensing for the purpose of ensuring, inter alia, the prevention of nuisances and annoyance, and compliance with the provisions of the laws relating to planning and building. Three classes of business are so regulated - those affecting public health, those affecting public safety and those requiring veterinary supervision. The industries listed in the first and second categories include most of the industries which are likely to cause pollution; power companies, however, are regulated by the Electricity Law, 1954. The licensing authority (normally the head of the local authority or a person empowered by him) may not issue a license without prior approval of the Ministry of Health, generally its District Sanitary Engineer. The latter is authorized to impose special conditions, to be fulfilled before the issue of the license or thereafter, which are calculated to further the objectives of the licensing. In this regard, the Health Ministry has a set of standard conditions for a large variety of industries. A perusal of these conditions shows a deep concern for the quality of effluent discharged from an industrial plant, but no regard at all for air-borne emissions, other than the stack height mentioned earlier. Theoretically the Minister has authority to order the installation of pollution abatement devices, but this power has not been exploited to any noticeable extent.

In this regard one should not lose sight of the mandatory provisions of the Abatement of Nuisances Law, 1961, known as the Kanovitz Law, after its sponsor. This statute will be discussed later, but for the moment, suffice it to say that sections 2 - 4 prohibit the causing of any considerable or unreasonable noise, smell or air pollution from any source whatsoever. In regard to the operational control of a plant through business licensing, section 9 of the Law directs that a business license under the Licensing of Business Law or a license required under any other statute shall be deemed conditional upon compliance with the provisions of the instant Law. Thus in determining whether to grant a business license, the issuing authority is required to examine as a pre-condition to its responding affirmatively to the application, whether the undertaking is likely to cause unreasonable air pollution. The reported decisions reveal one case only in which the refusal to grant a business license was premissed upon non fulfilment of this section: <u>Shimavitz vs Tel Aviv Municipality</u>, 24 P.D.(1) 302. A survey of the unreported decisions rendered in the Haifa area is now being undertaken by the writer and may disclose the true extent of the application of this section. One of the major limitations in the employment of section 9 to new license applications, in contrast to a renewal application of an existing undertaking, may be the fact that until the new plant is actually in operation it may be difficult to establish that its activities constitute a breach of statute in terms of the permissible limits of pollution. With regard to renewal applications, interviews conducted by the writer with a number of sanitary inspectors and engineers show that no new conditions are imposed unless the emission gives rise to complaints from the surrounding neighbours.

Israeli legislation in general has not attempted to prescribe or restrict the factors of production in relation to air pollution, as it has in water pollution control. Although the Commodities and Services (Control) Law, 1967, allows the various ministers to regulate and control the production, transportation, transfer, consumption, treatment or use of a commodity, such regulations may only be promulgated in order to maintain the supply and quality of an essential activity or product and to prevent speculation or public deception. In this regard, it would be feasable for the appropriate minister to regulate generally the quality of fuels to be used in industry, especially where the defence of the State or its economy is involved. An examination of the host of orders under this Law discloses no such controls as would affect industrial air pollution. The lone exception is in relation to motor vehicles. Even here there is no express mention of air pollution control, but only that the prototype of the vehicle must be submitted for examination to the Standards Institute.

Again, the one situation of direct control over factors of production is found in the operational regulation of the Tel Aviv power station. A decision of the Cabinet, expressed neither in the Law nor its regulations, requires that the plant use low sulphur fuel (less than 1%) under certain meteorological conditions and when the background pollution reaches a certain level. Cutbacks in operation may also be ordered. There seems to be no reason why similar provisions could not be made to apply to other industries which are known to pollute, under the Licensing of Business Law, and indeed, one such example of this type of regulation is applied in Toronto, Canada.

Factors of production having pollution characteristics may also be controlled by limiting their import to the State. The Import, Export and Customs Powers (Defence) Ordinance, 1939, authorizes the Minister of Finance to make such provisions as he thinks expedient for prohibiting or regulating the import and export of all or any goods. Here, too, there appears to be no bias in favour of encouraging the freedom to import pollution control devices or against the right to import goods which are generators of pollution. The unrestricted freedom to import diesel engines which existed until 1965, has been removed but for reasons unconnected with pollution control.

No grants or tax rebates are specifically designated for industrial air pollution abatement. Indeed incentives are not only lacking, but in some cases are even negative. Insofar as the Income Tax Ordinance [New Version] is concerned, there are no special provisions for accelerated depreciation for pollution control investment, and neither does investment in the introduction of changes in the systems, processes and methods of production receive special administrative encouragement. There is no evidence either of any special exemptions in favour of pollution abatement in the customs and excise duties payable under the Customs, Excise and Purchase Tax (Abolition of Special Exemption) Law, 1957. A perusal of the very extensive schedule annexed to this Law suggests that the reverse may be the case.

Thus, despite the express authority to control industrial air pollution by means of business licensing, efforts in this regard do not appear to have been extensive. With one exception, specification standards in the use of factors of production have not been employed, and neither has there been any noticeable use of taxation and other fiscal tools.

4. CONTROL EXTERNAL TO INDUSTRY

In the absence of specification standards for the prevention of air pollution, performance standards become essential. In this regard Israel has perhaps an excess of laws and regulations, most of which duplicate one another. It must be stated at the outset that protection of the environment in the broad sense, as has been expressed in American, Canadian and English legislation is not evident in Israel as yet. The thrust of all the legislation is in terms of public or private nuisance.

The Criminal Code Ordinance, 1936, section 199, makes penal the act of any person who voluntarily vitiates the atmosphere so as to make it noxious to the health of persons in general, dwelling or carrying on business in the neighbourhood or passing along the public way. Breach of the provision constitutes a misdemeanour, punishable by up to three years imprisonment, or by fine, and enforceable only by the Attorney General, and those empowered by him, such as municipal prosecutors.

This section has not been particularly effective, primarily because of the difficulty in proving a direct link between the act of the accused and the danger to health resulting therefrom. Secondarily, the objective of the Ordinance is not the maintenance of any particular standard of air quality; thus, pollution which is merely offensive to the senses does not constitute an infraction. Furthermore, the legal remedy does not directly encourage pollutors to improve industrial operating conditions, since the court is not empowered to impose conditions for improving the system.

In contrast with the purely punitive approach, a more constructive one is found in the Public Health Ordinance, 1941. This statute defines as a public nuisance, <u>inter alia</u>, any fireplace, furnace, chimney or the like which does not, so far as is practicable, consume the smoke or fumes arising from combustion, and emits them in a quantity such as to be injurious to health. The Ordinance exempts from liability any person who can show that the fireplace or furnace is constructed in such a manner as to consume as far as practicable, having regard to the nature of the industry, and that the furnace has been attended to satisfactorily.

Although this proviso is clearly directed to the protection of industrial interests, it does portray a crude attempt on the part of the Legislature to require the pollutor to operate his plant in accordance with the current state of the art. The difficulty with the section is that, in addition to imposing the normal burden on the sanitary inspector of proving that the emission could have been reduced practicably [economically?], it seems to grant protection to plants, which at the time of their construction, conformed to the standards of the time. Unfortunately, the section does not impose a clear duty to update processes or to introduce more modern equipment.

The administrative process behind the implementation of the Ordinance is esentially a partnership between the district offices of the Ministry of Health on the one hand, and industry on the other. Section 54 of the statute requires the authority to give the accused notice of the offence and to demand not only that it be remedied, but that the industrial concern take specific steps to prevent the situation from reoccurring. If the recommendations are not implemented, the authority is empowered to enter the premises, execute the work, and charge the cost thereof to the owner or occupier.

While such an approach may be appropriate with small firms, where management may be relatively unsophisticated in air pollution control, it is clearly much harder for the district sanitary engineer to make concrete suggestions in larger and more complicated enterprises, with whose detailed operations he is unfamiliar.

An important point to note is the manner in which the law is enforced. The district engineers of the Health Ministry, who are responsible for enforcement of this Ordinance, generally do not make the final decision to prosecute. This is left to the local municipal council concerned and the latter is given the opportunity of deciding whether or not to institute legal proceedings; not under the Public Health Ordinance, but under local sanitary bylaws enacted in accordance with the Municipalities and Local Councils Ordinances [New Version], 1968. In this way the Ministry of Health, improperly, it is submitted, shifts the responsibility for taking what might be a politically unpopular move from itself to the authority most directly 'interested.'

The standard sanitary bylaw, which significantly is approved not by the Ministry of Health but by the Ministry of Interior, contains a list of acts or omissions which are defined as sanitary nuisances. Among them is the absence of a smoke stack for which in the opinion of the local health inspector there is a need, or the presence of a stack which is broken or defective, emitting smoke in a manner likely to cause injury to health. A second nuisance is constituted by any furnace used in industry which, in the opinion of the local inspector, does not consume properly the fuel therein and causes an emission of smoke and gas to an extent which is likely to injure health.

It seems that, in comparison with the Public Health Ordinance, the local bylaws contain specification standards which the former does not; and also a kind of performance standard, namely the local inspector's opinion as to what constitutes a nuisance. On this latter point, the inspector's opinion is deemed to be decisive. Absent in the bylaw is the defence open to an accused in the Public Health Ordinance, that the furnace is constructed in accordance with the contemporary state of the art.

Thus, insofar as public control of air pollution is concerned, enforcement is left primarily at the local government level by means of municipal quasi-criminal legislation, whose execution is based on the relative opinion of the local inspector as to the potential hazard to health. In the past year, however, there has been a noticeable advance in the setting of absolute standards under the Abatement of Nuisance Law. Under this statute, as mentioned earlier, it is an offence to create unreasonable noise, smell, or smoke such as disturbs or is likely to disturb a person in the vicinity or a passer by. The Ministers of Interior and Health are required to issue enforcement regulations and are permitted to determine the degree to which the noise, smell or smoke become unreasonable. The Law does not specifically envisage clean air, neither does it appear to protect plant or animal life; it merely recognizes that, insofar as human activity is concerned, pollution up to certain limits, as defined by the Ministers, is permissable.

A further point which should be noted is the assumption implicit in this law that pollution is local, i.e. unreasonable emissions are defined in terms of a "person in the vicinity."

Judicial interpretation has fortunately taken a more realistic approach. In <u>Israel Electric Corporation vs Avisar</u>, 23 P.D.(2) 314, the Supreme Court held that the determination of whether the plaintiffs were "in the vicinity" (some 6 kilometres distant from the power station) ought not to be made on criteria mechanically applied, but on the basis of whether pollution about which a complaint has been made affects the plaintiff, even if such injury extends over a large area.

The responsible ministers have not been anxious to implement the Law; only under public pressure and court order have emission and ambient air standards been established. The standard of unreasonableness for emissions from "premises" is black smoke of a shade equal to or greater than Ringleman No. 2, for more than six minutes in any given hour. Such measures do not take into account gas content, volume of the emission, and in particular human error in making the comparison between the emission and the measuring chart. The ambient air standards, issued in 1971, define permitted concentrations of some six gases, and a soiling level over a given period. The difficulty with implementing these standards is the absence of any national or regional monitoring system; the law imposes no specific obligation to install one except in a case of the Tel Aviv power station.

Perhaps the most outstanding contribution of the Kanovitz Law is its combination of penal and civil sanctions. In general, air pollution has been considered a public nuisance, enforceable only by the Attorney General or by one who can show special injury. The statute now permits action by a private citizen without proof of such special injury. The Law also allows the court to introduce the civil remedy of injunction in criminal proceedings. However its provisions have not been thus invoked to any great extent for jurisdictional and procedural reasons. Municipal prosecutors prefer to rely on local bylaw enforcement for two reasons. First, municipal courts, which do have jurisdiction in matters of bylaws, planning, building, licensing, and public health have no jurisdiction to entertain a cause of action based on the Abatement of Nuisance Law. The latter can only be heard before a Magistrate's Court where, it is asserted, the proceedings are less summary and more dilatory. Secondly, it is argued, air pollution enforcement is in practice based not on systematic inspection, but on specific complaints largely because of a lack of skilled manpower. It is easier to show in a'complaint'situation a breach of the relative standard occurring under a bylaw than to demonstrate the absolute technical standards laid down in the regulations of the Kanovitz Law. Although the first criticism could be removed simply by amending order to the Municipal Courts Ordinance, the second requires a considerable investment in trained manpower and a national monitoring system, a network of staff and equipment which it is doubtful whether the national and local budgets can bear.

Private actions based on the Abatement of Nuisance Law have been relatively rare. Owing to the costs involved, the institution of legal action by organized citizens' groups has not yet become as accepted a weapon for abatement as it has in the United States. The Israeli pattern for protecting various interests is to establish a formal organization with government or quasi-public sponsorship. Such is the case of the Public Council for the Prevention of Noise and Air Pollution in Israel (Malraz). While such bodies are generally given limited grants from public funds, to the extent of supporting an organized nucleus of committed individuals, the amounts received barely cover the maintenance of a small secretariat. The concerned citizen is thus placed in a position of false security: he tends to believe that if a public organization exists for the protection of the environment and is supported by public funds, then such interest as he has in air pollution abatement is sufficiently protected without further financial commitment. This is unfortunately far from the truth, and only a concentrated effort on the part of those who are deeply involved in the issue of environmental quality can make pollution control an issue of greater political importance on the Israeli scene than it is at present.

AIR POLLUTION TRENDS IN TEL AVIV, ISRAEL

A.E. DONAGI,[1] J. MAMANE,[2] AND T. ANAVI[2]

[1]*Ministry of Health & School of Medicine, Tel Aviv University,* [2]*Ministry of Health*

1) Introduction

The first monitoring system installed in Israel is in the city of Tel Aviv. This system is composed of 9 fixed stations measuring SO_2, soiling, and wind speed and direction; as well as 4 stations (3 fixed and one mobile) that measure SO_2, NOx, CO, total hydrocarbons, soiling and wind direction and speed.

This system began operating in the summer of 1969. The available data of the years 1969-1970 were analyzed and possible trends of air pollution in Tel Aviv were investigated.

2) Sampling Sites

Fig. 1 shows the sampling sites in the city of Tel Aviv. The power station Reading D is situated in the north of the town; another, smaller power station, Reading C is located 1.5 km from Reading D. One should emphasize that the height of the stack of Reading D, which started to function in late 1970, is about 150 meters, while the height of the stack of Reading C is only about 10 meters.

The measuring stations are spread over the entire territory of Tel Aviv. The central monitoring station is located in northern Tel Aviv and called "Zafon".

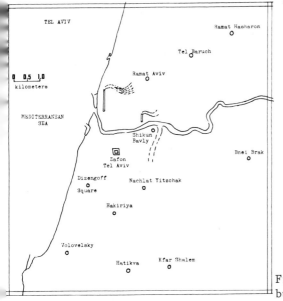

Fig. 1. Map of Network Sampling Stations

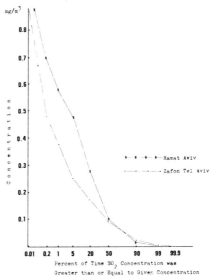

Fig. 2. Frequency of Distribution of Half an Hour Average SO_2 Concentration in Tel Aviv, 1970 (Zafon and Ramat Aviv)

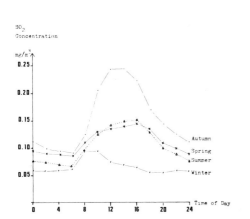

Fig. 3. Seasonal Daily SO_2 Concentration (1970) Tel Aviv

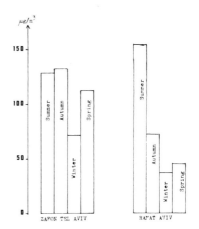

Fig. 4. Seasonal Variations of Sulfur Dioxide Concentrations in Tel Aviv (1970-1971)

In 1969 and 1970 two stations, "Zafon" and "Ramat Aviv" were functioning.

From the very beginning the emphasis was put upon measurement of SO_2, soiling index and NOx. Therefore, the forthcoming analysis refers mostly to the same pollutants.

Sulfur oxides were measured using either "Beckman" Acralyzer or Wösthoff "Ultragas-U3S" Analyzer. Both instruments are based on conductivity measurements. During 1969 and 1970 more than 75,000 valid half-hourly SO_2 readings out of a total number of 79,500 measurements were performed. This means that about 95% of the data is valid.

3) Frequency Distributions

Fig. 2 shows the frequency distribution of SO_2 half hourly averages in 1970, at the stations Zafon, and Ramat Aviv.

It can be seen that during 1% of the time the SO_2 concentrations were above 370 $\mu g/m^3$ in the Zafon station. This value is important because the Israeli standard refers to 99 and one percentes of the time. The corresponding value for Ramat Aviv is 598 $\mu g/m^3$. 50% concentration was about equal in both stations and its value was 94 $\mu g/m^3$. The annual average of Zafon station was 114 $\mu g/m^3$. This means that about 37% of the time, the SO_2 concentrations exceeded this value. In Ramat Aviv station the annual average was 154 $\mu g/m^3$. This means that during 38% of the time the readings were above this value.

4) The Diurnal Cycle

Fig. 3 shows the seasonal diurnal pattern of SO_2 concentration. The maximum values were measured at noon and early afternoon and the minimum values in the early morning.

It is suspected that the external temperature influenced the readings. This is supported by the fact that the highest readings during autumn, summer and spring were mostly obtained at noon time, while in winter, when the external temperature was low, the maximum concentrations were measured between 8 and 10 A.M.

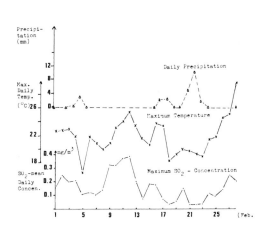

Fig. 5. Some Comparisons between Concentration, Temperature and Precipitation (Tel Aviv, Feb. 1970)

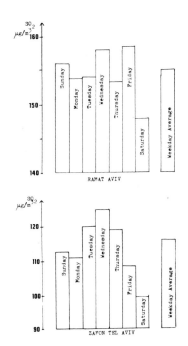

Fig. 6. Weekday and Saturday Sulfur Dioxide Averages. Tel Aviv 1970

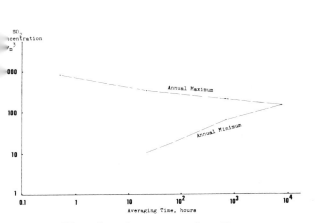

Fig. 7. Concentration Versus Averaging Time for SO_2. "Zafon" Tel Aviv 1970

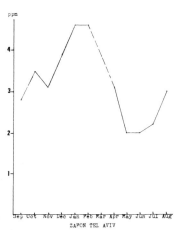

Fig. 8. Monthly Carbon Monoxide Levels (1970-1971) in Tel Aviv

5) Seasonal and Weekly Variations

The seasonal variation of SO_2 concentrations in Tel Aviv in 1970 is given in Fig. 4. It can be seen that the seasons with the highest SO_2 levels were low. This trend is not common in other countries. For example, in London and New York, the higher levels of SO_2 were measured in winter. It is believed that the climatic difference is one of the major reasons for this trend.

The consumption of fuel in the power stations is also higher in summer than in spring and autumn.

It was not possible in this case to find a correlation between fuel consumption and levels of air pollutants. For instance, in winter the consumption of fuel was very high but the SO_2 concentrations were low. This may be due to the good dilution in the air. Another factor that was investigated was the influence of the temperature and precipitation on the SO_2 levels (Fig. 5). It can be seen that, in February 1970, the days with relatively high precipitation had relatively low concentrations of SO_2 and vice versa. This is a kind of inverse relationship. It was also found that the days with maximum temperatures were the days with the maximum concentrations of SO_2.

From the synoptic point of view, it can be said that there is fairly good correlation between anticyclonic conditions and relatively high concentrations. On the other hand, stormy weather, which is connected with cyclonic conditions, causes a drastic decrease in the SO_2 concentrations.

Analysis of the potential effect of Saturdays on the SO_2 concentrations (Fig. 6) showed that the Saturday averages were about 10% lower than the average values for the other days of the week. This trend is probably a result of the decrease in industrial activities on Saturdays.

6) Time-Concentration Relationship

Fig. 7 gives a plot of the maxima concentrations, versus the averaging time, on a log-log paper. A straight line with a slope of about -0.2 was obtained. This figure is close enough to the values found by theoretical calculations for a similar line by other investigators such as Wipperman (-0.14) and Meade (-0.17) and scientists from the TNO (-0.24) in the Netherlands. This line is impor-

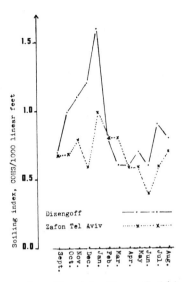

Fig. 9. Monthly Mean Soiling Index Levels (1970-1971) in Tel Aviv

Air Pollutants Location	SO_2 (p p m)	C O (p p m)	Soiling Index (COHS/1000 Lin. Feet)
Tel Aviv - 1970	0.044	3.1	0.80
Chicago - 1967	0.125	8.8	1.21
Philadelphia "	0.098	6.3	0.76
Cincinnati "	0.021	5.6	1.14
Washington "	0.048	4.8	1.01
St. Louis "	0.029	5.6	0.78
Denver "	0.005	7.6	0.95

Fig. 10. Levels of Some Air Pollutants in Tel Aviv and in the U.S.A. (Annual Averages)

tant because it gives the values of various maxima for any desired period of sampling time.

7) Trends of Pollution

Fig. 8 shows the monthly carbon monoxide concentrations for 1970-1971. The highest averages were found in the winter, in December, January and February. The lowest averages were measured in spring and in summer.

The soiling index, which can be used as an indication of smoke and particulates in the air, was measured in Tel Aviv during this period, at two stations: Dizengoff Square and Zafon. Soiling index monthly averages (Fig. 9) show the same trend as the CO concentrations. This means high values in winter and low values in summer. At the monitoring station located at Dizengoff Square, the averages are generally higher than in Zafon. In Dizengoff Square the density of motor vehicles is higher. It is interesting to observe that in spring and summer the levels are about equal and relatively low. A comparison of the trends for different air pollutants measured during August 1970 shows a marked parallelism between the CO and the soiling index trends. It is worth mentioning that on the 24-25 of the same month maximum daily concentrations for all the air pollutants measured were obtained. From the meteorological point of view, it was a sultry day with anticyclonic conditions and high temperatures in the coastal area of Israel.

In order to receive a picture of air pollution levels in Tel Aviv versus some cities in the U.S.A. a comparative table (Fig. 10) was prepared. This table shows that the levels of pollution in Tel Aviv are relatively low, for most pollutants. SO_2 is an exception, but perhaps the different measuring methods utilized may explain the reason for that difference. Despite the fact that this comparison is a rough one, yet it indicates that Tel Aviv is among the cities which have relatively low pollution.

In conclusion, it may be said that this trend-analysis cannot be considered as a definitive trend of air pollution in Tel Aviv. It refers only to the year 1970 and it gives mostly the direction of necessary research for the future, especially in regard to trends which are "unusual" in other parts of the world.

A DUAL-PURPOSE AIR POLLUTION ALERT AND IMPLEMENTATION SYSTEM FOR THE GREATER TEL AVIV AREA

A.E. Donagi [1], M. Naveh [1], A. Manes [2], M. Rindsberger [2], Y. Gat [3] and A. Friedland [3]

[1]*Ministry of Health, Israel* [2]*Ministry of Transport, Israel* [3]*Electric Power Company, Ltd., Israel*

Introduction

The first thermal power plant in Tel Aviv was constructed in the 1930's, about 2 miles north of the town. The capacity of the plant (Reading A) was 36 MW and the stack height reached 33m. Twenty years later, two 50 MW units (Reading B) were constructed at the same site, each unit connected to a 53 m. high stack. Two 20 MW units (Reading C) were added in the mid fifties, at a site located about 2 miles east of Reading A and B, close to a newly-built residential area.

The stack height of each of these units is about 10 m. At present these two units (Reading C) are not operated continuously at full load, but are maintained in order to provide backup at peak demand.

Since the 1930's, the city of Tel Aviv has expanded mainly to the north, reaching the power plant site (Reading A and B). During the years 1970-71, two large units (Reading D) having a nominal capacity of 214 MW each (240 MW overload capacity) were added to the first site. The height of the stack connected to these units is 150 m. This is a multiflue stack, the design and height of which were determined after an extensive study (1,2,3,4). The effluent of the two Reading B units are also released through the same stack.

The designed exit velocity of the flue gases of Reading D is 27.4 m/sec, and the average exit temperature is 145° C. Following the operation of Reading D, Reading A has been closed. The power plants use fuel No. 6 (Bunker C), residual petroleum oil containing about 3% sulfur and 0.01-0.03% ash. The daily consumption may reach 2500 metric tons, which means an emission of 150 tons/day of sulfur oxides, and 250-750 kgs/day of particulates.

In addition to the power plant, which is the only large and tall single point source, there are many low level sources, such as motor vehicles, industry, workshops, domestic heating, etc. As a matter of fact, the planned Reading D plant drew wide public attention as a possible air pollution hazard and became a governmental issue. The final decision to construct the plant was made by the government, which also appointed a special committee to set up an air pollution alert and implementation system for metropolitan Tel Aviv. This system is described in the following sections.

Objectives

The main purpose of the system is to ascertain that air pollution levels in the region remain within the limits specified by the appropriate Israeli Ambient Air Quality Standards (Table 1).

Table 1: Some Israeli Ambient Air Quality Standards

Pollutant	Max. ½ hr. Average	Max 2 hr. Average	Max. 24 hr. Average	Max. 8 hr. Average
SO_2	0.3 ppm	-	0.1 ppm	-
CO	30.0 ppm	-	-	10.0 ppm
NO_x	0.5 ppm	-	0.3 ppm	-
Soiling Index	-	2.0 COH/1000 lin.ft.	1.0 COH/1000 lin. ft.	-

Note: The requirement is that these values will not be exceeded during 99% of the time. The half-hour values could be exceeded for 1% of the time but shall not be

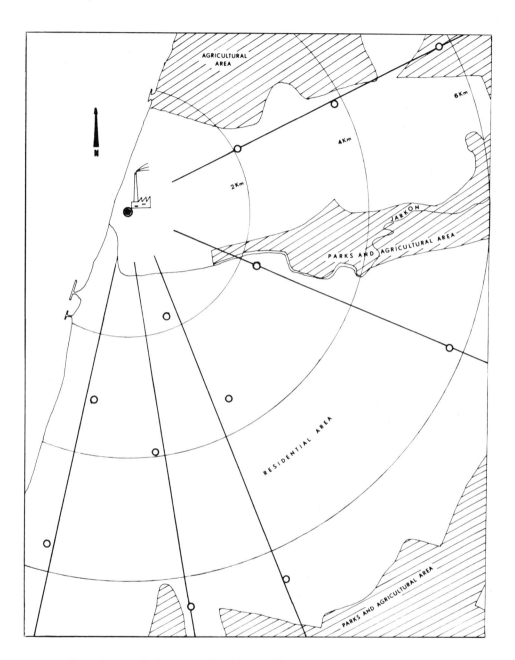

Fig. 1. Tel Aviv-Jaffa Air Pollution Monitoring Network

more than twice the specified values. The 24 hr. values could also be exceeded for 1% of the time, but shall not exceed the specified values by more than 50%.

The air pollution monitoring network, together with the meteorological network, supply continuous information on the air quality of the region and on the pollution potential. This information is mandatory for control and research purposes.

Description of the System

Sporadic air quality measurements in Tel Aviv were made since 1964. A monitoring network consisting of 12 fixed monitoring stations and one mobile unit has been set up, and gradually put into routine operation. Measurements of SO_2 and other pollutants commenced about 18 months prior to the operation of the first unit of Reading D. The siting for the monitoring network was selected in a way to enable discrimination between the contribution of the power plant and the influence of other pollution sources. The monitoring stations have been located at the intersections of three arcs, having the radii of 2, 4 and 6 km, with five directional lines (Fig. 1). In addition, the sites were selected in such a manner that they represent various typical sectors of the town. These are "pure" residential area, heavy traffic zone, industrial area, and mixed industrial - residential zone. The stations are equipped with the following instrumentation:

a) Sulfur Dioxide - continuous monitors with recording and integrating facilities in all stations.

b) Soiling Index - cumulative 2 hourly measurements--installed in six of the stations, including the mobile unit.

c) NOx, CO, Hydrocarbons - continuous monitors with recorders - installed in four of the stations, including the mobile unit.

d) Wind (direction and speed) - continuous recording at 6 stations, including the mobile unit.

Routine meteorological observations, including upper air measurements, are performed by the Israel meteorological service at Bet Dagan, located about 5 miles southeast of Tel Aviv. The monitoring network and the meteorological measuring system will be expanded in the near future. It will include town measurements of wind, temperature, humidity and solar radiation.

Alert and Implementation Procedure

The alert and implementation system is designed to deal with the power plant (a single and tall point-source), as well as with numerous different lower sources.

The system is based on a combination of special meteorological forecasts and results obtained from actual air quality measurements in the region. Special meteorologicalforecasts are issued by the Meteorological Service on a routine basis twice a day. These forecasts take in consideration the existence of dispersion conditions for either high or low sources separately. When adverse dispersion conditions, which may lead to a rise in the ground level pollutant concentrations are predicted to continue for at least thirty-six hours, a "Forecast" status is declared by the Meteorological service. This could be either "Meteorological Forecast for Power Plants" or "Meteorological Forecast for Background Pollution".

Following these Forecasts an "Alert Status" (for the Power Plant or for low-level sources) may be declared by the Health Authority, taking in consideration the actual pollutant concentrations measured. In the event that the predicted adverse conditions are verified by the actual air quality measurements, the meteorological service is requested to issue special forecasts, more frequently, up to four times a day, and if necessary to make some special upper air soundings.

There are four Operational Statuses which are designated "Alert", "Operational A", "Operational B" and "Emergency Status for Power Plants". A detailed step by step procedure, with the passage from one stage to another is given in Fig. 2.

The basic parameters, which determine the passage from one stage to another, are the sulfur-dioxide dosage (concentration multiplied by length of time) or a combination of the sulfur dioxide dosage and the soiling dosage.

Status	Col No.	Dosage limit requiring declaration of a new status				Time Interval Considered (hrs.)	Forecast** (hrs.)	Additional Conditions	Resulting new "status" to be forecast	Operations Required at the "Status"
		SO_2 ppm-hrs	Soiling Index Cohs/1000 ln.ft.	CO ppm-hrs	NO_x ppm-hrs					
Meteorological Forecast for Power Plants	1	-	-	-	-	-	36		"Alert" status for power plants	Surveillance of monitoring stations at least once every six hours
"Alert" status for Power Plants (Operational Status)	2	5.0	10.0	-	-	24	12	Reading of instruments indicate that the pollution results from Reading B and D	"Operational A" status	Shutdown of Reading C Increase of buoyancy of Reading B and D plume
	3	2.0	2.5	-	-	6	12			
	4	3.0	-	-	-	6	12			
"Operational A" Status	5	7.5	10.0	-	-	24	12	- " -	"Operational B" status	Mandatory usage of low sulfur oil in Reading D
	6	3.0	2.5	-	-	6	12			
	7	4.5	-	-	-	6	12			
"Operational B" status	8	12.5	20.0	-	-	24	12	- " -	"Emergency" status for Power Plants	Consultations on special action and on notice to the public, to traffic, and to industry
	9	7.5	-	-	-	6	12			

*The power plant set up in Tel Aviv is composed of three 12 megawatt units = Reading A (to be closed upon operation of Reading D), two 50 MW units (Reading B), two 20 MW units (Reading C) and two 240 MW units (Reading D).

**Duration of period of forecast of adverse meteorological conditions.

Fig. 2. Status Criteria and Implementation Procedure for a High Air Pollution Alert and Warning System in Tel Aviv*

Status	Col No.	Dosage limit requiring declaration of a new status				Time Interval Considered (hrs.)	Forecast* (hrs.)	Additional Conditions	Resulting new "status" to be forecast	Operations Required at the "Status"
		SO_2 ppm-hrs	Soiling Index Cohs/1000 ln.ft.	CO ppm-hrs	NO_x ppm-hrs					
Meteorological forecast for background pollution	10	-	-	-	-	-	36		Alert status for background pollution	Surveillance of monitoring stations at least once every six hours
"Alert" status for background pollution (Warning Status)	11	As in columns 2, 3, and 4					12		"Warning A" status	Shutdown of Reading C
	12	-	-	180	-	6	12			
	13	-	-	-	2	3	12			
"Warning A" status	14	As in columns 5, 6 and 7					12		"Warning B" status	Implementation procedure for warning status is being worked out at present
	15	-	-	300	-	6	12			
	16	-	-	-	3	3	12			
"Warning B" status	17	As in columns 8 and 9					12		"Emergency" status for background pollution	Implementation procedure for warning status is being worked out at present
	18	-	-	-	5	3	12			
Termination of status	19	-	-	-	-	-	0			

*Duration of period of forecast of adverse meteorological conditions.

Fig. 3. Status Criteria and Implementation Procedure for a High Air Pollution Alert and Warning System in Tel Aviv

The second part of the system deals with the lower air pollution sources. The appropriate procedure relavant to this situation is given in Fig. 3 and is similar to the procedure adopted for the power plant. However, there are some differences which are listed below:

1) Dosages of CO and NOx are taken into consideration, besides those of SO_2 and soiling.

2) Instead of "Operational Status" the various statuses are called "Warning Status" and there are "Alert", "Warning A", "Warning B", and "Emergency Status for Background Pollution".

The implementation procedure for warning status has not been completed. In the meantime, the necessary steps in each case are issued by the Health authorities if the need arises.

Preliminary Experience

Since the system has been put into operation, about fifty "Forecast" situations for low-level sources were issued, but only one "Forecast" for the power plant was declared. No "Operational" or "Warning" statuses were declared as yet.

The highest SO_2 30 min. mean concentration measured was 0.73 ppm (1905 $\mu g/m^3$) and the second highest was 0.32 ppm (805 $\mu g/m^3$). The average monthly concentrations of SO_2 measured at all 12 fixed monitoring stations varied between 20 to 40 $\mu g/m^3$, and at no station exceed 60 $\mu g/m^3$. Fig. 4 gives the monthly fuel consumption at the power plants. Fig. 5 gives the SO_2 ground level concentrations. Additional data and results are given in Ref. 5 and 6.

Summary

A dual-purpose Air Pollution Alert and Implementation System for the greater Tel Aviv area was described. This system takes into consideration potential high air pollution concentrations resulting from both a single tall and big point-source as well as numerous smaller and lower sources spread all over the city.

The operation of the power plant complex did not as yet cause any severe air pollution episodes in the

11. NATIONAL AND LOCAL PROBLEMS AND SOLUTIONS

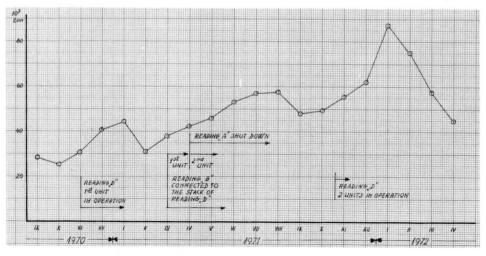

Fig. 4 - Monthly Fuel Consumption of the "Reading" Power Stations

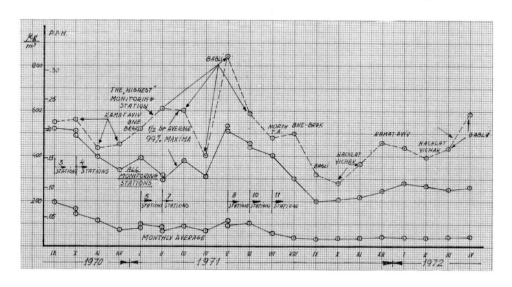

Fig. 5 - SO_2 Concentrations Measured in the Tel Aviv Area (Results from 12 monitoring stations)

region. However, the system described is described is designed to provide effective means of control if conditions leading to such an episode may occur.

It was found that an efficient pollution monitoring network, with a meteorolgical warning system, can function effectively at a relatively low cost through cooperative effort of various governmental and private institutions.

Bibliography

1) Rindsberger, M. and Manes, A.: Investigation of Potential Air Pollution in the Area of the Planned Reading D Power Station in Tel Aviv, Israeli Meteorological Service, 1967.

2) Wachs, A.W.: Height of Stacks for the Reading D Power Plant, Technion, Israel Institute of Technology, Haifa, 1967.

3) Brasser, L.J., McCormack, R. and Stern, A.C.: Report to the Israeli Ministry of Health on the Problems Associated with Air Pollution from the Proposed 214 MW Units, Tel Aviv, 1968.

4) Sporn, P. and Smith, M.E.: Report to the Prime Minister of Israel Regarding Reading D Power Plant Complex, 1968.

5) Donagi, A.E.: Air Pollution in Israel - to be published in the Proceedings of the 1st Internat. Mtg. Pollution: Eng. & Sci. Solutions.

6) Donagi, A.E. et al.: Air Pollution Trends in Tel Aviv - to be published in the Proceedings of the 1st Internat. Mtg. Pollution: Eng. & Sci. Solutions.

POLITICS AND POLLUTION: POLITICAL SOLUTIONS TO A DETERIORATING ENVIRONMENT

Ivan Völgyes

University of Nebraska - Lincoln

The waters of Lake Baikal are as polluted as those of Lake Erie. Smoke chokes people in Budapest and Tokyo, Mexico City and New York. Automobile pollution has destroyed the clean mountain air of small towns in provincial Italy. These are the symptoms of the industrial age; industrialization is destroying humanity regardless of race, religion or sex, ideology or belief. Industrialization, modernization and urbanization have begun to exhibit serious tendencies of failure just when the lofty dream of bringing health and abundance to the people of the world seems within reach.

The ills of society, of course, are not new. Pollution has been with man at least since he began to live like a civilized being. The Jerusalem of Biblical days was a virtual dungheap, but was still the cleanest city of the ancient world. Dirt in the ports of Greece was so bad that one had to climb over the rubble to see the ships come and go. The streets of Rome were rat-infested, smoke from cooking fires darkened the villas of Pompey. Paris's grime, soot and dirt was far worse in the sixteenth century than it has been since, and the great fire in London destroyed a decayed inner-city far more filthy and polluted than it was before the anti-pollution efforts began in the 1960s. The center of New York in the 1880s was almost as unhealthy to live in as it is today.

Until the second half of the twentieth century man had never really begun to clean up the environment. Although the list of people interested in the topic included Dante and Machiavelli, Erasmus and Kepler, Bentham and Marx as well as countless others, little was ever accomplished. All of the efforts trying to deal with pollution failed because those pursuing them had limited ways of exerting political pressure on the rulers of the time. Even today politics is the key to implementing technical solutions to the problems posed by a constantly deteriorating environment. Politics -- the relationship of man and society, government and people -- is the key to our ability to control pollution, and regardless of the political system, it is primarily political considerations which define the possible limits on the control of our physical environment.

The problem of pollution is handled differently in different political systems. For the purposes of classification three distinctive political systems will be identified: a. democratic, highly industrialized b. authoritarian, advanced, industrializing and c. underdeveloped, industrializing systems. There is no ideological bias in the omission of an authoritarian highly industrialized system from the classification; rather, so few examples of this type of society exist that the creation of a separate category was not warranted. East Germany alone could be included in such a classification.

Environmental deterioration has been most notable in the advanced developed, capitalist democracies of the West. The political means of solving the problems of the deteriorating environment in these countries can best be analyzed in terms of interest group politics. In each of these states the role of the government - to a lesser or greater degree -- is that of an arbiter among interest groups. The demand articulation of these interest groups ranges from the very significant to the minimal. Thus for example, in the United States the professional lobbies working for such giants as the United States Steel or Dupont are more likely to be able to muster political strength than the various groups fighting for environmental control measures. The legislators in the democratic countries frequently are likely to have re-election as a primary concern: they are likely to pay more attention to the local needs of running a factory and thereby keeping a high level of employment in their states than to the apparently less immediate needs of the citizenry for clean water or breathable air.

Legislators, however, are slowly being compelled by various citizens' lobbies to begin to pay attention to the environment which industrialization has destroyed. The high mercury content of polluted lakes threatens the children of legislators, of industrialists and bankers, as well as those of the workers and middle-class families living on the shores of the lake. It thus becomes their common concern to control the pollutants, and even if their efforts result in temporary loss in profits for the companies, public pressure has begun to make its mandatory that such controls be established.

In the U.S. the problem is complicated by the fact that stockholders demand profits and that the economic system is based on the notion of profit. The costs of environmental measures thus are tagged onto the price of a product causing prices to advance steadily. The argument advanced by the industries is "If you want cleaner air, water or non-polluting cars, you, the customer must pay for it." Implicit in this argument, unfortunately, is the assumption that there is a certain "natural right" of the producer to pollute the environment in exchange for producing his wares.

In the United States the political solutions to this problem imply a greater role for the federal government as the representative of all of the people and of the national interest. It now is expected to supply the needed impetus to change for the better. The government traditionally has adopted two methods of combatting the problem. On the one hand, it can supply subsidies to factories, thereby combatting the inflationary trend and, on the other hand, it can enforce federal standards of maximum allowed pollution. The first approach saddles the nation with higher taxes, thereby forcing all the people of the United States to pay for profits to only a small percentage of the citizenry; this approach puts the burden of paying for the cleaning up of the environment back on the populace. The second approach, enforcing strict federal standards, also poses difficulties. First, federal standards, as written by Congress or legislatures, always are

results of compromise. They are never optimum solutions,
and at best are "arrangements". Congress can never
view the problem of ecological deterioration solely from
the point of view of allocating scarce resources for
purposes which show no clear POLITICAL profit for its
members. Second, even when established, federal
standards have been difficult to rigidly enforce.
Industrial pollution in West Virginia mines, for example,
has not been controlled in spite of the fact that federal
legislation has been enacted since 1969. Large steel
plants, chemical firms or paper mills can get around
such legislation with relative ease, and the enforcement
would require so many federal agents at various jobs that
even the burgeoning civil service in the United States
would not be able to undertake such a task.

Another significant problem with pollution in the
United States relates to differences of opinion
concerning national priorities. While the prosperous
middle class views the problem of pollution as a primary
concern, this view is not shared by the residents of the
inner-cities and ghettoes of that country. To the
residents of these areas, the problem of pollution is not
a primary concern. The question of unbreathable air
comes secondary to the building of new hospitals. The
problem of industries belching smoke is second to having
industries which will find jobs for the unemployed and
pay decent wages. To the ghetto dwellers, cleaning up
Lake Erie is secondary to cleaning up the rat-infested
urban dwellings of Chicago.

Other industrial democracies have environmental
problems similar to those of the United States. Smog
in the city of London was cleared up, but air pollution
as a result of automobile exhaust fumes remains a
serious concern. Manchester, Birmingham, Glasgow, the
hundreds of medium-sized steel cities of the United
Kingdom, are still huddled under an umbrella of grey,
acrid smoke, and the control devices of the government
are not forced with as great a vigor as possible. In
Western Europe the polluted Rhine is nearly behond
salvaging as factories are still pouring their waste
into that river. The industries of the Ruhr valley spew
smoke over the country-side, but the Wirtschaftswunder
is progressing and it will become a wonder if the
wirtschafts can continue as the population becomes
barely able to breathe the air. Economic progress

production and development have been a prime concern in post-war Germany. The ecological threat has had to take second place to the need for rebuilding.

Italy's choked cities with their overwhelming noise and automobile pollution are subjects of endless jokes. One almost feels that the jokes are the only alternatives available to Italians because the scope of the automobile problem seemingly defies solution. The rivers of the North and the waters of the sea are beginning to produce uneatable fish because the mercury content is approaching unacceptable limits for human consumption.

France, too, is experiencing the same type of problems. Recent efforts at establishing automobile exhaust emission controls are not being carefully enforced; on wet, winter days the air of Paris in the air of Paris the results are plainly visible on the wide boulevards of the city. Málraux cleaned up the magnificent buildings which were covered with soot, but in merely five years a new cleanup job is needed. At Tours the factories pollute the Loire to an extent that is shocking, but the law enforcement agencies have not found factory activities contrary to the letter of the law. Although a new Ministry of Environment has been established its activities have not yet significantly affected the polluted producing areas.

Japan poses one of the greatest headaches to environmental concern. The rapid industrialization -- both between 1870 and 1940 and following 1949 -- had created a serious inbalance in the country. In Tokyo, street vendors are selling oxygen, in Nagasaki grey smoke stenches the thin air, while in places like Misawa, or Sapporo the environment seems relatively pleasant and clean. The Japanese government, too, has been unable to act decisively against the pollutants: economic concerns and a fear of unemployment or staggering inflation have stopped it short of serious control measures. A near fanatical drive to produce more and to live with "more things" overwhelms the Japanese urban dwellers' desire to live in a clean and healthy environment. The near universal acceptance of the norm that along with its blessing of a higher, more luxurious standard of living industrialization also brings choking, dirty air and polluted waters apparently have been acknowledged by many Japanese citizens to be the price of "progress".

The other industrialized democratic states are only slightly better off, although all of them groping for a solution to automobile pollution, smoke control and water pollution. In each of the countries the idea of having a modern, industrialized economy is a primary concern; the need for environmental controls have had to be secondary. The tragic point, however, is that many of the people themselves are not aware of being poisoned daily. Most of them accept the fact that the price of industrialization is pollution and they are meekly looking toward their governments to minimize pollution while maximizing their standards of living.

In all of these industrialized democracies some control measures are being implemented and a greater and greater consciousness is emerging among the people to turn to the government for more effective controls. Government is a more powerful instrument vis-á-vis industry in these states than in the United States. The national concerns are less diffuse, the inner-cities are not slums and the governments are beginning to under-take effective programs to combat the pollutants. Specifically within the Common Market a new, common framework of reference exists and perhaps within this framework some action may be undertaken in the 1970s to create a better environment to live in.

The problem of pollution and the implementation of control measures hinges on political solutions in all of the industrialized democracies. The question of the fine balance between the powerful industries and the governments, or between the economic and ecological interests of the states concerned can only be reached through political means. Governments and opposition parties alike must begin to incorporate in their programs the problem of pollution and means of reversing deterioration of the environment if any effective action is to be taken by these states. In this regard, popular pressure will be mandatory in combatting the influence of the powerful industrial interest groups.

In the category of authoritarian, industrializing systems we are dealing with a group of states whose governments exercise a strong control over the policy, where industry is, if not state-owned, at least state-controlled. Supposedly, thus, we are concerned with states where the government is relatively free to deal with the industries as it pleases, and therefore is able to insist on the quality and character of the environment. This generalization ought to apply clearly to the USSR and to the Communist countries as well as to a number of states with strong authoritarian regimes. Yet, there exists such a complex situation in these states that such generalizations are not warranted. Particularly in the Communist countries the governments have come to power with a mandate to industrialize, to lead the people to the promised land of a better standard of living. In other words, the ideologies have promised the material abundance, which exists in capitalist countries. The image of the good life in capitalist states remains extremely powerful: progress means industrialization; it demands more and more factories, hydro-electric plants, and eventually it means having private automobiles be they Polski Fiats or Zhigulis.

Thus, the governments have placed tremendous national efforts on industrialization and the result has been pollution of the air and destruction of the rivers and lakes. Modernization became the key to the success of these regimes and environmental concerns have been placed near the bottom of the political concerns. Lake Baikal is nearly ruined from pollution, the Danube is infested with 145 million tons annual industrial waste, oxygen dispensing stations have been set up in Budapest during the winter, and the factories of Galati and Nowa Huta and the other industrial centers of the Communist states belch smoke with little evidence of government or popular concern. Defense requirements demand ever-growing amounts of iron and steel, while the scarce quantity of consumer-merchandice must be maintained to keep the population happy. A sardonic Hungarian bitterly remarked, "In pollution we have already reached the Western capitalists."

Yet, the political control of production and pollution are vastly different from the practices of the industrialized Western democracies. In the industrialixing dictatorial systems it is the state, the government which pollutes and scarcely limits its productive activities. Specifically where single party systems rule all power to control the environment is theoretically vested in the regime's hands. The rulers

are free to decide whether to pollute or not to pollute.
That freedom, of course, is vastly encumbered by the
demands of the population to keep up with the luxuries
of the developed countries. Few are willing to pay the
price of decreased availability of consumer goods and a
lessened tempo of industrialization even if that would
mean a cleaner air. Coal heating is slowly being
replaced with gas heat in some places, but the gas and
the coil costs more money initially and other segments
of the industry must pay for it. The two-stroke engine
<u>Trabants</u> are still automobiles, and however much they
contribute to pollution, they are still "better" than the
uncomfortable, crowded buses.

Within some of the ruling hierarchies in these
systems during the late 1960s, however, a few voices of
dissent are beginning to be heard. The pollution
of Lake Baikal had become a national scandal in the USSR
and some of the younger economists and cityplanners have
begun to question the utility of headlong industrialization. Primitive ecological interest group articulation has begun to be noticeable in Hungary, Yugoslavie
and even within the Soviet economic elite. But so far
in all of these states a national commitment to severely
limit the pollution is missing, and in spite of the
governments' ability to carry out wholesale reforms a
commitment to such ecological reform is clearly missing.

The underdeveloped industrializing states generally
are ruled by small groups of people; whether they are
wedded to an ideology is not really critical in this
examination. They, too, regardless of the type of
system they select as a model for their regimes, attempt
to attain the all-embracing goal of modernization. And
here, too, the model they attempt to reach is the
industrial model of the capitalist states. It is a kind
of irony, that frequently the very regimes which come to
power on the basis of anticapitalist programmes attempt
to imitate the system they profess to despise. But
this irony is comprehensible if we understand that the
capitalist states have come close to achieving that
which has been one of the most elusive material goals of
mankind: abundance and an end to starvation and
deprivation. In this respect, then, it is no wonder
that the methods and means through which their success
was measured - i.e. industrialization - is being
emulated even in the underdeveloped states.

In these states, however, there is no effective counterweight to the power of the central government, no interest-groups even of such unarticulated types as exist in the USSR, which may raise voices in protest against environmental deterioration. Since the system of command from the central powers through the military to the local administration enforces industrialization regardless of its high price, little or no possibility exists to proceed cautiously and to guard the quality of the environment. The Volta Redonde, a hydro-electric dam in Brazil, or the Aswan dam in Egypt were built with small concern for the deteriorating environment and the saving of Abu Simbel was little more than a symbolic nod toward the outrage of the world community at the loss of an art treasure. The strip mining in Chile or the steel-mills sprouting up on the Latin American subcontinent are intended to improve the standard of living regardless of the high cost to the precious environment.

Philosophically we may argue about whether or not it is possible to save man from himself and to save the environment from man's greed of exploitation. But we are not dealing here so much with philosphy as with man's simple need to <u>survive</u>. Unless political means and methods are found to deal with the problem of the deterioration in our environment we may destroy not just the very quality of life to which all the progress of mankind is directed, but the political systems as well as human life itself.

SOME ECONOMIC, SPATIAL, SOCIAL AND POLITICAL INDICATORS OF ENVIRONMENTAL QUALITY OF URBAN LIFE

Chester Rapkin
Columbia University

I. THE NEED FOR URBAN INDICATORS

In a nation which is turning increasingly urban, the quality of the urban environment constitutes a major test of the level of its well-being as a society. The 243 metropolitan areas of the United States now comprise 69 percent of the nation's population and are more than ever the center of its economic and cultural life. Although planners and social scientists have long questioned the low priority given to enhancing the urban environment, growing segments of the public share their concern, recognizing glaring deficiencies in their surroundings. The national discussion about priorities in government expenditures is clearly tied to this realization and seeks to redirect government effort toward such pressing areas of domestic need as the improvement of urban development and housing, the health of the people, crime prevention, recreation and education, and environmental pollution, among others.

Without a systematic program to eliminate our slums and reconstitute our cities, the urban ambience has and will continue to lag behind less important components in the social system. Clearly, there is no question among informed persons that such a program is urgently required; rather, the current dilemma is one of formulating an effective and workable program that will contribute to the meeting of social needs. Even though political forces show signs of turning public policy in this direction, government planners lack tools for measuring the magnitude of public needs, as well as their changing characteristics over time.

At present, it is difficult to judge which problem areas are worsening, which are stable, and which are correcting themselves through regulation or through the market process, and it is thus difficult to prescribe the needed level of government intervention for each. Information is also lacking on the relationship among the different problem areas, and the likely effects that a particular treatment of one area may have on the others. In short, there is a need for more precise measures depicting social needs, indicating their direction of change, and evaluating the consequences of different types and degrees of intervention.

To be useful to policy-makers, however, these measures must be susceptible to expression in terms of explicitly stated social goals. Measuring the gap between the present level of well-being and the national goal in each problem area would be the preliminary purpose of government efforts. Regular readings of the extent to which the gap is being reduced could serve to evaluate the effectiveness of government programs. It has been particularly difficult to obtain a reasonable degree of agreement on national social goals because of wide divergence in views among the various economic, political, and class groups in our society. In the most part we have not been successful in moving beyond general statements to quantified standards,[1] except in certain specialized fields like housing, air pollution, and open space.

Economic indicators, developed by social scientists during the 1930's, offer a basis for conceptualizing measures of one dimension of urban environmental quality. These statistics have been refined, so that we now have accurate rates of employment, levels of living costs, and industrial production totals for the nation and for metropolitan areas. Census data furnish corresponding information on the demographic and housing aspects of environmental indicators. Nevertheless, these social statistics cannot as yet be combined into a consolidated system of accounts as can economic data. National income accounts summarize the status of the economy by aggregating related data into the gross national product, revealing whether the economy has grown or declined in balance. The many components of national income can be combined by utilizing a basic unit of measure--the dollar; and even though a vast variety of phenomena are incorporated, the framework is sensitive enough to reveal even a mild contraction or a modest advance.[2] Such measurements as rates of growth or decline allow analysts to assess the effectiveness of particular policies.

For all their usefulness, however, national or regional income accounts are by no means complete measures of the status of the urban environment. They do not include the social costs of production; and in fact by adding all goods and services together, actually consider both the products of a smoke-spewing factory and the resultant expense of combatting soot in the adjacent areas as contributions to national output. More importantly, national income figures touch only tangentially on most of the essential elements of an adequate urban environment, such as decent shelter, suitable community facilities and services, family stability, personal security from crime, enriching educational and cultural opportunities, swift and safe transportation, responsive government subject to local participation in decision-making, and an atmosphere of social justice.

A system of social accounts that combines these factors in a significant manner is still a distant prospect. Many of the statistics necessary for such a tally are gathered only when required by the administrative needs of agencies, and some areas of concern, such as alienation, may not even lend themselves to statistical measurement consistent with respect for privacy.[3] Moreover, a common unit of measurement is lacking among social indicators. Values play a crucial role in determining the quality of life, and measures cannot be uniformly reduced to a unit of currency, as in the case of national income statistics. It should be recognized, however, that not even all economic indicators are additive, and some must be evaluated separately because they are measured in other types of units, e.g., percent of labor force unemployed, or proportion of households earning less than half the median income. An added complication is the fact that useful statistics for social indicators are available from a wide variety of sources at varying intervals, rendering it extremely difficult to put together a comprehensive set of indicators on a regular basis. There is little in the theory of social science to indicate even a desirable frequency with which structural changes in the social pattern should be observed. Finally, social indicators still lack an organizing concept expressing basic cause and effect relationships.[4] Without a complete understanding of the environmental system in cities, including the indirect effects of individual social forces, discrete indicators cannot be weighted and added together to yield an aggregate index of urban quality. Rather than seek the unattainable goal of a perfect reading of social conditions, it would be well to adopt an empirical attitude, seeking to refine existing measures and to understand their implications better, while devising new proxy indicators of environmental factors not presently quantifiable.

Most of these observations would apply to social indicators, whether nationwide in scale or merely concerned with a single metropolitan area. It is important, however, to understand that urban indicators do have some basic differences from national social indicators.[5] Because they pertain to built-up areas limited in size, they must reflect the needs of a highly interdependent population. In general, urban dwellers are much more dependent on public services and facilities than are the rural and small-town population. The demands and strains of high density living in a climate of intense competition also severely test family stability and the psychological balance of individuals. The pace of social change is quickest in cities, and traditional forms of social organization tend to break down faster, leaving people uncertain of acceptable norms. Social pathology is the by-product of this process, and demands upon the schools, the courts, the police, the hospitals, and the welfare groups increase as people turn to impersonal public agencies for functions formerly performed by themselves, their families, or their neighbors. Thus, urban indicators must cover many areas, such as health care, crime prevention, cultural level, minority status, political participation, transportation, housing adequacy, land planning, and aesthetic satisfaction.

A fundamental purpose of a system of measures of urban environmental quality is to make comparisons among metropolitan areas possible. Each metropolitan area should be evaluated in the same categories of measurement, employing the same units as all others. The result would be an array of numbers--one for each category--which, while not subject to aggregation, could be compared category-by-category with all other metropolitan areas. A recent publication of the Urban Institute[6] has adopted this approach in analyzing the relative status of metropolitan Washington, D.C.

Urban indicators should be oriented to public policy questions. As comparative measures, their most basic use will be to guide the allocation of federal assistance funds among metropolitan areas. For the first time, it will be possible to employ objective criteria indicating the relative need of regions in each policy area. This use of urban indicators takes on special importance as the federal government considers the form which revenue sharing should take. It should be recognized, however, that even though the distribution of federal funds is increasingly becoming subject to metropolitan planning considerations, most assistance is allocated directly to state and local governments, rather than to metropolitan-wide entities. This means that sub-metropolitan indicators are needed as well, so that the central city and major suburbs can be evaluated vis-a-vis their competitors for federal funds across the country.[7] And, as Perloff has pointed out,[8] the indicators should also distinguish between residential and non-residential neighborhoods, and among types of communities like ghetto areas and middle class neighborhoods.[9] Acquiring complete data for such areas of limited size is extremely time-consuming and expensive, except when census tract information happens to coincide with the area in question. But at the scale of the central city or major suburb, such data might not by any more difficult to obtain than for the metropolitan area as a whole; in fact, the data needs of municipal agencies may generate more local data than are available on a metropolitan scale.

As indicators for each metropolitan area become available for different years, they will also be useful in disclosing which regions are meeting their problems successfully, which are stumbling along, and which are not making any headway. Government will be able to evaluate the effectiveness of its programs by checking its rate of progress in solving problems in various policy areas. Local effort will, of course, play a large part in determining the success or failure of government initiatives, and federal decision-makers will want to adjust grants-in-aid payments so that there is a maximum utilization of local resources, even if the absolute need for environmental improvement is great. Demonstrated progress will require a careful federal evaluation of the need for additional funds, depending upon the amount of local momentum and commitment, in addition to the relative status of the different policy areas and their effect upon each other. Lack of progress or retrogression may signal the need for additional federal assistance, or may indicate local disinterest or absence of effort. Clearly, simple measures of need will aid in the allocation of federal assistance, but they can never constitute more than a portion of the decision process when so many difficult-to-gauge variables are involved, and when the allocation of huge sums of public funds is at stake.

II. THE SOCIAL AND BEHAVIORAL CONTEXT

The scope of this paper differs from other discussions of the urban environment inasmuch as it is not concerned with the quality of such profoundly basic elements as water, air, and sound. Contamination of water and air has a deleterious effect upon the physical health of the individual, and if severe enough can even be lethal. Their effect is direct, indisputable, and universally accepted. The problems of noise contamination are somewhat different, being more in the realm of psychology and aesthetics than the other two, but also having direct physical effects. Once the intensity of the sound waves goes past the tolerance threshold, the function and structure of the hearing mechanism is affected and psychological disturbances are generated. In the case of all three, the problem of quality is concerned with the physical character and nature of the elements and the possibility of adverse consequences largely on the physical health of individuals in the community. It is certainly true that society would cease functioning abruptly if air and water became too contaminated for ready use. But long before this would happen, its consequences on the physical health of individuals would have become the overwhelming preoccupation of mankind.[10]

The elements with which we are concerned in this paper are also, but not entirely, physical in nature. Rather than the natural components of the environment, they involve man-made artifacts-- the homes, work places, public buildings, recreation areas, open spaces, streets and highways, and the arrangement of all of these components within the urban pattern. But in addition to the physical elements, the urban environment also subsumes a far more significant social component having to do with the way in which men have organized themselves to live in large numbers in densely concentrated geographic areas. Just as the urban physical structure has abundant variations, so does the nature of the social organization within which urban dwellers function.

How much of the social system is attributable to the entire society of which the urban area is a part, and how much of it is a function of the city itself is difficult to determine. With time, this distinction diminishes in importance as the nature of the world society becomes increasingly urbanized. What is important to explore is the extent to which the quality of the urban environment (now defined to exclude problems of contamination of water, air, and sound) influences health, behavior, and personality. Proshansky et al have summarized the two diametrically opposite psychological approaches that have sought to explain the relationship between the individual and his environment. The objective approach, epitomized by Watsonian behaviorism and developed in the more recent work by Skinner, holds that the individual responds in a fixed and discoverable manner to identifiable stimuli. In contrast, the phenomenological approach, developed by Kaffka and expanded by Lewin, holds that behavior is not a response to the stimuli of the objective world, but issues from that world transformed into an inner world by an inherently cognizing organism. Building on these extreme views, Proshansky et al assert that an individual's perceptions and behavior are not independent of the social milieu of which he is a part. The authors maintain that "from a theoretical point of view, there is no physical environment apart from human experience and social organization...and the individual's response to his physical world is never determined solely by the properties of the space and events that define it..."[11] They illustrate this point by explaining that "crowding," for example, is not simply a matter of the number of persons in a given space. Being crowded depends on many factors, including the activity going on, the people participating in it, previous experience involving numbers of people, gender of the individual, and the extent to which bodily proximity is acceptable in the given society.

The interaction between the quality of the physical environment and the social context is seen quite clearly if we accept the fact that there would be no housing problem if there were no housing standards. By and large, in the more economically advanced nations of the world, the concern about the quality of housing, work places, and other elements in the urban physical structure stems very little, at this stage of our progress, from the fact that the deficiencies constitute a menace to the physical health of the population. This is not to say that there are no slums in advanced nations like the United States, or that rat bite, lead poisoning, and pulmonary diseases are unknown, but rather that we have attained a reasonably high standard of housing, and that even slums have a level of facilities and amenities unknown in developing countries or in the western world as recently as a generation or two ago. In essence, our concern about the physical quality of our structures stems from the commentary that it offers on the equity of the distribution of the nation's economic products, on the social symbolism of housing, on matters concerned with privacy, and on amenity.

Housing standards, in effect, represent the community consensus of what each family in a given society should have if larger social goals are to be met. Over time, housing standards have continued to move upward, responding to rising incomes and advancing technology. In this process, the definition of a satisfactory housing unit has moved far beyond the level of concern with health and safety, and has concentrated more on convenience and amenity, as basically sound structures became more readily accessible to larger and larger sectors of society.

We would indeed be remiss if we did not attempt to place the effect that environment has on shaping behavior and personality in some larger perspective. What proportion of an individual's makeup is accounted for by external environment generally, and by the urban environment specifically? Among the great influences that shape man are his genetic baggage, the family in which he was reared, its composition and the relationships among its members, traumas of various sorts that he may have experienced, biochemical imbalances, or other physical pathologies that influence his life's experiences both inside and outside of his family. These experiences not only include interpersonal relationships, but also such matters as his social status, country or place of origin, and its cultural heritage and geographic location, as well as his opportunities for growth and development, his actual achievements and his feelings about them. Placed in this setting, it would seem that the urban environment qua physical structure would have a minor effect indeed as a formative influence on the totality of the individual, an intuition buttressed by the fact that there is considerable similarity in human behavior patterns the world over.

It has been commonplace, if less than scientific, to attribute the stress of modern life in substantial measure to the characteristics of the urban environment. Among the more recent studies which attempt to investigate the validity of this popular belief are explorations by the Regional Plan Association. The RPA maintains that high density has led to superficial, over-routinized experiences and sensory overload, which contributes to fatigue and frustration.[12] This is particularly apparent in "proxemic problems," what Ardrey, Hall, and others have termed "territoriality."[13] They postulate a "body buffer zone," which if violated produces anxiety. In some individuals the violation of this space causes great agitation. Kinzel has pointed out that many imprisoned for crimes of violence are far more sensitive to physical closeness than those jailed for property crimes.[14] The RPA also maintains that density at times stimulates competitive impulses, even among total strangers, which often may be gratuitous or excessive. This, in turn, intensifies the discomfort of being crowded, to which the impulse was an original response.

Urban society has also generated a substantial degree of migration and mobility, which impels a family to find a new home or new place of work, new friends, new shops, new neighborhood, new schools, if it wishes to improve its living or working conditions. For a middle-class family, such movement is dislocating indeed, but for poor families, where the rate of mobility is considerably higher, the dislocation is more profound, mitigated in part by the fact that they frequently tend to stay within the same general community. The typical worker, and more particularly the member of the poorest sector of society, depends upon his house and neighborhood not only to provide shelter and amenity, but also for safety and haven.[15]

All of these intrinsic characteristics of a large urban concentration tend to generate emotional pressures that frequently rise to intolerable levels and take their toll on the mental health of the urban population. Srole et al,[16] in a study of the apparently most stable population group of a 200-block area of mid-Manhattan, found that only 19 percent of this group showed no symptoms of mental illness; 59 percent gave evidence of mild to moderate illness, and 23 percent showed marked or severe mental illness.

These findings tended to support the suggestions that the tensions of the great metropolis were the root cause for the unbalance, until Leighton et al demonstrated that roughly the same incidence was to be found in a small rural area of Nova Scotia.[17] With mental illness so prevalent, one is impelled to ask, "if everybody is crazy, is anybody crazy?" In other words, has pathological behavior become the social norm? Has psychopathology become the accepted order of the day? To state the question in this way makes it sound ridiculous, but it does make one shudder when a prominent psychiatrist states that "the behavior of mental patients varies not in kind but in degree from that of normal individuals."[18]

Psychiatric studies by Faris and Dunham, Hollingshead and

Redlich, and Langner and Michael[19] have consistently revealed a high correlation between social class and mental illness, which provides a link between emotional stability and societal form. It is not astonishing to find that the lowest class, deprived of status, recipient of a minimum of the economy's product, and virtually devoid of power, is most susceptible to emotional difficulties leading to mental illness in their struggle for survival in every-day life.

Since each society attempts to produce the type of person that conforms with its norms, an obvious test of its effectiveness is the degree to which its environmental composition militates for or against its objectives and indeed produces people whose behavior patterns are substantially in accord or at variance with those that are acceptable and desirable. To the extent that the society produces criminality, i.e., presumably volitionally antisocial behavior, it has failed in its objectives. To the extent that it produces derangement among its people, that is, involuntary behavior of an unacceptable nature, it has indeed distorted itself violently in a manner that defeats its ultimate purposes. By the same token, a measure of success may be found in the degree to which society has reduced or eliminated the frictions and irritations of daily life, has provided acceptable alternatives to destructively aggressive behavior, and has bred a people capable of coping with the inevitable interactions, the stresses and tensions, as well as the major and minor emergencies that constitute the components of human existence wherever man finds himself.

III. A PROPOSAL FOR URBAN INDICATORS

Social reporting on the urban environment as a measure of the level of well-being and the effectiveness of public programs is closely related to the systems analysis approach in urban planning. Both use mathematical methods to evaluate progress in moving toward an explicitly stated goal.[20] In systems analysis, outcomes of alternative courses of action are compared in terms of some criteria of desirability, so that the optimum solution can be chosen. The Planning-Programming-Budgeting System (PPBS), the systems technique now used throughout the federal government, first proved itself useful in reorganizing the budget of the Defense Department. Its effective use in the programs of civilian agencies is more limited because federal agencies usually share authority with state and local governments, as well as with the private sector. Rational policy-making for domestic needs requires a deeper analysis of the many forces affecting the course of society, and social indicators are a logical extension of the systems approach. Through the quantification of strategic variables they attempt to bring a higher degree of precision to policy analysis.

Urban indicators should be conceived in clusters relating to broad policy areas. Not only is it uncomplicated to think of measures according to demographic, social, economic, political, and spatial changes, but indicators for each of these areas carry with them an implicit goal structure.[21] While related to each other, the goals for each form natural analytical boundaries. As we discuss each area, its goal orientation will emerge, even though our purpose here is not to set performance targets. Such numerical criteria will have to be determined administratively later, but they will undoubtedly be based upon existing standards in many areas of need. Once indicators for different years become available, the change registered in specific policy areas would provide the basis for a program review in each, directed toward deciding whether resources might be used more productively in other ways. This policy analysis, based upon the demonstrated effectiveness of programs in attaining stated goals, would go beyond the purpose of indicators and would call for a PPBS treatment on a government-wide scale.

We propose, then, that urban indicators be structured in the five major analytical areas mentioned above, so that they are closely tied to substantive problems and avoid merely recounting the expenditures and activities of government.[22] Some of these areas, particularly the demographic and economic, are already subject to sophisticated measures, while others, like social and political indicators, still require more understanding even to devise adequate proxy measures. Our intention here is to summarize the existing level of measurability in each area and suggest new directions for social research in quantifying urban problems.

Demographic Measures

Population counts are the oldest indicator of the state of the nation. They were considered so important that they were required decennially in the original Constitution. The census provides us with basic data on number of persons for the nation as a whole and for small subareas by age, sex, and race.[23] Over the years, the scope of the census has been broadened considerably beyond a simple headcount, and information is now also available on many aspects of demographic change. Rates of national increase (births minus deaths), household formation rates, and the proportion of the population comprised of elderly, minority, and impoverished persons all provide detailed information on the shifting pattern of need in metropolitan areas.

Population growth and decline, as well as changes in its composition, are the most fundamental factors in determining needed government programs and facilities. Demographic data from census to census indicate the magnitude and direction of change, and paint the basic picture of need in each region. Intercensal demographic estimates have become available in recent years, allowing policy-makers even more flexibility in monitoring change. And the expansion of the census into housing, personal income, and business activity has laid the groundwork for measurement in the other proposed policy areas. Nevertheless, additional information is needed on migration flows, neighborhood segregation, and household characteristics.[24] If demographic measures are to maintain currency, more frequent censuses must be undertaken, and it is therefore essential that proposals for a mid-decade census be adopted. If this is not possible, more complete intercensal estimates will be required, especially for subareas of metropolitan regions where rapid change is taking place.

Economic Measures

After population data, economic measures are our most complete component of urban indicators. There are two main categories of economic data that must be considered in rating the well-being of a metropolitan area: the income and assets of the individual, family and household, and the status of the region's economy.

Income is the most meaningful measure of a person's economic status, and median figures are reported by the census for regions and their subareas. The income distribution, which can be computed from available information, sheds light on the degree of social stratification in an area, and the relative hardships experienced by the underprivileged. Data on occupational status are also essential in establishing a region's level of well-being, especially for minority groups. Census data permit not only an examination of status at specific moments in time, but also indicate social mobility over time. Employment rates and proportion of households receiving welfare payments are additional important indications of the local economic conditions for which data are available. More complete information, however, is needed to determine the level of personal assets in local areas.

Economic base analysis offers a tested conceptual framework for rating the relative strength of metropolitan areas as productive units. This type of analysis divides the local economy into its constituent parts, disclosing the degree of importance of each, and their individual advances or declines, These data are useful in providing measures of the level of employment opportunity present in a metropolitan area and its general growth outlook. A region's economic role in the national scene can also be analyzed from data showing whether it imports or exports capital. Cost of living data are an essential overall evaluation of a region's economic impact on the individual household. Additional study is needed to determine the likely effect on minorities of trends pointed up by these measures.

Spatial Measures

Since measuring of pollution has been more than adequately covered at this conference by others, the spatial components of our proposed urban indicators will concentrate on housing, community facilities, and aesthetic satisfaction.

Housing has both an individual and community dimension. The options of individual households are limited by rent levels, the availability of mortgages, prevailing housing conditions, and their own household size. Their ability to satisfy their needs will depend upon the general availability of housing in the region and the rate of housing production. If the supply is limited, crowding and lack of privacy may result, and people may be forced to settle for housing below the national standard set by the federal government. The census and other data record these characteristics regularly, and their figures would go far in indicating the adequacy of public and private efforts to shelter the population.[25] The prevalence of discrimination remains difficult to gauge, however, and more research is required to test such techniques as the "segregation index" proposed by the Taeubers.[26]

Urban planning has devised yardsticks to measure the need for schools, parks, playgrounds, hospitals, museums, and other community facilities.[27] These standards are based upon local population, from the neighborhood to the region-wide scale, but they must be modified according to the special demographic characteristics of each area. For example, an area with an extensive elderly population would not require the typical allocation of playground space, but may need a greater than average number of hospital beds. Census data can be used to modify standards for local population needs as subarea goals are set.

The adequacy of local transportation facilities is a major spatial indicator of urban quality. Criteria of measurement are the choice of modes, speed, amenity, and comprehensiveness of coverage. Most of these factors would require the development of new measurements, but one helpful overall indicator might be the average time spent in the journey to work. At the present time, regional transportation studies make it possible to compare the efficiency of local movement systems by utilizing their data on the amount of time spent in the home-to-work trip.

The aesthetic quality of a metropolitan area has only recently become a vital public concern. Design amenities in the townscape are essential to the enrichment of life and serve in themselves to create an awareness of physical beauty. On a more functional level, an orderly environment is necessary if the individual is to be able to orient himself as he moves from place to place in our extensive metropolitan areas. Analytical tools to measure urban design quality are conspicuously lacking, save for intuitive judgments of experts. Lynch, Appleyard, and Myer[28] have supplied us with the basic terminology for a more precise system of measurement, but the few existing case studies must be expanded into an extensive body of research if progress is to be made in this area.

Social Measures

Social aspects of the urban environment are perhaps the most difficult factors to measure, and most often must be dealt with in terms of their symptoms. For example, rates of mental illness, suicides, and drug addiction are typically used to indicate the prevalence of anomie and inability of individuals to adjust to urban pressures.[29] Public order is usually measured by a variety of crime rates including robberies. Family stability is determined by the proportion of female-headed households and the rate of divorce and separation. Educational and cultural levels are indicated by census readings of average number of years completed in school and by national educational honors awarded to residents of the area in question. The social status of minorities is rated according to their representation among the high officials of government and business and in the professions. These surrogate indicators may have to serve temporarily until we can better grasp the interrelationships in the social sphere of the urban environment. Intensive research into more sophisticated measures of our social status deserves high priority in the social sciences.

Political Measures

Although in part codified by law, the urban political process is extremely difficult to measure qualitatively, largely because of the considerable amount of discretion given to administrators in determining procedures and priorities. There are four major areas, though, by which we can begin to measure the adequacy of government. First, the responsiveness of government to public needs can be considered a function of its expenditures per capita for a selected list of high priority purposes. Urban governmental effort, mentioned earlier, can be measured by the degree to which municipal income depends on taxation and the extent to which it is supported by grants-in-aid from higher levels of government. Second, public participation in government can be judged by the proportion of the eligible electorate which has registered and votes. In addition, there is a need for a broad-ranging measure of the strength of citizens groups and public hearings in the decision-making process. Third, local commitment to social justice is associated with laws forbidding discriminatory practices, and the effectiveness of the mechanisms that provide for their enforcement. Fourth, concern as to the future condition of the physical environment can be evaluated according to the quality of the area-wide planning process in the region. If it has an able technical staff supported by strong political backing, there is a good chance that the planning process will contribute to physical development characterized by high standards of spatial arrangement and satisfaction of social needs.

As with the proposed social indexes, these political indicators are still many years away from a level of sophistication comparable to demographic and economic measures. But they can be of use by stimulating additional thought on the subject of social reporting; in their estimates of need they are capable of alerting the general public and its political leadership to the magnitude of the environmental challenge that lies ahead in our cities.

IV. CONCLUSION

At this point in the development of urban indicators, it is still too early to expect an aggregate measure of urban environmental quality. In former years, social scientists were eager to develop the aggregate entity which is now known as national income, so that a summary statistical measure of the economy would be available. Only after this achievement was accomplished did we realize that the most useful part of the national income accounts were its components. Readings on the changes in the status of important industries and in the components of consumer expenditures became crucial indicators for policymakers. Similarly, with social indicators we should not be too frustrated by the great difficulty in arriving at an aggregate figure. Some of the measures proposed here have already begun to tell us whether given sectors of the urban environment meet current social needs, or whether they are past the possibility of adapting to emerging life styles. As one example, simple figures showing whether size and type of households are coordinated with the characteristics of the standing stock of dwelling units can offer an early warning of growing need for investment in housing. In terms of the perceptions to be gained through social reporting, the sum of the parts is clearly much greater than the whole.

Footnotes

[1] For example, see Goals for Americans, the report of the President's Commission on National Goals, Spectrum, 1960.

[2] Mancur Olson, Jr., "The Purpose and Plan of a Social Report," The Public Interest, No. 15, Spring 1969, p. 86.

[3] Eleanor Bernert Sheldon and Wilbert E. Moore, "Monitoring Social Change in American Society," Indicators of Social Change, edited by Sheldon and Moore, Russell Sage Foundation, 1968.

[4] Harvey S. Perloff, "A Framework for Dealing With the Urban Environment: Introductory Statement," The Quality of the Urban Environment, ed. Perloff, Resources for the Future, 1969, p. 16.

[5] Statement by Nathan Glazer, hearings before the Ad Hoc Subcommittee on Urban Growth of the Committee on Banking and Currency on the Quality of Urban Life, Washington, D.C., November 17, 1969, p. 400.

[6] Martin V. Jones and Michael J. Flax, The Quality of Life in Metropolitan Washington, D.C.: Some Statistical Benchmarks, The Urban Institute, Washington, D.C., March 1970.

[7] For an example of a set of urban indicators applying to a central city, see The Quality of Life in Urban America, New York City: A Regional and National Comparative Analysis, Vol. I and II, ed. John Berenyi, Office of Administration, Office of the Mayor, New York City, May 1971.

[8] Perloff, op. cit., pp. 17-18.

[9] For an example of indicators applying to neighborhoods within a city, see Urban Level-of-Living Index, Community Renewal Program, Department of City Planning, Pittsburgh, Pa. 1965.

[10] George L. Engle has classified the external injuries that strain the individual into four general categories: physical injury, insufficient food, water, air, etc., infection by pathogenic organisms, and psychological stress. Psychological Development in Health and Disease, W.B. Saunders, Philadelphia, 1962.

[11] Harold M. Proshansky, William H. Ittelson, Leanne G. Rivlin "The Influence of the Physical Environment on Behavior: Some Basic Assumptions," Environmental Psychology, Man and His Physical Setting. Holt, Rinehart and Winston, 1970, pp. 8 and 28. The works by Watson, Kaffka, and Lewin are referred to in Proshansky et al. The latest study by B.F. Skinner is entitled Beyond Freedom and Dignity, Alfred A. Knopf, New York, 1971.

[12] Regional Plan Association, "Human Responses to Patterns of Urban Density," New York, August 20, 1969.

[13] Robert Ardrey, The Territorial Imperative, Dell Publishing Co., New York, 1966. Edward T. Hall, The Hidden Dimension, Doubleday & Co., New York, 1966.

[14] Augustus F. Kinzel, "Body-Buffer Zone in Violent Prisoners," presented at American Psychiatric Association Annual Meeting, May 1969.

[15] Lee Rainwater, "Fear and House as Haven in the Lower Class," AIP Journal, Jan. 1966; Marc Fried, "Grieving for a Lost Home," Leonard J. Duhl (ed.), The Urban Condition: People and Policy in the Metropolis, Basic Books, 1963, and "Transitional Functions of Working Class Communities: Implications for Forced Relocation," Mildred B. Kantor (ed.), Mobility and Mental Health, Charles C. Thomas, 1963.

[16] Leo Srole et al, Mental Health in the Metropolis: The Mid-Town Manhattan Study, Vol. I, McGraw-Hill Book Co., New York, 1962. Thomas S. Langner and Stanley T. Michael, Life Stress and Mental Health: The Mid-Town Manhattan Study, Vol. II, Free Press of Glencoe, 1963.

[17] D.C. Leighton et al, The Sterling County Study, Basic Books, 1963.

[18] James P. Page (ed.), Approaches to Psychopathology, Temple University Publications, 1966, p. 6.

[19] R.E.L. Faris and W.H. Dunham, Mental Disorders in Urban Areas-An Ecological Study of Schizophrenia and Other Psychoses, University of Chicago Press, 1939. August B. Hollingshead and Fredrick C. Redlich, Social Class and Mental Illness-A Community Study, John Wiley & Sons, 1958. Langner & Michael, op.cit.

[20] Olson, op. cit. pp. 93-95.

[21] For a general discussion of an alternative urban indicator system divided along individual, family, and institutional lines, see Daniel P. Moynihan, "Urban Conditions: General," The Annals, Vol. 371, May 1967, pp. 159-177.

[22] Olson, op. cit., p. 90.

[23] For a full discussion of the relevance of demographic information to social indicators, see Conrad Taeuber, "Population: Trends and Characteristics," Indicators of Social Change, op. cit., pp. 27-73.

[24] For an extensive examination of information needed for a system of demographic indicators, see Grace Milgram and Robert W. Ponte, Demographic Indicators of Metropolitan Problems: A Suggested Framework for the General Guidance of H.U.D. Program Emphasis, report prepared for U.S. Dept. of Housing and Urban Development by Institute of Urban Environment, Columbia University, New York, October 1969.

[25] For a discussion of additional data needs for housing indicators, see Chester Rapkin and Robert W. Ponte, Indicators of Metropolitan Housing and Residential Construction Needs; A Suggested Framework for the Allocation of Future H.U.D. Assistance, report prepared for U.S. Dept. of Housing and Urban Development by Institute of Urban Environment, Columbia University, New York, June 1970.

[26] Karl E. Taeuber and Alma F. Taeuber, Negroes in Cities, Aldine Publishing Co., Chicago, 1965.

[27] F. Stuart Chapin Jr., Urban Land Use Planning, University of Illinois Press, Urbana, 1965.

[28] Kevin Lynch, The Image of the City, MIT Press, 1960; Kevin Lynch, Donald Appleyard, and John R. Myer, The View From the Road, MIT Press, 1964.

[29] Jones and Flax, op. cit., p. 49.

TERATOLOGICAL HAZARDS DUE TO PHENOXY HERBICIDES AND DIOXIN CONTAMINANTS

Samuel S. Epstein

Case Western Reserve University School of Medicine, Cleveland, Ohio

1. Teratogenicity as a Public Health Hazard

Potential hazards posed by environmental pollutants and drugs include toxicity per se, carcinogenicity, mutagenicity, and teratogenicity. Although teratogenic effects of various agents have been recognized for several decades, it was only as a reaction to the thalidomide episode that a requirement for teratogenicity data became established in 1962.

Teratology is the study of congenital malformations. These are generally defined as structural abnormalities, recognizable at or shortly after birth, which can cause disability or death. Less restrictedly, teratology, also includes microscopical, biochemical, and functional abnormalities of prenatal origin.

Congenital malformations pose incalculable personal, familial, and societal stresses. The financial cost to society of one severely retarded child, computed on the basis of specialized training and custodial care alone, approximates to $250,000 (Oberle, 1969); this estimate excludes further costs due to deprivation of earnings. In the absence of a comprehensive national surveillance system, the precise overall incidence of congenital malformations is unknown and has been variously estimated as ranging between 3-7% of total live births.

Three major categories of human teratogens have been identified -- viral infections, x-irradiation, and chemicals, e.g., thalidomide, mercury, and diethylstilbestrol.

2. Methods for Teratogenicity Testing

Teratogenic effects of chemicals and other agents should of course be prospectively identified in experimental animals, rather than in human beings following accidental or unrecognized exposure. Test agents should be administered to pregnant animals during active organogenesis of their developing embryos; administration after organogenesis is complete is also required to test for transplacental carcinogenicity. Shortly before anticipated birth, embryos should be harvested by caesarean section and examined. Parameters to be considered in test and concurrent control animals should include the incidence of abnormal litters, the incidence of abnormal fetuses per litter, the incidence of specific congenital abnormalities, the incidence of fetal mortality, maternal weight gains in pregnancy, and maternal and fetal organ/body weight ratios. Additionally, some pregnant animals should be allowed to give birth in order to identify abnormalities that may otherwise manifest only in the perinatal period or even in adult life.

Chemicals and their known metabolites should be administered to 2 or more mammalian species, under various nutritional conditions, during active organogenesis, and by a variety of routes reflecting possible human exposure. Of interest in this connection is the lack of available data on teratogenicity testing by the respiratory route; respiratory exposure is particularly important for pesticide aerosols and vapors. Agents should be tested at higher dose levels than might be anticipated in humans following high level accidental exposure, as well as following extensive low level exposure. This is essential to attempt to reduce the insensitivity of conventional test systems, based on very small numbers of animals, compared with the millions of humans at presumptive risk (HEW Commission, 1969; Epstein, 1970). To illustrate this further, let us assume that at human exposure levels a pesticide induces teratogenic effects or cancer in as many as 1 out of 10,000 humans, then the chances of detecting this in test groups of less than 50 rats or mice exposed at these actual levels would be very low. Indeed, many more than 10,000 rats or mice, depending on their spontaneous incidence of teratogenic effects or cancer, would be required to demonstrate one statistically significant event, if we assume that rats and humans have similar sensitivity to the teratogen or carcinogen being studied; of course, humans may be less or may be more sensitive than test animals. Meclizine for example, is teratogenic in the rat, but not apparently in the restricted number of humans studied (King, 1965; Yerushalamy & Milkovich, 1965). With thalidomide conversely, the lowest effective human teratogenic dose is 0.5 mg/kg/day; corresponding values for the mouse, rat, dog, and hamster are 30, 50, 100 and 350 mg/kg/day (Kalter, 1968). Thus, humans are 60 times more sensitive than mice, 100 times more sensitive than rats, 200 times more sensitive than dogs, and 700 times more sensitive than hamsters. Clearly, attempts to determine a safe level for thalidomide, based on animal teratogenicity data, would clearly expose humans to significant teratogenic hazards. Accordingly, it is routine practice to test for teratogenicity and carcinogenicity at a range of concentrations, including those higher than human exposure levels, and extending to maximally tolerated doses (MTD). Even at MTD levels, administered to mice from day 7 of life until sacrifice at 18 months, less than 10% of the 140 pesticides tested in the recent Bionetics study were shown to be carcinogenic (HEW Commission Report, 1969).

The Advisory Panel on Teratogenicity (HEW Commission Report, 1969) states unambiguously ... "Pesticides should be tested at various concentrations including levels substantially higher than those to which the human population are likely to be exposed". The report also emphasizes the insensitivity of standard test systems imposed by the relatively insufficient numbers of litters conventionally tested. The report further states ... "Thus, compounds showing no increase (in birth defects) cannot be considered non-teratogenic".

Epidemiological surveys of human populations may provide post hoc information on geographical or temporal clusters of unusual types or frequencies of malformations following exposure to undetected or untested teratogens in the environment. However, logistic considerations, quite apart from inadequate current surveillance systems, limit the utility of this approach. It should be emphasized that no major known human teratogen, such as X-rays, German measles, mercury, or thalidomide, has been identified by retrospective epidemiological analyses, even in industrialized countries with sophisticated medical facilities. Prospective epidemiologic surveys on agents previously shown or suspected to be teratogenic, by experimental studies or by retrospective population surveys, are clearly inappropriate.

3. The Bionetics Studies on Teratogenicity of 2,4,5-T

Bionetics Research Laboratories, Inc., of Litton Industries, under a contract from the National Cancer Institute, tested 48 pesticides, including 2,4,5-T and related phenoxy herbicides for teratogenic effects during 1965-68. Although the Bionetics studies were originally designed for purposes of large-scale screening, 2,4,5-T was tested more extensively than any other pesticide. The Bionetics Research Laboratory Report has never been publicly released, and in fact was only made available to the Advisory Panel on Teratogenicity of Pesticides of HEW Secretary Finch's Commission on Pesticides and Their Relationship to Environmental Health -- which inter alia was charged with evaluating all data on 2,4,5-T and making appropriate regulatory recommendations -- after the Chairmen indicated their intention to otherwise dissolve the panel. A revised and detailed statistical analysis of the Bionetic data was summarized by the Advisory Panel on Teratogenicity of Pesticides (HEW Commission Report, 1969).

2,4,5-T was tested on repeated occasions from 1965-1968 in 3 strains of mice and in one strain of rats by subcutaneous and/or oral administration, over a dose range from 4.6 - 113 mg/kg. The total numbers of litters tested at each dose level, by each route in all strains and species, excluding C3H mice in which only 1 litter was tested, are listed in Table I.

TABLE I

Test Animals	Dose mg/kg	Number of Litters in Test Groups	
		Oral	Subcutaneous
BL 6 mice	21.5	--	6
	46.4	6	--
	113.0	12	35
BL6/AK mice	113.0	--	13
AK mice	113.0	7	14
Sprague Dawley rats	4.6	8	--
	10.0	7	--
	21.5	8	--
	46.4	8	--

As can be seen, the bulk of the data was obtained with BL 6 mice. Due to control variability, the BL 6 data have been considered for 3 time intervals (Table II).

TABLE II*

Dose mg/kg	% Abnormal Fetuses (BL 6 mice)				
	Subcutaneous administration				Oral administration
	Prior to Sept.'66	Sept.-Nov.'66	Nov.'66-Aug.'68	Undated	Nov.'66-Aug.'68
21.5	7	--	--	--	--
46.4	--	--	--	--	37**
113.0 (Exper.1)	--	74*	31	--	67*
113.0 (Exper.2)	--	--	26*	--	--
113.0 (Exper.3)	--	--	--	78*	--

*2, 4, 5-T administered from days 6-14 or 9-17, and mice sacrificed on day 18 of pregnancy.
** Statistically significant increase compared with controls

TABLE III*

Subcutaneous Dose mg/kg	% Abnormal Fetuses (BL 6/AK mice)
113.0	39**

* 2,4,5-T administered from days 6-14, and mice sacrificed on day 18 of pregnancy.
** Statistically significant increase compared with controls.

TABLE IV*

Dose mg/kg	% Abnormal Fetuses (AK Mice)		
	Subcutaneous administration		Oral administration
	Prior to Nov. '66	Nov. '66- Aug. '68	Nov. '66 - Aug. '68
113.0	31**	36**	49**

* 2,4,5-T administered from days 6-15, and mice sacrificed on day 19 of pregnancy.
** Statistically significant increase compared with controls

TABLE V*

Oral Dose mg/kg	% Abnormal Fetuses (Sprague Dawley Rats)
4.6	39**
10.0	78**
21.5 (Exper. 1)	90**
21.5 (Exper. 2)	92
46.4 (Exper. 1)	100**
46.6 (Exper. 2)	75

* 2,4,5-T administered from days 10-15, and rats sacrificed on day 20 of pregnancy
** Statistically significant increase compared with controls

Major abnormalities induced in mice were cleft palates and cystic kidneys, and in rats, cystic kidneys, and gastro-intestinal haemmorrhages; increased fetal mortality was generally concomitant with these abnormalities. It is of particular interest that 39% abnormal embryos with cystic kidneys were seen in rats even at at the lowest dose tested; thus, the no effect level was not reached at 4.6 mg/kg. With reference to the rat data, the Bionetics report states (Bionetics, 1969): "It seems inescapable that 2,4,5-T is teratogenic in this strain of rats when given orally at the dosage schedules used here. These findings lend emphasis to the hazard implied by the results of studies on mice". This report further states: "These results imply a hazard of teratogenesis in the use of this compound. The problems of extrapolation preclude definition of the hazard on the basis of these studies, but its existence seems clear".

More refined and appropriate additional statistical analyses of these data were presented and discussed by the Advisory Panel on Teratogenicity (HEW Commission Report, 1969). As quoted below, these analyses are clearly confirmatory of the original conclusions of the Bionetics report on the teratogenicity of 2,4,5-T. "Tested more extensively than other pesticides, 2,4,5-T was clearly teratogenic as evidenced by production of statistically increased proportions of litters affected, and increased proportions of abnormal fetuses within litters in both DMSO and honey for both C57BL/6 and AK mice. In particular, cleft palate and cystic kidneys were significantly more prevalent. In addition, a hybrid strain resulting from a C57BL/6 female and AK male showed significant increases in anomalies, in particular cystic kidney, when administered at 113 mg/kg of body weight in DMSO .... Additionally, 2,4,5-T was tested in Sprague-Dawley rats. When given orally at dosages of 4.6, 10.0 and 46.4 mg/kg, from days 10 through 15 of gestation, an excessive fetal mortality, up to 60 percent at the highest dose, and high incidence of abnormalities in the survivors was obtained. The incidence of fetuses with kidney anomalies was three-fold that of the controls, even with the smallest dosage tested".

4. Recent Studies on Teratogenicity Testing of Relatively Pure 2,4,5-T

In view of the fact that the Bionetics study was conducted with a sample of 2,4,5-T which was subsequently shown to contain high concentrations, 27 ppm, of a tetrachlorodioxin contaminant, testing has been recently repeated by Dow Chemical Co., NIEHS, and the FDA, with relatively pure samples containing less than 1 ppm of this particular dioxin. The results of these 3 studies were presented at an FDA conference of 2/24/70; the Dow Chemical Co. data were additionally presented at the 9th annual meeting of the Society of Toxicology, Atlanta, 3/17/70. As can be seen from the data summarized below, purified 2,4,5-T is teratogenic in 3 species -- rats, mice, and hamsters.

a. Dow Chemical Co. Studies

2,4,5-T with 0.5 ppm dioxin was tested in pregnant rats by repeated oral administration at doses of 1, 3, 6, 12, and 24 mg/kg; the maximal dose tested was only 24 mg/kg. No embryo deaths or weight losses were noted within the limited dose range tested.

However, at 24 mg/kg there was a 7-fold increase in the incidence of fetuses with defective ossification of the 5th sternebra, which was inexplicably discounted; poor sternebral ossification was noted in 4/103 control fetuses, and in 29/103 fetuses of 2,4,5-T treated groups (Emerson et al, 1971). Defective sternebral ossification has been described in the rat as an expression of the teratogenic effects of drugs such as protamine zinc insulin and tolbutamide (Lichtenstein et al, 1951; Dawson, 1964).

b. Studies at The National Institute of Environmental Health Sciences (NIEHS)

Using the purest sample of 2, 4, 5-T, made available by Dow Chemical Co., teratogenic effects were induced in Swiss-Webster mice. Cleft palates were noted at dose levels of 150 mg/kg, and scattered abnormalities at 100 mg/kg; the cleft palate incidence in control mice was essentially zero. Subsequent studies showed that 2 relatively pure samples of 2, 4, 5-T were teratogenic in 3 strains of mice, but not in the rat, producing cleft palates and kidney anomalies (Courtney & Moore, 1971).

c. Studies at The Food and Drug Administration (FDA)

Hamsters were injected with 5 doses of 100 mg/kg/day of various batches of purified 2, 4, 5-T between days 6-10 pregnancy. In one of these studies, there was a 66% incidence of mortality in 50 fetuses. Of the surviving fetuses, 17% had congenital abnormalities -- crooked tail, missing limb, and defect in skull fusion. No data were presented on possible effects induced by doses less than 100 mg/kg. Of additional interest was the report, also presented at the same conference, that purified 2,4-D produced a 22% incidence of congenital abnormalities in hamsters at a dose level of 100 mg/kg/day.

5. Stability and Persistence of 2, 4, 5-T

Use of 2, 4, 5-T at rates of U.S. recommended application result in measurable residues in water, air, plants and animals which persist for relatively short periods (PSAC, 1971); residues may, however, persist as long as 3 months in soil, during which time they are potentially available for transport to non-target areas. Most water contamination in the U.S. is probably due to an aerial drift from target areas; residues of 70 ppb have been reported in Oregon surface waters (Tarrant & Norris, 1967). Much higher residues are likely following very heavy aerial applications, as in Vietnam (PSAC, 1971). Ecological and teratological considerations apart, the genetic implications of such persistent residues have been indicated by recent reports on in vivo chromosome breakage in early embryogenesis of Drospohila melanogaster (Darving & Sunner, 1971).

6. Dioxins
a. Toxicity

Rabbit ear skin is highly sensitive to dioxins, repeated application of which can produce chloracne, as a cumulative manifestation of local toxicity. Approximately 0.3 μg of the tetra isomer will produce a positive response; ">10 µg on a (surface) wipe sample indicates acute hazard (to man)". (Silverstein, 1970).

The acute oral LD_{50} of tetradioxin in male guinea pigs is 0.5-1.0 µg/kg, and in male and female rats, 22.5 and 45 µg/kg, respectively. Feeding dioxin-containing chicken edema factor diets produced cumulative toxicity in monkeys (Allen & Carstein, 1967). Storage of hexa, hepta, and octa isomers, as identified by GLC, has been reported in chickens and rats fed chicken edema factor diets (FDA, 1970). Chronic administration of 2,4,5-T or 2,4-D to dogs produces toxicity with gastrointestinal haemorrhages, suggestive of cumulative dioxin effects (Drill & Hiratska, 1953).

b. <u>Teratogenicity and Embryotoxicity</u>

The tetradioxin isomer was fed to Sprague Dawley rats between days 6-15 of pregnancy, over a dose range from 0.03 to 8.0 µg/kg/day. There was a marked increase in resorption sites at the 2 µg level. Gastrointestinal haemorrhages occurred dose-dependently over a range from 0.125 to 8 µg. Additionally, at the 0.125 µg/kg level there was a decrease in male fetal weights (Sparschu et al, 1971). It should be emphasized that cystic kidneys were not seen at the 0.125 µg/kg dose of the tetra isomer or at even higher levels. In the Bionetics study, 2, 4, 5-T at 4.6 mg/kg, containing 25 ppm of the tetradioxin isomer equivalent to 0.124 µg/kg, produced a 30% incidence of congenital abnormalities with cystic kidneys. There is thus a clear discrepancy between the teratogenic effects of 2, 4, 5-T containing 25 ppm dioxin, and the effects of the equivalent concentration of the same dioxin. It is, however, conceivable that this discrepancy may reflect synergistic interactions between dioxin and 2, 4, 5-T.

Recent NIEHS studies have confirmed the teratogenicity of tetradioxin, in 3 strains of mice and one strain of rats, producing renal abnormalities and cleft palates (Courtney & Moore, 1971); dioxin effects were not potentiated by concomitant administration of 2, 4, 5-T.

c. <u>Inadequacies of Toxicological Data</u>

There are no data in the available literature on the carcinogenicity or mutagenicity of any positional isomers of dioxins. Recent studies on the teratogenicity of dioxins have been largely restricted to the 2, 3, 7, 8-isomer; there are no available teratogenicity data on most of the other positional isomers. There are no experimental data on behavioral or psychopharmacological effects due to dioxins; this would be of interest in view of the possible psychiatric effects described in humans exposed to dioxins (Kimmig & Schulz, 1957). There are no data in the available literature on any toxicological studies on any dioxin isomer following acute or chronic administration by inhalation.

The extreme inadequacy of toxicological data on any dioxins clearly precludes consideration of potential human hazards due to exposure in air, food or water, and consideration of possible safety margins.

d. Dioxins as Contaminants of 2, 4, 5-T, and Other Phenoxy Herbicides, and Polychlorophenyls

Currently used 2, 4, 5-T formulations contain about 5% of known impurities, largely polychlorophenols. Analytic data on a sample of 2, 4, 5-T (Dow data, production batch 120449) are given in Table VI.

2, 4, 5-T is synthesized from 1, 2, 4, 5-tetrachlorobenzene which is reacted with methanol and caustic soda under high temperature and pressure to yield the sodium salt of 2, 4, 5 trichlorophenol; tetradioxin is probably formed at this stage by condensation of 2 molecules of 2, 4, 5-trichlorophenol. The aqueous trichlorosodium phenoxide is next reacted with chloroacetic acid under mildly alkaline conditions and the product is then acidified with sulfuric acid to produce 2, 4, 5-T. Polychlorophenols can be easily converted to corresponding dioxins (Cowan, 1970) in processing of fats to feeds and food, during rendering, treatment of soapstock and synthesis of phenoxy herbicides and a wide range of other biocides. As discussed below, pyrolysis of 2, 4, 5-T and Silvex produces high dioxin yields (Buu-Hoi et al, 1971; Buu-Hoi, 1972).

TABLE VI

Composition of 2, 4, 5-T Formulation	Percent
2, 3, 7, 8-tetrachlorodibenzo-p-dioxin	(0.5 ppm)
2, 6-dichlorophenoxyacetic acid	< 0.02
2, 5-dichlorophenoxyacetic acid	0.42
2, 4-dichlorophenoxyacetic acid	0.05
2, 3, 6-trichlorophenoxyacetic acid	0.55
2, 4, 6-trichlorophenoxyacetic acid	< 0.1
Bis-(2, 4, 5-trichlorophenoxy) acetic acid	0.4
3 isomers of dichloromethoxyphenoxyacetic acid	?.9
2, 4, 5-trichlorophenol	0.23 (max.)
sodium chloride	0.035
2, 4, 5-trichlorophenoxyacetic acid	Balance

There are no available data on the presence and concentration of the more than 60 positional isomers of dioxin, other than the 2, 3, 7, 8-tetrachlorodioxin isomer, in this batch of 2, 4, 5-T, or in other batches produced for food crop or other purposes in the U.S., and for export or use in Vietnam. In view of the relatively high concentration of polychlorophenolic impurities in 2, 4, 5-T, it is likely that a wide range of dioxins are also present. 2, 4-D and other phenoxy herbicides are similarly chemically uncharacterized. The higher positional dioxin isomers, hexa, hepta and octa, have been identified in 2, 4-dichlorophenol, a precursor of 2, 4-D.

e. Dioxins as Pyrolytic Products of 2, 4, 5-T and 2, 4, 5-TP

Apart from the presence of dioxins in polychlorophenols, heating of polychlorophenols will produce additional and very high yields of dioxin. Illustratively, heating 5 g of pentachlorophenol at 300°C for 12 hours yielded 1.5 g of the octadioxin isomer (Cowan, 1970). While improved production techniques may well reduce the levels of polychlorophenols and the levels of the 2, 3, 7, 8-dioxin isomer, apart from other isomers, in 2, 4, 5-T and other phenoxy herbicides, the degree to which this is practical does not yet appear to have been clearly defined.

More recently, high yields, ca 10%, of the tetradioxin have been reported following pyrolysis of purified 2, 4, 5-T (Buu-Hoi et al, 1971); additonally, yields in the order of 1% were found following pyrolysis of Silvex, 2, 4, 5-TP (Buu-Hoi, 1972).

f. Stability and Persistence of Dioxins

The extent of usage of 2, 4, 5-T and other phenoxy herbicides on food crops, and for other purposes in the U.S. and abroad, dictates the scale of resulting environmental contamination with 2, 3, 7, 8-dioxin and other isomers. The following data (Agricultural Economic Report, 1968) are illustrative:

TABLE VII

PRODUCTION AND USAGE ON CROPS OF PHENOXY HERBICIDES IN 1964 IN THE U.S.

Phenoxy Herbicides	Production (lbs.)	Use (lbs.) on crops in 48 States
2, 4-D	54,366,000	29,687,000 (i.e., 63%* of total production)
2, 4, 5-T	12,963,000	979,000 (i.e., 13%* of total production
MCPA (Methylchlorophenoxyacetic acid)		1,454,000
Other		684,000
		32,804,000

* The balance was largely used for brush control in rangelands, pastures, and rights-of-way, in aquatic weed control and in forestry.

These data reflect deliberate applications of phenoxy herbicides to crops, and ignore unintentional crop contamination following the more extensive application of herbicides for brush control or other purposes. There are no available data on the extent of such unintentional contamination. It is, however, well known that phenoxy herbicide dusts may drift for miles, even on non-windy days, following routine application (Federal Register, 1969); the concentration of phenoxy herbicides in the air in Washington in 1964 reached a maximum of 3.4 $\mu g/m^3$, with an average of 0.045 $\mu g/m^3$ (Bamesberger & Adams, 1966). As indicated in Table VIII, these figures probably underestimate the proportional concentration of atmospheric dioxins, in view of their high stability relative to phenoxy herbicides.

TABLE VIII

CALCULATED DIOXIN CONTAMINATION PER ACRE FOLLOWING 2, 4, 5-T APPLICATION

Application 2, 4, 5-T (lbs./acre)	Contamination with 2, 3, 7, 8-Dioxin (mg/acre)	
	Based on 0.5 ppm Dioxin	Based on 25 ppm Dioxin
2.5 (domestic use)	0.57	28
25.0 (export use)	5.70	284

The above figures for export use should be adjusted to reflect varying concentration of 2, 4, 5-T in different formulations. These calculations are based on 2, 3, 7, 8-dioxin alone, and as a contaminant alone, and ignore additional contamination due to other dioxin isomers, and due to any dioxins formed by pyrolysis. The high concentration of polychlorophenolic impurities in 2, 4, 5-T, approximately 5%, apart from other sources of polychlorophenols, may result in extremely high yields of dioxins. As mentioned previously, heating of 5 g of pentachlorophenol at 300°C for 12 hours resulted in a yield of 1.5 g of the octadioxin isomer. Combustion of shrub, brush, timber, or other materials exposed to phenoxy herbicides or other polychlorophenols, may thus liberate high concentrations of dioxins in the atmosphere. Whether the high yields of dioxin produced by pyrolysis of 2, 4, 5-T and 2, 4, 5-TP (Buu-Hoi et al, 1971; Buu-Hoi, 1972) originate in the herbicide itself or in its chlorophenolic impurities has not been determined.

It is thus of interest to examine the data on stability and persistence of dioxins in the environment. Tetradioxin is known to be heat stable up to 800°C; there are, however, no available data on the heat stability of other dioxin isomers. There are also no available data on the stability and persistence of the 2, 3, 7, 8- and other dioxin isomers in water, crops, milk, and animal or human tissues. No biological degradation of dioxin has been noted in soil over a 5 month period (Kearney et al, 1971); similarly, radioactivity can be quantitatively recovered from leaves of plants grown under fluorescent lighting for 3 weeks after application of labelled dioxin (Inensee & Jones, 1972). Most importantly, there are no available data on the possible accumulation and transmission of 2, 3, 7, 8- and other dioxins in the food chain -- air, soil and water⟶ plants, brush and crops ⟶ fish, birds and cattle -- with attendant accumulation in man. The heat stability of the tetraisomer, the general lipid solubility of dioxins, and their cumulative toxicity in experimental animals, all serve to enhance the possibility of food chain transmission of the various isomers.

7. Ecological Effects of 2, 4, 5-T

Massive defoliation in Vietnam, with particular emphasis on destruction of mangrove forests and associations and upland forest trees with subsequent invasion by persistent bamboo and coarse grass, crop destruction, destruction of rubber and hardwood trees, and nutrient dumping with possible risks of subsequent soil laterization, has been recently reported (Constable & Meselson. 1971; Westing, 1971) effects on animal, bird and fish life have not yet been adequately documented.

Losses of 250 reindeer and 40 abortions in a herd of 600 occurred in North Sweden in April, 1970, following grazing of animals in a conifer forest that had been aerially sprayed on 7/12/69 (Erne & Nordkvist, 1971). Grazing foilage contained 10 ppm of 2, 4, 5-T and 25 ppm of 2, 4-D; liver and kidneys of dead reindeer contained 0.45 - 0.5 ppm of 2, 4-D and 0.05 - 0.1 ppm of 2, 4, 5-T. A similar episode in cattle accidentally sprayed with 2, 4, 5-TP has been recently reported in Socorra, New Mexico (Adang, 1971).

Apart from the very high toxicity of dioxins to chickens, resulting in chick edema disease (see below), phenoxy herbicides, and even more so dioxins, are highly toxic and teratogenic to the chick embryo (Verrett, 1970); additionally, spraying partridge and pheasant eggs with 2, 4-D produced lethality and teratogenic effects. including paralysis and skeletal anomalies (Lutz-Ostertag & Lutz, 1970).

8. Chronological Summary of Scientific Regulations and Related Events Relevant to 2, 4, 5-T and Dioxins
a. Occupational and Toxicological Data on Dioxins

2, 4, 5-T was first registered for use as a herbicide by Amchem Products, Co., Ambler, Pa. on 3/2/48. Discovery of dioxins has originated from two unrelated sources. Firstly, it was reported in 1957 that certain batches of chicken feed were toxic, producing chick edema disease. The chick edema factor was traced to toxic fats, and later identified as a tetradioxin by X-ray crystallography

(Wooton, 1967). Occupational reports on occurence of chloracne in 2, 4, 5-T workers date back to an accident in a Monsanto plant in March, 1949 (Kettering Laboratory, Unpublished Reports). Chloracne was subsequently recorded in German 2, 4, 5-T workers, and tetradioxin was identified as the causal agent (Kimmig & Schulz, 1957). An outbreak of 60 cases of chloracne in a Dow Chemical plant in 1964, following a change in synthetic reaction conditions in attempts to increase production, forced the closing of the facility (Dow, 1971). Dow informed other 2, 4, 5-T manufacturers of their problem in 1965, and in the same year opened a new plant capable of producing 2, 4, 5-T and 2, 4, 5-trichlorophenol with less than 1 ppm of dioxin. The occupational toxicology of dioxins has been recently reviewed (Poland et al, 1971), with particular reference to induction of chloracne and porphyria cutanea tarda in a series of 73 male employees in a herbicide plant; some of these employees had been previously reported to have chloracne 7 years earlier (Bleiberg, et al, 1964).

b. Bionetics Study on 2, 4, 5-T

The original experimental data on the teratogenicity and fetotoxicity of 2, 4, 5-T were obtained by Bionetics Research Laboratories, under contract to the National Cancer Institute in July 1966, and these experiments were amplified and completed by August, 1968; these findings were, however, not publicly released until October 1969, a delay characterized as "inordinate" by a Presidential Advisory Committee (P.S.A.C., 1971). Low doses of 2, 4, 5-T and some of its esters produced birth defects in mice and rats, and in rats "no effect" levels were not reached at even the lowest dose tested, 4.6 mg/kg.

c. Report of Secretary Finch's Commission

On the basis of a detailed evaluation of the above data, the Teratogenicity Panel of Secretary Finch's Commission on Pesticides and Their Relationship to Environmental Health, HEW, December, 1969 unanimously recommended, "The use of currently registered pesticides to which humans are exposed and which are found to be teratogenic by suitable test procedures in one or more mammalian species should be immediately restricted to prevent human exposure. Such pesticides in current use, include ..... 2, 4, 5-T".

d. Appointment of a Herbicide Assessment Commission

The American Association for the Advancement of Science (AAAS) appointed a Herbicide Assessment Commission under the chairmanship of Dr. Matthew Meselson on 12/28/69. The Commission spent August and September of 1970 in Vietnam, and subsequently released a preliminary report including background material on 12/30/70 at the AAAS Annual Meeting, which was subsequently published in the Congressional Record (Meselson et al, 1971). Their final report is anticipated by the summer of 1972.

e. Claims by Industry that the Bionetic Data were Artefactual

The belated discovery of high concentrations of a tetradioxin contaminant, ca. 27 ppm, in the sample of 2, 4, 5-T tested by

Bionetics Laboratories suggested the possibility that the reported teratogenic and toxic effects could in part be due to the contaminant rather than to the herbicide per se. It was also suggested that formulations with lower concentrations of the dioxin posed no potential teratogenic or other hazards.

f. Articles on 2, 4, 5-T by Mr. Whiteside in The New Yorker

An article, "A Reporter at Large: Defoliation", dealing with the use of defoliants in Vietnam, toxicological and teratological effects of 2, 4, 5-T and dioxins, and some aspects of the regulatory and political background appeared in The New Yorker on 2/7/70. This article, the first of a series, is generally regarded as meticulous scientific journalism and excited extensive public interest culminating in Congressional hearings some two months later.

g. Hearings Before Senator Philip Hart, April, 1970

Senator Philip Hart, Chairman of the Subcommittee on Energy, Natural Resources and The Environment". Available toxicological data on 2, 4, 5-T and on dioxins were presented by various witnesses including Dr. Jacqueline Verrett, Dr. Arthur Westing, Mr. Harrison Wellford and Dr. Samuel Epstein. These hearings proved to be a potent stimulus for subsequent regulatory restrictions.

h. Restrictions on Use of 2, 4, 5-T

On 4/20/70 the Secretary of the USDA, in conjunction with the Surgeon-General, announced suspension as an "imminent danger" to public health of the use of liquid formulations of 2, 4, 5-T around the home and of uses of all formulations in lakes, ponds, and ditchbanks. USDA also cancelled the registration of 2, 4, 5-T in granular formulations around the home and recreation areas. and the use of 2, 4, 5-T on food crops. However, the failure to suspend these latter uses meant that 2, 4, 5-T could be freely used on food crops pending the outcome of lengthy cancellation proceedings and litigation. The Department of Defense also announced restrictions in military use over Vietnam, although it appears that such restrictions were not enforced until much later.

i. Appeal by Industry against Restrictions and Appointment of EPA Advisory Committee

Under Section 4C of the Federal Insecticide, Fungicide, and Rodenticide Act of 1947, 2 manufacturers of 2, 4, 5-T, Dow Chemical Co. and Hercules, Inc., in May of 1970, petitioned for an Advisory Committee to review the need for the imposed restrictions on 2, 4, 5-T.

The National Academy of Sciences was accordingly requested to supply a roster of individuals, from whom an Advisory Committee was then appointed by the Secretary of Agriculture in circumstances that have not been adequately defined. This Committee first met on February 1 and 2 or 1971, and submitted its secret report to EPA on May 7, 1971.

At public hearings on May 13, 1971, before the Pesticide Board of the Commonwealth of Massachusetts, following the imposition of an emergency ban on aerial applications and other uses of 2, 4, 5-T, representatives of the National Agricultural Chemicals Association and of Dow Chemical Co., both referred to the report of the Advisory Committee as a National Academy of Science (NAS) Committee. In subsequent correspondence with Dr. Handler, President of the NAS, he clearly corrected this misstatement, "This committee is not a committee of the National Academy of Sciences or the Natural Research Council, and its advice is in no way subject to our usual review procedures". A representative of the National Agricultural Chemicals Association subsequently admitted to the use of "loose" terminology in misrepresenting the Advisory Committee as a National Academy of Sciences Committee.

j. President's Science Advisory Committee Report of the Panel on Herbicides

The main thrust of this report (PSAC, 1971), was the critical need for more data on teratogenicity thresholds and dioxin accumulation in food; the report made no recommendations to remove existing restrictions on 2, 4, 5-T usage.

k. Circuit Court of Appeals Instruction to the EPA

In response to a challenge by a coalition of consumer and environmental groups -- who decided not to wait for the conclusions of the EPA secret Advisory Committee -- against USDA's refusal to take further action against 2, 4, 5-T, the Federal Court of Appeals for the D.C. circuit in a ruling on 1/7/71 held that the Government failed to consider the "risk of injury to farm workers or others who might be exposed to the chemical by virtue of its use on food crops". This decision reopened the public health question of uses of 2, 4, 5-T.

l. Decision by EPA Not To Suspend Further Uses of 2, 4, 5-T

On 3/18/71 EPA announced that it would not suspend further uses of 2, 4, 5-T pending the decision of the Advisory Committee.

m. Recommendations of the Advisory Committee

The Advisory Committee report was "leaked" shortly after its submission in secrecy to EPA on 5/7/71. Nine of the 10 members made the following recommendations:

All restrictions on the use of 2, 4, 5-T should be removed; residues of 2, 4, 5-T must be <0.1 ppm in food and in drinking water; existing stocks of 2, 4, 5-T with 0.5 ppm of tetradioxin may be used, except domestically; future formulations of 2, 4, 5-T must contain 0.1 ppm of tetradioxin; household formulations of 2,4, 5-T must carry a label warning of dangers to pregnant women; and finally, further research should be undertaken to study whether dioxins accumulate in the soil and food chains.

One dissenting member issued his own strongly worded minority report, which was not only highly critical of the majority report, but which argued that most existing restrictions on 2, 4, 5-T be maintained.

n. Critiques of the Advisory Committee Report

A detailed critique of the report's contents and a discussion of its discrepancies and ommissions was published in Nature on 6/25/71 by their Washington correspondent.

On 7/14/71 the Committee for Environmental Information held a Press Conference in Washington severely criticizing the conduct and recommendations of the Advisory Committee. The critique was supported by a wide range of independent scientists including Drs. John Edsall, Arthur Galston, Matthew Meselson, Barry Commoner, Jeremy Stone and James Crow. Additionally, the legal and scientific implications of the report were specifically reviewed and criticized by Mr. Harrison Wellford of the Center for the Study of Responsible Law, and also by Dr. Samuel Epstein, respectively, who made the following points:

"The Advisory Committee conferred with 22 outside persons including 6 representatives of the 2 petitioners. With the exception of a legal representative from the Center for the Study of Responsive Law, the committee conferred with no scientists, such as Drs. Galston, Meselson and Verrett, who had expressed concern on the teratogenic hazards of 2, 4, 5-T and who were critical of removal of restrictions on its use, and with no representatives of environmental or ecological groups; the committee, in fact, refused to accept evidence from the Environmental Defense Fund. The decision to exclude evidence from the environmentalists was apparently taken unilaterally by the Chairman of the Committee, although the majority of the Committee are alleged to have expressed a contrary opinion at the first meeting".

"The Advisory Committee, with one dissenting member, takes the view that, in spite of many admittedly unanswered questions, the public may be exposed first and the experiments performed later".

"In the Committee's brief discussion of the Bionetics studies, no reference is made to the rat data, where dose response relationships were established and no effect levels were not reached at the lowest dose tested, 4.6 mg/kg. -- Much of the report is based on unpublished data and reports which have not been subject to customary independent scientific review, and whose validity must, therefore, be open to question. -- Several recent experiments demonstrating the teratogenicity of "pure" 2, 4, 5-T have been discounted by the Advisory Committee on the grounds that dosages used, ca. 100 mg/kg, were too high. Yet, there are other studies that show definite teratogenic and fetotoxic effects at lower dose levels. Additionally, in the experiments conducted by Dow Chemical Co., 2, 4, 5-T at concentrations of 24 mg/kg, with ca. 0.5 ppm dioxin contaminants, produced an approximately 7-fold increase in incidence of sternebral ossification defects in rats as compared with controls. It should also be pointed out that much of the data reviewed by the Committee were based on tests at relatively high dose levels, and that there is a serious inadequacy of data at the purportedly more critical low dose levels. -- The

claims by the committee that there are threshold levels for the
teratogenic effects of 2, 4, 5-T and dioxins are completely
unsubstantiated. Again, with the exception of the dissenting
member, the Committee ignores the insensitivity of teratogenicity
testing in animals, as a function of small test sample size
compared with the millions of humans at presumptive risk. --
It becomes a <u>reductio ad absurdum</u> to discount the validity of such
high dose data, when low dose data statistically cannot possibly
demonstrate effects. -- Apart from its unwarranted assumptions
on the existence of threshold effects for the teratogenicity of
2, 4, 5-T, the Committee ignores the real possibility that humans
may be more sensitive to the teratogenicity of 2, 4, 5-T than
rodents. Indeed this is the case with thalidomide. -- So that,
even if pure 2, 4, 5-T was teratogenic to rodents only at doses
as high as <u>ca</u>. 100 mg/kg, and there is little evidence that this,
in fact, is the case, this would not preclude the possibility
of serious potential human teratogenic hazards."

"The Chairman of the Advisory Committee, a well-known
teratologist, performed some rather unusual experiments with
small numbers of animals and no concurrent controls, which in
fact, showed fetotoxic trends at the lowest dose tested, 20 mg/kg;
these effects are unaccountably ignored in the report. Commenting
on these experiments, another committee member is alleged to have
stated that the Chairman "Would never have published these data".
However, data that do not merit publication should clearly not
be cited in support of decision making processes."

"The Committee fails to point out a serious error in a
Hercules Corporation-sponsored study, which was reported to it,
where no effects were claimed with pure 2, 4, 5-T at 113 mg/kg
doses in mice, when in fact the dosage mistakenly used was 10 mg/kg.
A subsequent repeat test at correct dosage levels confirmed
teratogenic effects, i.e., an 11% incidence of cleft palates in
mice with a Dow formulation, and a 1% incidence with a Hercules
formulation; the Advisory Committee made no comment on this
important and striking discrepancy. When challenged on this,
the Committee Chairman is alleged to have charitably indicated
that he would rather not air a mistake".

"The Committee concludes that 2, 4, 5-T will rapidly break
down in the environment, and ignores the possibility that certain
bound forms in plants may not be detected by currently used
techniques, without initial acid extraction". For these reasons.
the Presidential Advisory Committee (PSAC, 1971) emphasized that
"unless an effort is made to release the bound residues, the
standard analytical scheme does not provide a suitable basis for
estimating the total residue concentration". "The Committee
states that further research should be undertaken to study whether
dioxin accumulates in the soil and food chains. Paradoxically,
the Committee nevertheless concurs that it is "virtually impossible"
for anyone to be exposed to toxic doses with currently produced

2, 4, 5-T. This conclusion is based on the unsubstantiated belief that dioxin does not accumulate in the air, water, plants, or in the food chain; it is, however, admitted that the tetradioxin is highly persistent and can persist for over a year in soil. The Committee's confidence is also based on the belief that dioxins are poorly soluble in oils, and thus would not accumulate in body fat. This is directly contradicted by available information on dioxins, which are causally responsible for chick edema disease and which are found in certain batches of cotton seed oil; feeding of chick tissues can transmit this disease to rats and monkeys".

"Levels of 0.6 ppb of tetradioxin produce ca. LD_{80} effects in guinea pigs, and 0.125 ppb are embryotoxic in rats. Yet, the Committee concludes, in the total absence of cited information, that humans cannot be exposed to these levels. The Committee also appears to ignore the fact that these levels are well below current analytic sensitivity limits of 50 ppb, or at best 10 ppb. Such analytic procedures are grossly insensitive, compared to the acute toxicological effects of dioxin, which approaches that of the botulinus toxin in biological activity".

"The Committee does not consider problems of carcinogenicity or mutagenicity due to tetradioxins, and fails to mention that there are no data on these important problems. -- The Committee does not consider any other of the ca.60 chloroisomers of dioxins which do not appear to have yet been looked for in 2, 4, 5-T, and does not consider the possibility of their carcinogenic, mutagenic or teratogenic effects. -- The Committee does not consider possibilities of production of any dioxins from pyrolysis of the 5% trichlorophenol impurities present in current formulations of 2, 4, 5-T; the 5% impurities are dismissed as toxicologically insignificant. -- The Committee does not consider problems of chronic inhalation exposure to dioxins or 2, 4, 5-T".

"The Committee's analysis of the Vietnam data on possible herbicide effects can only be regarded as poorly comprehensible. -- The Committee states that "any attempt to relate birth defects or stillbirths to herbicide exposure is predestined to failure". -- Such an opinion was considered "extraordinary" by Dr. Matthew Meselson, Chairman of the AAAS Herbicide Commission, who maintains that properly designed surveys can possibly establish such relationships. Also, the Committee viewpoint is not apparently shared by the Department of Defense, who have recently commissioned a National Academy of Sciences Report on this very subject. A Committee member is recently alleged to have confessed that the Committee statement was an "overdramatic and rather poor way of putting it". The dissenting Committee member claims that the report evinces "an unjustified certainty" that the Vietnam data exclude the possibility of teratogenic defects. -- It is possible that the AAAS report of increased stillbirth rates in Vietnam could have been an artefact, unrelated to herbicide exposure, as indeed the AAAS Commission carefully points out while recommending the critical need for further study in Vietnam; but, as the Minority Committee Report states: "Factors could just as easily have worked to hide a large stillbirth rate than to spuriously create one". The Committee appears to ignore this possibility".

"The Advisory Committee is totally unresponsive to its charge to discuss alternatives to the use of 2, 4, 5-T and to assess the economic benefits and penalties attendant on the cancellation or suspension of its registrations. Such data are available in the open literature. Illustrative is the USDA Agricultural Economic Report No. 199, March 1971, entitled "Restricting the Use of 2, 4, 5-T; Costs to Domestic Users". This report states that the economic costs to domestic users of banning 2, 4, 5-T in 1969 would have been only $52 million, providing all other herbicides could still be used, and $172 million if other phenoxy herbicides were also restricted".

Dr. Epstein concluded as follows:

"The present restrictions on 2, 4, 5 T should be maintained until the underlying scientific questions on its experimental toxicity can be properly resolved to the satisfaction of the open and disinterested scientific community. High priority should be accorded to such experiments, particularly to problems of dioxin accumulation in food chains, and to proper investigations of teratogenic dose-response relationships of 2, 4, 5-T, its commercial formulations and variants, including Silvex, and dioxins. Urgent consideration should be directed to the impropriety of decision-making processes, on issues of critical environmental and public concern, conducted in secrecy and in the absence of formal representation of qualified scientific and legal representatives of public interest and other concerned groups. The statements of Mr. Ruckelshaus that the public-at-large cannot be continuingly exposed to potential and serious hazards until adequate research has resolved outstanding problems are to be warmly commended. It is recommended that this philosophy be uniformly applied in all pending and future regulatory practice".

o. EPA Rejection of the Advisory Committee Report

On 8/9/71, Mr. Ruckelshaus, Administrator of EPA, announced that he would not accept the recommendation of his Advisory Committee to remove all restrictions on the use of 2, 4, 5-T, and instead ordered a public hearing in response to a request made by industry prior to the Advisory Committee Report. Such a decision is unprecedented in the history of U.S. regulatory agencies.

Subsequent to this decision, Dow Chemical Co. and other manufacturers of 2, 4, 5-T on 11/2/71 requested a public hearing on this matter. Dow then proceeded to directly challenge Mr. Ruckelshaus' failure to accept the advice of his Committee and asked for a statement in explanation. Mr. Ruckelshaus on 11/4/71 responded with the following 10-point statement justifying his action:

i   "A contaminant of 2, 4, 5-T -- tetrachlorodibenzoparadioxin (TCDD, or dioxin) -- is one of the most teratogenic chemicals known. The registrants have not established that 1 part per million of this contaminant -- or even 0.1 ppm -- in 2, 4, 5-T does not pose a danger to the public health and safety.

ii   There is a substantial possibility that even "pure" 2, 4, 5-T is itself a hazard to man and the environment.

iii   The dose-response curves for 2, 4, 5-T and dioxin have not been determined, and the possibility of "no effect" levels for these chemicals is only a matter of conjecture at this time.

iv   As with another well-known teratogen, thalidomide, the possibility exists that dioxin may be many times more potent in humans than in test animals.

v   The registrants have not established that the dioxin and 2, 4, 5-T do not accumulate in body tissues. If one or both does accumulate, even small doses could build up to dangerous levels within man and animals, and possibly in the food chain as well.

vi   The question of whether there are other sources of dioxin in the environment has not been fully explored. Such other sources, when added to the amount of dioxin from 2, 4, 5-T, could result in a substantial total body burden for certain segments of the population.

vii   The registrants have not established that there is no danger from dioxins other that TCDD, such as the hexa- and heptadioxin isomers, which also can be present in 2, 4, 5-T, and which are known to be teratogenic.

viii   There is evidence that the polychlorophenols in 2, 4, 5-T may decompose into dioxin when exposed to high temperatures, such as might occur with incineration or even in the cooking of food.

ix   Studies of medical records in Vietnam hospitals and clinics below the district capital level suggest a correlation between the spraying of 2, 4, 5-T defoliant and the incidence of birth defects.

x   The registrants have not established the need for 2, 4, 5-T in light of the above-mentioned risks. Benefits from 2, 4, 5-T should be determined at a public hearing, but tentative studies by this agency have shown little necessity for those uses of 2, 4, 5-T which are now at issue".

Subsequent to this statement (12/2/71), Dow sued EPA in Arkansas on the implicit grounds that Mr. Ruckelshaus had acted capriciously and arbitrarily, and requested an injunction to postpone public hearings until the Courts had ruled.

p.   Petition by EDF and Center for Responsive Law

On February 16, 1972, the Environmental Defense Fund (EDF) and Mr. Harrison Wellford of the Center for Responsive Law refiled petition against the USDA for total suspension of all uses of 2, 4, 5-T.

9. References

Adang, P.J.
    Personal Communication 11/19/71
Agricultural Economic Report No. 131
    Economic Research Service, U.S.D.A., Washington, D.C. 19
    Quantities of Pesticides used by Farmers in 1964
Allen, J.R.
    Am. J. Vet. Res: $\underline{28}$, 1513, 1967
Bamesberger, W.L., and Adams, D.R.
    In, Organic Pesticides in the Environment
    Advan. Chem. Ser: $\underline{60}$, 1966
Bionetics Research Laboratories
    Unpublished Report, 1969
Bleiberg, J., Wallen, M., Brodkin, R., and Applebaum, I.
    Arch. Derm. $\underline{89}$, 793, 1964
Buu-Hoi, N.P., Saint-Ruf, G., Bigot, P., and Mangane, M.
    C.R. Acad. Sci. Series D: $\underline{273}$, 708, 1971
Buu-Hoi, N.P.
    Personal Communication, 1972
Constable, J., and Meselson, M.
    Sierra Club Bulletin: $\underline{56}$, 4, 1971
Courtney, D.K., and Moore, J.A.
    Toxicol. Appl. Pharmacol: $\underline{20}$, 396, 1971
Cowan, J.C.
    U.S.D.A. Memo, Possible sources of polychlorodi-
    benzodioxins in fats. 2/12/70
Darving, L., and Sunner, M.
    Hereditas: $\underline{68}$, 115, 1971
Dawson, J.E.
    Diabetes, $\underline{13}$, 527, 1964
Dow Chemical Co.
    Report to PSAC, 1971
Drill, V.A., and Hiratzka, T.
    Ind. Hygiene & Occupat. Med: $\underline{1}$, 61, 1953
Emerson, J.L., Thompson, D.J., Strebing, R.J., Gerbig, C.G., and Robinson, V.B.
    Fd. Cosmet. Toxicol. $\underline{9}$, 395, 1971
Epstein, S.S.
    Nature: $\underline{228}$, 816, 1970
Erne, K., and Nordkvist, M.--Nat. Swedish Vet. Inst.
    Communication 308/70/KA, 5/20/70
FDA
    Unpublished data, 1970
Federal Register, 5/21/69
Inensee, A.R., and Jones, G.E.
    J. Agric. Fd. Chem. in Press, 1972
Kalter, H.
    Teratology of the Central Nervous System
    University of Chicago Press, 1960

Kearney, P.C., Isensee, A., Helling, C.S., Woolson, E.A., and Plimmer, J.R.
    Abstr. 161st National Meeting
    Am. Chem. Soc. March, 1971
Kettering Laboratory
    Unpublished Reports, Communication to PSAC, 1971
Kimmig, J., and Schulz, K.D.
    Dermatologica, 115, 540, 1957
King, C.
    J. Pharm. Exp. Therap. 147, 391, 1965
Lichtenstein, H., Guest, G.M., and Warkany, J.
    Proc. Soc. Exp. Biol. & Med: 78, 398, 1951
Lutz-Ostertag, Y., and Lutz, M.H.
    C.R. Acad. Sci: 271: Series D, 2418, 1970
Meselson, M.S., Westing, A.H., Constable, J.D., and Cook, R.E.
    Preliminary Report of the Herbicide Assessment Commission of The American Association for the Advancement of Science
    Hearings by Senator E. Kennedy, Chairman, Sub-Committee to Investigate Problems Connected with Refugees and Escapees, of the U.S. Senate Committee on the Judiciary, 92nd Congress, 4/21/71
Oberle, M.W.
    Science: 165, 991, 1969
Poland, A.P., Smith, D., Metter, G., and Possick, P.
    AMA Archives Environ. Health: 22, 316, 1971
President's Science Advisory Committee (PSAC)
    Report of the Panel on Herbicides: Report on 2,4, 5-T
    Office of Science and Technology, Washington, D.C., 3/71
Report on the Secretary's Commission on Pesticides and Their Relationship to Environmental Health: Advisory Panel on Teratogenicity of Pesticides, HEW, December, 1969
Silverstein, L.C., Dow Chemical Co. Memo.
    Safe Handling of Tetrachlorodibenzo-p-dioxin (TCBD) in the Laboratory: 2/7/70
Sparschu, G.L., Dunn, F.L., and Rowe, V.K.
    Fd. Cosmet. Toxicol: 9, 405, 1971
Tarrant, R.F., and Norris, L.A.
    Proc. Symposium on Herbicides in Vegetation Management
    Oregon State University, 1967
Verrett, J.
    Testimony Before the Subcommittee of Energy, Natural Resources and the Environment of the U.S. Senate Committee on Commerce, 91st Congress, 4/15/70
Yerushalamy, J., and Milkovich, L.
    Am. J. Obs. Gynae: 93, 553, 1965
Westing, A.H.
    Bioscience: 21, 893, 1971
Wooton, J.C.
    Chem. & Eng. News: 45, 10, 1967

STRATEGY FOR MAINTAINING ENVIRONMENTAL QUALITY IN DEVELOPING TECHNOLOGICAL SOCIETIES

Anthony Peranio

Center for Environmental Engineering Faculty of Civil Engineering, Technicon, IIT, Haifa

INTRODUCTION

In the event of outbreak of large scale war or series of wars in which even a small percentage of the nuclear weapons presently stockpiled by the major powers are unleashed, we may reasonably expect Man's disappearance from the Earth. This could happen in a matter of just several more generations. If we believed that such destruction is to come about, then meetings of this sort----with such optimistic titles: Pollution: Engineering and Scientific Solutions----can only be viewed as excercises in futility.

Most of us wish to believe that Man's demise will not come with a Bang! There is a very good chance, though, that it will come with a whimper (and a last gasp for a breath of decent air). Or, as one ecologist put it, "...when the last beetle eats the last blade of grass."

Putting aside the choice of rapid destruction, there is the possibility that over some more protracted period (say from 200 years to over several thousand years) Man's activities will have sufficiently harmful effects upon the environment making human life impossible.

Finally, we also have the possibility that Man's term on Earth can be extended for tens of thousands, even hundreds of thousands, or more years. However, the hour is late and we have already caused irreversible changes in our environment that will cut short the life possibilities for present and future generations. Each new assault upon the environment cuts down more of these life possibilities. But, before we become lulled into believing that the pollution crisis is really an exercise in philosophy for <u>future</u> generations, permit me to put my argument starkly: We are <u>not</u> discussing some nebulous damage to life and environment. Rather, we are talking about the serious harm that is occurring to the majority of us---- yes, those of us sitting in this room---- whereby as much as 5 or 10 years may be taken off our life spans, where we may become the premature victims of dreadful diseases, and where we may be forced to live in environments of such low quality that life loses its taste and meaning for us.

It is <u>already</u> happening. Let us take a few Israeli examples Pollution of the entire coast line by oil and tar so that our ability to enjoy the beaches and sea has been seriously lessened; serious pollution of the coast off Tel Aviv by sewage; lesser, but growing, sea pollution at other locations; destruction of rivers such as the Kishon, Yarkon, and that at Hadera mainly by industrial effluents; increasing salinity and contamination of our underground water supplies by salts (including increasing nitrogen content), detergents, and other undesirable contaminants; endangering the integrity of Lake Kinneret's water by indiscriminate discharge of many types of contaminants either directly into the lake or into the surroundings; virtually uncontrolled discharge of all industrial pollutants into either sewers, water courses, or the air, so that the environment in and around the large cities has become dangerously polluted; overloaded or non-existant sewage treatment facilities---- a good portion of Jerusalem sewage flows, untreated, down the wadis and into the sea; a stench emanates from the sewage lagoons; there is a pile up of solid wastes; composting is not entirely successful; many communities have burning garbage dumps, plastics included; virtually uncontrolled vehicular pollution with its disturbing health effects; increasing noise and its harmful effect on human health and well-being brought about by increased air and ground traffic, intensified industrial activity, and crowded city life; and the attack on scenic sites, air corridors, parks, and other green areas by highways, airports, general construction, high-rise buildings----just to mention the most glaring cases. Others would add to this list, the increase in crimes of violence, crime in general, and civil disturbances-----all of which point to a reduction in the quality of the social environment.

Since this is an international meeting of experts, I need not stress the point that although the list of examples cited is strictly Israeli, it is far from unique. On the contrary, it is most likely that each visitor can point to almost identical parallels in his own country and may be able to add some "horror" stories of his own. The obvious questions in our minds at this point are: What can be done to stop the environment from degrading even further? Is there some strategy that we can employ?

THE BASIS OF THE PROBLEM

For the sake of completeness, permit me to sketch in what appears to be the basis of the decline in environmental quality, before going on to strategies and solutions.

Consensus establishes the root of the pollution problem in the great and rapidly increasing needs for power, fuel, food, water, a great variety of manufactures, and the very air itself, needed both by Man and his machines. In attempting to meet these needs, large quantities of heat, solid and particulate matter, liquids, gases, and vapors must perforce be rejected to the environment as waste. And here is the rub: how can this be done with minimal negative impact both on Man and his environment? Before considering this question in detail, let us take Israel as an example, again, in several selected statistics. By the year 2000, it is expected that in Israel: electrical capacity will increase about 10 times from the present 1500 to 16,000 megawatts; population will increase about 2½ times from 3.3 million to about 8 million persons; water consumption will double from 1.5 to 3.0 billion cubic meters per year (which implies that perhaps half of this total will have to come from desalination plants); and, it is expected that the number of road vehicles in use will quadruple to reach a total of about 1.5 million by the year 2000.

By themselves, these statistics are difficult to evaluate in terms of what might happen to the quality of Israel's environment unless some careful planning and execution of projects is carried out. For example, with regards to electrical power, the experts have announced plans for large electric power complexes of around 3000 megawatts each (5 or 6 such) all sited along Israel's coastline! If this concept is realized it will virtually commit all of Israel's population to being crowded into the coastal strip; a situation wherein about 85 or 90% of the population will be crowded into about 4 or 5% of the land area of Israel (pre-1967). As in a Greek tragedy, we can see the forces at work grinding to the inevitable sorry end: electric power is produced near the coast since it needs water for cooling, and the coast is populated with people and is thus the center of electrical load. Because we have concentrations of workers, electric power, water, and an established transportation grid, industry will then site itself where the action is; in and around the established crowded cities. This requires more electricity, people, and industry, and so on. The process is a most familiar one to all visitors from abroad, and is a classic description of megalopolis formation. Once again, even though a specific Israeli example, it can be applied to the pattern of urban growth now taking place almost universally. It is a process whereby newly developing and expanding industries sited in and around centers of population find tnemselves in already polluted environments. To prevent pollution levels from rising and the making of further inroads upon environmental quality, proper strategies must be devised and applied.

STRATEGIES FOR PRESERVING ENVIRONMENTAL QUALITY

A three-concept strategy is suggested:

Planning

Long range plans must integrate such needs, for example, as those for electricity and fuel production, water supplies, transportation, sewage and refuse disposal, recreation areas, and areas for living where suitable environmental quality can be maintained.

Placement

On the basis of integrated and long range plans, newly developing and expanding industry can be sited sufficiently removed from centers of population without undue economic hardships.

Prevention

Use must be made of "maximal feasible control technology" on all new industry and other sources of pollution. Established sources of pollution must be brought under control during a phasing out period. Under this category we must include the reduction of waste to a minimum by converting and recycling by-products of manufacturing and energy production.

Clearly, at every stage, precious supplies of land, minerals, water, and air must be used judiciously and only after thorough planning.

NEGATIVE FORCES

Few people would argue against implementation of the three-concept strategy: planning - placement - prevention. The problem, of course, is putting such a strategy into effect. Let us consider just several of the pitfalls:

Difficulty in Forecasting

The extreme and rapid changes brought about by technological innovation makes it difficult to predict future needs with accuracy. This shortcoming is compounded by the fact that future conditions themselves depend upon the implementation of programs being planned; programs which may or may not ever be completed.

Getting Planners Together

Although we have come to realize that a multidisciplinary and "systems" approach is required for arriving at adequate solutions to our major problems, getting experts from different disciplines to speak a common language is another matter. Even when they do understand one another, rivalry and competition between groups prevent sorely required cooperation and integration of plans. Sometimes mere physical distance and poor administrative delineation of command and jurisdiction prevent this cooperation. Let us take another Israeli example: the placement of electric power plants. Plans for electric power production should certainly be integrated with master plans for transportation, industrial expansion, population dispersal, water production and utilization, and tourism----just to name a few. To plan the placement of electrical power plants on the basis of the cheapest and most expedient location, using power engineering criteria alone, is surely the path to environmental chaos. Yet, this is the apparent policy of the Electric Corporation when it plans 5 large generating installations of 3000 MW each spread up and down Israel's coast line.

Vested Economic Interest

Although vested interest may be capable of long range planning, it is generally incapable of or unwilling to make long range <u>integrated</u> plans. A large bus cooperative, for example, would not be willing to abide other types of public transportation, even though by so doing, great savings in fuel, and cost might be obtained. The managers of this same cooperative would be unable to "understand" that by moving the majority of people around in subways, high speed trains, and other electric conveyances instead of diesel-engined buses, it would be possible to improve greatly the quality of the air we breathe. Vested economic interest looks out for itself and is ready to compromise the best interests of the nation at large, in order to obtain immediate and selfish gains.

Money Problems

It is the nature of private capital to seek out the areas of investment that bring the greatest and most immediate return. To use money in long range plans costs dearly, if indeed, the necessary capital can be found at all. For example, finding money for setting up industry for manufacturing many commodities (including obvious and wasteful luxuries) that are consumed soon after manufacture is a relatively easy process as long as reasonable profits are obtained concurrently with investment. Finding $1 billion for a long range power and water system is a **virtually** impossible task, even though over a period of years the long range investment may bring exceptional returns.

Lack of Societal Goals

The list of problems besetting long range and integrated planning would be incomplete without considering the fact that in most countries of the world today there is a lack of clearly defined societal goals. What leader, for example (**exce**pt in a pure authoritarian state) can say with certainty what exactly is desired in his nation in twenty, fifty, or a hundred years from today. And, even if such a convinced leader existed, we can be certain that there might be wide discrepancy between his aims and those of his political henchmen, not to mention the wide gap between the leader's party and his opposition. How then can long range and integrated plans be devised (never mind implemented) when political leaders come and go in periods of only several years, and the ideology of the party in power shifts from one pole to the other? And, if as occurs in many countries as a matter of course, and nearly always in every country at one time or another, greedy and criminal persons use graft and other corruption to direct funds into special channels, what chance is there that long range and integrated plans will ever come to fruition?

THE ROLE OF PUBLIC AWARENESS

In this closing section I will outline the nature of what is believed to be the key element in any strategy for attempting to maintain environmental quality in developing technological societies: public awareness of the nature of the environmental crises. Without a public aware of the dangers to health and life inherent in a continuation of the present processes that degrade our environment, then it is almost certain that the negative forces mentioned above will succeed in reducing the world to a highly undesirable state as far as Man's existence is concerned. Economic motivation will be paramount, and political and other social forces will act to produce a world in which profits have been maximized and the environment has been reduced in quality to a point that is just acceptable. And here is the key concept: if the majority of people in a given society tolerate a certain quality of their environment, no matter how dangerous or undesirable this may be for them, then it can be assumed that it is this environment that will be obtained in that society. No real-life politician will agitate the people under his rule with disquieting thoughts of "improving environmental quality" unless they are already restless and complaining, and wish for some improvement. We must face the fact that in order to prevent the polluting of our environment requires Herculean and costly efforts. No group of citizens unaware of the dangers of the crisis will tolerate the increased taxes and other economic burdens and sacrifices implied in a program consisting of implementing long range and integrated plans for environmental protection.

But, let the citizens become aware of these dangers, commit themselves to making changes, and then take action to change the state of affairs, and we will witness a miraculous change for the better. As public pressure grows, politicians will begin to compete with one another as champions of the cause of environmental quality. The story is an old one and well told by Machiavelli. In general, the wise and experienced politicians try to give the public what the majority believes it wants and needs, especially when in doing so, public health and welfare are enhanced. When the public is aware of and vocal in expressing its needs, the politician then knows that he may reasonably call upon the citizens to make the sacrifices required of them to improve the situation. Witness the acceptance of heavy defense burdens by citizens of nations throughout the world. Most are convinced that without their own armies and navies to defend them they will be destroyed by their enemies. Therefore, there are no politicians in power today calling for abolition of armies and navies (no matter what their political complexion). To do so would be political suicide. Similarly, once the desire for environmental protection reaches the same level of intensity as the desire for national protection, we will witness a vigorous attempt to correct matters. In closing I will present a final Israeli example.

About 11 years ago a broad anti-pollution law was passed. It called upon the Ministers of Health, Interior, and Transport, and several others to implement the law by issuing suitable standards and regulations. Since that time and up to just recently, there were no air quality standards in force, and no emissions standards for industrial pollutants. In a particular and obvious case, the Portland cement plant in Haifa was seriously polluting the air with the emission of about 100 tons of cement dust per day for several years. The authorities had not succeeded in legal prosecution of the company even though the Minister of Health and various of his subordinates had warned of the health and environmental danger involved.

Over a period of years, public awareness of the pollution problem has been growing in Israel. Every day now items appear in the press describing attacks upon health and environment. In the past two years, this awareness has grown considerably, and is largely responsible for an encouraging High Court decision that was issued recently based on an order nisi requested by a citizen's group, the Israel Public Council for the Prevention of Noise and Air Pollution (MALRAZ). In the judgement, the Minister of Health was called upon to issue standards for the emission of dust, make these effective within 6 months' time and apply them to the cement company. If the cement company fails to comply, the Minister is ordered to use his legal power not to renew the plant's operating license.

Such a judgement could not have been obtained two years ago. Today, it has real meaning to a large number of people on all levels of life in Israel.

Thus, we have seen examples of how long range integrated plans encorporating a strategy of Planning - Placement - Prevention are required if environmental quality is to be improved in technological societies. The negative forces at work inhibiting and preventing the implementation of the required plans are powerful ones. A key counter-force is a wide spread publice awareness of the consequences of the environmental crisis. A knowledgeable and organized citizenry can do much to encourage political leaders to take the required action. Without this public awareness and commitment, we may expect a serious worsening in environmental quality in the coming decades.

ENVIRONMENT'S MOST DANGEROUS PEST: MAN

Donald C. Royse

Washington University, St. Louis, Missouri

The title of this paper suggests a breadth which is misleading and an indictment of man which needs further explanation. It may not even be an appropriate title as the list of issues with which I will deal are considerably less inclusive than a thorough listing of environmental problems might be. The indictment of man, however, at this point in history seems justified. Virtually every thoughtful person agrees that the environment is being affected in deletrious ways by the collective actions of man, and man collectively must be held accountable for his actions.

There was a time when man's understanding of his effect on his environment was based more on superstition than on any knowledge of the physical sciences. For instance, there is no evidence that the early inhabitants of the Tigris and Euphrates river valleys understood that it was their methods of cultivation which gradually changed the fertile valleys into an agriculturally non-productive desert. There is no reason, then, to expect that corrective action would have been taken to avoid this destruction of productive land. But ignorance can no longer be claimed as an excuse. We know that sections of the Imperial Valley in California have become salinized through the repeated irrigation of the soils and accompanying deposits of salts as water evaporates. We can no longer excuse it. Such information is increasingly available for any who will take the time to inform himself. There is, of course, a very great difference between a situation in which only a relatively few scientists and concerned citizens have a meaningfull understanding of environmental problems and one in which the entire population is well informed and willing to reorder priorities in response to his knowledge. Further, what one man sees as an environmental problem, another may view as a beautiful indication of achievement. It is hard to convince a less developed nation that the black smoke of its new factory is a problem.

Silent Spring by Rachel Carson, The Closing Circle by Barry Commoner, The Population Bomb by Paul Ehrlich, are only three of many works recently written which solicit our concern for environmental problems. What is the response? It is, of course, too early to say and it is not easy to predict the impact today's leaders in the environmental movement will have on future generations.

You probably are aware of the fact that there is no unanimity of opinion about which aspects of the environmental problems are the most important. Environment Magazine recently published contrasting articles by Paul Ehrlich and John Holdren on one side and Barry Commoner and his associates on the other which contain in combination most of the key variables found on anyone's list of environmental problems: population, affluence, faulty technology and pollution. It is the relative effects of these and the way that they interact and also the attitudes of private and public decision-makers and citizens that create the arguments between Ehrlich and Commoner, and also between others concerned about environmental problems and their solutions.

If one looks briefly at the four variables mentioned: population, affluence, faulty technology and pollution, what are some of the central issues which they raise, and what should be one's response both as a professional and as a humanitarian?

POPULATION

Population control has long been at the center of Ehrlich's concern with the environment but has been viewed by Commoner and others as misdirected attention. In Ehrlich's and Holdren's review of Commoner's book, The Closing Circle, the argument is presented as follows: "Commoner's unidimensional approach is exemplified by his oft repeated analogy[1] that pressing for population control 'is equivalent to attempting to save a leaking ship by lightening the load and forcing passengers overboard'. Needless to say, if a leaking ship were tied up to a dock and passengers were still swarming up the gangplank, a competent captain would keep any more from boarding while he manned the pumps and attempted to repair the leak."[2]

However, Commoner's point is well taken. The population crusaders, by their lack of real attention to existing social inequities which would be intolerable even if there were zero population growth, are left open to attack by the poor and their advocates, the third world and others interested in social reform.

In broader terms, population, even if it were to grow only at a very modest rate, presents a potential problem if in coming years the poorer countries of the world industrialize and develop the philosophy of consumerism now existing in the more industrialized countries of the earth. To state it in different terms, tremendous problems are going to arise as the "less-developed" portions of the world's population approach the levels of per capita energy consumption currently existing in the "more-developed" countries.

A look at some estimates of the energy reserves of the world and estimates of energy required to produce the goods and services at the level presently existing in the developed countries, shows that enormous demands will be made on the assumed reserves. For example, S. R. Eyre points out that "world reserves of petroleum - the mineral upon which world industry (particularly its transport sector) leans so heavily - are running out very rapidly. Evaluations from a range of estimates indicated quite clearly that the oil reserves of the U.S.A. (excluding Alaska) had been reduced by half by about the year 1968, and that only 10 percent will remain by about the year 1990. Equivalent estimates of world resources show that they will have been halved by a point in time somewhere between 1988 and A.D. 2000 with 90 percent exhaustion somewhere between the years 2020 and 2030."[3]

Petroleum supplies 75% of the total energy requirements of the U.S. and if the underdeveloped countries develop similar patterns of energy consumption in future years, the implications for the provision of energy are sobering. The thrust, it seems to me, of the population problem is that patterns of consumption of world resources are tragically lopsided. As Eyre points out "...although a large percentalge of the world's industrial raw materials are extracted in underdeveloped countries, these materials are largely consumed by the manufacturing industries of Europe and North America."[4] He gives several examples which show the imbalance; for instance, the developed countries (Europe, the U.S.S.R. and U.S.) with a population of 850 million consumed over 75% of the world copper while the underdeveloped countries (Asia, Africa and Latin America) with a population of three times as large consumed only 25%. Figures for tin are similar and for aluminum the figures are even more striking with the western developed countries consuming 7 million out of 7,415,000 short tons in 1965. Japan consumed 300,000 leaving 115,000 tons for the underdeveloped countries. Discussion later in the paper of the higher energy consumption required for the production of aluminum shows that the imbalance is compounded.

Considering these facts it is possible to define an environmental problem in terms of excessive population. A solution to the problem so defined would logically be population control. There are, of course, other ways to define the problem such that other solutions seem equally logical. As Commoner points out "Population control (as distinct from voluntary, self-initiated control of fertility), no matter how disguised involves some measure of political repression, and would burden the poor nations with the social cost of a situation - overpopulation - which is the current outcome of their previous exploitation, as colonies, by the wealthy nations."[5]

AFFLUENCE

Ehrlich holds that affluence is one of the factors which explains part of the increase in environmental problems and cites the per capita increases in automobiles, telephones, ranges, water heaters, air-conditioners, refrigerators, and clothes dryers in support of his point. Increased per capita expenditure in housing is another factor which is included in Ehrlich's list. The environmental impact of these affluence items is manifested in the form of increased resource consumption.

Commoner states that "...what I have tried to express in my writings, is that regardless of the myth created by advertising propaganda and the consequent cultural attitudes, true affluence ought to be measured by the actual consumption of goods that in fact contribute to human welfare."[6] In explaining this point he maintains that per capita consumption has not risen alarmingly but that the impact of modern production methods and the goods and services which they spawn have had greater deleterious effect on the environment. As an example, while it is true that the per capita expenditure in housing has risen from $149 in 1946 to $235 in 1960 (constant dollars), it does not necessarily follow that we are better housed. Nor, more importantly do we know without asking further questions about the consumption patterns for the materials used in the construction and the operation and maintenance demands of the building whether or not there is greater environmental impact. Later in the paper I will discuss some of the effects of current practices in architecture and construction on the environment.

Ehrlich's list of affluence items, however, is hard to refute as indicative of a particular life style and an attitude about comfort and consumption which is prevalent in the more developed countries, particularly the U.S. Following is the table of Ehrlich's affluence items (Environment Magazine, April 1972):

TABLE 1

AMENITIES AND AFFLUENCE IN THE POSTWAR UNITED STATES

Item	Initial Value		Final Value		Percent Increase
Automobiles in use per capita*	0.208	(1940)	0.416	(1968)	100
Telephones in use per capita**	0.165	(1940)	0.540	(1968)	227
Automatic heating units in use per capita†	0.042	(1946)	0.133	(1960)	217
Ranges in use per capita††	0.242	(1951)	0.305	(1960)	26
Water heaters in use per capita‡	0.122	(1950)	0.177	(1960)	45
Percentage of households with air conditioners‡‡	0.2%	(1948)	13.6%	(1960)	6,700
Refrigerators in use per capita‡‡‡	0.145	(1946)	0.277	(1960)	91
Clothes dryers per capita‡‡‡	0.0013	(1949)	0.053	(1960)	4,000

*U.S. Department of Commerce, *Statistical Abstract of the United States*, U.S. Govt. Printing Office, Washington, D.C., p. 544, 1970.

**Ibid., p. 491.

†Landsberg, H., L. Fischman, and J. Fisher, *Resources in America's Future*, Johns Hopkins University Press, Baltimore, p. 733, 1963.

††Ibid., p. 742. ‡Ibid., p. 744. ‡‡Ibid., p. 747. ‡‡‡Ibid., p. 748.

The meaningful measure against which to judge the environmental impact of affluence items, I think all would agree, is threefold: 1) The depletion of a non-regenerative natural resource; 2) The consumption of energy in its manufacture and maintenance; 3) The environmental impact of the fuel used in its operation. There is another dimension which should enter into a consideration of the environmental impact of affluence items and that is the expected life of the item. Recycling is one response to the realization that as goods wear out and are replaced more resources are consumed unless there is some way to reutilize those used in their original manufacture. There is, of course, energy consumed in the reprocessing or reclaiming of resources which diminishes the benefit, or more accurately, shifts the burden from one resource to another - the one used in the generation of power for the recycling.

We have simply developed a throw away mentality and a lifestyle (some call it affluence) which places greater demand on the environment than has been true in the past or than is true for the less developed countries. The transportation sector, which I will discuss later, is one such example of consumption where the demands on the environment is most dramatically seen. Again, the demands are going to be most strongly felt when the gap in available goods and services between the more and the less developed nations is narrowed, especially if our model of consumption is followed in the transportation sector.

FAULTY TECHNOLOGY AND POLLUTION

One of the points which Barry Commoner makes very effectively in The Closing Circle is that technological developments in the production sector have in the past 20 years or so given us a system in which the emission of pollution has grown faster than the output of goods. In Commoner's words "While two factors frequently blamed for the environmental crisis, population and affluence, have intensified in that time (i.e., following World War II), these increases are much too small to account for the 200 to 2,000 percent rise in pollution levels since 1946. The product of the two factors, which represents the total output of goods (total production equals population times production per capita), is also insufficient to account for the intensification of pollution. Total production - as measured by GNP (Gross National Product) - has increased 126 percent since 1946, while most pollution levels have risen by at least several times that rate."[6]

If one looks at a specific major polluter, the automobile, and relates the contribution to pollution and the energy used in its manufacture with its contribution to a desired goal - movement of people - the point is well made. In 1937, 179 kilowatt hours of electricity were used to produce one automobile; in 1967 the figure had risen to 708 kilowatt hours - an increase of 3.95 times in the manufacture of one automobile. While pollution emissions are now regulated - at least in part - the contribution to atmospheric pollution with all controls operating will still be considerable. But more importantly the fuel consumed by the transportation sector represents a major environmental impact. "The transportation of people and goods required 16,5000 trillion BTU in 1970, equal to 24% of total U.S. energy consumption."[7]

In one sense it can be argued that this is less an immediate example of "faulty technology" than a socio-political priority set by the American public for a high powered, individual unit of transportation. It is, however, a problem amenable to technological solution assuming the individual nature of the transportation system must be maintained. In terms of efficiency of the gasoline engine to deliver power it has been estimated that one-tenth of the raw fuel is delivered to the wheels of the car. Similar estimates for cars powered by electric power show that their energy-system efficiency is approximately 20 percent.[8]

In addition to the aspects of waste due to faulty technology, there is evidence that solutions to pollution problems resulting from new technology often fall much more heavily on the poor. One typical example illustrates the point. "The fight to clean up air pollution in New Jersey provides another type of example. Many environmentalists supported the proposal that automobiles be made to pass exhaust emission tests in order to stay on the road just as they are now required to pass safety tests. This support was presumably based on the recognition that automobile exhaust contributes considerably to air pollution. The estimated cost of bringing a car up to standard was in the tens of collars. Clearly the burden of this approach falls most heavily on the poor, despite the fact that less stringent standards were set for older cars. This proposal requires the poor to pay disproportionately for the solution to a problem which is rooted in Detroit's technological failure and in the absence of adequate public transportation. If action on air pollution cannot wait for the coming of a sane and diversified transportation system, then perhaps we should have the grace to recognize that since society in general would be the beneficiary of lower exhaust emission, then the public in general should pay for it within the framework of a progressive tax structure, rather than taking it out of the hides of those least able to pay......The challenge to the environmental movement is clear. Either elevate social justice to an essential criterion for the solution of environmental problems or earn the enmity of the poor for being a movement of narrow social vision."[9]

The main purpose of this incomplete discussion of technology and pollution is really a repetition of a plea by others (not all, by any means) that this aspect of the environmental movement has several interrelated facets: 1) Technology which produces a given level of goods and services may, through changes in the production system, have greater environmental impact; 2) Production technology which even if "clean" in terms of pollution may be much more consumptive of fuel; 3) Solution to environmental pollution may inequitably require those least able to pay to bear the brunt of the cost.

ROLE OF ARCHITECTURE AND PLANNING

The more general issues discussed so far in this paper have relevance to different profesionals in the terms of their own professions. Several implications stand out for the fields of architecture and planning.

Architecture as a profession is only beginning to address itself to the environmental problems we have been discussing. When I was studying architecture 12 years ago there was simply no emphasis on these issues. Population was certainly not considered relevant to the profession; affluence was popularly discussed but, since in a real sense its existence was essential to the well being of the profession there was no consideration of its impact on either architecture or the environment; technology was discussed in terms of strength of materials, evolution of new materials and construction techniques but seldom in terms of energy transfers required in the manufacture of a given material or impact of the use of different materials on available resource banks; pollution was recognized not as particularly relevant to the field but as a potential real threat to the whole society exemplified primarily by radioactivity fears associated with the nuclear fission.

When I studied planning some 5 years ago the public awareness had certainly been raised but there was a similar lack of interpretation of the problem into the training of a planner so as to be subsequently useful.

What has changed in the education of architects and planners today? Not as much as one would hope. There are only a few professors of architecture or professional architects who devote their time to questioning the impact of architecture and the construction industry on the environment.

Let us look at some of the possible areas of impact. One of the major concerns which has developed focuses on the use of energy: energy required to turn raw materials into construction materials, energy required in the construction process and energy required in the operation of buildings. In the first section of this paper I referred several times to energy consumption as one of the major problems upon which most environmentalists agree.

In the narrower limits of the construction industry the proportion of the G.N.P. which it provided, 10 percent, gives an indication of its energy demands. Of all the electricity produced in the U.S. the construction industry consumes 7.5% of the total; this is 22% of all electricity used by the industry. The projected growth pattern for industry calls for energy use which is growing much more rapidly than the projected growth curve of the population. Why is the energy consumption in industry increasing at a faster rate than that of the population?

In an important paper presented to the AAAS Philadelphia Meeting in 1971, Richard Stein presented some facts which give some answers to the question.[10] He has placed them in the following areas to which I have added a brief explanation:

1) The Design of Buildings - The code required safety factors and standard design practice result in structures which are built to stand loads 5 to 6 times greater than those expected. Further, construction practices designed to save labor costs often result in failure because bracing is removed too soon and engineers often overdesign to avoid such failures.

2) Design of Lighting - Lighting engineers have consistently raised the assumed necessary foot candles for adequate illumination from 20 in classrooms in 1952, 30 in 1957 and 60 in 1971. This is typical for other use areas, some of which go to 90 foot candles, and is based on a study by A. R. Blackwell that efficiency in the performance of visual tasks is in direct proportion to the foot candle level of diffuse undifferentiated light achieved in a space. There is substantial contraditory evidence best exemplified in the Westinghouse studies on factors affecting work productivity.

3) Electric Heating - Electric heating is inherently inefficient but its use has been pushed by the utilities by advertising and preferential rates for "all-electric homes" and other large consumers of electricity. Electric heating is also popular because of its low initial installation costs and because it is considered to be "clean" at the point of "delivery". However, because of its inherent inefficiency it is more polluting at its source of generation.

Another somewhat more complicated practice related to the ease of administering large buildings contributes to greater consumption of energy in heating, cooling and lighting. That is the total sealing of a building such that heating and cooling is dependent solely on mechanical systems - regardless of the conditions outside. I have no figures at my disposal and in some instances this kind of control may be necessary - in many cases, however, it is not and the energy consumption capacity of the urbanizing world scarcely needs this kind of boost.

Building materials vary in their requirements of energy to be transformed from their raw state into construction materials. I cannot list many materials now with respect to their rates of energy consumption in manufacture but some are well known: for example, aluminum takes much more power in its production than does steel; and concrete requires more than wood (and wood is a regenerative resource). More research into this aspect of the construction industry is needed and that information should be made available to students and professionals who would be affected.

One final comment, or really a suggestion of further research to be done, concerns the evolving patterns of urbanization in the U.S. Urbanists, such as Jane Jacobs, and most architects and planners have on the grounds of esthetics, life style and wasteful use of land, decried the rapid suburbanization of our metropolitan areas. Nationally, suburbs have grown approximately four times faster than their city centers and land is being absorbed in this suburbanization at a rate double or triple that of the developed central areas.

I find it difficult to condemn suburbia on the grounds of life style and esthetics listed above but this density of development does absorb great quantities of land. Objectively, however, when one considers that approximately 74% of the U.S. population is living on less than 5% of the land, and further when one considers government policy to hold land out of production in order to avoid surpluses, this problem in the foreseeable future does not appear to be a serious one. There are other implications, however, which do seem serious.

The present pattern of suburbanization can only be effectively served by a transportation system which is highly mobile and individualized, i.e., the one that has developed - the private automobile. When one looks at the statistics on the energy demands of such a system - 24% of the total U.S. energy consumption - the insanity of the system becomes apparent.

Mobility is essential for a developed industrial (or postindustrial) society but a land use pattern that virtually rules out more efficient modes is hard to support. Some comparisons between various modes of transportation illustrate the problem:

		Passenger Miles per gallon	BTU passenger miles
1.	Buses	125	1090
2.	Railroads	80	1700
3.	Automobiles	32	4250
4.	Airplanes (averaged total)	14	9700

Within urban areas other modes can be considered:

Energy-Efficiency for Urban Passenger Traffic[a]

		Vehicle-miles Gallon	Passengers Vehicle	Passenger-miles Gallon	BTU[b] Passenger-mile
1.	Bicycles[c]	-	-	756	180
2.	Walking[c]	-	-	450	300
3.	Buses	5.35	20.6	110	1240
4.	Automobiles	14.15	1.9	26.9	5060

[a]Data from Statistical Abstract (1970) for 1965 and Automobile Facts and Figures (1971).

[b]Assuming 136,000 BTU/gallon.

[c]Efficiencies for walking and bicycling computed as follows: an excess of 225 calories/hour (893 BTU/hr.) is required for moderate walking or bicycling, from Rice (1970). Assuming 5 mph by bicycle and 3 mph by foot yields the values given above.[11]

It becomes clear that the patterns which have developed in most American cities is one of the most energy expensive systems that could be devised. Remembering the energy demands which will be made on world supplies and more cogently the massive subsidies which are built into the automobile oriented system at the expense of the relatively poor, one wonders how long this particular inequity will be allowed to exist.

Other interesting research which could make more forceful the argument for residential patterns alternative to the predominant single family detached house one, would be an analysis of the overall energy demands for the single family house alternatives as compared with various forms of multiple housing. The analysis should look at construction costs (assuming similar materials in both cases), heating and cooling costs of both (it is obvious that there is more exposed exterior space in the single family detached house than in multiple housing units), and costs which go into the maintenance of both. I do not suggest that the results would be compelling but in combination with the implied transportation costs and the duplication of all the paraphernalia required to maintain it, the single house would appear to present quite a different energy consumption figure from various forms of multiple housing.

In conclusion, I do not feel I have covered adequately an extremely broad topic and I would perhaps have been well advised to narrow the topic considerably, but I do feel that there is a need to at least attempt to place one's professional concerns in the context of what in this case is clearly a much larger problem. Man has created social, economic and consumption patterns which have very real implications for his continued existence on this earth. If man is not to become a pest exhibiting the patterns typical of pests which the doom-sayers of the movement predict, he must as citizen and as professional consider the impacts of his action.

BIBLIOGRAPHY

1. The Closing Circle, review by Ehrlich and Holdren of book by Barry Commoner in "Environment", Vol. 14, No. 3, pages 4-5, with Commoner's response.

2. Given verbally before the President's Committee on Population Control and the American Future, at the meeting of the American Association for the Advancement of Science in December 1970, on page 255 of The Closing Circle.

3. S. R. Eyre, "Man, the Pest", New York Review of Books, Vol. XVII, No. 8, page 24.

4. S. R. Eyre, Op. Cit., page 22.

5. <u>The Closing Circle</u>, op. cit., page 52.

6. Barry Commoner, <u>The Closing Circle</u>, Knopf, New York, 1971, page. 140.

7. Eric Hirst, <u>Energy Consumption for Transportation in the U.S.</u>, Oak Ridge National Laboratory, Oak Ridge, Tennessee.

8. D. P. Grimmer and K Luszczynski, <u>Lost Power</u>, in "Environment", Vol. 14, No. 3, page 22.

9. Daniel Kohl, unpublished review of "In Defense of People" by Richard Neuhaus, April 1972, page 15.

10. Richard G. Stein,"Architecture and Energy", Paper, AAAS Philadelphia Meeting, December 28-29, 1971.

11. Eric Hirst, <u>Energy Consumption for Transportation in the U.S.</u>, National Laboratory, Oak Ridge, Tennessee.

ENVIRONMENTAL EDUCATION AS A MEANS OF CREATING AN AWARENESS OF POLLUTION BY TOMORROW'S YOUTH

Phillip Bedein

*Bureau of Air Quality and Noise Control
Pennsylvania Department of Environmental Resources*

Many ecologists and environmentalists are concerned about the future of "space ship earth" especially since the controversial book "Limits to Growth" which was written by a group of scientists from M.I.T. and has just been published. A computer was used to determine what environmental conditions will exist in the future. The input for the computer was on conditions existing today in the world. The conclusion was that doomsday will arrive within a hundred years unless drastic action takes place now to improve our environment.

Based on the results of the computer, a chart in the book shows that a gradual decline in natural resources and in food per capita is beginning now. The decline will become steeper commencing around the year 2025. The total amount of environmental pollution will decrease slightly until 1990 and then there will be a steep climb reaching its peak in 2050. The industrial output will rise rapidly until after 1990 when it will begin to parallel the upward curve of the amount of pollution. The prognosis is clear. Due to the great amount of air and water pollution and the small amount of food per capita, the world's population which has been increasing rapidly will begin to decrease sharply within the next hundred years.

One must remember, however, that the input to the computer is based on available technology with no major changes expected during the next 100 years.

Accordingly, we have time in preventing this doomsday prediction. Clearly, the major avenue to accomplish this is to create an awareness of the environment through the education of our children from the age of six and even earlier, and then continuing their education up even to adulthood. Only in this manner will these youngsters as adults feel that it is their responsibility to be committed to the improvement of the environment in their communities.

Although "Earth Week" is beginning to lose much of its emotional impact, environmental education is not a passing fad. The world's environmental crises with their accompanying threats to human survival mandate that man begin to understand his role in the overall scheme of existance or predictably create his own demise during the 21st century as some people predict.

We must decide early in life whether we want to live in harmony or in conflict with our environment. By educating our youngsters regarding environmental problems, we will be able to create an awareness that will enable them to accept the challenge and the responsibility of the community.

During the summer of 1970, a conference was held at the Smithsonian Institute's Belmont Center in Virginia. This conference was sponsored by the Educational Facilities Laboratories of New York City and was attended by nationally recognized authorities in disciplines related to environmental education. The participants included architects, planners, government leaders and educators.

A report of discussions that took place was summarized by the educational facilities laboratories in the following four statements:
1. Affecting change must be the absolute essential of effective environmental education.
2. Educators must become familar with facilities that can best contribute to effective environmental education.
3. An interdisciplinary approach should be the methodology of instruction. The student should have environmental encounters and experiences. He should interconnect a thematic strand through many aspects of a subject.
4. Major capital expenditures are not necessary to mount effective programs in environmental education. An effective successful program can use not only existing school plants and sites but also existing community resources.

Although ecology and environmental education mean different things to different people, the participants did agree on what environmental education is and what it is not.

Environmental Education is:
1. A new approach to teaching about man's relationship to his environment.
2. An integrated process dealing with man's natural and manmade surroundings.

3. Experience - based learning using the total human, natural, and physical resources of the school and surrounding community as an educational laboratory.
4. An interdisciplinary approach which relates all subject areas to a whole earth.
5. Oriented toward survival in an urban society.
6. Life centered and oriented toward community development.
7. An approach for developing self reliance in responsible, motivated members of society.
8. A rational process to improve the quality of life.
9. Geared toward developing behavior patterns that will endure throughout life.

The participants also agreed on what environmental education was not:
1. An 100% conservation of nature study.
2. A cumbersome new program requiring vast outlays of capital and operating funds.
3. A self-contained course to be added to the already overcrowded curriculum.
4. Merely getting out of the classroom.

Following upon this premise, the environmental education program would be divided into elementary, junior high, and senior high school levels. In the elementary school, environmental education would involve students working on small group projects rather than in individual projects. The projects might become more sophisticated as the student becomes older.

Environmental encounters can focus on a basic resource such as solid waste, air, water, as well as upon community environmental problems such as waste disposal, housing and recreation. These encounters can be a series of experiences that focus the attention of young people on the relationship of the economic, ecological, social, and political realities of living. Environmental programs can be designed to be topical and relevent to the particular needs of an individual school as well as to the community. For example, in the community that I reside, a decision has been reached to construct a controversial middle school after several years of emotional arguments. Typical of many built-up suburbs, the community has an increasing number of senior high school students and a decreasing number of children entering elementary schools. Now that the green light has been given, the local community officials will not issue a permit to install an "on-site" sewage plant or a septic tank while the State will not allow the use of the sewer lines which dump into an overcrowded water treatment plant. There is no doubt that someone is going to have to give in. Young adults could have made the question: "to build or not to build a middle school", a relevant project. I personally feel that they would not have done any worse than the adult community did. Another group of high school students currently are involved in finding the flood plains of their community and publicizing their location so that no unsuspecting homeowners will build there. At the same time they will research and suggest to the community concerning what to do about the flood plains.

During the years the student spends in elementary school, the basic science courses should play an important part in his environmental education so that he can become familiar with the environment. Children are not aware of the interrelationships of elements interrelated, or even of how their subjects in school are related to the environment. For example, the idea of field trips into the parks is not simply to identify plants, animals and insects, but to create an attitude to promote an understanding of how man uses and abuses nature around him. Children must learn to make value judgements regarding the environment, and to appreciate what impact their own decisions and actions can have.

The children in the lower grades can begin to study the general laws of nature which represent the logical sequence of learning. Variety, and similarity, patterns (leaves on a treee), interaction and interdependence (the ant colony), continuity and change, and adaption and evolution.

The emphasis should be on the children engaging in activities not always in the classroom but beyond the classroom into the community where the students are lead into the exploration and discussion of interacting forces - e.g. how communities are dependent upon soil, water, air, trees and birds; how plant life which includes vegetables and flowers go through their "life" cycle.

The children should be taken to the zoos and a discussion should be led by the teacher on animal survival. Children should be taken beyond the classroom into real world. All day trips to the state parks, science museums and public gardens, plus weekend trips in the spring and fall should be conducted where the emphasis should be on nature. If the school has a summer program, most of this should take place outside of the classroom.

The class should have as projects the raising of small pets like rabbits and mice, birds and fish and insects like worms, snails and ants.

Throughout his education, the student should be learning of the important part man plays in the environment and the necessity of cooperation among men to survive and to obtain results.

At the highschool level, many students are turned off by science. An informal approach in which students limit their scientific vocabularly and teachers feel comfortable because they do not require any rigorous scientific education. The approach should interweave taxonomical classification and open-ended research into all environmental learning so that students recognize that man and his environment are related to and dependent upon each other.

For example, in the study of history or anthropology the students would learn how the prehistoric animals lived and the part the environment played in the extinction of these animals. Additional study could focus on the non polluted water system of the Romans which permitted them to prosper while the polluted drinking water in Europe during the dark ages caused sickness and death, for many people and animals.

In America, the students should study why Thomas Jefferson and his fellow patriots orientated America toward an agricultured economy never dreaming that America would become a highly industrialized nation.

In both English and Social Studies, man and his environment can be studied by the reading of literature and investigating man's relation to the environment and his fellow man, and to himself.

We must understand why events took place and why we are moving so fast that we must question where are we going. Man is at the crossroads and must make a decision as to the future of man and determine how to live in harmony with the environment nature has offered. Nature and the environment will live long after man will have vanished.

Our thinking must change and man must learn to move away from the cities. Cities throughout the ages seem to hold a golden promise of instant wealth and glamour but rarely make man happy and content. Cities have always been places of paradoxes - problems of crime and poverty, yet, endless sources of inspiration to writers and artists. Man can be lonely and yet he can feel the pulse of humanity. He must be made to realize that he is part of the problem and part of the solution. He must either rebuild our old cities or build new ones that are planned with designated locations for residential, commercial, industrial and recreational areas.

It is said that 20th century man has replaced God with the dollar bill. An attempt must be made to arouse the interest and concern of the students in the problems of pollution. They must be aware that the threats are serious, that the losses are critical, and that need for action is urgent. The students must gain insight into the economics of pollution. It is difficult for them and for adults to answer affirmatively that pollution is threatening our existence especially if you live in a residential suburban area. The students must explore the problems of measuring the costs of pollutions and also the alternatives for controlling pollution. For example, rather than make the expensive anti-pollution devices for autos or installing a engine design, maybe it would be better to ban autos in urban areas and to create an efficient public transportation system. However, if this is done, how would the auto, oil and steel industries be effected. Is there a "limit to growth"? If so, where can people go to find jobs.

Admiral Rickover stated a few years ago "in the brief span of time - a century or so - that we have had a science based technology, what use have we made of it? We have multiplied inordinately, wasted irreplaceable fuels and minerals and perpetrated incalculable and irreversible ecological damage".

The philosophy of every business is that growth is necessary, and with growth, we have an increase in the amount of pollution which can be in the form of dirty air, dirty water or solid waste. The solid waste problem is approaching the critical stage where we just don't know what to do with our waste. Mr. Russell Baker in the New York Times suggested that we drop the accumulated trash on our enemy. This would solve our solid waste problem, save money by not using bombs, and devastate their land with our solid waste. Unfortuneately, someone might interpret the Geneva Convention and discover that it is illegal to dump solid waste from an airplane even during a war on enemy territory.

The students could obtain a copy of State's and the local community regulations on pollution control to learn the effectiveness of the regulations. They should be encouraged to meet with the community's zoning board and planning commission to find out what areas are set aside for industrial plants, shopping centers, homes, and apartments. Research should be done on traffic patterns, public transportation systems, sewage lines and the availability of utilities lines.

A small private school, Akiba Hebrew Academy, located just outside of Philadelphia has launched a successful program for several years to create a student awareness of the problems of the community. Since 95% of the school's graduates go to college upon graduation, restlessness occurs among the students after the middle of April when the students have been notified as to what colleges have accepted them. In order to graduate from this school, each student must select a community agency and work there from 4 to 6 weeks. Once a week in the morning, the students meet at the school to discuss their project. A report on their project is required by each student on their assigned project at the end of their work session. The students do not receive any money although some of them may get expense money to pay for lunch and transportation.

In the State of Tennessee an environmental awareness workshop has been established. One of these activities has been setting up an activity in which students and adults partitipate in a simulation experience as members of regional planning commission.

At these environmental workshops, each participant is given background of a community. Each member receives information on the type of person he is to assume. The instructor serves as the chairman of the group and explains the guidelines and restrictions under which the participants must work during the activity.

Problems are presented to the group for approval or rejection. The problems affect the life of at least one Planning Commission member in some manner. The member is expected to interact with the problem that is consistent with the role that has been assigned to him. The problem might be that of deciding where to locate a new agricultural - chemical factory. General and technical information is available and in addition, each participant has some "inside" information that governs his actions or his relations with his fellow board members.

Through these simulations, the workshop participants get opportunities to look at environmental problems from a multidisciplinary viewpoint. They are made aware of the pressures that are brought to bear on people who are responsible for making the decisions that affect the environment. Each problem has been designed to demand consideration of social, political, economic, and natural environmental impacts of the decision made by the group.

After completion of the simulation, the Chairman conducts a critique. A review of the actions of the participants provides insight into the impact of the decisions that the Planning Commission has made.

Another program has been successful in several states. This is an air monitoring project where high schools students have installed hundreds of air sampling stations throughout the state. One goal was to measure the amount of dust sedimentation and compare the amounts collected by other sample stations. After a length of time the samples were sent back to a central lab where they were analyzed. This concept is not new as monitoring devices to measure dust was first used in 1878 in London. One of the interesting results was that during the winter, large amounts of calcium chloride which is used to salt the roads was found in air samples.

In my talk, I have discussed environmental problems such as pollution. The solutions are not always scientific, but are sometimes of economical, political and social concern. By taking the youngsters out of the classroom, we can better acquaint them with the problems of the environment and make them realize they are needed to help determine the proper solution.

On June 5, the first United Nations Conference on the Human Environment will be publicized on four stamps. The official announcement of the stamp notes, "This is the first time in its history that mankind faces not merely a threat but an actual worldwide (environmental) crisis involving all living creatures, all vegetable life, the entire system in which we live, all nations whether large or small."

POLLUTION AND PUBLIC INFORMATION

TERRI AARONSON AND DANIEL H. KOHL

Environment Magazine St. Louis, Missouri

    The topics discussed here this morning may have been familiar ones to you. But how many ordinary citizens with non-technical backgrounds, do you think are aware of the health aspects of pesticides or of low-level radiation? Surely many citizens (and I speak of Americans in particular) are aware that DDT causes problems; but how many citizens can intelligently give you the pros and cons of banning DDT? And DDT is perhaps the best publicized environmental pollutant.

    Is there a means, though, by which a citizen <u>can</u> become well informed about environmental problems? The answer, some people believe, lies in scientists taking the initiative for translating their knowledge into laymen's terms. Is that worth the trouble? Does a citizen really need to understand that a certain percentage of mercury consumed by eating fish will actually lodge in the brain? Must your Aunt Millie understand that even natural background radiation can cause some amount of genetic damage?

    The answers to these and other questions like them can be derived from the fact that if a citizen in a democratic society is to be a part of the decision-making process, he must understand the questions before him.

    Environmental matters that governments the world over must deal with are not simply scientific or engineering problems. They are complex social, economic, and political issues that should be in the domain of the public -- not of the scientists and engineers alone. Whether emission standards for automobiles should be set at 0.4 grams of hydrocarbons per mile or 0.2 grams per mile is not simply a technical question. Citizens in free societies should have the opportunity to decide if the extra health benefits afforded by the lower emission standard are worth the extra monetary cost of a car meeting those strict standards. The part of the scientist in cases such as this is to disseminate information, not make decisions. A committee of the American Association for the Advancement of Science in a report on the social responsibility of the scientist put it this way:

> "In sum...the scientific community should, on its own initiative, assume an obligation to call to public attention those issues of public policy which relate to science, and to provide for the general public the facts and estimates of the effects of alternative policies which the citizen must have if he is to participate intelligently in the solution of these problems. A citizenry thus informed is, we believe, the chief assurance that science will be devoted to the promotion of human welfare."[1]

From a commitment by a group of scientists to the principles stated by the AAAS committee has grown the "science information movement." Several hundred scientists and engineers are now members of the sixteen committees for scientific information in as many cities in the U.S. Their work is coordinated by the Scientists' Institute for Public Information in New York. This "movement" has participated in numerous areas of concern, beginning with the consequences of atmospheric testing of nuclear weapons. In each case the scientists involved have attempted to bring objective scientific information to the attention of the public. The means of transmitting the information have varied -- from speakers at meetings, to Congressional testimony, to Environment magazine -- but let us stress again, that the concern of scientists involved in this movement is to provide information, not advocate solutions. Scientists and engineers acting in a professional capacity should restrict their actions to providing information. In their capacity as private citizens, however, they can, and perhaps should, be more concerned with "action-oriented" projects. The aim of the science information movement, however, is to separate objective science and political activity.

The way in which science information works is perhaps best described by citing specific instances in which it has led to changing a presumably set course of action. Let me begin by what is now a kind of "classic" example of science information providing the background for citizen participation.

Bodega Head is a beautiful promontory jutting out into Bodega Bay, situated about fifty miles north of San Francisco,

California. In 1958 the Pacific Gas & Electric Company (PG&E) began buying up land on Bodega Head with the aim of building a power plant there. Plans were initially delayed, however, by a stubborn land owner who refused to sell her property. During the slow judicial procedure of declaring the utility's right of eminent domain, local citizenry became aware of PG&E's intentions and began to organize. The opposition to a power plant on Bodega Head was founded in conservationist, as opposed to environmental, grounds -- a power plant would not be compatible with the recreational uses foreseen for the area by the citizens of Bodega Head; and the University of California was considering a marine biology laboratory there that would certainly not be compatible with a thermally-polluting plant.

Opposition to the plant became focused, though, by an announcement by PG&E that the proposed plant would be nuclear.

The site on Bodega Head, chosen by PG&E was only 1,000 feet from the San Andreas fault, the most active source of earthquakes in the United States. Officials of PG&E maintained, however, that a nuclear plant could be built that would safely survive the worst foreseeable earthquake in that area. The officials did hedge somewhat by admitting that there might be some damage to the reactor core even though the building would remain intact. Presumably the building would be able to contain any amount of radiation that would be released in case of a major reactor accident following disruption by an earthquake.

Citizens of Bodega Head were not convinced by PG&E's assurances. Organized into the Northern California Association to Preserve Bodega Head and Harbor, the citizenry amassed an impressive amount of scientific information. The impact of this information was provided by the testimony of independent scientists, technical reports by individual geologists, and by publication in Environment[2] pointing out the important test of public policy concerning nuclear safety that was taking place at Bodega Head. In spite of Environment's small circulation at that time, newspapers throughout the country picked up the story and printed the information.

Opposition to the Bodega power plant grew when, during site preparation, PG&E discovered a fault running directly beneath the reactor core. In spite of this and increasing evidence of the geologic unsuitability of the site, the utility continued to insist that a reactor could safely be situated there. By the fall of 1964, however, it became clear that the AEC would almost certainly disapprove the proposed reactor. On October 30, 1964 Robert H. Gerdes, then president of PG&E was quoted in the San Francisco Chronicle as stating, "We would be the last to desire to build a plant with any substantial doubt existing as to public safety."[3]

So saying, he announced that plans for the reactor on Bodega Head had been dropped.

Whether citizen action, backed by expert science information was actually <u>the</u> causative factor in the dropping of the Bodega Head plans is not likely ever to be known; but it is probable that without citizen intervention, plans for Bodega Head would have advanced much more quickly, especially if the opposition had not been fueled by accurate scientific information.

The case of Bodega Head was only the first in a long line of proceedings involving nuclear reactors -- proceedings that often take years (although the process in the U.S. will soon be streamlined somewhat as a result of rule changes that are now being made). In spite of the annoyance of delayed utility schedules, Americans have had an opportunity to take part in decisions affecting their lives. They have been able to participate, though, only because they have had access to information that enabled them to discuss the situations at hand.

The second case that I would like to mention this morning deals with the herbicide 2,4,5-T. In this case scientific information played a part in affecting a decision, but this instance also highlights the importance of the availability of information to scientists themselves.

In the spring of 1970 the registration of 2,4,5-T was suspended for some uses and cancelled for others (the difference being that cancellation is a weaker regulatory move because if the action is appealed by a manufacturer, the substance in question can still be sold in interstate trade during the lengthy appeal process). The safety of 2,4,5-T has been questioned because of the chemical's ability to cause birth defects in laboratory animals. The cancellation orders were appealed by Dow Chemical Company and Hercules, Inc. As part of their rights under the U.S. Federal Insecticide, Fungicide, and Rodenticide Act, the manufacturers requested referral of the matter to an advisory committee. The committee was selected from a list of names provided by the U.S. National Academy of Sciences. The committee met during the winter of 1971 to consider the safety of 2,4,5-T and to write a report. In May, 1971 the report was submitted to William Ruckelshaus, administrator of the Environmental Protection Agency. The report recommended that the registration of 2,4,5-T be reinstated for all uses in question, but that a contaminant of 2,4,5-T, known as dioxin, be kept at a specified minimal level. The dioxin contaminant is thought by many to be responsible for the teratogenic effects of 2,4,5-T. On the other hand, some studies [4] definitely indicate that 2,4,5-T alone is capable of causing birth defects in laboratory animals. At any rate, the conclusions of the advisory committee on 2,4,5-T were not unanimous. There was one dissenting opinion -- that of Dr. Sterling. Sterling dissented, stating that our knowledge of 2,4,5-T is not sufficient to warrant its use around the home, recreation or similar sites, and on food crops.

Advisory committee reports in the past have been confidential at least until a final decision has been reached by the appropriate officials. However, a copy of the advisory committee report on 2,4,5-T came to the attention of the Committee for Environmental Information (CEI), publisher of Environment, before a decision had been made by Mr. Ruckelshaus on whether to follow the 2,4,5-T committee's recommendation or to take other steps regarding the registration of the herbicide. In accordance with the aims of the science information movement, CEI felt that the material contained in the advisory committee report should have been a matter of public record. Furthermore the members of CEI felt that the material contained in the advisory committee report was incomplete and that the committee had been overconfident in its assumptions. The deficiencies of the report were publicized by CEI at a press conference held in July of last year. Shortly after the press conference Mr. Ruckelshaus announced that restrictions on the use of 2,4,5-T would remain in effect. In making this move Mr. Ruckelshaus disregarded the recommendations of the advisory committee. It is much less likely that the EPA administrator would have done this if he had not known that a prestigious group of scientists had little confidence in the conclusions of the advisory committee's report.

What is most significant in this case, though, is not the decision on 2,4,5-T, which, by the way, has again been appealed -- this time by a request from the manufacturer for a public hearing. What is most significant is that within one week of the press conference on 2,4,5-T, CEI had word from the Environmental Protection Agency that henceforth all advisory committee reports would be made available to the public prior to decisions being made concerning the registration of the substance in question. In an announcement of this policy change last July 28, Mr. Ruckelshaus said, "These reports form an important part of the scientific evidence upon which administrative determinations as to the possible cancellation or suspension of registration of a pesticide will be based. As such, they should be made available to the public as rapidly as possible." 5

The awareness of the Environmental Protection Agency that scientific evidence is only background for social decisions is of major importance to the cause of science information, for without access to the technical information submitted to EPA, independent scientists are unable in turn to present the information to the public. In the case of 2,4,5-T it was merely fortuitous that we received the information and were then able to disseminate it. Hopefully, more information that was once held confidential will

be available to us in the future. Recent announcements by the
Food and Drug Administration that approximately 90 percent of the
information it receives will be available to the public are
definitely a step in the direction of supporting the aims of
science information.

To be fair, we should note that there are difficulties in
informing the public. There must be some aspect of an issue
that catches the public's eye -- something to perhaps make it more
personal. Would the citizens of Bodega Bay have actively opposed
a power plant if it were to be placed next to the town dump instead of in a spot of particular beauty? The answer is by no
means certain.

Sometimes information dissemination has not been immediately
effective. Issues have had to lie at rest until some major,
usually adverse, incident occurred. Scientists in the U.S. were
aware of the hazards of mercury pollution a full two years before
it became a public issue in 1971. The May 1969 issue of Environment
was devoted entirely to the problems of environmental pollution by
mercury. Why was the public unconcerned about mercury in 1969?
Probably because no one incident stood out that made people sit up
and take notice. Mercury as an issue in America had to wait until
a Canadian graduate student found excessively high levels of mercury in fish caught in Lake St. Clair. At that point the information that had been gathered by independent scientists became
invaluable in educating the public to the consequences of mercury
pollution.

PCBs -- polychlorinated biphenyls -- were the subject of an
article in Environment in January 1970.[6] PCBs were not a public
issue in America until some months later when Congressman
William Ryan of New York, having read the Environment article began
active opposition to the widespread use of these chemicals. And
it was not until the summer of 1971, when thousands of chickens
and eggs were found to be contaminated by PCBs, that many citizens
learned that PCBs even existed.

Our most recent case of an attempt to alert the public to a
newly discovered potential hazard concerns phthalate plasticizers.
Plasticizers in particular have been shown to leach or vaporize
out of the plastic, with unknown, but quite possibly hazardous,
effects on human health.[7] Certainly it is the right of the public
to know that the pleasant smell of a new car is actually vaporizing
plasticizer being inhaled with unknown health consequences.

Automobile manufacturers are aware of the problem -- primarily
because the vaporizing phthalates contribute to window fogging.
Awareness does not mean that phthalates are no longer used.
General Motors Corporation, for instance, recently requested that
its suppliers of polyvinyl chloride for car interiors delete

phthalates from their formulations -- ostensibly to cut down on window fogging.[8] It is possible, though, that GM's request -- and it is only a request -- was in anticipation of public recognition of possible health hazards from phthalates. Again, it is the purpose of those in the science information movement to make available all the information concerning phthalates that is known, plus, of course, detailed accounting of those potentially significant aspects of phthalates that are not known.

In reviewing the cases of Bodega Head, 2,4,5-T, mercury, PCBs, and phthalates -- all instances in which science information had a hand -- I think you will be aware that those people dedicated to science information have attempted to bring issues of public importance into the arena of public discussion. But in order to inform the public, scientists and others must have information available to them. We constantly need more and more data as we get deeper into the complexities of environmental matters. So far we have concentrated on bringing issues before the public. Now we need information about how to solve some of those problems that have been raised. Certainly it is our hope that engineers and other technically competent individuals will join those in the science information movement in telling the public of the technical means available or proposed that can aid in solving our environmental and other scientific problems.

NOTES

1. "Science and Human Welfare," report of the AAAS Committee on Science in the Promotion of Human Welfare, 1960.

2. Mattison, Lindsay and Richard Daly, "Bodega: The Reactor, the Site, the Hazard," Environment, April 1964.

3. San Francisco Chronicle, October 30, 1964.

4. Results of numerous studies on the teratogenicity of 2,4,5-T have been summarized in the "Report of the Science Advisory Committee on 2,4,5-T," submitted to the administrator of the Environmental Protection Agency, May 7, 1971.

5. Environmental Protection Agency, press release, July 28, 1971.

6. Risebrough, Robert with Virginia Brodine, "More Letters in the Wind," Environment, January-February 1970.

7. Shea, Kevin, "The New-Car Smell," Environment, October 1971.

8. Chemical and Engineering News, March 27, 1972, p. 7.

ARE INDUSTRY AND GOVERNMENT FULFILLING THEIR RESPONSIBILITIES FOR POLLUTION CONTROL?

Mitchell R. Zavon

Huntingdon Research Center

The effect on health of environmental pollutants has been difficult to prove in many instances. Many of those concerned with environmental pollution have made allegations unsupported by reliable data in their zeal to redo what they have viewed as a rapidly deteriorating environment. Frustrated by inability to prove an effect on human health, they have given up on man as a focus and have based decisions on the presumed effects on other species or on materials destruction.

When the advisor has no ultimate responsibility for the effects of his advice, the feedback effect is inadequate. It is very easy to make recommendations if the consequences of those recommendations are unlikely to affect the advisor in any material way. We have had a highly contagious epidemic going on for at least five years. This epidemic consists of destructive criticism on the part of well intentioned people. They forecast dire consequences for all mankind if their criticism goes unheeded. The disease syndrome in many instances includes painful cries of gloom and doom and occasional expostulations because a chemical compound has been newly discovered in sphagnum moss or mothers' milk. Recovery from the disease may be hastened by a grant for a study of the newly described "disaster" or sufficient newspaper headlines to make a celebrity of the "scientist" who first developed the "illness".

The increase in the number of non-responsible critics has the effect of making industries and governments cower in fear of adverse publicity. Industrial managers, fearful of the impact of allegations of adverse effect from their actions, fail to demand supporting data and fail to present the known facts and alternatives

to the procedures being criticized. Government managers, ever
sensitive to public criticism, hear only the loud clamor of a
noisy few and fail to ask whether those who yell the loudest are
representative of anyone other than themselves. Historians of
the future will undoubtedly study this epidemic of "environmental
concern" with the same ill-concealed amazement that we now reserve
for the study of earlier mass hysterias such as the Childrens
Crusade or the intense adulation of a singing group such as the
Beatles.

The hysteria which has prevailed worldwide does not result
in detailed consideration of real environmental problems which must
be properly managed if man's health is not to be seriously threatened. Proper environmental management and hysteria about the environment are quite different matters. The Christian Science
Monitor of December 18, 1970 (1) gave an example of the type of
illness which I have described. An officer of an organization
dedicated to preservation of the environment had urged from a
location 3000 miles distant, that the people of the State of Maine
stop cutting evergreens for Christmas trees and substitute artificial trees. This, presumably, to prevent excessive cutting of
the Maine forest. But properly managed forests are a renewable
resource and failure to cut timer can be wasteful and result in
more disease of trees in the forest. This is not proper management of the environment. It is an amateurish love affair with
what is thought to be natural.

THE PURPOSE OF INSTITUTIONS

Society develops institutions in response to the need for
organized effort in order to accomplish specific functions. Government, industry and universities are products of the need to organize
effort to fulfill societal needs. The industrial revolution, which
transformed life from a rural-agrarian society to a more central-
ized, production oriented society, created large scale industry for
more efficient utilization of resources. The factory which developed as an institution became the central focus of an industrially
oriented society.

Industrial organization is the product of its times. Rarely
the leader in social innovation, industry is frequently the stabilizing influence because the needs of producing for society's needs
requires stability to insure raw material flow and the availability
of labor. The goal and function of industry has always been production. The idea that profit is an end in itself serves to obscure
this fundamental fact. Industry follows the prevailing ethic of the
society which it serves. Raw materials are bought at the least
possible price. Labor is paid the lowest wage at which it can be

bought. Waste from the manufacturing process is disposed of at the least possible cost. This has been the ethic of our society until very recently. It continues to be the ethic today for significant portions of the world's society. For most of the period of industrial development the only real responsibility of industry has been to produce.

The ethic by which our society lives is beginning to change. The realization that the frontiers are gone, that man is inundating the earth and using its non-renewable resources at an ever increasing rate has forced the beginning of a new ethic. Simultaneously, we have developed the ability to control the ancient scourges of mankind, hunger and pestilence. We can now ask - and hope to answer - are we harming man by our own actions? Are we creating diseases of man made origin as we reduce or eliminate disease caused by microorganisms? The complexity of the questions we must answer requires large, organized effort and substantial monetary support. Only in recent years have these questions become of prime concern to governments. Only occasionally since the plague of 1665 in London have governments had to concern themselves with the total environment in which the populace lives. We now face a situation in which human health has become of major political importance and voices are raised to influence those political decisions. If decision is based largely on whose voice is loudest, scientific-political decisions will not be made in the best interest of the people of this planet.

MANAGEMENT

The direction of industry and of government is provided by managers who usually occupy their positions for comparatively short periods of time. Achievement is measured in periods as brief as a year. Five years is a long time in this scheme of measurement. The manager of an enterprise is expected to show results - fast.

In the United States the concept of the "profit-center" has dominated industrial management in recent years. Profit is the measure of performance. The manager of a factory, a sales office, a division of a larger company is expected to show that the part of the operation for which he is responsible shows a profit. His position depends on his ability to produce a profit. He cannot really be concerned with the overall profitability of the entire company or with its profitability five, ten, or twenty years hence. He must produce results today for that is the basis on which he is judged.

This situation holds true for many managers in government positions in the United States and is also true in large part for managers in industry and government in many other countries. If managers must show quick results there may not be time to fully determine the facts. Impatience has come to be looked on as a sign of the right kind of attitude. Forgotten is the fact that all motion is not progress.

ENVIRONMENT AND POLLUTION

The natural environment is disrupted by any system which imposes large numbers of one species in a limited area. Man disrupts the natural environment by industrialization, rationalization of agriculture and urbanization. Alteration in the natural environment occurs because of the needs of a complex human society. There is little evidence that the alterations are bad for human health. We require an environment in which man and the other creatures inhabiting the earth can continue to exist and in which the quality of man's life is such as to allow a full span of years in reasonable comfort and health.

It is overly simplistic to talk of "pollution" and the adverse effects of "pollution" without definition of the term. The words "pollution" and "ecology" have become useful to headline hunters and headline writers, but have been used loosely and so have lost meaning. If adverse health effects can be shown to have resulted from a change in the environment, they are given wide publicity. If no adverse effect is shown, the results of the study are generally ignored (2). This results in the misleading impression that all environmental changes are bad.

There are two basic problems which overshadow all others and result in a world where human problems are obscured and neglected. The first of these is the national, racial, religious, economic and human rivalries which have lead to continuous warfare throughout modern history. The second is the tremendous population growth of the past fifty years. Unless war and its disastrous effects on the environment, direct and indirect, can be prevented, our concern with environmental problems ultimately becomes an exercise in futility. Unless population growth is stopped, we only delay the inevitable destruction of our environment and exhaustion of the earth's natural resources.

Let us assume that solutions to these two problems are possible although little evidence is presently available that effective answers are in sight. If solutions are possible than we can discuss whether industry and government are fulfilling their responsibilities in solving pollution problems and controlling pollution. To explore the question we might first ask a few preliminary questions.

1. Is the quality of the environment on earth deteriorating?
2. Is life today worse than it was 100 years ago?
3. Is the air more "polluted" now than it was a century ago?
4. Is the water more "polluted" now than 100 years ago?
5. Is the health of man worse now, by any reasonable measure, than it was 100 years ago?

Whether or not conditions are worse now than they were 100 years ago, is immaterial, we can certainly improve the present conditions. Historical perspective is however helpful in understanding how far we have come.

My thesis is that neither industry nor government is fulfilling its responsibility in protecting the environment and enhancing the status of man. At the risk of oversimplification, I maintain that the failure is basically the result of the "profit center" concept. The need for quick solutions, for immediate answers, for maintaining a sense of crisis in order to get credit for solution of the "crisis" results in failure to undertake research which must be long term, failure to limit profit in the interest of long term profitability, and a use of non-renewable resource without regard to long range consequences.

Hardin (3) in his much quoted paper on the "Tragedy of the Commons" refers to the tragedy of overgrazing and lack of care of land which was used by all. He extends the application of this social institution to other areas such as water and air and Crowe (4) uses the simple mathematical model of the utility concept to illustrate. If each individual can add one sheep (kilo of waste could be substituted), he gains an increase of almost one in utility and the negative utility is only a fraction of minus one. Obviously the summation benefits the individual herdsman but all the herdsmen come to the same conclusion and we have overgrazing and destruction. We have seen a similar series of actions all over the world in the use of waterways and air for waste disposal. One factory on a water course may foul the river but the river's absorptive capacity may be sufficient to recover in a reasonable distance downstream. Even a second plant may be acceptable. But further industrialization overwhelms the "common" river and deterioration takes place.

When industry and government are separate and distinct institutions with government regulating industry, the institutional relationships are relatively simple. In today's society, industry may be a branch of government or largely controlled by government. Government may establish an industry and be a sole owner. The institutional lines are blurred. The manager, irrespective of whether in industry or government, too often has conflicting goals. In addition to the "profit-center" mentality, he may have conflicting objectives. He may be under pressure to produce more coal for a power plant but at the same time the only way he can achieve that goal in the time alloted is to resort to the environmentally destructive practice of strip mining. Too often the health of individuals or of the populace is relegated to secondary consideration as these difficult decisions are reached. Too often the persons who could advise on the health effects of an action are not included in the decision-making process by either industry or government.

When the earth's human population was a billion or less, we abused our waters and our air as much or more than we do now but the results were not as significant because the population was small compared to the present human population. Disease of man made origin was a minor factor. Present day methods of illness prevention were not available. Malaria, the great plague of mankind, had no cure. Levi could write (5) "In this region malaria is a scourge of truly alarming proportions; it spares no one and when it is not properly cared for it can last a lifetime. Productive capacity is lowered, the race is weakened, the savings of the poor are devoured; the result is a poverty so dismal and abject that it amounts to slavery without hope of emancipation ...."

Malaria remains the most important disease of man for which the cause and cure is known and for which effective methods of prevention are known. The use of DDT to control malaria has had revolutionary effects on the human population of large parts of the world. Yet, in response to accusations of harm from DDT, many responsible managers in industry and government have failed to insist on the same quality of scientific evidence of harm that would be required of a doctoral candidate at any first rate university. As a result of this failure, the continued use of this most beneficial of man's discoveries is seriously threatened.

Industry has continued to externalize its costs by relying on the social system of the "Common." In this tragedy it has been aided and abetted by governments which look at balance of payments rather than long term benefit for its people. It makes little difference whether we are talking of a capitalist society or a socialist society, industry and government have proceeded in similar fashion irrespective of the economic system under which they operate (6).

Industry has continued to run fearfully when in doubt and to advertise perhaps to a greater extent than to act. The need for reorienting ourselves away from the "Commons" approach is quite obvious. But industrial management has not yet fully recognized this fact and governments are only slowly becoming aware of this new reality.

In many respects, the environment in which we live may be no worse than it was 100 years ago and in some respects that environment may be better. The pall of wood and coal smoke which used to hang over most temperate climate cities in winter has disappeared to be replaced by the effluent from automobiles. The stench of sewerage has been eliminated in many areas. But all is not well. We cannot measure against perfection, for we don't know what perfection is. We must measure in relation to what is possible or what has been achieved.

We cannot allow industry to disperse its wastes into the common waterway or the common air without evaluation of cost to society versus alternative costs. At the same time we cannot afford to allow the myth of pollution to be perpetuated by those with no regard for the effects of their actions. Recently observation of bitumen in the oceans was given as evidence of widespread pollution. But this has been antedated by 2000 years. Boatmen in the centuries before 1 A.D. reported that "lumps of bitumen rose to the surface of the sea, the size of decapitated bulls" (7). Pollution is not new and to claim that it is new may add drama but doesn't help to decide priorities on a rational basis.

SOLVING OUR PROBLEMS

Man's continued existence and improvement in his state is the goal of many. In the United States any further improvement in the standard of living will in large part be dependent on the availability of electrical power. Any future improvement in the standard of living of the less affluent in our society will depend on the availability of electric power. More electric power production brings with it further need for control of effluent from power stations. Control of effluent must be meshed with optimization of resource use. If we fail to use resource allocation as the basic criterion for decision making, our efforts must ultimately fail, at great cost to all mankind.

Malaria, not air pollution, remains the most significant world health problem (8). If we are ever to solve our health problems they must be sorted out and given their proper priorities. Health, as I said earlier, has become a political question. It is no longer solely a scientific or medical matter. If the public is to participate in a meaningful way in ordering priorities and helping to solve health problems, it is up to the managers of industry and government to make health problems understandable. If environmental problems with health implications are to be understood by the public, these problems must be simplified. Hughes has said, "Complexity can induce the public to give up, if only out of sheer satiation" (9). Environmental problems are complex. They must be simplified in order to allow intelligent political decisions. If there are health effects from environmental pollution, the evidence must be made available. If there are no health effects the public should be informed so that the political decisions can be made without the threat of adverse health effects.

Neither industry nor government will be fulfilling its responsibility unless the scientists and engineers who work in these two institutions cease to behave like advocates for a cause and attempt to work together toward rational solutions to environmental problems. In some parts of the world such an approach is now the rule. Without such an approach the people are the ones who will suffer and human health and the human condition will not improve.

REFERENCES

1. Gould, J., Christmas Ornament. Christian Science Monitor 18 Dec. 1970.
2. Committee on Pest Control and Wildlife Relationships, Division of Biology and Agriculture, National Academy of Sciences - National Research Council Publication 920. National Academy of Sciences - National Research Council, Washington, D.C. 1963.
3. Hardin, G. Science 162, 1243 (1968).
4. Crowe, B. L. Science 166, 1103 (1969).
5. Levi, C. Christ Stopped at Eboli.
6. Goldman, M. I. Science 170, 37 (1970).
7. Allegro, J. M. The Sacred Mushroom and the Cross. Doubleday 1970 pp. 65, 66.
8. Pollack, H. and Sheldon, D. R. The Factor of Disease in World Food Problems. JAMA 212:4 598 (1970).
9. Hughes, T. L. On the causes of our discontents. Foreign Affairs 47:4 653 (1969).

INDUSTRIAL ZONING

A.E. Donagi AND I. Nizan
Israeli Ministry of Health

Environmental nuisances created by industrial plants and workshops are recognized as one of the severest problems in the field of industrial zoning. In 1661, John Evelin wrote(1) indignantly that London was wrapped "in Clouds of Smoake and Sulphur, so full of Stink and Darkness". He suggested that "all industries be moved to the leeward side of the city and that sweet-smelling trees be planted in the city itself". We may regard him as the father of industrial zoning.

Investigation of the situation in Israel and abroad shows the existing lack of appropriate environmental planning, as a result of which inhabitants are subjected to harmful effects due to industrial nuisances.

Today, too frequently, industrial zones are placed adjacent to residential neighborhoods. At times, although industrial zones and residential areas are built at some distance from one another, rapid urbanization and faulty planning result in their growing close together. In addition to industries, environmental nuisances are sometimes created by "services" such as solid and liquid waste disposal plants, incinerators, tank farms and airfields (where residential areas bordering on the flight path may be affected).

Compromises must be made between public health and welfare, on the one hand, and technical, economic, and political factors on the other. We may therefore regard zoning as an instrumental guide to private and public land use, taking into consideration the micrometeorological conditions, the nuisance producing potential of the industry and the nuisance control installations.

Three bodies are, or should be, interested in and affected by the choice of the plant's site:

1. The owner of the plant - chooses the site mostly according to economic factors--distance from raw materials, markets and manpower.

2. Part of the population - who regard the industry and the development of their area as a source of income.

3. The rest of the population - who, as a result of the zoning, may suffer from offensive industrial effects at their dwelling area.

Let us look at some solutions that other countries have found.

U.S.S.R.

According to the labor laws in the U.S.S.R., plants expected to discharge objectionable pollutants must be located downwind of nearby residential areas and be separated from them by "sanitary protection zones" (2,3). Those zones are divided into five categories: 1000 meters, 500 meters, 300 meters, 100 meters, and 50 meters.

Poland

The Council of Ministers authorized the chairman of the Central Water Utilization Bureau to establish the width of protective zones around sources of pollution. The categories are the same as in the U.S.S.R.(4).

France

Industries are classified, according to nuisance, into three classes(5,6,7). In the first class are the establishments which must be constructed outside urban areas and which are subject to authorization only after inquiry and examination. In the second class are those

accordance with the quality and quantity of its raw materials, products and by-products.

The relative value of each of the parameters, and the type of industry and its appropriate distance group, were based on specific knowhow, personal experience, technical references, and literature on the subject. Appropriate hearings were also organized.

The final results were published in 1971 as "Industrial Zoning in Israel"(8). This publication is divided into three parts. Each part is intended to give guidance to national, municipal and regional town planning committees.

The classification in each of the parts is in accordance with that of the Central Bureau of Statistics.

The guide in its present form is merely a beginning. Attempts at improving matters continue. The guide's main advantage is that it affords the possibility of initiating planning and is of assistance to town planners and economists. In spite of the fact that it was designated as a "Tentative List of Standards", this guide obtained a semi-legal status in Israel, following its official adoption by three various governmental ministries.

It should be reiterated that the publication is a preliminary work which needs completion by inclusion of various parameters which have not as yet been treated in full. Some of these are listed below:

1. The economic aspect - what is the land-use designation of the specific area? (amusement area, industrial area, development, etc.)
2. The topographic, meteorologic and climatic conditions.
3. Existence of nuisance control installations.
4. The effect of one industry upon another.

In spite of the various shortcomings and the incomplete basis for some of the values appearing in the tentative list of the sanitary protection zones, it has numerous advantages, and can serve as a useful tool for a better environmental design in the future. The major goals that already may be obtained by using these standards are the following:

Fig. 1: TENTATIVE CLASSIFICATION OF TRADES AND
INDUSTRIES INTO SANITARY PROTECTION ZONES

1. Abbatoirs, by-products, production	5
2. Abbatoirs, regional preparation and preserving of meat (over 50 heads/day)	5
3. Abbatoirs, regional, up to 50 heads/day	4
4. Abbatoirs, small animals, regional, (over 1000/day)	4
5. Abbatoirs, small animals, (up to 1000/day)	2
6. Abrasive stones, manuf.	2
7. Accumulators, manuf.	3
8. Aeroplanes and aircraft parts, manuf., assembly and repair	6
9. Airport, international	6
10. Airport, domestic	6
11. Aluminum, extrusion and profiling	3
12. Aluminum, reduction (electrolytic)	6
13. Aluminum sulfate, production	3
14. Ammonia, production	5
15. Ammonium nitrate production	5
16. Ammonium sulfate production	5
17. Ammunition, production	6
18. Animal corrals holding over 300 heads	5
19. Animal waste storage	4
20. Argon, compressed, production	3

which can be erected in urban areas after appropriate inquiry and authorization. In the third class are those which need only a declaration and have to follow general rules of air pollution prevention.

Israel

In Israel, an interministerial committee for industrial zoning was set up in 1965. The members of the committee represented the Ministries of Health, Interior, Commerce and Industry, and Agriculture.

The purpose of this committee was to determine guidelines for the approval of newly planned industrial plants and zones, as well as for enlarging existing plants. The purpose of the work of the committe was to search for the minimum distances between specific industries and neighboring residential areas; and to find a way to adjust the minimum distances to the diminutive size of the country, taking into account the overcrowding of the population in the shore area. The distances decided upon were 2000 meters, 1000-2000 meters, 500-1000 meters, 150-500 meters, 50-150 meters and less than 50 meters. Fig. 1 gives an example of the Tentative Classification of Trades and Industries into Sanitary Protection Zones. The first 20 items, alphabetically arranged, out of the 456 various items on the list are shown in the figure.

Each industry was examined from the viewpoint of the expected nuisances, quanitatively and qualitatively, according to the following categories:

1. Air pollution - dust, aerosols, gases, etc.
2. Noise - different disturbances at various frequencies.
3. Liquid wastes - surface and underground water pollution, river and sea pollution, etc.
4. Problems resulting from creation of solid wastes.
5. Possibility of fire hazard.
6. Possibility of explosions (gas and fuel tanks, explosives, etc.
7. Offensive odors.
8. Radiation hazards.
9. Other specific problems.

Throughout the evaluation of the expected nuisance, each specific industry was analyzed and categorized in

1. The protection of the public from the deleterious effects of noxious industries.
2. Prevention of uncontrolled growth of existing plants that can be a source of environmental nuisances.
3. Assistance in the preparation of a countrywide master plan, where specific noxious zones will be set up, in accordance with meteorological, climatic, geographic and social conditions.

Bibliography

1. Evelyn, J.: Fumifugium or the Inconvenience of the Aer and Smoake of London Dissipated, First published in 1661, repr. by Natl. Soc. for Clean Air, London, 1961.
2. Stern, A.C. (Ed.): Air Pollution, Vol. III, p. 64, Academic Press, New York, 1968.
3. Kettner, H., Bull. Inst. Wasser-, Boden-, Lufthyg. 12, 46 (1957).
4. Stern, A.C. (Ed.): Air Pollution, Vol. III, p. 825, Academic Press, New York, 1968.
5. Journal Officiel de la Republique Francaise; Law No. 61-842, 3rd August 1961.
6. Journal Officel de la Republique Francaise; Decree No. 63-963, p. 8539, 21 Sept. 1963.
7. Journal Officiel de la Republique Francaise; Legislation Nomenclature et Reglementation des Etablissements Dangereux Insalubres ou Incommodes, Journaux Officiels No. 1001, 1967.
8. Donagi, A.E. and Nizan, I.: Industrial Zoning in Israel, Min. of Health Publ. No. APR-20, 70 pp, 2nd. Ed., July 1971.

SUBJECT INDEX

aerosols, 54
  acid, 149
agricultural chemicals, 133
  fertilizers, 136
agricultural refuse
  burning, 152
air pollution
  atmosphere motion, 424, 453-467, 469-476, 511, 554
  aluminum smelting, 548
  caused by agriculture, 148
  fluid mechanical model, 77 ff
  forecasting, 415, 675, 734
  in Canada, 643
  in Israel, 664
  in South Australia, 618
  monitoring from satellites, 54 ff
  particulates, 564
  raman scattering, 439
  temperature inversions, 455
air quality testing
  airborne sensor systems, 415-432
  carbon monoxide, 405
  flouride compounds, 550
  hydrocarbons, 409
  instrumentation, 398-414
  nitrogen dioxide, 409
  ozone, 407
  particulate matter, 400, 420, 564
  standardization of methods, 410
  sulfur dioxides, 401, 666
aldehyde, 80 ff
amonia, 61
  amonium sulfate, 3
  free, 350
  scrubbing 229, 237, 245
Anopheles, 162
antimetabolites, 188
antimony, 99 ff
  organoantimony, 95
ash, 105
  fly ash, 198
  pluverised fuel ash, 136
aurora, 26
  artificial 27
arsenic, 186

Bahco-scrubbing, 259
Bilharziasis, 164

Bischoff process, 259
blackbody emission
    spectra, 60
boilers, 105 ff
    air heater fouling,
        112
    cleanliness, 105 ff
    coal fired, 124 ff
    cold end corrosion,
        112
    packaged oil, 129
boranes, 99 ff
boron, 136
    boron hydrides, 95

cadmium, 322
calcium, 193
    arsenate, 186
    polysulfide, 141
    sulfate, 198, 220
    sulfite, 198
carbon
    activated, 140, 315
    carbon dioxide, 1,
        54 ff, 193
    carbon monoxide, 2,
        54 ff, 149, 405
    dust and particulates,
        3, 118, 149
    hydrocarbons
        unburned gaseous,
            3, 409
        chlorinated, 186
catalysts, 282
    catalytic absorption,
        265
    catalytic oxidation,
        227
    catalytic reduction,
        281
chimney effect, 38
chlorine, 193
clouds
    cirrus, 7
    cumulus, 13
cocoa, 185
Colorado potato beetle, 186
cooling towers, 13, 19 ff,
    151

contrails, 3 ff
copper, 134, 322
    cupric oxide, 186
    fungicides, 134
    smelting industry, 220
correlation spectrometer, 65
cross flow absorber, 196

DDT, 138, 141, 186
death rate, 15
dirigible mirror, 38
dust
    carbon, 3, 118, 149
    cement, 149
    control, 122 ff

"ecotechnics", 177
emissions
    aircraft, 1
    crankcase, 153
    exhaust, 80 ff
energy consumption, 12
    crisis, 11
    solar, 48
erosion, 171
ethene, 149
explosives, 268 ff, 288 ff,
    298 ff, 304, 307
    incineration, 298 ff
    TNT, 288 ff, 307

fertilizers, 136
flourides, 149, 550
    compound toxicity, 552
flue gas, 105
    desulfurization, 224
fluorosis, 149
fog, ground, 13
forest burning practices,
    155
fuel
    additives, 105
    consumption, aircraft,
        3
    sulfur, 106

gaseous pollutants, 58ff, 149

SUBJECT INDEX

haloxydine, 140
heat island, 13, 38
heavy oil, 236
hycanthone (HC), 165
hydrogen sulfide, 65
hydrometallurgical pressure leaching, 219
Hypomagnesaemia, 136

incinerators, 358-367, 371
industrial chemicals, 95 ff
Infrared Interferometer Spectometer (IRIS), 65
insecticides, 142
   degradation by microorganisms, 141
   evaporation, 137
   organochlorine, 135, 142
   photochemical degradation, 139
insect pest situations, 182 ff
   control, 158 ff
   population models, 504 ff
inversion layers, 39, 455
ionosphere modification, 28
isobutylene, 80 ff

landscape degradation, 168 ff
lapse rate, 39
leaching, 140
lead arsenate, 186
   mining, 320 ff
   styphnate, 312
legislation
   against herbicides, 722, 761
   against pesticides, 721
   in Canada, 625, 633
   in Israel, 651-663
   noise, 572, 594
   U.S. Environmental Protection Agency, 607
   zoning, 774
lime, 192 ff
   lime-limestone fly ash, 225
   lime-limestone scrubbing, 192 ff

magnesium oxide, 106
   scrubbing, 226
manganese, 106
marble-bed absorber, 196
Medium Resolution Infrared Radiometer (MRIR), 71
Meisenheimer complexes, 272 ff
mercaptans, 149
mercury, 138
   fungicides, 134
   organometallic methyl-mercury, 95
methane, 61, 281
mine backfilling, 331
mineral acids, 310
molecular diffusion, 77
molecular sieve, 282
Multi-Spectral Scanner, 71

NASA Data Processing Facility (NDPF), 73
natural variety, 173
New Source Emission Standards, 122
nitrogen, 136
   nitric acid, 65, 279, 310
   plants, 278 ff
   nitrobodies in aqueous effluents, 288 ff
   organic, 350
   oxides, 2, 6, 80 ff, 149, 193, 278 ff, 409
noise pollution, 572-586
   control, 576, 594-601
   criteria, 573
   hearing loss, 574, 602
   measurement, 575
   motor vehicles, 587-593

non-ferrous smelting industry, 215 ff
nucleophiles, 141, 273

organophosphorus, 142
　esters, 186
oxidation ponds, odor control, 477
ozone, 54 ff, 80 ff, 149, 407

paladium, 282
Peppard Program, 304
peroxyacetyl nitrates, 81, 149
phytophagous insects, 183
phytotoxic effect, 173
picloram, 135, 140
pink waters, 272 ff, 288 ff
plants and vegetation
　contamination, 553, 556, 718
　destruction and damage, 553, 719
Plasmodium, 162
platinum, 282
plume in petrochemical plants, 119
pollution sources, 150
polychlorinated biphenyls (PCB) 95
potassium 136, 193
　potassium formate, 232
propellants, 268 ff, 298 ff, 304
Propylene, 80 ff

radioactive fission products, 429
radiosonde, 61
red waters 272 ff, 288 ff
resource recovery systems
　chemical, 369
　compost, 361
　economics, 369-376
　energy, 365-367
　heat, 359
　materials, 363
　mathematical analysis, 487-503
reverbatory furnace, 220

satellites, 61
　Earth Resources Technology Satellite, ERTS, 71
　Nimbus, 65, 70
　satellite sweeping, 32
schistosomiasis, 164
scrubbers, 195
　amonia, 229, 237, 245
　Bahco, 259
　Bischoff, 259
　double alkali, 230
　grid, 196
　magnesium oxide, 226
　molten carbonate, 231
　venturi, 195
septage, 346 ff
sewage sludge, 136
Sheffe's test, 338
sickle-cell trait, 162
simazine, 138, 141
sleeping sickness, 161
sociological considerations
　environmental education, 750
　expanding societies, 739, 768
　informing the public, 758
　lifestyle, 741-742
　measurement, 689
　national differences, 681, 695
　political, 680, 689, 765
　population centers, 739, 768
　priorities, 683, 734-736, 739

SUBJECT INDEX

unsupported allegations, 765
smog, photochemical, 78
sodium, 193
soil decontamination, 137
soil pollution, 133
solid wastes, 357-377
  in Nigeria, 649
soot, 3
Starfish nuclear explosion, 32
"sterile-male technique", 188
sugar cane pests, 185
styphnic acid, 310 ff
sulfur, 322
sulfur dioxide, 2, 54, 149, 173, 192 ff, 224, ff, 235 ff, 401, 666
sulfur dioxide removal
  absorption process, 255
  activated carbon process, 246, 248
  activated manganese oxide process, 245
  alkali absorption, 257
  amonia scrubbing, 229, 237, 245
    Bahco-scrubbing, 259
    Bischoff Process, 259
  char sorption process, 231
  double alkali scrubbing, 230
  Grillo-AGS-process, 261
  hydrodesulfurization, 236
  lime-gypsum process, 239, 242
  magnesium oxide scrubbing, 226
  metal oxide sorption, 232
  molten carbonate scrubbing, 231
  reduction process, 249
  sodium citrate process, 228
  Wellman-Lord Process, 228, 247
  zinc, oxide process, 238

sulfur oxides, 105, 193, 215, 280
  air quality standards, national, 216
  sources, 215 ff
sulfuric acid, 3, 112, 218, 225
  dilute, 244
sulfurous acid, 193

Teratogenicity
  of dioxins, 715, 761
  of phenoxy herbicides, 708, 761
  testing methods, 708
thermal plasma, injection of, 31
thermal pollution, 10, 19
toxic mists, 149
triazines, 135, 139
trypanasomiasis, 162
tsetse fly, 162, 172
turbulent contact absorber, 195
turbulent diffusion, 77

urban mortality, 15

vanadium
  corrosion, 105
  fuels, 106
  vanadates, 106
Van Allen belt, 26

waste water
  carbon absorption, 523, 526
  computer control, 529-547
  filtration, 523
  lime treatment, 380, 523
  ozone treatment, 380
  ptt control, 523, 527

removal of metals,
    380-385
sedimentation, 387-
    396, 523
water pollution
  distribution mechanisms,
    510-521
  in Canada, 635
  in South Australia,
    617

mercury, 763
salt content, 434-436
water vapor, 1, 54 ff
Whistlers, 30

zeolite, 282
zinc, 322
    venturi scrubber, 195